Operator Theory: Advances and Applications
Vol. 128

Editor:
I. Gohberg

Operator Approach to Linear Problems of Hydrodynamics

Volume 1: Self-adjoint Problems for an Ideal Fluid

Nikolay D. Kopachevsky
Selim G. Krein

Springer Basel AG

Nikolay D. Kopachevsky
Mathematical Department
Taurida National V. Vernadsky University
ul. Yaltinskaya 4
95007 Simferopol
Ukraine

e-mail: kopachevsky@tnu.crimea.ua

2000 Mathematics Subject Classification 76-02, 76M22, 76M99, 47N20, 47N99

A CIP catalogue record for this book is available from the Library of Congress, Washington D.C., USA

Deutsche Bibliothek Cataloging-in-Publication Data

Kopačevskij, Nikolaj D.:
Operator approach in linear problems of hydrodynamics / Nikolay D. Kopachevsky ;
Selim G. Krein - Basel ; Boston ; Berlin :
Birkhäuser, 2001
Vol. 1. Self-adjoint problems for an ideal fluid. - 2001
ISBN 978-3-0348-9525-5 ISBN 978-3-0348-8342-9 (eBook)
DOI 10.1007/978-3-0348-8342-9

Originally published by Birkhäuser Verlag, Basel - Boston - Berlin in 2001
Softcover reprint of the hardcover 1st edition 2001
Printed on acid-free paper produced from chlorine-free pulp. TCF ∞
Cover design: Heinz Hiltbrunner, Basel

ISBN 978-3-0348-9525-5

9 8 7 6 5 4 3 2 1 www.birkhasuer-science.com

Preface

The main topics presented in this book deal with methods from functional analysis applied to the study of small movements and normal oscillations of hydromechanical systems having cavities filled with either ideal or viscous fluids. The book is a sequel to and at the same time substantially extends the volume entitled "Operator Methods in Linear Hydrodynamics: Evolution and Spectral Problems," by N. D. Kopachevsky, S.G. Krein, and Ngo Zuy Kan that was published in 1989 by the Nauka publishing house in Moscow.

The present book includes several new problems on the oscillations of partially dissipative hydrosystems and the oscillations of visco-elastic or relaxing fluids.

The contents of this book do not overlap almost at all with the ones in the following volumes: "Mathematical Problems of the Motion of Viscous Incopressible Fluids," by O. A. Ladyzhenskaya, "The Dynamics of Bodies with Cavities Filled with Fluids," by N. N. Moiseev and V. V. Rumiantzev, "Navier-Stokes Equations," by R. Temam, and "Boundary Problems for Navier-Stokes Equations," by S. M. Belonosov and K. A. Chernous. Mainly, the contents of the present book rely on the authors' and their students' works.

We would like to express our gratitude to I. T. Gohberg and A. S. Markus, who encouraged us to publish the book and who offered many helpful suggestions. Our gratidude goes also to our colleagues T. Ya. Azizov, O. A. Ladyzhenskaya, N. N. Moiseev, A. D. Myshkis, F. L. Chernousko, V. I. Yudovich, and Ngo Zuy Kan with whom we had fruitful discussions during the process of obtaining some mathematical results and of writing the final version of the book.

We would also like to acknowledge the work of E. D. Volodkovich and E. A. Danilova who drafted a preliminary version of the English translation of the book. The staff in the Department of Mathematics and Theoretical Mechanics at the Voronezh Forestry Engineering University, Russia, and in the Department of Mathematical Analysis at the Simferopol State University, Ukraine, offered us their professional assistance during the time of writing the book. Their help is greatfully acknowledged.

Our particular gratitude goes to Mircea Martin and his wife, Larisa, for their professional and qualified work in finalizing the English translation and typesetting the monograph in TeX.

The Authors
Simferopol, Voronezh
October 20, 1995

While the book was prepared for press, its co-author, Selim Grigoryevich Krein, a dedicated teacher, an elder friend of mine, an outstanding mathematician and person, and a great specialist in functional analysis and its applications, passed away. In the process of investigating initial boundary value and spectral problems of mathematical physics and hydrodynamics with his students and colleagues, he used to inquire how the problem under investigation would look in the operator form. This book answers the question in the case of linear problems of hydrodynamics. S. G. Krein dreamed of this book being published, not only in Russia and in the former USSR, but also for the English-speaking readers as well. I hope they will not be disappointed by this two volume monograph.

Nikolay D. Kopachevsky
May 18, 2001

Table of Contents

Volume I

Preface v
Table of Contents
Volume I vii
Volume II xv
Introduction 1

Part I: Mathematical Foundations of Linear Hydrodynamics 5

Chapter 1: Operators on Hilbert Spaces

1.1 General Facts 7
1.1.1 The Concept of a Hilbert Space 7
1.1.2 The Space $L_2(\Omega)$ 8
1.1.3 Orthogonality. Projection onto a Subspace 9
1.1.4 Equivalent Norms 10
1.1.5 Linear Functionals. Riesz Theorem 10
1.1.6 Embeddings of Spaces. Riesz Theorem for Equipments 11
1.1.7 Orthonormal Systems and Bases 12
1.1.8 Bounded Linear Operators 12
1.1.9 Adjoint Operators 14
1.1.10 Self-Adjoint Operators 14
1.1.11 Self-Adjoint Compact Operators 15
1.1.12 Compact operators. s-numbers 16

1.1.13 Riesz Bases and p-Bases 17
1.1.14 Direct Sum of Subspaces. Invariant Subspaces 18
1.1.15 Eigen- and Associated (Root) Elements. Root Subspaces . . 19
1.1.16 Unbounded Linear Operators 20
1.1.17 Resolvent and Spectrum of a Linear Operator 21
1.1.18 Classification of Points in the Spectrum of a Linear Operator 22
1.1.19 Spectrum of a Self-Adjoint Operator. Weyl Theorem 23
1.1.20 Riesz Projections 24
1.1.21 Symmetric and Self-Adjoint Operators 25
1.1.22 Spectral Decomposition of Self-Adjoint Operators. Functions of Operators 26
1.1.23 Spaces with Degenerate Scalar Products. Seminorms 28
1.1.24 Equivalent Corrections of Seminorms 29

1.2 Sobolev Spaces 29
1.2.1 Finite Functions 30
1.2.2 Generalized Derivatives 30
1.2.3 The Definition of Sobolev Spaces 31
1.2.4 The Space $L_2^1(\Omega)$. Regions of the First Type 31
1.2.5 The Subspace $H_0^1(\Omega)$ 32
1.2.6 Embedding $H^1(\Omega)$ into $L_2(\Omega)$. Regions of the Second Type 34
1.2.7 The Trace Operator. Regions of the Third Type 34

1.3 Spaces With Indefinite Metrics 35
1.3.1 J-Spaces 35
1.3.2 Uniformly Definite Subspaces 37
1.3.3 J-Orthonormal Systems and Bases 37
1.3.4 Linear Operators on J-Spaces 38
1.3.5 Invariant Subspaces of J-Self-Adjoint Operators 39
1.3.6 Pontryagin Spaces 39
1.3.7 On Completeness and Basicity for the System of Root Elements of a $\boldsymbol{J}$-Self-Adjoint Operator 41

1.4 Eigenvalue Problems 43
1.4.1 Operator B is the Identity Operator 43
1.4.2 Operator B is Positive Definite 44
1.4.3 Positivity Condition for a Matrix Operator 47
1.4.4 Simplifying Equations with an Alternating Operator 48

1.4.5 Equations in Spaces with Indefinite Metrics 49

1.5 Evolution Equations in Hilbert Spaces 51

1.5.1 First Order Linear Differential Equations with Bounded Operator Coefficient 51

1.5.2 The Cauchy Problem for Equations with Unbounded Operators 52

1.5.3 Equations With a Negative Self-Adjoint Operator 53

1.5.4 Equations With a Dissipative Operator 54

1.5.5 Equations With Perturbed Operators 55

1.5.6 Stability 56

1.5.7 Nonhomogeneous Equations 57

1.5.8 Linear Differential Equations of the Second Order 58

1.5.9 Volterra Integral Equations 62

1.6 Spectral Theory of Operator Pencils 63

1.6.1 Eigen- and Associated Elements of an Operator Pencil . . . 63

1.6.2 Root Functions 65

1.6.3 Fredholm Holomorphic Operator-Valued Functions 66

1.6.4 Linear Pencils. Theorems on Completeness of the System of Eigen- and Associated Elements 66

1.6.5 Keldysh-Type n-Multiple Completeness 68

1.6.6 Spectral Factorization of an Operator Pencil 69

1.6.7 Completeness With Finite Defect of a System of Eigen- and Associated Elements of an Operator-Valued Function 70

1.6.8 Asymptotic Behavior of Branches of Eigenvalues 71

1.6.9 Self-Adjoint Operator Pencils 72

1.6.10 On Riesz Basicity of the System of Eigenelements of a Self-Adjoint Operator-Valued Function 73

1.6.11 Variational Methods for Investigating Continuous Operator-Valued Functions 75

1.7 Asymptotic Methods for Solving Evolution Equations With a Small Parameter Attached to the Derivative 76

1.7.1 Equations With a Small Parameter Attached to the Derivative 77

1.7.2 Splitting of a Homogeneous Equation 78

1.7.3 Solvability of Commutator Equations 79

1.7.4 Asymptotic Expansions of Solutions 82

1.7.5 The Special Case of a Splitable Operator Kernel 86
1.7.6 The Case of a Nonstationary Perturbation 87
1.7.7 Nonhomogeneous Equations 87
1.7.8 Eigenvalue Problems 90

1.8 A General Scheme for Solving Boundary Value Problems 93
1.8.1 Hilbert Pairs. Generating Operators 93
1.8.2 Hilbert Pairs Connected With the Spaces $H^1(\Omega)$ and $L_2(\Omega)$ 95
1.8.3 Hilbert Scale of Spaces. Space $E^{-1/2}$ 98
1.8.4 Self-Adjoint Extensions of Positive Definite Symmetric Operators. Generalized and Weak Solutions of Equations . . 99
1.8.5 Nonhomogeneous Boundary Value Problems 101
1.8.6 Spaces of Harmonic Functions 103
1.8.7 Embedding and Mapping. The Abstract Green Formula . . . 107

Chapter 2: Fundamental Spaces and Operators of Linear Hydrodynamics

2.1 Fundamental Spaces and Hydrodynamics Operators for an Ideal Fluid 109
2.1.1 Fields with Finite Kinetic Energy 110
2.1.2 Potential Fields 110
2.1.3 Divergence of Fields with Finite Kinetic Energy 111
2.1.4 The Space of Solenoidal Fields 112
2.1.5 Laplace Operator on the Space $H^1(\Omega)$ 112
2.1.6 Normal Component of a Field on the Boundary 113
2.1.7 Green Formula for the Laplace Operators. Harmonic Fields 114
2.1.8 Weyl Decomposition 115
2.1.9 Approximation by Smooth Fields 116
2.1.10 The Space of Velocity Fields for an Ideal Fluid in an Open Container 117
2.1.11 Systems of Nonmixing Fluids 118
2.1.12 Spaces of Velocity Fields for Systems of Nonmixing Ideal Fluids 124

2.2 Spaces and Hydrodynamics Operators for a Viscous Fluid 125
2.2.1 Forces of Internal Friction. Energy Dissipation 126
2.2.2 Divergence Operator. Solenoidal Fields 127
2.2.3 Vector Laplace Operator. Green Formula 128

2.2.4 Movement of a Viscous Fluid in a Closed Container. Korn Identity and Korn Inequality 129
2.2.5 Stokes Operator 131
2.2.6 Spaces of Velocity Fields for a Viscous Incompressible Fluid in an Open Container 132
2.2.7 Main Boundary Value Problems for the Fluid Movement in an Open Container 134
2.2.8 Spaces of Velocity Fields for a System of Viscous Fluids . . . 138

Appendix A: Remarks and Reference Comments to Part I
A.1 Chapter 1 143
A.2 Chapter 2 146

Part II: Motion of Bodies With Cavities Containing Ideal Fluids . . 149

Chapter 3: Oscillations of a Heavy Ideal Fluid in Stationary and Nonstationary Containers
3.1 Equations of the Motion of a Rigid Body with a Cavity Filled with an Incompressible Fluid 151
3.1.1 Basic Concepts of Kinematics 152
3.1.2 Equations of Motion for an Incompressible Fluid 153
3.1.3 Boundary Conditions 154
3.1.4 Motion Equations for a Gyrostate 155
3.1.5 Dynamics Equations of the System "Body + Fluid" With a Partially Filled Cavity 158
3.1.6 Transition to Undimensional Variables 163
3.2 Motion of an Ideal Fluid in a Closed Stationary Container 164
3.2.1 Basic Equations 164
3.2.2 Existence of Solutions 165
3.3 Small movements of an Ideal Fluid in an Open Immovable Container 166
3.3.1 Statement of the Problem and the Basic Equations 166
3.3.2 Projection of Euler Equations 167
3.3.3 Existence of Solutions 168
3.3.4 Proper Oscillations 170
3.4 Small Joint Movements of a Fluid and a Container 171
3.4.1 Statement of the Problem and the Basic Equations 171
3.4.2 Finding the Velocity Field and the Pressure 172

3.4.3 Defining the Motion Law of the Body 173
3.4.4 Zhukovsky Potentials 174

3.5 Small Joint Movements Around a Fixed Point of a Body and a Fluid Partially Filling the Cavity 174
3.5.1 Statement of the Problem and the Basic Equations 175
3.5.2 The Law of Full Energy Balance 176
3.5.3 Projecting Euler Equations 177
3.5.4 Kinetic Moment Equation 178
3.5.5 Investigating the Complete System of Motion Equations . . 179
3.5.6 Proper Oscillations 184
3.5.7 Solving the Evolution Problem 187

3.6 Oscillations of a System of Fluids in an Immovable Container . . . 189
3.6.1 Statement of the Problem 190
3.6.2 Orthogonal Projection Method 191
3.6.3 Transition to Operator Equation 193
3.6.4 Proper Osccilations 194
3.6.5 Small Movements of Stable Systems 196

Chapter 4: Problems on Oscillations of Capillary Fluids and Problems on Hydroelasticity in Immovable Containers

4.1 Oscillations of a Capillary Fluid in a Rigid Container 199
4.1.1 On the Equilibrium State 200
4.1.2 Statement of the Problem on Small Oscillations 201
4.1.3 Law of Energy Balance 203
4.1.4 Transition to Operator Equation 204
4.1.5 Properties of the Potential Energy Operator 205
4.1.6 Proper Oscillations 206
4.1.7 Instability Conditions of the System 208
4.1.8 Solvability of the Evolution Problem 208
4.1.9 Oscillations of a System of Capillary Fluids 209

4.2 Oscillations of Fluids in Containers With Elastic Ends 213
4.2.1 Solving the Static Problem 213
4.2.2 Formulation of the Problem on Small Oscillations 216
4.2.3 Energy-Preserving Law 217
4.2.4 Operator Equation of the Problem 219
4.2.5 Properties of the Operators of the Problem 220
4.2.6 Proper Oscillations 221

4.2.7 Evolution Problem 223
4.2.8 Oscillations of a Fluid in a Container With One Elastic End 224
4.2.9 Oscillations of a System of Fluids in a Container With Elastic Plane Partitions 226

4.3 Oscillations of a Fluid in a Partially Filled Container With an Elastic Bottom 232
4.3.1 Determining the Equilibrium State 232
4.3.2 Formulation of the Problem on Small Oscillations 234
4.3.3 On Solvability of the Evolution and Spectral Problems . . . 234
4.3.4 Oscillations of a Capillary Fluid in a Partially Filled Elastic Container 237
4.3.5 Systems of Heavy Fluids in a Container With Elastic Plates 241
4.3.6 Systems of Capillary Fluids in a Container With Elastic Ends 244
4.3.7 Compound Systems of Plate-Partitions and Nonmixing Fluids 246

Chapter 5: Other Operator Approaches to Hydrodynamics Problems of Ideal Fluids

5.1 Plane Problems on Proper Oscillations of a Heavy Fluid in a Channel. An Application of the Stream Function 249
5.1.1. Spectral Problem for the Stream Function 249
5.1.2. Properties of Nodal Lines of Stream Eigenfunctions 252
5.1.3. Estimates of Eigenvalues 256

5.2 Shallow Water Theory in Problems on Oscillations of Heavy Ideal Fluids in Bounded Regions 257
5.2.1. Formulation of the Problem with a Small Parameter 257
5.2.2. Asymptotic Solution in First Approximation 259
5.2.3. Formulas for Calculating Second Order Approximations . . 261
5.2.4. Plane Problems 263
5.2.5. Examples 264
5.2.6. Systems of Nonmixing Fluids 265

5.3 Oscillations of a System "Fluid-Gas" in a Bounded Region 270
5.3.1. Formulation of the Initial Boundary Value Problem 270
5.3.2. Formulation of the Spectral Problem 272

5.3.3. Transition to a System of Operator Equations 274
5.3.4 Theorem on Spectrum 276
5.3.5 Variational Principles for Eigenvalues 278

Chapter 6: Oscillations of an Ideal Rotating Fluid

6.1 Motion of Fluids in Rotating Containers 281
6.1.1 Statement of the Problem and the Main Equations 281
6.1.2. Existence of Solutions 282
6.1.3 Normal Oscillations 284

6.2 Motion of a Gyrostate Similar to Uniform Rotation
About a Fixed Axis 287
6.2.1 Statement of the Problem 287
6.2.2 Transition to the Evolution Equation in a Hilbert Space . . . 288
6.2.3 Properties of the Operators in the Problem 289
6.2.4 Existence of Solution to the Boundary Value Problem 289

6.3 Rotation of a Fluid in a Partially Filled Container 290
6.3.1 On the Equilibrium State 290
6.3.2 Statement of the Problem on Small Oscillations 292
6.3.3 Method of Orthogonal Projection 294
6.3.4 Properties of the Operators of the Problem 297
6.3.5 Systems of Nonmixing Fluids 299
6.3.6 Transition to an Operator Equation and Properties of the
Operators of the Problem 302

6.4 Solving the Initial Boundary Value Problem 310
6.4.1 Generalized Solution of the Operator Equation 310
6.4.2 Small Movements of Fluid in a Partially Filled Container . . 313
6.4.3 On the Structure of the Spectrum of a Vortical Operator . . 313
6.4.4 Classes of Free Movements 316
6.4.5 Free Movements of a System of Fluids 318

6.5 Self-adjoint Operator Pencils Generated by Problems on
Oscillations of a Rotating Ideal Fluid 321
6.5.1 The Main Operator Pencil 321
6.5.2 On the Spectrum of the Operator Pencil 322
6.5.3 Operator Pencils with Analytic Perturbations 323
6.5.4 Factorization of the Operator Pencil 325
6.5.5 Systems of Eigenelements With Defect Basicity 327

6.5.6 Double-Sided Estimates of Positive and Negative Eigenvalues . . . 328
6.5.7 On the Essential Spectrum of the Problem . . . 331
6.6 Proper Oscillations of a Rotating Fluid . . . 333
6.6.1 Surface and Internal Waves . . . 333
6.6.2 Properties of the Surface Waves . . . 335
6.6.3 On Existence and Properties of Internal Waves . . . 338
6.6.4 Oscillations of a System of Nonmixing Fluids . . . 341

Appendix B: Remarks and Reference Comments to Part II
B.1 Chapter 3 . . . 345
B.2 Chapter 4 . . . 347
B.3 Chapter 5 . . . 350
B.4 Chapter 6 . . . 350

Standard Reference Texts . . . 355

Bibliography . . . 357

List of Symbols . . . 375

Subject Index . . . 379

Table of Contents

Volume II

Preface
Table of Contents
Volume II
Volume I
Introduction

Part III: Motion of Bodies with Cavities Containing Viscous Incompressible Fluids

Chapter 7: Motion of Bodies with Cavities Completely Filled with Viscous Incompressible Fluids

7.1 Motion of Fluids Completely Filling a Cavity in a Stationary Body
7.1.1 Statement of the Problem and the Basic Equations
7.1.2 Reducing the Problem to a Differential Equation in a Hilbert Space. Existence of Solutions
7.1.3 Structure of the Spectrum of the Problem
7.1.4 Perturbation of the Stationary Motion of a Fluid
7.1.5 Small Movements of a Fluid in a Rotating Container

7.2 Small Movements of a Gyrostate around a Fixed Mass Center
7.2.1 Statement of the Problem and the Basic Equations
7.2.2 Transition to a Differential Equation in a Hilbert Space

7.2.3 Properties of the Translation Operator
7.2.4 Existence of Solutions of the Evolution Problem
7.2.5 Normal Oscillations

7.3 Rotating Motion of a Gyrostate
7.3.1 Statement of the Problem and the Basic Equations
7.3.2 Transition to a Differential Equation in a Hilbert Space
7.2.3 Properties of the Operators of the Problem
7.3.4 Normal Oscillations
7.3.5 Solvability of the Nonstationary Problem

7.4 Asymptotic Solutions for High Viscosity
7.4.1 Solving the Hydrodynamics Problem
7.4.2 Asymptotic Equations of the Motion of a Rigid Body
7.4.3 An Example

7.5 Oscilations of a Pendulum with a Cavity Completely Filled with a Viscous Fluid
7.5.1 Towards the Statement of the Problem
7.5.2 Transition to a System of Operator Equations
7.5.3 The Indefinite Metric Approach
7.5.4 Other Properties of Solutions of the Spectral Problem
7.5.5 On Riesz and p-Basicity of Modes of Dissipative Waves

7.6 Problems on Fluids Flowing through a Given Cavity
7.6.1 The Basic Equations
7.6.2 Application of the Abstract Scheme
7.6.3 Transition to Operator Equations in Orthogonal Subspaces
7.6.4 Theorem on Existence of a Generalized Solution

7.7 Convective Movements of Fluids in a Closed Cavity
7.7.1 Equations of Thermal Convection
7.7.2 Conditions of Mechanical Equilibrium
7.7.3 Final Statement of the Problem
7.7.4 Transition to an Operator Equation
7.7.5 Solvability of the Initial Boundary Value Problem
7.7.6 Normal Movements of a System Heated from Below
7.7.7 Normal Oscillations of a System Heated from Above
7.7.8 On the Transition of Eigenvalues to the Left Half-Plane in the Case of Heating from Below

Chapter 8: Motions of Viscous Fluids in Open Containers
8.1 Small Movements of Viscous Fluids in an Open Immovable Container
8.1.1 Classical Statement of the Problem
8.1.2 Auxilliary Boundary Value Problem
8.1.3 Generalized Solution of the Homogeneous Nonstationary Problem
8.1.4 Motions with Small Mass Forces
8.1.5 Equation of Energy Balance
8.1.6 Equation of Normal Oscillations
8.2 The Main Operator Pencil
8.2.1 Structure of the Spectrum of the Problem
8.2.2 Linearization of the Pencil
8.2.3 Mutual Relationships between Eigen- and Associated Elements of the Two Pencils
8.2.4 Transformation to a Nondegenerate Pencil
8.2.5 The Property of Two-Multiple Basicity
8.2.6 Spectral Factorization of the Overdamped Pencil. Separate Basicity
8.2.7 Double-Sided Inequalities for the Two Branches of Eigenvalues
8.2.8 The General Case
8.3 Normal Oscillations and the Spectrum of the Hydrodynamics Problem
8.3.1 General Properties of the Spectrum
8.3.2 Influence of Fluid Viscosity on the Structure of the Spectrum of the Problem
8.3.3 Properties of Surface and Internal Waves
8.3.4 On the Basicity of Modes of Normal Oscillations
8.4 Oscillations of a Heavy Rotating Fluid
8.4.1 Statement of the Problem
8.4.2 Transition to a System of Operator Equations
8.4.3 Solvability of the Nonstationary Problem
8.4.4 Equations of Normal Oscillations
8.4.5 Investigation of the Spectrum Problem
8.4.6 On the Completeness of Systems of Modes of Normal Oscillations of the Initial Problem

8.5 Asymptotic Solutions for High Viscosity
8.5.1 The Cauchy Problem
8.5.2 Normal Oscillations
8.5.3 Motions under Mass Forces
8.5.4 The Case of Rotating Viscous Fluids
8.6 Oscillations of a System of Nonmixing Fluids
8.6.1 Statement of the Problem on Small Oscillations
8.6.2 Transition to a System of Operator Equations
8.6.3 The Theorem on Existence of a Generalized Solution
8.6.4 Normal Oscillations of a System of Fluids
8.6.5 On the Stability of the Relative Equilibrium State
8.7 Motions around a Fixed Point of a Body with a Cavity Partially Filled with Fluid
8.7.1 Basic Equations
8.7.2 Transition to a System of Operator Equations
8.7.3 Auxilliary Results
8.7.4 Existence of Solution of the Boundary Value Problem
8.8 Normal Oscillations of a Pendulum Partially Filled with a Fluid (The Plane Problem)
8.8.1 Statement of the Problem
8.8.2 Transition to a Differential Equation in a Hilbert Space
8.8.3 Properties of Operator Coefficients of the Evolution Equation
8.8.4 Normal Oscillations. Properties of the Spectrum Problem
8.8.5 Theorem on Instability
8.8.6 On the Solvability of the Initial Boundary Value Problem
8.9 Convection in a Partially Filled Container
8.9.1 Statement of the Problem
8.9.2 Transition to a System of Operator Equations
8.9.3 Solvability of the Evolution Problem Equation
8.9.4 Normal Convective Movents. Reduction to an Operator Pencil
8.9.5 Dissipatively Thermal and Surface Waves under the General Law of Heat Transfer

8.9.6 Surface and Internal Waves for Heating From Above
8.9.7 Normal Oscillations for Heating from Below and for a Given Heat Flow on the Free Surface
8.10 Sufficient Conditions of Instability for Convective Movements of a Fluid
8.10.1 Transition to a Two-Parameter Pencil
8.10.2 On the Structure of the Kernels of Operator Coefficients
8.10.3 On the Existence of Eigenvalues in the Left Complex Half-Plane
8.10.4 Double-Sided Estimates for Eigenvalues
8.10.5 Derivation of a Sufficient Condition for Instability
8.10.6 Remarks
Chapter 9: Oscillations of Capillary Viscous Fluids
9.1 Statement of the Problem
9.1.1 Basic Equations and Boundary Conditions
9.1.2 Some Properties of Solutions of the Normal Oscillation Problem
9.1.3 On the Spectrum Structure of Normal Oscillations
9.2 Oscilations of Capillary Fluids in Arbitrary Containers
9.2.1 Transition to a System of Operator Equations
9.2.2 Normal Oscillations. Properties of the Operators of the Problem
9.2.3 Normal Oscillations of a Rotating Fluid
9.2.4 Normal Oscillations of a Nonrotating Fluid
9.2.5 The Matrix Structure of the Main Operator
9.2.6 On the Finiteness of the Number of Nonreal Eigenvalues
9.2.7 Heuristic Considerations. The Abstract Spectral Problem
9.2.8 Heuristic Considerations. Physical Conclusions
9.2.9 The Solvability of the Evolution Problem
9.3 The Inverse of the Lagrange Theorem on Stability
9.3.1 Formulating the Theorem
9.3.2 Auxilliary Propositions
9.3.3 The Principle of Changing Stability
9.3.4 Transition to an Equation with a Compact Operator
9.3.5 Application of Perturbation Theory

9.3.6 The Existence of Eigenvalues in the Left Complex Half-Plane for an Arbitrary Viscosity Value

9.4 Motions of a Rigid Body Containing a Cavity Filled with a Capillary Fluid under Conditions of Complete Low Gravity

9.4.1 Statement of the Problem

9.4.2 Transition to a System of Operator Equations

9.4.3 Normal Oscillations. Transition to an Equation with a Dissipative Operator

9.4.4 Properties of Normal Movements

9.4.5 The Existence of a Generalized Solution to the Nonstationary Problem

Appendix C: Remarks and Reference Comments to Part III

C.1 Chapter 7

C.2 Chapter 8

C.3 Chapter 9

Part IV: Small Oscillations of Complex Hydrodynamic Systems

Chapter 10: Oscillations of Partially Dissipative Hydrosystems

10.1 Statement of the Problem

10.1.1 The Classical Statement of the Problem

10.1.2 The Law of Full Energy Balance. Definition of a Generalized Solution

10.1.3 Normal Oscillations. Statement of the Problem

10.2 Studying an Initial Boundary Value Problem

10.2.1 Projections of Euler and Navier-Stokes Equations on Orthogonal Subspaces

10.2.2 Auxiliary Boundary Value Problems

10.2.3 Properties of Matrix Blocks and Their Physical Meanings

10.2.4 Theorem on Correct Solvability of the Initial Boundary Value Problem

10.3 Model Problem on Normal Oscilations of Partially Dissipative Hydrosystems

10.3.1 Statement of the Model Problem

10.3.2 Obtaining the Characteristic Equation

10.3.3 Studying the Characteristic Equation

10.3.4 General Conditions and Hypotheses on the Structure of the Spectrum of the Hydrodynamical Problem

10.4 Normal Oscillations of a Partially Dissipative Hydrosystem in an Arbitrary Domain

10.4.1 Transition to an Operator Pencil with Bounded Operator Coeficients

10.4.2 General Properties of the Spectrum

10.4.3 The Theorem on Spectrum Location

10.5 On the Completeness of the System of Modes of Normal Oscillations

10.5.1 Auxiliary Results

10.5.2 Theorem on Completeness. Keldysh Scheme Realization

10.5.3 On the Existence of Branches of Eigenvalues with a Limit Point at Infinity. Heuristic Arguments

10.5.4 Concluding Remarks

Chapter 11: Oscillations of Visco-Elastic and Relaxing Media

11.1 Visco-Elastic Fluids in Completely Filled Containers

11.1.1 On a Model of Visco-Elastic Fluid

11.1.2 Statement of the Initial Boundary Value Problem

11.1.3 On the Solvability of the Initial Boundary Value Problem

11.1.4 Normal Oscillations

11.2 Abstract Evolution and Spectral Problems Generated by Small Motions of a Visco-Elastic Fluid

11.2.1 Statement of the Problem. Transition to an Equation with a Dissipative Operator

11.2.2 On the Solvability of the Cauchy Problem for an Integro-Differential Equation

11.2.3 Spectral Problem. Transition to an Equation with Bounded Operators

11.2.4 Properties of Operator Coefficients of the Spectral Problem

11.2.5 Properties of Solutions of the Spectral Problem

11.3 Small Motions and Normal Oscillations of a Visco-Elastic Fluid in an Open Container

11.3.1 Mathematical Statement of the Problem
11.3.2 Transition to a System of Operator Equations
11.3.3 On the Solvability of the Initial Boundary Value Problem
11.3.4 Normal Oscillations. The main Operator Pencil

11.4 Multiple Basicity of the System of Eigen- and Associated Elements for the Problem on Normal Oscillations of a Visco-Elastic Fluid in an Open Container
11.4.1 The Linearization of the Pencil
11.4.2 Theorem on Basicity of the System of Root Elements of the Linear Pencil
11.4.3 Connection between the System of Root Elements of the Two Problems
11.4.4 Theorem on Basicity of Elements of a Special Form

11.5 Additional Properties of Solutions of the Spectral Problem
11.5.1 On the Existence of Different Branches of Eigenvalues
11.5.2 On the Location of Nonreal Eigenvalues in the Complex Plane
11.5.3 Multiple Basicity and p-Basicity. An Indefinite Metric Approach Using Krein Space Theory

11.6 Oscillations of Relaxing Fluids
11.6.1 Classical Statement of the Problem on Small Motions of a Relaxing Fluid
11.6.2 Transition to an Initial Boundary Value Problem for One Scalar Function
11.6.3 The Simplest Problem on Oscilations of a Relaxing Fluid
11.6.4 On the Solvability of the Cauchy Problem for an Abstract Integro-Differential Equation Connected with Small Motions of a Relaxing Fluid
11.6.5 Normal Oscilations of a Relaxing Fluid with Variable Medium Characteristics
11.6.6 Physical Conclusions

Appendix D: Remarks and Reference Comments to Part IV

D.1 Chapter 10

D.2 Chapter 11

Standard Reference Texts

Bibliography

List of Symbols

Subject Index

Introduction

The last decades have been characterized by the implementation of various functional analysis methods into the field of continuous medium mechanics. The founders of this process were H. Weyl, J. Leray, S. L. Sobolev, and K. O. Friedrichs. Later on, many mathematicians and mechanicians from all over the world contributed jointly to the development of this particular area.

The problems on the motion of solid bodies with cavities filled by fluids are classical. Their investigation originated with the works of G. Stokes, H. Helmholtz, H. Lamb, and N. E. Zhukovsky. During the last decades, these problems became modern once again due to advancements in aviation and tanker and space engineering. The present book provides a detailed description of some results obtained by using several operator theory methods in the study of the dynamics of bodies with cavities filled with fluids.

In this type of problems, one needs to use different linear operators such as the standard operators of vector analysis, the Stokes operator, the operators of kinetic and potential energy, translation operators, the Coriolis—or gyroscopic—operator, various projections, etc. The definitions and basic properties of all these operators, their interaction in the linearized problems of the dynamics of bodies filled with fluids, and a qualitative study of evolution equations and the corresponding spectral problems are among the highlights of the book.

The three fundamental directions of study pursued throughout the book are the following: To prove the existence of solutions of various initial boundary value problems, to investigate the properties of normal oscillations and the structure of their frequency spectrum, and to obtain asymptotic formulas. Though some of the sections in the book deal with specific problems such as, for instance, an analytic solution for cavities of a given form, however, many of these special problems are not explicitly addressed. The papers and monographs where these problems are studied in detail are listed in the reference section and additional bibliographic comments are included in appendices.

In this book, unlike in many other known books on applications of functional analysis methods to hydrodynamics problems, the authors have deliberately refrained from presenting the basic results obtained in the study of a certain problem as lemmas and theorems. They believe that the reader, particularly the mechanicians, will benefit more from a description of the problems studied in this book in the form of a free conversation, where the material is divided into sections and subsections. Such an approach, however, does not intend to abate the exactness or correctness of the results obtained and formulated along the way. An attentive reader will easily identify in every section some theorems and lemmas, their proofs, and also physical interpretations of the main properties of the studied problem.

The book has two volumes and is divided into four parts. The first volume, divided into two parts, deals with the study of hydrodynamical systems containing an ideal fluid and includes basically the description of self-adjoint problems.

The first part consisting of Chapters 1 and 2 is subsidiary. Chapter 1 presents all the prerequisits in functional analysis that are necessary to read the three other parts of the book. Many of the basic results are stated without proofs, yet some of them (for example, in Section 1.8) not included in standard functional analysis textbooks are covered in more detail. To the reader who is not sufficiently well acquainted with the methods of functional analysis we suggest to begin by reading Section 1.1 and eventualy to use Chapter 1 as a handbook.

Chapter 2 reflects to a certain extent the authors' viewpoint that the metrics in different spaces of fluid velocity fields should have a physical meaning. For an ideal fluid, such a metric is generated by the quadratic form of the kinetic energy, and for a viscous fluid, by the same form and the form of the velocity of the energy dissipation. All the other spaces of velocity fields (solenoidal, potential, satisfying some boundary conditions, etc.) are subspaces of the corresponding Hilbert spaces of fields with a finite velocity of energy dissipation. Dual spaces consisting, as a rule, of generalized functions (i.e., distributions) of finite order, are also considered. For all the basic operators of vector analysis one can find their largest domains of definition in the above mentioned spaces of fields.

It should be noted however that in hydrodynamics we have to deal naturally with domains with nonsmooth boundaries such as containers with edges, cavities partially filled with fluid, etc. Therefore, in many problems the authors will only look for and find generalized or weak solutions. Methods of improving the smoothness of generalized solutions are not addressed in our book.

The second part of the book studies problems on the motion of an ideal fluid in a stationary or moving container. In Chapter 3, we consider the problem of oscillations of a heavy ideal fluid. Chapter 4 deals with the problem of oscillations of a capillary fluid under conditions similar to weightlessness, and the problem of oscillations of a fluid in a container with an elastic bottom. In Chapter 5, we discuss several new problems based on an operator approach that is different from the approaches used in some other chapters of the book. Finally, in Chapter 6 we construct a theory of small oscillations of an ideal homogeneous fluid that partially fills a container and, in an unperturbed state, is uniformly rotating together with that container.

The second volume is divided into two parts as well, and addresses mainly nonself-adjoint problems on small oscillations of hydrodynamic systems containing a viscous fluid. In Part III, several problems on the motion of bodies with a cavity filled with a homogeneous viscous fluid are considered: Chapter 7 describes the case of completely filled cavities, Chapter 8 deals with cavities filled with a heavy fluid, and Chapter 9 is devoted to the study of cavities filled with a capillary fluid. Part IV of the book is concerned with the study of small oscillations of complex hydrodynamic systems. In Chapter 10, partially dissipative hydrosystems are studied for the first time. Chapter 11 is devoted to problems on small oscillations of a visco-elastic or relaxing fluid.

The cross-references throughout the book are specified by two- or three-number labels: the first number indicates the chapter, the second one refers to the the section, and the third one—if any—is the number of the subsection. Therefore, sections have two-number labels and subsections have three-number labels.

Part I

MATHEMATICAL FOUNDATIONS OF LINEAR HYDRODYNAMICS

This part of the book is designed as an auxiliary intended to provide the necessary prerequisites from functional analysis. The presentation is focused on various issues related to evolution operator equations in Hilbert spaces and spectral problems for such equations. The readers acquainted in part with this kind of problems are advised to take a brief look at the general ideas outlined in Chapter 1 and illustrated to a greater extent in the existing literature on this subject. Section 1.3 and Chapter 2 require a more attentive and detailed study.

Chapter 1

Operators on Hilbert Spaces

This chapter gathers some facts and outlines results from the theory of Hilbert spaces and operator theory. Particular attention is paid to a few special, yet well-known issues, such as the theory of operator pencils and asymptotic methods.

1.1 General Facts

This section informally introduces the basic concepts pertaining to the theory of Hilbert space operators. Emphasis is placed on several theoretical facts that are used in other parts of the book. Most of this basic material is stated without proofs.

1.1.1 The Concept of a Hilbert Space

Let E be a linear space over $\mathbb{C}$, the field of complex numbers. A *scalar product* on E is a mapping that assigns to every ordered pair of elements x, y in E a complex number (x, y), called the *scalar product of x and y*, with the following properties:

1. $(x, y) = \overline{(y, x)}$;
2. $(x_1 + x_2, y) = (x_1, y) + (x_2, y)$;
3. $(\lambda x, y) = \lambda(x, y)$, where λ is a complex number;
4. $(x, x) \geq 0$ for each x, and $(x, x) = 0$ only when $x = 0$.

Whenever E is a linear space over $\mathbb{R}$, the field of real numbers, one implicitly assumes that each scalar product (x, y) is a real number.

Properties 1–4 lead to some simple corollaries:

1. $(x, y_1 + y_2) = (x, y_1) + (x, y_2)$;
2. $(x, \lambda y) = \bar{\lambda}(x, y)$;
3. $|(x, y)| \leq \sqrt{(x,x)} \cdot \sqrt{(y,y)}$ (**Buniakowsky-Schwarz Inequality**).

Using the scalar product on E, one defines a *norm* in this space by setting

$$\|x\| := \sqrt{(x,x)}, \qquad x \in E.$$

The norm has the following properties:

1. $\|x\| \geq 0$ and $\|x\| = 0$ only when $x = 0$;
2. $\|x + y\| \leq \|x\| + \|y\|$ (**Triangle Inequality**);
3. $\|\lambda x\| = |\lambda| \, \|x\|$.

Employing the norm, one can introduce the important concept of convergence on E as follows. A sequence $\{x_n\}$ in E converges to an element x_0—in symbols, $x_n \to x_0$ as $n \to \infty$—if and only if the numerical sequence $\{\|x_n - x_0\|\}$ converges to 0, that is, $\|x_n - x_0\| \to 0$ as $n \to \infty$.

Furthermore, a sequence $\{x_n\}$ in E is called *fundamental*, if $\|x_n - x_m\| \to 0$ as $n, m \to \infty$. The space E is called *complete* provided every fundamental sequence in E converges to an element of this space. A complete infinite-dimensional space E with a scalar product is called a—complex or real—*Hilbert space*. Every infinite-dimensional linear space with a scalar product which is not complete can be completed in a standard way to a Hilbert space.

A set of elements of a Hilbert space E is called *closed*, if it contains the limits of all its convergent sequences.

A closed linear subset in E is referred to as a *subspace* of E.

A set is called *relatively compact*, provided from any of its sequences one can select a convergent subsequence.

1.1.2 The Space $L_2(\Omega)$

Let Ω be a region in the n-dimensional space $\mathbb{R}^n$, where $n = 2, 3$. The space $L_2(\Omega)$ consists of all equivalence classes of real-, or complex-valued measurable functions $u(x)$ on Ω, such that $|u(x)|^2$ is integrable. In the case of real-valued functions, the standard scalar product on $L_2(\Omega)$ is introduced according to the formula

$$(u, v)_{L_2(\Omega)} := \int_\Omega u(x)v(x)d\Omega.$$

For complex-valued functions one sets

$$(u, v)_{L_2(\Omega)} := \int_\Omega u(x)\overline{v(x)}d\Omega.$$

The space $L_2(\Omega)$ is complete, so it is a Hilbert space in both cases.

1.1.3 Orthogonality. Projection onto a Subspace

Two elements, x and y, of a Hilbert space E are said to be *orthogonal*, if $(x, y) = 0$. An element $x \in E$ is called orthogonal to a set $G \subseteq E$, if $(x, y) = 0$ for any $y \in G$. Finally, two subspaces G_1 and G_2 are called orthogonal, provided each element of G_1 is orthogonal to every element of G_2.

Let L be a subspace of E. The collection of all elements of E that are orthogonal to L forms a subspace $M = L^{\perp}$, called the *orthogonal complement* to L. The two subspaces M and L have only the zero element in common.

One of the most important properties of a Hilbert space is the following **Projection Theorem**:

If L is a subspace of a Hilbert space E, then every $x \in E$ has a unique representation as

$$x = y + z,$$

where y is in L and z is orthogonal to L. The element y is called the projection of x into L.

Thus, each element of E splits up into a sum of two elements, one in the subspace L, and the other in M, the orthogonal complement to L. In other words, the whole space E is split into a direct sum consisting of the subspace L and its orthogonal complement M. Hence, we can write

$$E = L \oplus M.$$

If the orthogonal complement to L has finite dimension $n \geq 1$, then it is said that the subspace L has *codimension*, or *defect*, n.

A linear subset $L \subseteq E$ is called *dense* in E, provided each element of E is the limit of a sequence of elements from L.

A set $G \subseteq E$ is called a *complete set* in E, if all the possible linear combinations of elements from G form a dense set in E. In this regard, the Projection Theorem has an important corollary:

A set $G \subseteq E$ is complete in E if and only if there is no nonzero element in E that is orthogonal to the subset G.

A set of elements of E is called *minimal*, provided each element of that set does not belong to the closed linear hull of the subset formed by the other elements.

1.1.4 Equivalent Norms

The scalar product and the associated norm can be introduced on a linear space E in many different ways. Two norms $\|\cdot\|_1$ and $\|\cdot\|_2$ on E are called *equivalent*, provided there are two positive constants c_1 and c_2 such that

$$c_1\|x\|_2 \leq \|x\|_1 \leq c_2\|x\|_2, \qquad x \in E.$$

One should notice that two equivalent norms give rise to the same topology on E. Consequently, all the topological properties connected with convergence, closure, density, or compactness, are the same for both norms.

The following result, which is just a possible form of the **Banach Theorem**, provides a test for equivalence.

Let $\|\cdot\|_1$ and $\|\cdot\|_2$ be two norms on a linear space E. If E is complete relatively to each of them and

$$\|x\|_1 \leq c_2\|x\|_2, \qquad x \in E,$$

then the two norms are equivalent.

1.1.5 Linear Functionals. Riesz Theorem

Suppose that E is a Hilbert space. A *functional* l on E is a mapping that assigns a complex or real number $l(x)$ to each element x in E. A functional l is called *linear*, if

$$l(\alpha x + \beta y) = \alpha l(x) + \beta l(y)$$

for any x, y in E and all real or complex numbers α, β.

A linear functional is said to be *bounded*, provided there exists a constant c such that

$$|l(x)| \leq c\|x\|, \qquad x \in E.$$

The smallest value of c that can be used in this inequality is denoted by $\|l\|$ and is referred to as the *functional norm*. Bounded linear functionals on E form a linear space relatively to the standard operations of pointwise addition and multiplication by numbers. This space is called the *dual space* to E and is denoted by E^*.

The fundamental **Riesz Theorem** formulated below provides a simple description of dual spaces.

Every bounded linear functional $l(x)$ on a Hilbert space E can be represented in terms of the scalar product of E as

$$l(x) = (x, u), \qquad x \in E,$$

where u is an element of E uniquely determined by l and satisfying $\|l\| = \|u\|$.

According to Riesz Theorem, the dual space E^* to a Hilbert space E can be identified with the space E itself. For example, one may consider the Hilbert space $L_2(\Omega)$ introduced in 1.1.2. In this case, for each bounded linear functional $l(v)$ on $L_2(\Omega)$ there is a unique function $u(x) \in L_2(\Omega)$, such that

$$l(v) = \int_\Omega v(x)\overline{u(x)}\mathrm{d}\Omega$$

for any v in $L_2(\Omega)$.

1.1.6 Embeddings of Spaces. Riesz Theorem for Equipments

Suppose that E and F are two Hilbert spaces. One says that the space F is *embedded* in the space E, provided F is a linear subset of E and the inequality $\|x\|_E \leq a\|x\|_F$ is satisfied for every $x \in F$, where a is a certain constant. This circumstance is indicated by using the notation $F \hookrightarrow E$. If, in addition, F is dense in E, then one says that F is *densely embedded* in E. Whenever the bounded sets in F are relatively compact in E, one says that F is *compactly embedded* in E.

Assume now that $F \hookrightarrow E$ and let $(\cdot, \cdot)$ be the scalar product on E. Suppose $y \in E$ is fixed and set $y(x) := (x, y)$ for each $x \in F$. In this way one gets a linear functional on F. Since

$$|y(x)| = |(x, y)| \leq \|x\|_E \cdot \|y\|_E \leq a\|x\|_F\|y\|_E$$

for every $x \in F$, it follows that this functional is bounded. Therefore, each element $y \in E$ can be identified with a functional from F^*. If F is dense in E, then different elements of E generate different functionals in F^*. Under the identification mentioned above, E is dense in the space F^*. If the opposite were true, there would exist a nonzero element $z \in F$, such that $(z, y) = 0$ for every $y \in E$. In particular, $(z, z) = 0$, so $z = 0$ and one reaches a contradiction.

Therefore, whenever l is a bounded linear functional on E, there exists a sequence $\{y_n\}$ of elements of E, such that

$$l(x) = \lim_{n\to\infty} y_n(x) = \lim_{n\to\infty} (x, y_n).$$

This formula makes it possible to extend the scalar product on E in the case when one multiplier belongs to F and the other to F^*.

Specifically, if $x \in F$ and $l \in F^*$ one sets

$$(x, l) = l(x) = \lim_{n\to\infty} (x, y_n).$$

In effect one gets that each functional $l \in F^*$ can be represented in terms of the scalar product on E. This result is referred to as the **Riesz Theorem for Equipments.** To make a point, one needs to recall that $F \hookrightarrow E \hookrightarrow F^*$. Such chains of embeddings are called *equipments* of the space E. It is implicit in the definition of an equipment $F \hookrightarrow E \hookrightarrow F^*$ that F is densely embedded in E and E is densely embedded in F^*.

1.1.7 Orthonormal Systems and Bases

A set of elements $\{e_\alpha\}$ of E is called an *orthonormal system*, provided $(e_\alpha, e_\beta) = 0$ for $\alpha \neq \beta$ and $(e_\alpha, e_\alpha) = 1$, that is, $(e_\alpha, e_\beta) = \delta_{\alpha\beta}$. For any element x in E, the numbers (x, e_α) are called *Fourier coefficients* of x corresponding to the orthonormal system $\{e_\alpha\}$. The series $\sum_\alpha (x, e_\alpha)e_\alpha$ is called a *Fourier series* for x. Fourier coefficients satisfy the **Bessel inequality**

$$\sum_\alpha |(x, e_\alpha)|^2 \leq \|x\|^2.$$

An orthonormal system $\{e_\alpha\}$ which is a complete set in E is referred to as an *orthonormal basis* for E. If $\{e_\alpha\}$ is an orthonormal basis, each element x in E can be expanded as a Fourier series in the unique way

$$x = \sum_\alpha (x, e_\alpha)e_\alpha,$$

and the **Parseval identity**

$$\|x\|^2 = \sum_\alpha |(x, e_\alpha)|^2$$

is true.

A Hilbert space E is called *separable*, provided there is a countable orthonormal basis for it. For instance, the space $L_2(\Omega)$ defined in 1.1.2 above is separable.

1.1.8 Bounded Linear Operators

Let E and F be two Hilbert spaces. An operator A from E to F is a mapping that assigns to each element x in E a certain element y in F. The correspondence between x and y is indicated by writing $y = Ax$.

An operator A is called *linear*, if

$$A(\alpha x_1 + \beta x_2) = \alpha A x_1 + \beta A x_2 \tag{1.1}$$

for any $x_1, x_2 \in E$ and any numbers α, β.

A linear operator is called *bounded*, provided there is a constant c, such that

$$\|Ax\|_F \leq c\|x\|_E$$

for all $x \in E$. The smallest value of c such that this inequality holds for every x is called the *operator norm*, and is denoted by $\|A\|$.

The space of all bounded linear operators acting from E to F is denoted by $\mathcal{L}(E, F)$. In the special case when $E = F$, instead of $\mathcal{L}(E, F)$ we may use the simpler notation $\mathcal{L}(E)$, or just $\mathcal{L}$.

Suppose now that A is an operator from E to F. The subset of F consisting of all the elements $y = Ax$, where x ranges over E, is called the *range* of A and we denote it by $\mathcal{R}(A)$. The range of a linear operator is a linear subset—possibly not closed—of F.

Assume next that A is a linear operator from E into F and $0 \in E$ is the only element of E transformed by A into $0 \in F$. An operator with this properties defines a one-to-one correspondence between E and $\mathcal{R}(A)$. Under such assumptions, there exists an algebraic *inverse operator* to A, denoted by A^{-1}, that assigns to every element $y \in \mathcal{R}(A)$ its preimage $x \in E$. In other words, if $y = Ax$, then $x = A^{-1}y$. From Banach Theorem, we get the following result:

If a bounded linear operator A acting from a Hilbert space E to a Hilbert space F is one-to-one and the inverse operator A^{-1} is defined on the whole space F, then A^{-1} is a bounded linear operator.

In the case when such an operator exists, the spaces E and F are called *isomorphic*.

A bounded linear operator A from E into F is called *isometric*, provided it preserves the norm, that is,

$$\|Ax\|_F = \|x\|_E,$$

for all $x \in E$. Each isometric operator has an inverse, which is also isometric. An isometric operator that maps a Hilbert space E *onto* itself is called *orthogonal* or *unitary*, if E is a real or a complex Hilbert space, respectively.

A linear operator A from E to F is called *compact* or *completely continuous*, provided A transforms any bounded subset of E into a relatively compact set in F. There is an equivalent definition of a compact operators that uses weakly convergent sequences and yields an easier test for compactness. To state it let us recall that a sequence $\{x_n\}$ in E is said to converge weakly to x_0 if $\lim_{n\to\infty} (x_n, y) = (x_0, y)$ for any $y \in E$. Each weakly convergent sequence is bounded and from every bounded set in E one can select a weakly convergent sequence. It is not difficult to prove that whenever a sequence $\{x_n\}$ in E converges weakly to x_0 and a second sequence $\{y_n\}$, also in E, converges in norm to y_0, then

$$(x_n, y_n) \to (x_0, y_0) \qquad \text{as } n \to \infty.$$

A linear operator from E to F is compact if and only if it transforms every weakly convergent sequence in E into a sequence in F that converges in norm.

From the previous assertions one easily infers that whenever A is a compact linear operator on E and the sequence $\{x_n\}$ in E converges weakly to x, then

$$(Ax_n, x_n) \to (Ax, x) \qquad \text{as } n \to \infty.$$

1.1.9 Adjoint Operators

Let A be a bounded linear operator acting from E to F. Suppose that $y \in F$ in the scalar product $(Ax, y)_F$ is fixed and $x \in E$ is arbitrary. In this setting, to each $y \in F$ one associates a functional l on E defined as

$$l(x) := (Ax, y)_F, \qquad x \in E.$$

Since l is a bounded linear functional, by the Riesz Theorem there exists a unique element $u \in E$ such that

$$l(x) = (x, u)_E, \qquad x \in E.$$

Thus, every $y \in F$ corresponds to an element $u \in E$. This mapping from F to E is called the *adjoint operator* to A and is denoted by A^*. Combining the last two equations with $u = A^*y$ one gets

$$(Ax, y)_F = (x, A^*y)_E \qquad x \in E,\ y \in F.$$

The operator A^* turns up to be a bounded linear operator from F to E. Moreover, A^* and A have the same norm.

1.1.10 Self-Adjoint Operators

A bounded linear operator A mapping a Hilbert space E to itself is called *self-adjoint*, provided A coincides with its adjoint operator, that is, $A = A^*$, or

$$(Ax, y) = (x, Ay), \qquad x, y \in E.$$

Orthogonal projections onto subspaces of E are important examples of self-adjoint operators. To be specific, let L be a subspace in E. By the Projection Theorem in 1.1.3, every element $x \in E$ has a uniquely determined projection y into L. Setting $y = P_L x$, one gets an operator P_L on E, called the *orthogonal projection, or orthogonal projector,* onto L. It turns out that P_L is self-adjoint and its norm equals 1.

According to the previous definition, $(P_L)^2 = P_L$. Conversely, every self-adjoint linear operator P, which satisfies $P^2 = P$, is the orthogonal projector onto its own range.

To each linear operator A on E one associates a bilinear form (Ax, y) called the *bilinear form* of A, and the corresponding quadratic form (Ax, x), called the *quadratic form* of A. The quadratic form of a self-adjoint operator is a real-valued functional, even in complex Hilbert spaces. This property can be used to define special types of self-adjoint operators as follows.

A self-adjoint operator A is called *nonnegative*—in symbols, $A \geq 0$—if $(Ax, x) \geq 0$ for every $x \in E$; A is called *positive*—in symbols, $A > 0$—if $(Ax, x) > 0$ for any $x \neq 0$ in E; finally, A is called *positive definite* or *uniformly positive*—in symbols, $A \gg 0$—whenever there exists $a > 0$ such that $(Ax, x) \geq a\|x\|^2$ for all $x \in E$. In the case when the inequality $(Ax, x) \geq b\|x\|^2$ holds true for a certain real number b and each $x \in E$, the operator A is said to be *bounded from below* by b; under these assumptions it follows that $A - bI \geq 0$, where I is the identity operator, $Ix := x$, $x \in E$.

If A is a bounded linear operator acting from E to F, then A^*A is a self-adjoint nonnegative operator on E and AA^* has the same properties as an operator on F.

1.1.11 Self-Adjoint Compact Operators

If a self-adjoint operator A is compact, then the whole space E can be split up into the orthogonal sum of two subspaces, $E = E_0 \oplus E_1$. These subspaces have the following properties:

1. Operator A turns into the zero-operator on the subspace E_0, that is,

$$Ax_0 = 0 \qquad x_0 \in E_0;$$

2. The subspace E_1 has an orthonormal basis $\{e_k\}_{k=1}^{\infty}$, $e_k = e_k(A)$, consisting of *eigenelements* of the operator A. These elements satisfy the equations

$$Ae_k = \mu_k e_k \quad \mu_k \neq 0, \qquad k = 1, 2, \ldots.$$

The numbers $\mu_k = \mu_k(A)$ are called *eigenvalues* of the operator A.

Thus, any $x \in E$ can be represented as

$$x = x_0 + \sum_k (x, e_k)e_k, \tag{1.2}$$

while

$$Ax = \sum_k \mu_k (x, e_k) e_k. \tag{1.3}$$

This important result is known as the **Hilbert-Schmidt Theorem.**

According to the previous equations, $(Ae_k, e_k) = \mu_k$ and, therefore, the eigenvalues are real. Moreover, $(Ax, x) = \sum_k \mu_k |(x, e_k)|^2$. Consequently, a compact self-adjoint operator is nonnegative if and only if its eigenvalues are nonnegative.

The properties of self-adjoint compact operators outlined above are generalizations of properties of integral operators with a symmetric kernel. The following equation is a generalization of some integral equations,

$$x - \lambda Ax = y.$$

If $1/\lambda$ does not coincide with any of the eigenvalues of operator A, then this equation can be solved by the formula

$$x = y + \sum_k \frac{\lambda \mu_k}{1 - \lambda \mu_k} (y, e_k) e_k.$$

If $1/\lambda$ does coincide with one or several μ_i, then a solution exists if and only if y is orthogonal to the corresponding elements e_i. That solution can be found using the same formula as above, with the sum extended only over indices k such that $1/\lambda \neq \mu_k$.

To conclude, one notices that an orthogonal projection P_L can be compact if and only if the subspace L is finite-dimensional. Such projections have a finite number of nonzero eigenvalues and all of them are equal to 1.

1.1.12 Compact operators. s-numbers

Compact operators acting in a Hilbert space E have the following properties:

1. Any linear combination of compact operators is a compact operator, and

2. The product (from the left or from the right) of a compact and bounded operator is a compact operator.

Let A be a compact operator on E. According to the previous remarks, the operator A^*A is compact. As it was mentioned in Section 1.1.10, this operator is nonnegative and self-adjoint. Its positive eigenvalues can be listed in descending order, $s_1^2 \geq s_2^2 \geq s_3^2 \geq \cdots$, according to their multiplicity. The corresponding eigenelements are denoted by f_k, $k = 1, 2, 3, \ldots$. The numbers $s_k = s_k(A)$ are called the *s-numbers* of operator A. According to the definition,

$$A^*Af_k = s_k^2 f_k.$$

If $Af_k = s_k g_k$, then $A^* g_k = s_k f_k$. Expansions (1.2) and (1.3) have the following forms

$$x = x_0 + \sum_k (x, f_k) f_k, \quad Ax_0 = 0, \quad Ax = \sum_k s_k (x, f_k) g_k.$$

Compact operators are classified according to the decreasing rate of their s-numbers. The symbol $\mathfrak{S}_p$, $p \geq 1$, will denote the set of all compact operators that satisfy the following condition

$$(\|A\|_p)^p := \sum_k (s_k(A))^p < \infty.$$

The class $\mathfrak{S}_p$ is a linear space. Multiplication from the left or from the right of an operator from $\mathfrak{S}_p$ by any bounded operator results in an operator in the class $\mathfrak{S}_p$. If $p < q$, the embedding $\mathfrak{S}_p \hookrightarrow \mathfrak{S}_q$ holds true.

If $A \in \mathfrak{S}_p$, $B \in \mathfrak{S}_q$ and $1/p + 1/q < 1$, then the product $AB \in \mathfrak{S}_r$, where $1/r = 1/p + 1/q$. Finally, if $A \in \mathfrak{S}_p$, then for each $k \in \mathbb{N}$ the operator A^k belongs to $\mathfrak{S}_{p/k}$ and $\|A^k\|_{p/k} \leq (\|A\|_p)^k$.

Operators in the class $\mathfrak{S}_1$ are called *nuclear*; operators from $\mathfrak{S}_2$ are called *Hilbert-Schmidt operators*. The class of all compact operators is usually denoted by $\mathfrak{S}_\infty$. If $A \in \mathfrak{S}_p$ $(p < \infty)$, then one says that A has a finite order.

1.1.13 Riesz Bases and p-Bases

A system of elements $\{\varphi_k\}_{k=1}^\infty$ is called a *basis* for a separable Hilbert space E, provided any $x \in E$ has a unique representation

$$x = \sum_{k=1}^\infty a_k \varphi_k,$$

where the series converges in the norm of the space E.

If $\{e_k\}_{k=1}^\infty$ is an orthonormal basis in E and Q is a bounded linear operator which has a bounded inverse operator, then the system $\{Qe_k\}_{k=1}^\infty$ is a basis in E. Indeed, for each $x \in E$ the element $Q^{-1}x$ can be expanded as a Fourier series in the unique way

$$Q^{-1}x = \sum_{k=1}^\infty (Q^{-1}x, e_k) e_k.$$

Hence

$$x = \sum_{k=1}^\infty (Q^{-1}x, e_k) Qe_k,$$

and this representation is unique.

A basis of this type is called equivalent to an orthonormal basis, or in short, a *Riesz basis.* If $Q = I + T$ where $T \in \mathfrak{S}_p$, then the basis $\{Qe_k\}_{k=1}^{\infty}$ is called a *p-basis.*

A basis is called *unconditional,* if any permutation of its elements yields a new basis. A basis $\{f_k\}_{k=1}^{\infty}$ is a Riesz basis if and only if it is an unconditional basis, and the sequence $\{f_k\}_{k=1}^{\infty}$ is almost normalized, that is,

$$\inf_k \|f_k\| > 0, \qquad \sup_k \|f_k\| < \infty.$$

Whenever one has an expansion

$$x = \sum_{k=1}^{\infty} c_k f_k$$

corresponding to a Riesz basis $\{f_k\}_{k=1}^{\infty} = \{Qe_k\}_{k=1}^{\infty}$, from the Parseval identity for the orthonormal basis $\{e_k\}_{k=1}^{\infty}$ one gets

$$\|Q^{-1}\|^2 \sum_{k=1}^{\infty} |c_k|^2 \le \|x\|^2 \le \|Q\|^2 \sum_{k=1}^{\infty} |c_k|^2.$$

1.1.14 Direct Sum of Subspaces. Invariant Subspaces

One says that the space E is decomposed into a direct sum of its subspaces L and M, in symbols,

$$E = L \dot{+} M, \tag{1.4}$$

provided every element $x \in E$ can be represented as $x = y + z$ with $y \in L$ and $z \in M$ in an unique way. The operator P that maps each element x into its component y is called a *projection.* This operator is bounded and $P^2 = P$. Conversely, every bounded operator P for which $P^2 = P$ corresponds to the decomposition into a direct sum

$$E = PE \dot{+} (I - P)E. \tag{1.5}$$

Each bounded linear operator acting on E can be represented as a matrix. First, one represents every element $x \in E$ as a column vector $(y; z)^{\mathrm{t}}$, where y and z are the components of x in the direct sum (1.5). In effect, each operator A on E corresponds to a 2×2 operator matrix,

$$A = \begin{pmatrix} A_{11} & A_{12} \\ A_{21} & A_{22} \end{pmatrix}. \tag{1.6}$$

Here one sticks to the standard rules used to multiply a column vector by a matrix. Specifically,

$$A_{11} = PAP, \quad A_{12} = PA(I-P), \quad A_{21} = (I-P)AP, \quad A_{22} = (I-P)A(I-P),$$

where P is the projection corresponding to the decomposition (1.5).

A subspace L is called an *invariant subspace* of a bounded linear operator A, if for any $y \in L$ the element Ay also belongs to L.

If the subspace L in the decomposition (1.4) is invariant, then the matrix that represents the operator A is

$$A = \begin{pmatrix} A_{11} & A_{12} \\ 0 & A_{22} \end{pmatrix}. \tag{1.7}$$

Moreover, if both L and M are invariant subspaces, then the matrix representation is diagonal,

$$A = \operatorname{diag}(A_{11}; A_{22}) := \begin{pmatrix} A_{11} & 0 \\ 0 & A_{22} \end{pmatrix}. \tag{1.8}$$

In this case the study of the operator A is reduced to the examination of its restrictions A_{11} and A_{22} to the subspaces L and M.

The decomposition of a space into a direct sum of n subspaces and the representation of an operator as an $n \times n$ matrix can be introduced in quite a similar way.

One should notice that the orthogonal decomposition considered in Section 1.1.3 is a special case of the decomposition (1.4).

1.1.15 Eigen- and Associated (Root) Elements. Root Subspaces

A nonzero element $x \in E$ is called a *root element* corresponding to an eigenvalue $\lambda_0 \in \mathbb{C}$ of an operator $A \in \mathcal{L}(E)$, provided

$$(A - \lambda_0 I)^n x = 0 \tag{1.9}$$

for some $n \in \mathbb{N}$.

If $x = x_0$ and $n = 1$, then x_0 is an eigenelement of the operator A that corresponds to the eigenvalue λ_0. If $x = x_1$ in (1.9), $n = 2$, and $(A - \lambda_0 I)x \neq 0$, then it is said that x_1 is a first associated element to the eigenelement x_0. In the same way one can define a second, a third and all the subsequent associated elements. This process yields a chain $x_0, x_1, \dots, x_m, \dots$ (finite or infinite), which consists of linearly independent elements and is called a *Jordan chain* corresponding to the eigenelement x_0. The maximum length of such a chain for a given x_0, that is, the total number of nonzero elements, is called the *multiplicity* of the element x_0.

The linear hull of all root (i.e., eigen- and associated) elements of the operator A corresponding to the same eigenvalue λ_0 forms the *root lineal* $L_{\lambda_0}(A)$; its dimension $\nu_{\lambda_0} = \nu_{\lambda_0}(A)$ is called the *algebraic multiplicity* of the eigenvalue λ_0. If $\nu_{\lambda_0} < \infty$, the root lineal $L_{\lambda_0}(A)$ is closed, and in this case one can speak about the *root subspace* that corresponds to the eigenvalue λ_0. Any root subspace $L_\lambda(A)$ of an operator A is invariant relatively to A.

The *eigensubspace* $\mathcal{Z}_{\lambda_0}(A)$, that is, the linear hull of all the eigenelements of an operator A corresponding to the eigenvalue λ_0, is a part of $L_{\lambda_0}(A)$, that is, $\mathcal{Z}_{\lambda_0}(A) \subseteq L_{\lambda_0}(A)$. The dimension $\alpha_{\lambda_0} \leq \nu_{\lambda_0}$ of the subspace $\mathcal{Z}_{\lambda_0}(A)$ is called the *eigenmultiplicity* of the eigenvalue λ_0. The eigenelements of a self-adjoint operator have no associated elements, hence $\alpha_{\lambda_0} = \nu_{\lambda_0}$

1.1.16 Unbounded Linear Operators

The vast majority of linear differential operators we will be dealing with are not bounded as operators from a Hilbert space E to a Hilbert space F. As a rule, they are not even defined for all the elements of the space E. In essence, one deals with an operator A that has the linearity property (1.1), but is defined only for elements of some linear subset $\mathcal{D}(A) \subseteq E$, which is called the *domain of definition* of this operator. If $\mathcal{D}(A)$ in dense in E, then it is said that the operator A is *densely defined* in E. If, under this condition, the operator A is bounded in $\mathcal{D}(A)$, that is, there exists a constant c such that

$$\|Ax\|_F \leq c\|x\|_E \qquad x \in \mathcal{D}(A), \tag{1.10}$$

then A can be continuously extended to the whole space E in the following way: for each $x \in E$ one first selects a sequence $x_n \in \mathcal{D}(A)$ such that $x_n \to x$, and then sets

$$Ax := \lim_{n\to\infty} Ax_n.$$

The limit exists, and according to inequality (1.10) Ax is correctly defined. Thus, one gets a bounded operator defined on the whole space E.

A more complicated issue is to determine whether an operator is closed or not. A linear operator A is called *closed*, if from $x_n \to x$ and $Ax_n \to y$ it follows that $x \in \mathcal{D}(A)$ and $y = Ax$.

It turns out that whenever a closed operator is defined on the whole space E, then that operator is bounded.

A linear operator A_1 is called an *extension* of a linear operator A, provided $\mathcal{D}(A_1) \supset \mathcal{D}(A)$ and $A_1x = Ax$ when $x \in \mathcal{D}(A)$. An operator A_2 is called a *restriction* of a linear operator A on a subset $\mathcal{D} \subset \mathcal{D}(A)$, in symbols, $A_2 = A \mid \mathcal{D}$, if $\mathcal{D}(A_2) = \mathcal{D}$ and $A_2x = Ax$ for each $x \in \mathcal{D}$.

A linear operator A that is not closed admits a closed extension if and only if from $x_n \to 0$, $x_n \in \mathcal{D}(A)$, and $Ax_n \to y$, it follows that $y = 0$. If A has closed extensions, then there is a smallest one, which, in turn, is a restriction of any other extension. This smallest extension is called the *closure* of the operator A and is denoted by $\bar{A}$. If a closed operator A has an inverse, then the inverse operator A^{-1} is also closed.

1.1.17 Resolvent and Spectrum of a Linear Operator

Let G be a region of the complex plane $\mathbb{C}$ and suppose that every point λ in this region corresponds to a bounded linear operator $A(\lambda)$ acting on a complex Hilbert space E. In this case one says that an operator-valued function is defined on G; this function takes its values in $\mathcal{L}(E)$. Usually, such an operator-valued function $A(\lambda)$ is called a *pencil.*

An operator-valued function $A(\lambda)$ is called *holomorphic* at a point $\lambda_0 \in \mathbb{C}$, provided the limit

$$\lim_{\lambda \to \lambda_0} (A(\lambda) - A(\lambda_0))/(\lambda - \lambda_0) =: A'(\lambda_0)$$

exists. The limit here is to be understood in the sense of convergence in the operator norm on the space $\mathcal{L}(E)$. Actually, it would be sufficient if the limit exists for all elements $x \in E$ in the sense of weak convergence.

Many concepts and theorems for ordinary holomorphic functions hold true for holomorphic operator-valued functions. Further, they will be used without special explanation.

The resolvent of a linear operator is one of the most important operator-valued functions. Let A be a closed linear operator acting on a complex Hilbert space E. One considers the equation

$$Ax - \lambda x = y, \qquad \lambda \in \mathbb{C}. \tag{1.11}$$

If this equation has a unique solution for each $y \in E$, then one says that λ is a *regular point* of the operator A. This property is equivalent to the existence of a bounded inverse operator $(A - \lambda I)^{-1}$, defined on the whole space E and called the *resolvent* of the operator A. The set $\rho(A)$ of all the regular points of an operator A is called the *resolvent set* of the operator A. This set is open and the resolvent is a holomorphic operator-valued function on the set $\rho(A)$.

The complement $\sigma(A) := \mathbb{C} \setminus \rho(A)$ to the resolvent set $\rho(A)$ in the whole complex plane is called the *spectrum* of the operator A. The spectrum $\sigma(A)$ is closed. The spectrum of a bounded operator is always bounded and not empty. Obviously, it is located within the circle $|\lambda| \leq \|A\|$.

When $|\lambda| > \|A\|$, the resolvent is defined by the Neumann series

$$(A - \lambda I)^{-1} = -\frac{1}{\lambda}\left(I + \frac{1}{\lambda}A + \frac{1}{\lambda^2}A^2 + \dots\right). \tag{1.12}$$

As expansion (1.12) shows, the resolvent of a bounded operator A has a zero of the first order at infinity. If the operator A is unbounded, then its resolvent has a singular point at infinity. The set that consists of the spectrum of an unbounded operator and the point at infinity is called the *extended spectrum* of that operator.

The spectrum of a compact operator consists of the denumerable set of its nonzero eigenvalues μ_k and, possibly, zero. Moreover, zero may be the only accumulation point, if any, of the spectrum of a compact operator.

The spectrum of a self-adjoint operator is located on the real axis.

The radius $r(A)$ of the smallest circle with center at zero that contains the spectrum of an operator A is called the *spectral radius* of operator A.

The set $W(A)$ of all the numbers (Ax, x), where $\|x\| = 1$, is called the *numerical range* of an operator $A \in \mathcal{L}(E)$. This set is bounded and convex, but, in general, it is not closed.

All the eigenvalues of an operator A belong to $W(A)$. The entire spectrum $\sigma(A)$ of an operator A may not be located within $W(A)$, but it lies within the closure of this set.

1.1.18 Classification of Points in the Spectrum of a Linear Operator

The existing mathematical literature outlines different ways of classifying points in the spectrum $\sigma(A)$ of an operator A. This section briefly reviews the well-known coarse classification of such points.

The set of the points $\lambda \in \sigma(A)$ for which $A - \lambda I$ is not one-to-one is called the *point spectrum* of the operator A and is denoted by $\sigma_{\mathrm{p}}(A)$. In other words, $\lambda \in \sigma_{\mathrm{p}}(A)$ if and only if $Ax = \lambda x$ for some nonzero $x \in E$, that is, $\sigma_{\mathrm{p}}(A)$ is the set of all eigenvalues of the operator A.

One says that $\lambda \in \sigma(A)$ is a point in the *continuous spectrum* $\sigma_{\mathrm{c}}(A)$ of the operator A provided:

1) the operator $A - \lambda I$ is one-to-one;
2) the set $(A - \lambda I)E$ is dense in E and $(A - \lambda I)E \neq E$.

One says that $\lambda \in \sigma(A)$ is a point in the *residual spectrum* $\sigma_{\mathrm{r}}(A)$ of the operator A if:

1) the operator $A - \lambda I$ is one-to-one;

2) the set $(A - \lambda I)E$ is not dense in E.

The spectral sets $\sigma_{\mathrm{p}}(A)$, $\sigma_{\mathrm{c}}(A)$, and $\sigma_{\mathrm{r}}(A)$ are disjoint and

$$\sigma(A) = \sigma_{\mathrm{p}}(A) \cup \sigma_{\mathrm{c}}(A) \cup \sigma_{\mathrm{r}}(A). \tag{1.13}$$

It turns out that if A is a self-adjoint operator, then $\sigma_{\mathrm{r}}(A) = \emptyset$.

The point spectrum $\sigma_{\mathrm{p}}(A)$ includes a subset $\sigma_{\mathrm{d}}(A)$, which is called the *discrete spectrum* of the operator A. The eigenvalues $\lambda \in \sigma_{\mathrm{d}}(A)$ are called *normal eiganvalues* of the operator A. By definition, $\lambda \in \sigma_{\mathrm{d}}(A)$ provided:

1) λ is an isolated point of the spectrum $\sigma(A)$;
2) the root subspace $L_\lambda(A)$ corresponding to λ is finite-dimensional;
3) the whole space E can be decomposed into a direct sum of subspaces

$$E = L_\lambda(A) \dot{+} N_\lambda(A), \tag{1.14}$$

where $L_\lambda(A)$ is as in 2) and $N_\lambda(A)$ is an invariant subspace of the operator A on which $A - \lambda I$ is invertible.

It should be noticed that the direct sum decomposition (1.14) is unique.

1.1.19 Spectrum of a Self-Adjoint Operator. Weyl Theorem

It has already been mentioned that the residual spectrum of a self-adjoint operator A is the empty set, that is, $\sigma(A) = \sigma_{\mathrm{p}}(A) \cup \sigma_{\mathrm{c}}(A)$. For such a self-adjoint operator, in some cases it is more convenient to use a different decomposition of $\sigma(A)$ into disjoint subsets.

Let $\sigma_{\mathrm{d}}(A)$ be the discrete spectrum of the operator A. One introduces the set $\sigma_{\mathrm{e}}(A) := \sigma(A) \backslash \sigma_{\mathrm{d}}(A)$, which is called the *essential spectrum* (*accumulation spectrum*, *limiting spectrum*) of the operator A. Obviously,

$$\sigma(A) = \sigma_{\mathrm{e}}(A) \cup \sigma_{\mathrm{d}}(A), \quad \sigma_{\mathrm{e}}(A) \cap \sigma_{\mathrm{d}}(A) = \emptyset. \tag{1.15}$$

The essential spectrum $\sigma_{\mathrm{e}}(A)$ is a closed set, which contains:

1) all the points of the continuous spectrum $\sigma_{\mathrm{c}}(A)$;
2) the set $\sigma_{\mathrm{p}}^{\infty}(A)$ of all the eigenvalues with infinite multiplicity;
3) any eigenvalue from the closure $\overline{\sigma_{\mathrm{c}}(A)}$.

Thus,

$$\sigma_{\mathrm{e}}(A) = \sigma_{\mathrm{c}}(A) \cup \sigma_{\mathrm{p}}^{\infty}(A) \cup (\overline{\sigma_{\mathrm{c}}(A)} \cap \sigma_{\mathrm{p}}(A)). \tag{1.16}$$

In every component interval from the open set $\mathbb{R} \setminus \sigma_e(A)$, the spectrum of A can have only isolated eigenvalues of finite multiplicity, that is, it is discrete; these eigenvalues may accumulate only at the endpoints of such component intervals.

The next result, which is formulated in terms of the operator A, provides a test for points in $\sigma_e(A)$: a point λ belongs to $\sigma_e(A)$ if and only if there exists a *Weyl sequence* for A at λ, that is, a sequence $\{x_k\}_{k=1}^{\infty}$ in E that converges weakly to zero (in particular, an orthonormal system), such that

$$\inf_k \|x_k\| > 0, \qquad (A - \lambda I)x_k \to 0 \quad \text{as } k \to \infty. \tag{1.17}$$

It turns out that the essential spectrum $\sigma_e(A)$ of a self-adjoint operator A is stable under compact perturbations; the precise result is provided by the **Weyl Theorem**:

If $A = A^* \in \mathcal{L}(E)$ and $V = V^* \in \mathfrak{S}_\infty$, then the operator $B := A + V$ has the same essential spectrum as A, that is, $\sigma_e(A) = \sigma_e(B)$.

1.1.20 Riesz Projections

Let Γ be a simple or composite rectifiable closed curve, or boundary that bounds a region $G_\Gamma \subset \mathbb{C}$ and its completely located within the resolvent set $\rho(A)$ of an operator $A \in \mathcal{L}(E)$. Therefore, the resolvent $R(\lambda) = (A - \lambda I)^{-1}$ is a holomorphic operator-valued function on Γ.

Using the positive orientation on Γ one defines the operator

$$P_\Gamma := -\frac{1}{2\pi i} \oint_\Gamma R(\lambda) \mathrm{d}\lambda \in \mathcal{L}(E). \tag{1.18}$$

This operator, expressed as a *Riesz integral* and referred to as a *Riesz projection*, has a number of remarkable properties:

1. P_Γ is a projection that commutes with A and, therefore, both summands in the direct sum decomposition

   $$E = L_\Gamma \dot{+} N_\Gamma, \quad L_\Gamma := P_\Gamma E, \quad N_\Gamma := (I - P_\Gamma)E$$

   are invariant subspaces of the operator A;
2. The spectrum of the restriction $A \mid L_\Gamma$ of the operator A to the subspace L_Γ is located within the region G_Γ;
3. The spectrum of the restriction $A \mid N_\Gamma$ of the operator A to the subspace N_Γ is located outside of the closure $\bar{G}_\Gamma$;

4. If Γ_1 and Γ_2 are two different boundaries with the properties mentioned above, and the regions G_{Γ_1} and G_{Γ_2} have no common points, then for the corresponding projections,
$$P_{\Gamma_1} P_{\Gamma_2} = P_{\Gamma_2} P_{\Gamma_1} = 0.$$

Hence, if the spectrum of the operator A has only a finite number of points $\{\lambda_k\}_{k=1}^n$ in G_Γ, then

$$P_{\Gamma_1} = \sum_{k=1}^{n} P_{\lambda_k}, \quad P_{\lambda_k} P_{\lambda_j} = 0, \qquad k \neq j,$$

where P_{λ_k} $(k = 1, 2, \ldots, n)$ are projections onto the invariant subspaces $P_{\lambda_k} E$ that coincide with the root subspaces $L_{\lambda_k}(A)$ of the operator A for each eigenvalue $\lambda = \lambda_k$.

The Riesz projections P_Γ can be used to characterize eigenvalues in the discrete spectrum $\sigma_{\mathrm{d}}(A)$ of an operator A. To be specific, a point λ_0 is in $\sigma_{\mathrm{d}}(A)$ if and only if λ_0 is an isolated point of $\sigma(A)$ and the corresponding projection

$$P_{\lambda_0} = -\frac{1}{2\pi i} \oint_{|\lambda - \lambda_0| = \delta} R(\lambda) \mathrm{d}\lambda \tag{1.19}$$

is finite-dimensional.

Riesz integrals allow the introduction of a certain natural class of functions of bounded operators by means of an analogue to the Cauchy integral formula. To make the point, assume that $f(\lambda)$ is a holomorphic function in an open neighborhood of the spectrum $\sigma(A)$ of an operator A, and let Γ be a boundary, which lies in that neighborhood and contains the spectrum $\sigma(A)$. Then, by definition,

$$f(A) := -\frac{1}{2\pi i} \oint_\Gamma f(\lambda) R(\lambda) \mathrm{d}\lambda.$$

In particular, if $f(\lambda) = \lambda^k$ $(k \in \mathbb{N})$, then, as one can easily check,

$$A^k = -\frac{1}{2\pi i} \oint_\Gamma \lambda^k R(\lambda) \mathrm{d}\lambda;$$

this result just proves that the introduced definition is meaningful.

1.1.21 Symmetric and Self-Adjoint Operators

A linear operator A with a dense domain of definition $\mathcal{D}(A)$ in a Hilbert space E is called *symmetric* provided

$$(Ax, y) = (x, Ay) \tag{1.20}$$

for any $x, y \in \mathcal{D}(A)$. If $\mathcal{D}(A) = E$, then (1.20) is the definition of a self-adjoint operator. In the general case, it is not true that the two classes of operators are the same.

If A is a linear, densely defined unbounded operator, then an adjoint operator A^* can be defined in the following way. Suppose y is an element of E such that (Ax, y) is a bounded linear functional of x on $\mathcal{D}(A)$. This functional can be continuously extended to a bounded functional on the whole space E, and, by Riesz theorem, there is a unique element u which satisfies

$$(Ax, y) = (x, u).$$

One proceeds by setting $A^*y = u$. Such elements y form the domain of definition $\mathcal{D}(A^*)$ of the operator A^*. Thus,

$$(Ax, y) = (x, A^*y) \tag{1.21}$$

when $x \in \mathcal{D}(A)$, $y \in \mathcal{D}(A^*)$.

A comparison of (1.21) and (1.20) shows that the operator A^* is an extension of the symmetric operator A. An unbounded operator A is called *self-adjoint*, if $A = A^*$.To test for self-adjointness, one has to show that the inequality

$$|(Ax, y)| \le c\|x\|, \qquad x \in \mathcal{D}(A),$$

leads to $y \in \mathcal{D}(A)$ and

$$(Ax, y) = (x, Ay).$$

A symmetric operator always admits a symmetric closure. A self-adjoint operator is closed.

1.1.22 Spectral Decomposition of Self-Adjoint Operators. Functions of Operators

Let A be an unbounded self-adjoint operator with a domain of definition $\mathcal{D}(A)$. This operator generates a family of orthogonal projections P_λ $(-\infty < \lambda < \infty)$, which have the following properties:

1. $P_\lambda \le P_\mu$ when $\lambda < \mu$;
2. $P_\lambda x$ is a left continuous function of λ for any $x \in E$;
3. $P_{-\infty} = \lim_{\lambda \to -\infty} P_\lambda = 0$, $P_{+\infty} = \lim_{\lambda \to +\infty} P_\lambda = I$.

An element $x \in E$ belongs to $\mathcal{D}(A)$ if and only if

$$\int_{-\infty}^{\infty} \lambda^2 \mathrm{d}(P_\lambda x, x) < \infty, \tag{1.22}$$

and for such an element

$$Ax = \int_{-\infty}^{\infty} \lambda \mathrm{d}P_\lambda x. \tag{1.23}$$

If the operator A is semibounded from below, that is, $(Ax, x) \geq b(x, x)$, then $P_\lambda = 0$ for $\lambda < b$ and

$$Ax = \int_{b}^{\infty} \lambda \mathrm{d}P_\lambda x.$$

If the operator A is bounded, and $m(x, x) \leq (Ax, x) \leq M(x, x)$, then

$$Ax = \int_{m}^{M} \lambda \mathrm{d}P_\lambda x$$

and the integral converges for every $x \in E$.

In virtue of the properties 1–3, the function $(P_\lambda x, x)$ is an increasing, left continuous function of bounded variation, and the integral (1.22) is a Stieltjes integral. The operator integral (1.23) is also defined in a natural way. If A or A^{-1} is a compact operator, then the integral (1.23) equals a sum of the form (1.3).

If $f(\lambda)$ is a continuous function on an interval that contains the spectrum of a self-adjoint operator A, then the operator $f(A)$ can be defined by

$$f(A) := \int_{-\infty}^{\infty} f(\lambda) \mathrm{d}P_\lambda.$$

The so defined operator may be bounded or unbounded, and its domain of definition consists of all x satisfying

$$\int_{-\infty}^{\infty} |f(\lambda)|^2 \mathrm{d}(P_\lambda x, x) < \infty.$$

If the function $f(\lambda)$ is real, then the operator $f(A)$ is self-adjoint. If the function $f(\lambda)$ is bounded, then the operator $f(A)$ is also bounded, and $\|f(A)\| = \sup|f(\lambda)|$, where λ ranges over the spectrum of the operator A. The resolvent $(A - \lambda_0 I)^{-1}$ at a regular point λ_0 can be represented as

$$(A - \lambda_0 I)^{-1} = \int_{-\infty}^{\infty} \frac{1}{\lambda - \lambda_0} \mathrm{d}P_\lambda.$$

Suppose next that A is a positive definite operator, that is, $(Ax, x) \geq a(x, x)$, for each x in $\mathcal{D}(A)$, where $a > 0$. One defines powers of the operator A with arbitrary real exponents by setting

$$A^\alpha := \int_a^\infty \lambda^\alpha \mathrm{d}P_\lambda.$$

If A is unbounded, then each of its positive powers is unbounded, and all the negative powers are bounded. In the case when A is a compact operator, one gets—see (1.3)—

$$A^\alpha x = \sum_k \mu_k^\alpha (x, e_k) e_k.$$

In a similar way, one can define the operators $\exp(-tA)$, $\sin(tA)$, and $\cos(tA)$, where A is a positive definite operator and t is a real parameter. Such functions of operators will be extensively used in Section 1.5.

1.1.23 Spaces with Degenerate Scalar Products. Seminorms

There are many situations when, as a substitute for a scalar product on a linear space E, one needs to use a functional (x, y), $x, y \in E$, that satisfies the first three properties of a scalar product (see Section 1.1.1), and a weak form of the fourth property, namely, $(x, x) \geq 0$ for each x in E. In other words, one could have $(x, x) = 0$ for some nonzero elements x. Such a functional is called a *degenerate* scalar product on E.

The associated functional $\|x\| = \sqrt{(x, x)}$, $x \in E$, is referred to as a *seminorm* on E. It has all the properties of a norm except the first property, which in its new form merely states that $\|x\| \geq 0$ for each x in E. Therefore, a seminorm may be equal to zero for some nonzero elements of E.

Let N be the collection of all the elements z in E such that $\|z\| = 0$, where $\|\cdot\|$ is a seminorm. If $z \in N$, then for each $x \in E$ one has $\|x\| = \|x + z\|$. Indeed, according to the triangle inequality,

$$\|x\| = \|x\| - \|z\| \leq \|x + z\| \leq \|x\| + \|z\| = \|x\|. \tag{1.24}$$

From this property one easily derives that N is a linear subset in E.

Sometimes, in order to use a Hilbert space, it is convenient to deal with the *quotient space* $\hat{E} = E/N$ of space E. The elements of $\hat{E}$ are residue classes of E modulo N, that is, sets of the form $\{x + N\} = \hat{x}$. The linear space operations on $\hat{E}$ are defined in a natural way; the norm on E is introduced by setting $\|\hat{x}\| = \|x\|$. If space E is complete (the same definition as in 1.1.1), then $\hat{E}$ is a Hilbert space.

1.1.24 Equivalent Corrections of Seminorms

Using the same notation as above, assume that N is finite-dimensional. To proceed, one first takes a scalar product $(\cdot,\cdot)_1$ on N with the associated norm $\|\cdot\|_1$, and a projection P, $P^2 = P$, that maps the space E onto its finite-dimensional subspace N. Next, one defines a scalar product on E,

$$(x,y)_{\mathrm{I}} := (Px, Py)_1 + (x,y),$$

and the corresponding norm

$$\|x\|_{\mathrm{I}}^2 = \|Px\|_1^2 + \|x\|^2. \tag{1.25}$$

These scalar product and norm have all the required properties on the whole space E. If E is complete relatively to the initial seminorm, then it is complete, and, therefore, E is a Hilbert space relatively to the norm (1.25).

There are some other natural ways of introducing a scalar product and norm on N. Using a different norm $\|\cdot\|_2$ on N one can introduce a new norm on E by setting

$$\|x\|_{\mathrm{II}}^2 := \|Px\|_2^2 + \|x\|^2.$$

This second norm is equivalent to the norm $\|\cdot\|_{\mathrm{I}}$ if the following inequality holds true

$$\|Px\|_2 \le c\|x\|_{\mathrm{I}}$$

for some $c > 0$ and for all $x \in E$.

If $\|x\|_3$ is another norm on N such that $\|Px\|_3 \le c_1\|x\|_{\mathrm{II}}$ for each $x \in E$, then the norm on E defined as

$$\|x\|_{\mathrm{III}}^2 := \|Px\|_3^2 + \|x\|^2$$

is equivalent to the norm $\|x\|_{\mathrm{I}}$.

The following result is quite useful.

If F is a subspace of E closed in the norm $\|\cdot\|_{\mathrm{I}}$ and satisfying $F \cap N = \{0\}$, then the functional $\|\cdot\|$ yields a norm on F equivalent to the norm $\|\cdot\|_{\mathrm{I}}$ on F.

1.2 Sobolev Spaces

Sobolev functional spaces play an important part in addressing many problems of mathematical physics. The whole theory of these spaces constitutes the subject matter of several monographs. This section introduces only some basic information about properties of functions from the Hilbert space $W_2^1(\Omega) = H^1(\Omega)$, and properties of its subspaces. Traces of such functions on the boundary of a region are also considered. It should be pointed out that the norms in these spaces usually have direct physical meaning, which is connected with energy characteristics of the considered system.

1.2.1 Finite Functions

Further, Ω will denote a bounded region, or domain, in the three-dimensional space $\mathbb{R}^3$, that is, a bounded, open, connected set; $\partial\Omega$ will denote the boundary of this region.

If $\varphi(x)$ is a function defined in Ω, then the closure of the set that consists of all the points $x = (x_1, x_2, x_3)$ such that $\varphi(x) \neq 0$, is called the *support* of this function. A function, which has continuous partial derivatives of all orders and a support located within Ω, is called *finite* in Ω. One defines the extension of a finite function on the whole space $\mathbb{R}^3$ by setting it equal to zero out of the region Ω. The resulting function is infinitely differentiable in $\mathbb{R}^3$.

Functions finite in Ω form a linear set, which lies and is dense in the Hilbert space $L_2(\Omega)$.

1.2.2 Generalized Derivatives

If $\alpha = (\alpha_1, \alpha_2, \alpha_3)$ is an integer multiindex and $|\alpha| = \alpha_1 + \alpha_2 + \alpha_3$, then D^α denotes the partial derivative $D^\alpha = \partial^{|\alpha|}/\partial x_1^{\alpha_1}\partial x_2^{\alpha_2}\partial x_3^{\alpha_3}$. If $\varphi(x)$ is a finite function, and $f(x)$ is a smooth function in Ω, then the following formula of integration by parts holds,

$$\int_\Omega \varphi(x) D^\alpha f(x) d\Omega = (-1)^{|\alpha|} \int_\Omega D^\alpha \varphi(x) f(x) d\Omega.$$

This identity is the base of the definition of a generalized derivative. Specifically, let $f(x)$ and $g(x)$ be two locally summable functions in Ω (i.e., they are summable in any subregion, which is located, including its closure, in Ω), and assume that for any finite function $\varphi(x)$ in Ω the following equation is satisfied,

$$\int_\Omega \varphi(x) g(x) d\Omega = (-1)^{|\alpha|} \int_\Omega D^\alpha \varphi(x) f(x) d\Omega.$$

Then the function $g(x)$ is called the *generalized derivative* of order α for the function $f(x)$ in the region Ω, in symbols, $g(x) := D^\alpha f(x)$.

Functions, which have generalized derivatives of order α, form a linear set, and the operation D^α is linear on this set. The usual formula for the derivative of a product is true,

$$\frac{\partial}{\partial x_i}(f_1 \cdot f_2) = \frac{\partial f_1}{\partial x_i} f_2 + f_1 \frac{\partial f_2}{\partial x_i},$$

as long as one assumes that the right-hand side exists and is locally summable (for example, belongs to $L_2(\Omega)$).

If all generalized derivatives of the first order for some function are equal to zero in Ω, then that function is a constant in Ω.

If a sequence of functions $\{f_n\}$, $f_n \in L_2(\Omega)$, converges to a function f, and the generalized derivatives $D^\alpha f_n$ exist and converge to a function g in $L_2(\Omega)$, then $g = D^\alpha f$.

1.2.3 The Definition of Sobolev Spaces

One says that a function $f(x)$ from $L_2(\Omega)$ belongs to the class $H^l(\Omega)$, provided $f(x)$ has generalized derivatives of order up to l and all such derivatives are in $L_2(\Omega)$. The scalar product in the class $H^l(\Omega)$ is introduced according to the formula

$$(f,g)_{H^l(\Omega)} := \int_\Omega \sum_{|\alpha|\le l} D^\alpha f(x) D^\alpha g(x) \mathrm{d}\Omega,$$

and the class $H^l(\Omega)$ becomes a Hilbert space. In Sobolev's notation, $H^l(\Omega) = W_2^l(\Omega)$.

We will be mostly interested in the space $H^1(\Omega)$; according to the previous formula, the norm in this space is given by

$$\begin{aligned}\|f\|^2_{H^1(\Omega)} &= \int_\Omega |f(x)|^2 \mathrm{d}\Omega + \int_\Omega \sum_{k=1}^{3} \left|\frac{\partial f}{\partial x_i}\right|^2 \mathrm{d}\Omega \\ &= \|f\|^2_{L_2(\Omega)} + \|\ |\nabla f|\ \|^2_{L_2(\Omega)}. \end{aligned} \tag{2.1}$$

1.2.4 The Space $L_2^1(\Omega)$. Regions of the First Type

Let $L_2^1(\Omega)$ be the collection of all summable functions in Ω for which the generalized derivatives of the first order exist and belong to the space $L_2(\Omega)$. One introduces a degenerate scalar product and a seminorm in this space by setting

$$(f,g)_{L_2^1(\Omega)} := \int_\Omega \nabla f \cdot \nabla g \mathrm{d}\Omega, \quad \|f\|^2_{L_2^1(\Omega)} = \int_\Omega |\nabla f|^2 d\Omega.$$

The space on which this seminorm equals zero is one-dimensional and consists of constant functions.

The next question arises naturally: Will a function from $L_2^1(\Omega)$ belong to $L_2(\Omega)$? This fact takes place for a wide class of regions Ω, called *regions of the first type*. It turns out that for such regions, the **Poincare inequality** is true,

$$\|f\|^2_{L_2(\Omega)} \le \left|\frac{1}{\mathrm{mes}\Omega}\int_\Omega f(x)\mathrm{d}\Omega\right|^2 + C(\Omega)\|f\|^2_{L_2^1(\Omega)} \qquad f \in L_2^1(\Omega), C(\Omega) > 0. \tag{2.2}$$

The operator P that assigns to each function $f(x)$ the mean value of that function in the region Ω, that is, $Pf := (\mathrm{mes}\,\Omega)^{-1}\int_\Omega f(x)d\Omega$, is a projection from the whole space $L_2^1(\Omega)$ onto the subspace of constants. Therefore, the right-hand side of (2.2) is the square of the norm $\|\cdot\|_{\mathrm{I}}$ defined in Section 1.1.24, namely,

$$\|f\|^2_{\mathrm{I}} = \left|\frac{1}{\mathrm{mes}\,\Omega}\int_\Omega f(x)\mathrm{d}\Omega\right|^2 + C(\Omega)\|f\|^2_{L_2^1(\Omega)}. \tag{2.3}$$

From the facts mentioned above and from inequality (2.2) it follows that this norm is equivalent to the following one,

$$\|f\|^2_{H^1(\Omega)} = \|f\|^2_{L_2(\Omega)} + \|f\|^2_{L^1_2(\Omega)},$$

which coincides with the norm of the Sobolev space $H^1(\Omega)$. Thus, for regions of the first type the space $L^1_2(\Omega)$ with the norm (2.3) is isomorphic to the Sobolev space $H^1(\Omega)$.

As an example of regions of the first type one may consider a region that is star-shaped relatively to an inner point, that is, a region Ω such that the interior of any interval connecting that point with a point in $\bar{\Omega}$, is located within Ω. Naturally, the union of a finite number of regions of the first type is also a region of the first type.

One notices that for regions of first type the collection of all infinitely differentiable functions from $H^1(\Omega)$ is dense in $H^1(\Omega)$.

There is a subspace of $H^1(\Omega)$ that is used more often in applications; this subspace consists of all functions in $H^1(\Omega)$ with a mean value equal to zero, and is denoted by H^1_Ω. Evidently, this space does not contain nonzero constants, so the norm in this space is equivalent to the *Dirichlet norm* (taking into account some comments made in Section 1.1.24),

$$\|f\|^2_{H^1_\Omega} := \int_\Omega |\nabla f|^2 \mathrm{d}\Omega. \tag{2.4}$$

1.2.5 The Subspace $H^1_0(\Omega)$

This subspace plays an important role. The subspace $H^1_0(\Omega)$ of the space $H^1(\Omega)$ is defined as the closure in $H^1(\Omega)$ of the linear set of all finite functions in Ω. If Ω_1 is a region which contains Ω, then any function from $H^1_0(\Omega)$ can be extended as the zero function in $\Omega_1 \setminus \Omega$; the resulting function will belong to the space $H^1_0(\Omega_1)$.

For two functions $f(x) \in H^1(\Omega)$ and $g(x) \in H^1_0(\Omega)$ the following formulas of integration by parts are true,

$$\int_\Omega \frac{\partial f(x)}{\partial x_i} g(x) \mathrm{d}\Omega = -\int_\Omega f(x) \frac{\partial g(x)}{\partial x_i} \mathrm{d}\Omega, \qquad i = 1, 2, 3;$$

these formulas can be obtained by taking the limit in the formulas that define the generalized derivatives $\partial/\partial x_i$.

For every function $f(x) \in H^1_0(\Omega)$ the next integral representation is valid,

$$f(x) = \frac{1}{4\pi} \sum_{i=1}^{3} \int_\Omega \frac{(x_i - y_i)}{|x - y|^3} \frac{\partial f(y)}{\partial y_i} \mathrm{d}\Omega_y. \tag{2.5}$$

This equation leads to some important facts. Each function $f(x) \in H_0^1(\Omega)$ is defined almost everywhere; any changes of its values in a set of measure zero do not affect this function as an element of the space $H_0^1(\Omega)$. Formula (2.5) indicates a choice of one representative in each class of functions equivalent in this sense, which is uniquely defined at every point $x \in \Omega$ by the right-hand side of the equation (2.5).

Furthermore, from formula (2.1) it follows that the spaces $H^1(\Omega)$ and, therefore, $H_0^1(\Omega)$ are embedded in the space $L_2(\Omega)$. The representation (2.5) shows that the embedding of $H_0^1(\Omega)$ in $L_2(\Omega)$ is compact. Indeed, the kernels of the integral operators involved in this representation have a weak singularity and, therefore, generate compact operators from $L_2(\Omega)$ to $L_2(\Omega)$. If a set of functions is bounded in $H_0^1(\Omega)$, then their first derivatives form a bounded set in $L_2(\Omega)$, and, by (2.5), the functions by themselves form a compact set in $L_2(\Omega)$.

Since the space $H_0^1(\Omega)$ is always densely embedded in $L_2(\Omega)$ (finite functions are dense in $L_2(\Omega)$), there is an equipment of $L_2(\Omega)$ that corresponds to this embedding, namely,

$$H_0^1(\Omega) \hookrightarrow L_2(\Omega) \hookrightarrow (H_0^1(\Omega))^*.$$

Further on, the space $(H_0^1(\Omega))^*$ will be denoted by $H^{-1}(\Omega)$.

One should observe that the space $H_0^1(\Omega)$ does not contain nonzero constants. Indeed, if $c \in H_0^1(\Omega)$, then there is a sequence of finite functions $\varphi_n(x)$ in Ω such that $\|c - \varphi_n(x)\|_{H^1(\Omega)} \to 0$ as $n \to \infty$. From formula (2.1) for the norm in $H^1(\Omega)$ it follows that $\|c - \varphi_n\|_{L_2(\Omega)} \to 0$ and $\|\partial\varphi_n/\partial x_i\|_{L_2(\Omega)} \to 0$, $i = 1, 2, 3$. Let now $p(x)$ be an arbitrary smooth function. According to the integration by parts formula

$$\int_\Omega \frac{\partial\varphi_n}{\partial x_i} p \mathrm{d}\Omega = -\int_\Omega \varphi_n \frac{\partial p}{\partial x_i} \mathrm{d}\Omega.$$

By taking the limit in this equation ($n \to \infty$), one gets

$$0 = \int_\Omega c \frac{\partial p}{\partial x_i} \mathrm{d}\Omega.$$

Finally, by setting $p \equiv x_i$, one easily concludes that $c = 0$.

Every region Ω is enclosed within some region Ω_1 of the first type (for example, as Ω_1 one may take a sphere of a large radius). From Section 1.1.24 it follows that the norm on the space $H_0^1(\Omega_1)$ is equivalent to the norm $(\int_{\Omega_1} |\nabla f|^2 \mathrm{d}\Omega)^{1/2}$. Hence, the norm (2.1) is equivalent in the space $H_0^1(\Omega)$ to the norm

$$\|f\|_{H_0^1(\Omega)} := \left(\int_\Omega |\nabla f|^2 \mathrm{d}\Omega\right)^{1/2} \tag{2.6}$$

Further on, one will assume that the norm in the space $H_0^1(\Omega)$ is introduced according to this law.

1.2.6 Embedding $H^1(\Omega)$ into $L_2(\Omega)$. Regions of the Second Type

As mentioned above, $H^1(\Omega)$ is always embedded into $L_2(\Omega)$. For various applications, it is useful to know when this embedding is compact. It turns out that this is the case for a wide range of regions, which will be called *regions of the second type*. Every region Ω that satisfies the so-called "*cone condition*" can be considered as an example of a region of the second type. The "cone condition" states the existence of a fixed circular cone whose vertex can be placed at any point in $\bar{\Omega}$ and which can be turned around in such a way that in its new position the whole cone lies within $\bar{\Omega}$.

1.2.7 The Trace Operator. Regions of the Third Type

The restriction of a function continuous in $\bar{\Omega}$ to the boundary $\partial\Omega$ is called a *trace*.

For a certain class of regions, one can introduce a linear operator that acts boundedly from $H^1(\Omega)$ to $L_2(\partial\Omega)$. This operator maps each function continuous in $\bar{\Omega}$ into its trace on $\partial\Omega$. Such an operator is called a *trace operator*, and is denoted it $\gamma_{\partial\Omega}$. The trace operator equals zero on each finite function in Ω, and, in virtue of its boundedness, it equals zero on the subspace $H_0^1(\Omega)$. It is said that a region Ω is a region of the third type provided the trace operator $\gamma_{\partial\Omega}$ exists, is compact, and equals zero only on functions from $H_0^1(\Omega)$.

A Lipschitz region is an example of a region of the third type. For every point on the boundary $\partial\Omega$ of such a region one can find a neighborhood of that point and an orthogonal coordinate system $Oy_1y_2y_3$ such that the equation of the part of $\partial\Omega$ located in this neighborhood is of the form $y_3 = g(y_1, y_2)$, where g is a function satisfying the Lipschitz condition $|g(y) - g(z)| \leq c|y - z|$. Each Lipschitz region satisfies the "cone condition" and, therefore, it is a region of the second type also. Almost all the regions filled with fluid that will be considered further are supposed to be Lipschitz regions.

The existence of the trace operator makes it possible to introduce some new equivalent norms in the space $H^1(\Omega)$. From the boundedness of this operator one gets

$$\|\gamma_{\partial\Omega} f\|_{L_2(\partial\Omega)} \leq c\|f\|_{H^1(\Omega)}. \tag{2.7}$$

Here, the left-hand side does not equal zero on constants, thus, according to Section 1.1.24, the norm defined by the formula

$$\left\{\|\gamma_{\partial\Omega} f\|^2_{L_2(\partial\Omega)} + \|\ |\nabla f|\ \|^2_{L_2(\Omega)}\right\}^{1/2} \tag{2.8}$$

is equivalent to the norm of the space $H^1(\Omega)$ (regions of the first type). According to the Cauchy-Buniakowsky inequality,

$$\left|\int_\Gamma \gamma_{\partial\Omega} f \mathrm{d}\Gamma\right| \leq \sqrt{\mathrm{mes}\Gamma}\left(\int_\Gamma |f|^2 \mathrm{d}\Gamma\right)^{1/2} \leq \sqrt{\mathrm{mes}\Gamma}\ \|f\|_{L_2(\partial\Omega)},$$

where $\Gamma \subseteq \partial\Omega$ is a part of the boundary $\partial\Omega$ that has a positive surface measure. The functional in the left-hand side does not equal zero on constants, so the norm in the space $H^1(\Omega)$ is equivalent to the norm

$$\left\{ \left| \int_\Gamma \gamma_{\partial\Omega} f \mathrm{d}\Gamma \right|^2 + \| \, |\nabla f| \, \|^2_{L_2(\Omega)} \right\}^{1/2} . \tag{2.9}$$

Now, if one considers the subspace $H^1_\Gamma(\Omega)$, which consists of all functions from $H^1(\Omega)$ satisfying

$$\int_\Gamma \gamma_{\partial\Omega} f \mathrm{d}\Gamma = 0, \tag{2.10}$$

then this subspace does not contain any nonzero constants, and the norm of the space $H^1(\Omega)$ is equivalent in this subspace to the Dirichlet norm

$$\|f\|^2_{H^1_\Gamma(\Omega)} := \int_\Omega |\nabla f|^2 \mathrm{d}\Omega. \tag{2.11}$$

1.3 Spaces With Indefinite Metrics

This section provides some information on the geometry of spaces with indefinite metrics. There are also some facts from the theory of linear operators acting on such spaces. Operators of this kind appear in many problems of mechanics and theoretical physics.

1.3.1 J-Spaces

Let E be a Hilbert space with a scalar product $(\cdot, \cdot)$ and let

$$E = E_+ \oplus E_- \tag{3.1}$$

be an orthogonal decomposition of E into two subspaces E_+ and E_-. Furthermore, let P_+ and P_- be the orthogonal projections corresponding to this decomposition and let J be an operator defined as

$$J := P_+ - P_-. \tag{3.2}$$

This operator J is self-adjoint and unitary at the same time, that is,

$$J = J^* = J^{-1}, J^2 = I. \tag{3.3}$$

A new "scalar product" is introduced in the space E by setting

$$[x, y] := (Jx, y).$$

It can be used to produce an indefinite metric, or *J-metric*, with the help of the quadratic form

$$[x, x] = (Jx, x). \tag{3.4}$$

Two elements x and y from E satisfying $[x, y] = 0$ are called *J-orthogonal*, in symbols, $x[\perp]y$. In a similar way, one defines the concepts of J-orthogonality of an element x to a set $M \subseteq E$ and J-orthogonality of two sets N and M. To be specific, $N[\perp]M$ if and only if $[x, y] = 0$ for all $x \in N$ and $y \in M$.

$N^{[\perp]}$ denotes the J-orthogonal complement to a subspace $N \subseteq E$, that is the set of all $y \in E$, which are J-orthogonal to N. The set $N^{[\perp]}$ is always a subspace. It turns out that N and $N^{[\perp]}$ may have a nonzero intersection. In the case when $N \cap N^{[\perp]} = \{0\}$, the direct sum $N \dot{+} N^{[\perp]}$ may not coincide with E, but it is dense in E. The direct sum of two J-orthogonal subspaces N and M is denoted by $N[+]M$. If

$$N[+]N^{[\perp]} = E,$$

then the subspace N is called *J-orthocompletable.*

The initial subspaces E_+ and E_- are J-orthocompletable because in this case $E_- = E_+^{[\perp]}$ and

$$E = E_+[+]E_-.$$

In regard to the J-metric (3.4), every element x from E is either positive ($[x, x] > 0$), or negative ($[x, x] < 0$), or neutral ($[x, x] = 0$). A linear subset $L \subseteq E$ is called a *positive* (*nonnegative*) *lineal*, if for all its elements x, except zero, $[x, x] > 0$ ($[x, x] \geq 0$). *Negative, nonpositive*, and *neutral lineals* are defined in a similar way.

A lineal $L \subseteq E$ is called *nondegenerate*, provided the equation $[x_0, y] = 0$, which is true for a certain $x_0 \in L$ and all $y \in L$, implies $x_0 = 0$. In the opposite case, L is called a *degenerate lineal*, and the element x_0 is called an *isotropic element* of the lineal L. The totality L_0 of all isotropic elements of L is called the *isotropic lineal* for L. For every $x_0 \in L_0$ the equation $[x_0, x_0] = 0$ is true, that is, L_0 is a neutral lineal.

For subspaces in the space E, the concepts of *positive* (*nonnegative*), *negative* (*nonpositive*) and *neutral subspaces* are introduced as above. A nonnegative (nonpositive) subspace is called *maximal*, if it does not lie within a larger nonnegative (nonpositive) subspace. Every nonnegative (nonpositive) subspace can be extended to a maximal subspace.

Positive and negative subspaces (lineals) are also called *definite* subspaces (lineals).

If N is a maximal nonnegative subspace, then $N^{[\perp]}$ is a maximal nonpositive subspace.

A nonnegative (nonpositive) subspace L is maximal if and only if $P_+L = E_+$ ($P_-L = E_-$).

Maximal definite subspaces can be described by means of the so-called *angular operators.* If L is a maximal nonnegative subspace in E, then

$$L = \{x \in E : x = x_+ + x_- = x_+ + Kx_+, x_+ \in E_+\},$$

where K is an angular operator acting from E_+ to E_-. The norm of the operator K is less than or equal to 1. For a maximal nonpositive subspace M in E, a similar representation takes place,

$$M = \{x \in E : x = Qx_- + x_-, x_- \in E_-\}$$

with an angular operator $Q : E_- \to E_+$, $\|Q\| \leq 1$.

1.3.2 Uniformly Definite Subspaces

If L is a positive lineal, then the form

$$[x, y]_L := [x, y] \qquad x, y \in L \tag{3.5}$$

can be considered as a new scalar product in L. Generally speaking, L is not complete relatively to this product. In this case, the completeness or incompleteness of L in the norm $[x, x]_L^{1/2}$ is not connected with its closedness in the norm $\|x\|$. The norm $[x, x]_L^{1/2}$ is dominated by the norm $\|x\|$, because

$$\begin{aligned} [x, x] &= (J(x_+ + x_-), (x_+ + x_-)) = (x_+ - x_-, x_+ + x_-) \\ &= \|x_+\|^2 - \|x_-\|^2 \leq \|x_+\|^2 + \|x_-\|^2 = \|x\|^2 \, (x \in L)\,. \end{aligned} \tag{3.6}$$

If L is a subspace, that is it is closed in the norm $\|x\|$, then the completeness of L in the norm $[x, x]_L^{1/2}$ (in virtue of (3.6) and Banach Theorem) amounts to the equivalence of the norms $[x, x]_L^{1/2}$ and $\|x\|$, that is, to the estimate

$$[x, x] \geq a\|x\|^2, \qquad a > 0, x \in L. \tag{3.7}$$

In this case, the positive subspace (lineal) L is called *uniformly positive* or *regular*.

Uniformly negative lineals and *subspaces* with the internal norm $(-[x, x])^{1/2}$ are defined in a similar way. Uniformly positive and uniformly negative subspaces (lineals) are called *uniformly definite subspaces (lineals)*.

A maximal subspace is uniformly definite if and only if the norm of the corresponding angular operator is less than 1.

1.3.3 J-Orthonormal Systems and Bases

A system of elements $\{e_k\}$ in a J-space E is called *J-orthonormal*, provided $[e_i, e_j] = \pm\delta_{ij}$. Such a system is a union of two disjoint sets, $\{e_k : [e_k, e_k] = 1\}$ and $\{e_k : [e_k, e_k] = -1\}$. Their closed linear hulls will be denoted by L_+ and L_-, respectively.

A J-orthonormal system $\{e_k\}$ is called *complete*, provided there is no such an element $x \neq 0$ from E, which is J-orthogonal to this system. If the system $\{e_k\}$ is complete, then the closure of the set $L_+[+]L_-$ coincides with the whole space E.

It turns out that a complete J-orthonormal system of elements $\{e_k\}$ is a Riesz basis in E if and only if $L_+[+]L_- = E$, that is $L_\pm$ are maximal uniformly definite subspaces.

1.3.4 Linear Operators on J-Spaces

As long as in a J-space E

$$[x, y] = (Jx, y), \qquad x, y \in E,$$

for each operator $A \in \mathcal{L}(E)$ there exists the *J-adjoint* operator A^+ defined by the equation

$$[Ax, y] = [x, A^+ y], \qquad x, y \in E. \tag{3.8}$$

The J-adjoint operator A^+ is related to the usual adjoint operator A^* by the formula

$$A^+ = JA^*J. \tag{3.9}$$

An operator $A \in \mathcal{L}(E)$ is called *J-self-adjoint*, provided

$$(JA)^* = JA. \tag{3.10}$$

In particular, the operator J is an example of a J-self-adjoint operator. J-self-adjoint operators have a number of remarkable properties. For a later use, one needs to point out the following:

1. The spectrum $\sigma(A)$ of a J-self-adjoint operator A is symmetric relatively to the real axis.
2. The spectrum of a J-self-adjoint nonnegative operator ($JA \geq 0$), that is, an operator such that
$$[Ax, x] = (JAx, x) \geq 0, \tag{3.11}$$
is real.
3. If $\lambda \in \sigma_{\mathrm{d}}(A)$ for a J-self-adjoint operator A, then $\bar{\lambda} \in \sigma_{\mathrm{d}}(A)$ also, that is, $\bar{\lambda}$ is a normal eigenvalue. Moreover, the root subspaces $L_\lambda(A)$ and $L_{\bar{\lambda}}(A)$ are J-skew-connected, that is, none of these subspaces contains a nonzero element J-orthogonal to the other subspace.
4. If $\{\lambda, \mu\} \subset \sigma_{\mathrm{p}}(A)$ and $\lambda \neq \bar{\mu}$, then the root lineals $L_\lambda(A)$ and $L_\mu(A)$ and,

therefore, their closures are J-orthogonal. In particular, if $\lambda \neq \bar{\lambda}$, then $L_\lambda(A)$ is a neutral lineal.

5. If A is a J-self-adjoint operator and

$$P_+ A P_- \in \mathfrak{S}_\infty, \tag{3.12}$$

then the nonreal spectrum $\sigma_{\text{non}}(A)$ of A consists of normal eigenvalues, i.e.,

$$\sigma_{\text{non}}(A) \subset \sigma_{\text{d}}(A). \tag{3.13}$$

1.3.5 Invariant Subspaces of J-Self-Adjoint Operators

One of the central problems in the theory of linear operators acting on a J-space is the existence of invariant subspaces and, in particular, of maximal nonnegative and maximal nonpositive invariant subspaces for J-self-adjoint operators. The main result stated below gives sufficient conditions for the existence of such subspaces.

Suppose that condition (3.12) is true for a J-self-adjoint operator $A \in \mathcal{L}(E)$, and assume that the nonreal spectrum $\sigma_{\text{non}}(A)$ of A is somehow decomposed into a union,

$$\sigma_{\text{non}}(A) = \Lambda \cup \bar{\Lambda}, \tag{3.14}$$

of two disjoint parts Λ and $\bar{\Lambda}$, which are symmetric relatively to the real axis. Then there exist the maximal nonnegative and nonpositive subspaces L_+ and L_-, which are invariant for A and, in addition, satisfy the spectral conditions

$$\sigma_{\text{non}}(A \mid L_+) = \Lambda, \quad \sigma_{\text{non}}(A \mid L_-) = \bar{\Lambda}. \tag{3.15}$$

In particular, compact J-self-adjoint operators in a J-space always have maximal nonnegative and maximal nonpositive subspaces with the properties (3.15).

If an operator $A \in \mathcal{L}(E)$ is J-nonnegative, then $\sigma(A) \subset \mathbb{R}$. Actually, in this case the following property takes place: for each J-nonnegative operator $A \in \mathcal{L}(E)$ there exist two maximal J-orthogonal subspaces L_+ and L_-, which are invariant relatively to the operator A, and satisfy the spectral conditions

$$\sigma(A \mid L_+) \subset [0, \infty), \quad \sigma(A \mid L_-) \subset (-\infty, 0]. \tag{3.16}$$

1.3.6 Pontryagin Spaces

A Pontryagin space, usually denoted by Π_κ, is an important, special, and from a historical perspective the first example of an infinite-dimensional Hilbert space E

with an indefinite metric. For such a space, the rank of indefiniteness is finite, that is,

$$\min\{\dim E_+, \dim E_-\} =: \kappa < \infty. \tag{3.17}$$

For the sake of convenience, one will assume that

$$\kappa = \dim E_-, \tag{3.18}$$

hence P_- is a κ-dimensional orthogonal projection and $\dim E_+ = \infty$.

For a space Π_κ, the following additional facts take place.

1. All nonpositive lineals L_- in the space Π_κ are finite-dimensional and, therefore, they are closed, $\dim L_- \leq \kappa$. All κ-dimensional nonpositive subspaces are maximal nonpositive and vice versa. Any nondegenerate subspace in Π_κ is orthocompleteable. In particular, the uniform definiteness of definite subspaces follows from this fact.
2. A J-self-adjoint operator $A \in \mathcal{L}(E)$ acting on $E = \Pi_\kappa$ does not have residual spectrum. Such an operator has at least one eigenvalue, and the entire spectrum of the operator $A = A^+$ lies on the real axis, except for no more than κ pairs of nonreal eigenvalues (with regard to their multiplicity), which are located symmetrically relatively to the real axis.
3. Since condition (3.12) holds true automatically for Π_κ, due to the finite dimension of the space E_-, for every J-self-adjoint operator A on Π_κ there are invariant subspaces L_+ and L_- with the properties (3.14), (3.15). The subspace L_+ can be chosen in such a way that the linear hull $L_\Lambda(A) \subset L_+$ of all the root lineals corresponding to eigenvalues λ with $\operatorname{Im}\lambda > 0$ has the dimension $\dim L_\Lambda(A) \leq \kappa$. A similar assertion is true for the root lineal $L_{\bar{\Lambda}}(A)$, corresponding to eigenvalues λ with $\operatorname{Im}\lambda < 0$.
4. Let $\lambda \in \mathbb{R}$ be an eigenvalue of an operator $A = A^+$ on the space Π_κ. It turns out that the root lineal $L_\lambda(A)$ can be decomposed into a direct sum,

$$L_\lambda(A) = L'_\lambda(A)[+]L''_\lambda(A),$$

where the subspace $L'_\lambda(A)$ is positive and wholly consists of eigenelements of the operator A; $L''_\lambda(A)$ is a finite-dimensional invariant subspace of A. Thus, the associated elements of such an operator A may occur only in the finite-dimensional subspace $L''_\lambda(A)$. The multiplicities of the eigenelements corresponding to the eigenvalue λ do not depend on the choice of the subspace $L'_\lambda(A)$.

Properties 3 and 4 lead to the conclusion that for every J-self-adjoint operator

A, the space Π_κ admits a J-orthogonal decomposition

$$\Pi_\kappa = L[+]M$$

into two invariant subspaces of A, such that L is no more than 2κ-dimensional and M is a uniformly positive subspace; therefore, A has a complete J-orthonormal system of eigenelements in M corresponding to real eigenvalues.

1.3.7 On Completeness and Basicity for the System of Root Elements of a J-Self-Adjoint Operator

The following criteria and sufficient conditions for completeness and basicity are generalizations to the case of J-spaces of the well-known Hilbert-Schmidt Theorem for compact self-adjoint operators acting on a Hilbert space E.

In a Pontryagin space Π_κ, the criterion can be formulated in a simpler way.

The system of root elements of a J-self-adjoint operator $A \in \mathfrak{S}_\infty$ is complete in Π_κ, if and only if the root subspace $L_0(A)$ corresponding to the number $\lambda = 0$ is nondegenerate.

In the case of completeness, there exists a Riesz basis in Π_κ, which consists of root elements of the operator A.

For J-self-adjoint operators acting on a J-space E, the criteria and the theorems on completeness and basicity for root elements look more complicated. Therefore, they will be formulated for the cases of both bounded and unbounded J-self-adjoint operators. Preliminarily, one introduces some definitions.

A subspace L is said to be invariant relatively to an operator A, provided the lineal $L \cap \mathcal{D}(A)$ is dense in L and $A\,(L \cap \mathcal{D}(A)) \subseteq L$.

One says that a J-self-adjoint operator A belongs to the class $\mathcal{H}$, in symbols, $A \in \mathcal{H}$, provided A has at least one maximal nonnegative invariant subspace and each of these subspaces can be decomposed into the direct sum of a finite-dimensional neutral subspace and a uniformly positive subspace.

A J-self-adjoint operator A is in $\mathcal{H}$ if and only if there exists at least one maximal nonpositive invariant subspace for this operator, and each of these subspaces can be decomposed into the direct sum of a finite-dimensional neutral subspace and a uniformly negative subspace.

A J-self-adjoint operator A belongs to $\mathcal{H}$, provided it has at least one maximal nonnegative or nonpositive invariant subspace, and the angular operator of that subspace is compact.

If a J-self-adjoint operator A belongs to $\mathcal{H}$, then the set $\mathbb{C} \setminus \mathbb{R}$ consists of the regular points of the operator A and no more than a finite number of its normal eigenvalues.

One says that an operator A, which has at least one regular point, commutes with an operator B, provided $B\,(A-\lambda I)^{-1} \supset (A-\lambda I)^{-1}\,B$ for every $\lambda \in \rho(A)$. It is said that a J-self-adjoint operator B belongs to the class $\mathcal{K}(\mathcal{H})$, if there exists a J-self-adjoint operator $A \in \mathcal{H}$ that commutes with B.

A J-self-adjoint operator B belongs to $\mathcal{K}(\mathcal{H})$ if and only if it has at least one maximal nonnegative invariant subspace that can be decomposed into the sum of a finite-dimensional neutral subspace and a uniformly positive subspace.

A J-self-adjoint operator B belongs to $\mathcal{K}(\mathcal{H})$, if it has at least one maximal nonnegative invariant subspace with a compact angular operator.

A real number λ is called a critical point for a J-self-adjoint operator A, provided $\ker(A-\lambda I)$ is degenerate, that is it contains nonzero isotropic elements. The set of critical points of an operator A will be denoted by $s(A)$.

The closed linear hulls of eigen- and root elements of an operator A are designated by the symbols $\mathcal{M}_0(A)$ and $\mathcal{M}(A)$, respectively.

The main theorem on completeness and basicity of the system of root elements of a J-self-adjoint operator acting on a space E can now be formulated as follows.

Let A be a J-self-adjoint operator from the class $\mathcal{K}(\mathcal{H})$, and suppose that its spectrum $\sigma(A)$ has no more than a countable set of accumulation points. Then:

a) $\dim(E/\mathcal{M}(A)) \le \dim(E/\mathcal{M}_0(A)) < \infty$;
b) $\mathcal{M}_0(A) = E$ *if an only if* $s(A) = \emptyset$ *and* $L_\lambda(A) = \ker(A-\lambda I)$ *for every* $\lambda \neq \bar{\lambda}$, *where* $L_\lambda(A)$ *is the root subspace of operator* A *that corresponds to the eigenvalue* λ;
c) $\mathcal{M}(A) = E$ *if and only if the closure of the linear hull of the set* $\{L_\lambda(A) : \lambda \in s(A)\}$ *is a nondegenerate subspace;*
d) *if* $\mathcal{M}_0(A) = E$ *(respectively,* $\mathcal{M}(A) = E$*), then there exists in* E *an almost* J*-orthonormal (that is,* J*-orthonormal up to a finite defect) Riesz basis that consists of eigenelements (respectively, root elements) of the operator* A;
e) *if* $\mathcal{M}_0(A) = E$, *then the existence in* E *of a* J*-orthonormal Riesz basis consisting of eigenelements of the operator* A *is equivalent to the spectral condition* $\sigma(A) \subset \mathbb{R}$;
f) *the aforementioned bases can be chosen as* p*-bases if and only if the operator* A *has an invariant maximal nonnegative subspace with the angular operator from the class* $\mathfrak{S}_p$.

In particular, the above statement holds true for the case when A is a bounded or compact operator acting on E. To be specific, for a compact operator A either $s(A) = \emptyset$, or $s(A)$ consists of the single point $\lambda = 0$, and in condition c) one should take the set $L_0(A)$ instead of the closure of the linear hull of the set $\{L_\lambda(A) : \lambda \in s(A)\}$.

1.4 Eigenvalue Problems

The theoretical investigation of oscillations of mechanical nondissipative systems with a finite or infinite number of degrees of freedom entails the problem of finding nonzero solutions of the equation

$$Bu = \lambda Au, \tag{4.1}$$

where A is an operator connected with the kinetic energy of the system and B is an operator related to the potential energy of the considered system. As a rule, A is a positive operator and B is self-adjoint possibly with a nonzero kernel. In this section, one examines a number of typical situations, which will make it possible eventually to address a great deal of specific problems. Equation (4.1) is considered in a Hilbert space E, operator A is assumed to be positive and compact, and B is a self-adjoint operator.

If $u \neq 0$ is a solution of equation (4.1), then $(Bu, u) = \lambda(Au, u)$, hence λ is a real number.

1.4.1 Operator B is the Identity Operator

In this case one gets the following equation,

$$u = \lambda Au. \tag{4.2}$$

Since the positive operator A is compact from the results mentioned in Section 1.1.11 it follows that there is an orthonormal basis $\{e_k\}_{k=1}^{\infty}$ for the space E consisting of the eigenelements $e_k = e_k(A)$ of the operator A; these eigenelements correspond to the eigenvalues $\mu_k = 1/\lambda_k$ of the operator A. All the numbers λ_k are positive. Usually, they are listed in nondecreasing order, $0 < \lambda_1 \leq \lambda_2 \leq \ldots \leq \lambda_k \leq \ldots$, and $\lambda_k \to \infty$ as $k \to \infty$.

Each eigenvalue is counted for a certain number of times, which equals the multiplicity of that eigenvalue; multiplicity in its turn, equals the number of linearly independent eigenelements corresponding to that eigenvalue (this number is always finite). From the orthogonality of the basis $\{e_k\}_{k=1}^{\infty}$ ($(e_i, e_j) = 0$ when $i \neq j$), in virtue of equation (4.2) one gets $(Ae_i, e_j) = 0$, $i \neq j$, that is, the system of elements $\{e_k\}_{k=1}^{\infty}$ is also orthogonal in the scalar product (Au, v).

For any $x \in E$ the following expansions are valid,

$$x = \sum_{k=1}^{\infty} c_k e_k, \qquad Ax = \sum_{k=1}^{\infty} \frac{c_k}{\lambda_k} e_k, \qquad (Ax, x) = \sum_{k=1}^{\infty} \frac{| c_k |^2}{\lambda_k}.$$

Let now y be an element orthogonal to the first eigenelements $e_1, e_2, \ldots, e_{n-1}$. Then

$$y = \sum_{k=n}^{\infty} a_k e_k, \quad Ay = \sum_{k=n}^{\infty} \lambda_k^{-1} a_k e_k.$$

Therefore,

$$(Ay, y)/(y, y) = \left(\sum_{k=n}^{\infty} \lambda_k^{-1} | a_k |^2 \right) / \left(\sum_{k=n}^{\infty} | a_k |^2 \right) \leq \lambda_n^{-1} = \mu_n.$$

This inequality becomes an equality when $y = e_n$, hence

$$\mu_n = \lambda_n^{-1} = \max_{y \perp \{e_1, \ldots, e_{n-1}\}} \left[(Ay, y) / (y, y) \right].$$

On the other hand, it turns out that the maximum value of the functional $(Az, z)/(z, z)$ on the orthogonal complement to any $(n-1)$-dimensional subspace M_{n-1} is at least μ_n. Thus, the formula

$$\mu_n = \lambda_n^{-1} = \min_{M_{n-1}} \max_{z \in M_{n-1}} \left[(Az, z)/(z, z) \right] \tag{4.3}$$

holds true, where the minimum is considered over all the $(n-1)$-dimensional subspaces M_{n-1} in E. This formula expresses the so-called *minimax principle.*

It should be noticed that to find the minimum above, one can use not all the $(n-1)$-dimensional subspaces, but any family of these subspaces that includes the subspace formed by the linear hull of the elements $e_1, \ldots, e_{n-1}$.

The minimax principle has an important corollary:

If A_1 and A_2 are two compact positive operators and the inequality $(A_1 z, z) \leq (A_2 z, z)$ holds true for all $z \in E$, then $\mu_n(A_1) \leq \mu_n(A_2)$ and, therefore, $\lambda_n(A_1) \geq \lambda_n(A_2)$ for every $n = 1, 2, \ldots$.

1.4.2 Operator B is Positive Definite

One starts by making the substitution $B^{1/2}u = v$ in equation (4.1). In virtue of the uniform positiveness of B, hence of $B^{1/2}$, there exists the bounded inverse operator $B^{-1/2}$, so $u = B^{-1/2}v$. Replacing u by $B^{-1/2}v$ in equation (4.1) and applying the bounded operator $B^{-1/2}$ to both sides, one gets

$$v = \lambda B^{-1/2} A B^{-1/2} v. \tag{4.4}$$

The product of a compact and a bounded operator is again a compact operator. Furthermore, by the self-adjointness of operator $B^{-1/2}$ and positivity of operator A,

$$\left(B^{-1/2} A B^{-1/2} v, v\right) = \left(A B^{-1/2} v, B^{-1/2} v\right) > 0, \qquad v \neq 0,$$

that is, the operator $B^{-1/2} A B^{-1/2}$ is positive. Thus, the equation (4.4) is of the type considered in the previous section and, therefore, there exists a sequence of numbers $0 < \lambda_1 \leq \lambda_2 \leq \cdots \leq \lambda_k \leq \cdots$, with $\lambda_k \to \infty$ as $k \to \infty$. The nonzero solutions of equation (4.4) correspond to each of these numbers:

$$\lambda_k B^{-1/2} A B^{-1/2} e_k = e_k.$$

As long as the element from the left side of this equation belongs to the domain $\mathcal{D}(B^{1/2})$, the element from the right side also belongs to this domain, that is, $e_k \in \mathcal{D}\left(B^{1/2}\right)$. Then

$$B^{1/2} e_k = \lambda_k A B^{-1/2} e_k.$$

By denoting $u_k = B^{-1/2} e_k$, one gets $u_k \in \mathcal{D}(B)$ and

$$B u_k = \lambda_k A u_k. \tag{4.5}$$

The elements $\{e_k\}_{k=1}^{\infty}$ form an orthonormal basis, that is, $(e_k, e_j) = \delta_{kj}$, and every $u \in E$ has the expansion

$$u = \sum_{k=1}^{\infty} c_k e_k, \qquad c_k = (u, e_k).$$

Writing e_k in terms of u_k and using equation (4.5), one obtains

$$\left(B^{1/2} u_k, B^{1/2} u_j\right) = (B u_k, u_j) = \lambda_k (A u_k, u_j) = \delta_{kj}. \tag{4.6}$$

Thus, the system $\{u_k\}$ is orthonormal in both the scalar products (Bu, v) and (Au, v). Let $u \in \mathcal{D}(B^{1/2})$. The series that represents the element $B^{1/2}u$ according to the system $\{e_k\}$ is given by

$$B^{1/2} u = \sum_{k=1}^{\infty} c_k e_k, \qquad c_k = \left(B^{1/2} u, e_k\right). \tag{4.7}$$

Applying the bounded operator $B^{-1/2}$ to both sides, we have

$$u = \sum_{k=1}^{\infty} c_k u_k, \qquad c_k = \left(u, B^{1/2} e_k\right) = (u, Bu_k). \tag{4.8}$$

Since $\mathcal{D}(B^{1/2})$ is dense in E, it follows in particular that the system $\{u_k\}$ is complete in E. Furthermore, from the Parseval identity applied to (4.7) and from (4.8) one gets

$$\|B^{1/2}u\|^2 = \sum_{k=1}^{\infty} |c_k|^2 = \sum_{k=1}^{\infty} |(u, Bu_k)|^2.$$

If $u \in \mathcal{D}(B)$, then

$$(Bu, u) = \|B^{1/2}u\|^2 = \sum_{k=1}^{\infty} |(u, Bu_k)|^2. \tag{4.9}$$

Applying the bounded operator A to both sides of expansion (4.8), we get

$$Au = \sum_{k=1}^{\infty} c_k Au_k.$$

By using the orthogonality relations (4.6), formulas (4.8), and the equations (4.5), it follows that

$$(Au, u) = \sum_{k=1}^{\infty} |c_k|^2 (Au_k, u_k) = \sum_{k=1}^{\infty} |(u, Bu_k)|^2 \lambda_k^{-1}. \tag{4.10}$$

Equations (4.9) and (4.10) show that the forms of kinetic and potential energy (Au, u) and (Bu, u) are reduced to a sum of squares by means of the same system of elements.

Finally, it should be noticed that in the case when operator B is bounded, the system $\{u_k\}$ forms a Riesz basis.

From the minimax principle applied to equation (4.4), it follows that

$$\lambda_n^{-1}(A; B) = \min_{M_{n-1}} \max_{z \perp M_{n-1}} \left[\left(B^{-1/2}AB^{-1/2}z, z\right) / (z, z)\right].$$

Minimizing over subspaces $M_{n-1} \subset \mathcal{D}(B^{1/2})$ and making the substitution $B^{-1/2}z = v$, one obtains

$$\lambda_n^{-1}(A; B) = \min_{L_{n-1}} \max_{v \perp L_{n-1}} \left[(Av, v) / \left(B^{1/2}v, B^{1/2}v\right)\right], \tag{4.11}$$

where L_{n-1} is an arbitrary $(n-1)$-dimensional subspace. (Here, the following fact was used: if $z \perp M_{n-1}$, then $B^{-1/2}x \perp B^{1/2}M_{n-1}$, that is, $v \perp L_{n-1}$.)

If the operator B is bounded, then

$$\lambda_n^{-1}(A; B) = \min_{L_{n-1}} \max_{v \perp L_{n-1}} \left[(Av, v) / (Bv, v)\right].$$

The inequalities $\lambda_n(A_1; B) \geq \lambda_n(A_2; B)$, $n = 1, 2, \dots$ for $A_1 \leq A_2$ hold true for the equations

$$Bu = \lambda A_1 u, \quad Bu = \lambda A_2 u.$$

1.4.3 Positivity Condition for a Matrix Operator

Further, one will be dealing with the situation when the space E is decomposed into the orthogonal sum of two of its subspaces:

$$E = E_1 \oplus E_2.$$

Each bounded linear operator A acting on the space E can be represented as a matrix, which corresponds to this decomposition (see Section 1.1.14). If P_1 and P_2 are the operators of orthogonal projection onto E_1 and E_2 respectively, then for $u \in E$ we have $u = P_1 u + P_2 u =: v + w$, and

$$Au = \begin{pmatrix} A_{11} & A_{12} \\ A_{21} & A_{22} \end{pmatrix} \begin{pmatrix} v \\ w \end{pmatrix},$$

where $A_{ij} = P_i A P_j$.

It is easy to show that operator A is self-adjoint only when operators A_{11} and A_{22} are self-adjoint and $A_{21} = A_{12}^*$.

If A is positive then the operators A_{11} and A_{22} are positive. Indeed, if $u = v \in E_1$, then $(Au, u) = (A_{11}v, v) > 0$. The fact that $A_{22} > 0$ can be proved in a similar way.

Assume now that A is positive and the operator A_{11} has a bounded inverse operator in the space E. The property of positivity of operator A amounts to

$$\begin{aligned}(Au, u) &= (A_{11}v + A_{12}w, v) + (A_{21}v + A_{22}w, w) \\ &= (A_{11}v, v) + 2\mathrm{Re}\,(v, A_{12}w) + (A_{22}w, w) > 0 \qquad (u \neq 0).\end{aligned}$$

Isolating a perfect square in this sum, one obtains

$$\begin{aligned}&(A_{11}^{1/2}v + A_{11}^{-1/2}A_{12}w, A_{11}^{1/2}v + A_{11}^{-1/2}A_{12}w) \\ &\qquad + (A_{22}w, w) - \left(A_{21}A_{11}^{-1}A_{12}w, w\right) > 0.\end{aligned}$$

Let $w \neq 0$ be an arbitrary element in E_2 and set $v = -A_{11}^{-1}A_{12}w$. The first term equals zero, hence one gets the inequality

$$\left(\left(A_{22} - A_{21}A_{11}^{-1}A_{12}\right)w, w\right) > 0. \tag{4.12}$$

Thus, from the positivity of the operator A and taking into account the condition that the operator $A_{11} = P_1AP_1$ is invertible, one derives the positivity of the operator $\tilde{A} := A_{22} - A_{21}A_{11}^{-1}A_{12}$.

It should be noticed that in the case when the space E_1 is finite-dimensional, the positivity of the operator A_{11} implies its invertibility.

Finally, if operator A_{11} has an unbounded inverse operator, then because the previous reasoning was made for a fixed element w, inequality (4.12) is valid for those w on which the operator $A_{21}A_{11}^{-1}A_{12}$ is defined, that is, in the case when $A_{12}w \in \mathcal{D}\left(A_{11}^{-1}\right)$.

1.4.4 Simplifying Equations with an Alternating Operator

First, one considers the case when zero does not belong to the spectrum of operator B. Take the operator $|B|$. If zero does not belong to the spectrum of B, then $B = J|B|$, where the spectrum of operator J consists only of two numbers, $+1$ and -1, and this operator commutes with $|B|$ and $|B|^{1/2}$. One next makes the substitution $|B|^{1/2}u = v$ in equation (4.1). Then

$$Jv = \lambda|B|^{-1/2}A|B|^{-1/2}v, \tag{4.13}$$

or

$$Jv = \lambda Cv, \tag{4.14}$$

where $C := |B|^{-1/2}A|B|^{-1/2}$ is a self-adjoint positive compact operator.

The whole space E can be decomposed into a direct sum of invariant subspaces of B, corresponding to the positive and negative parts of its spectrum: $E = E_+ \oplus E_-$. Then, equation (4.14) can be written down in the matrix form

$$\begin{pmatrix} I & 0 \\ 0 & -I \end{pmatrix}\begin{pmatrix} z \\ w \end{pmatrix} = \lambda\begin{pmatrix} C_{11} & C_{12} \\ C_{21} & C_{22} \end{pmatrix}\begin{pmatrix} z \\ w \end{pmatrix}. \tag{4.15}$$

Now, let zero be an isolated point of the spectrum of operator B. The subspace E_0, on which operator B turns into zero, corresponds to this isolated point. The space E can be decomposed as $E = E_0 \oplus \tilde{E}$. Equation (4.1) turns into a system

$$\begin{aligned} 0 &= \lambda\left(A_{11}v + A_{12}w\right), \\ \tilde{B}w &= \lambda\left(A_{21}v + A_{22}w\right), \end{aligned} \tag{4.16}$$

where $v \in E_0$, $w \in \tilde{E}$, and $\tilde{B} := B \mid \tilde{E}$ is the restriction of operator B to $\tilde{E}$.

Assume that operator A_{11} has a bounded inverse operator in the space E_0. Then, from the first equation for $\lambda \neq 0$ one obtains $v = -A_{11}^{-1} A_{12} w$. By substituting this expression into the second equation, one gets an equation in the space $\tilde{E}$, namely,

$$\tilde{B}w = \lambda \tilde{A}w \left(= \lambda \left(A_{22} - A_{21}A_{11}^{-1}A_{12}\right) w\right). \tag{4.17}$$

On the space $\tilde{E}$, zero does not belong to the spectrum of operator $\tilde{B}$, hence (4.17) can be transformed into an equation of the type (4.15) by means of the above mentioned substitution.

To conclude, one notices that the solutions of system (4.16) corresponding to $\lambda = 0$ fill the subspace E_0; an orthonormal basis can be selected from these solutions.

1.4.5 Equations in Spaces with Indefinite Metrics

Going back to equation (4.14), one may introduce an indefinite metric on E by setting

$$[u, v] := (Ju, v).$$

Since (4.14) is the eigenvalue problem for the J-positive compact operator JC, that is,

$$JCv = \mu v, \quad \mu = \lambda^{-1}, \quad C = |B|^{-1/2} A |B|^{-1/2}, \tag{4.18}$$

the results obtained in Section 1.3 can be applied to equation (4.18).

If B is a semibounded self-adjoint operator with exactly κ eigenvalues on the negative semiaxis and $0 \in \rho(B)$, then, obviously, $J = J_\kappa = P_+ - P_-$, $\dim P_- = \kappa$ and (4.18) is an equation in a Pontryagin space Π_κ. Since $C > 0$, problem (4.18) does not have neutral eigenelements. The whole space E splits up into a J_κ-orthogonal sum

$$E = L_+[+]L_- \tag{4.19}$$

of positive and negative maximal invariant subspaces L_+ and L_- of $J_\kappa C$; $\dim L_- = \kappa$, and each subspace is uniformly definite.

In the subspace L_+ operator $J_\kappa C$ has the J_κ-orthogonal system of eigenelements $\left\{e_k^+\right\}_{k=1}^\infty$, corresponding to the eigenvalues $\mu_k^+ = 1/\lambda_k^+$, with $\mu_k^+ \to 0$ as $k \to \infty$. In the subspace L_- operator $J_\kappa C$ has exactly κ negative eigenvalues $\left\{\mu_k^-\right\}_{k=1}^\kappa$ and the J_κ-orthogonal system of eigenelements $\left\{e_k^-\right\}_{k=1}^\kappa$.

Since $0 \notin \sigma(J_\kappa C)$ and $L_0(J_\kappa C) = \{0\}$, the criterion stated in Section 1.3.7 shows that the union of the two systems of eigenelements $\{e_k^+\}_{k=1}^\infty$ and $\{e_k^-\}_{k=1}^\infty$ forms a Riesz basis in the space E.

If $B = B_+ - B_-$, and $B_+ \gg 0$, $\dim B_\pm = \infty$, then in (4.18) $J = P_+ - P_-$, $\dim P_\pm = \infty$, that is, one deals with the eigenvalue problem for the J-positive operator JC in a J-space. L_+ and L_- in (4.19) are positive and negative maximal invariant subspaces, $\dim L_\pm = \infty$. In this case, in the subspace L_+ operator JC has the J-orthogonal system of eigenelements $\{e_k^+\}_{k=1}^\infty$, $[e_k^+, e_k^+] = 1$, and the system of eigenvalues $\{\mu_k^+\}_{k=1}^\infty$, $\mu_k^+ > 0$, $\mu_k^+ \to 0$, $k \to \infty$. Similarly, in the subspace L_- operator JC has the J-orthogonal system of eigenelements $\{e_k^-\}_{k=1}^\infty$, $[e_k^-, e_k^-] = -1$, and the system of eigenvalues $\{\mu_k^-\}_{k=1}^\infty$, $\mu_k^- < 0$, $\mu_k^- \to 0$, $k \to \infty$. Not every system $\{e_k^\pm\}_{k=1}^\infty$ is complete in $L_\pm$. However, the union of the systems $\{e_k^+\}_{k=1}^\infty$ and $\{e_k^-\}_{k=1}^\infty$ results in a complete system in the space E.

Employing the inverse substitution $e_k^\pm = |B|^{1/2} u_k^\pm$ for the solutions $e_k^\pm$ of problem (4.18), one comes to the conclusion that $u_k^\pm \in \mathcal{D}(B)$, and

$$Bu_k^\pm = \lambda_k^\pm A u_k^\pm, \qquad \lambda_k^\pm = 1/\mu_k^\pm, \tag{4.20}$$

that is, $\{u_k^\pm\}_{k=1}^\infty$ are the eigenelements for problem (4.1). The conditions of J-orthonormality for $e_k^\pm$ lead to the conditions of orthogonality

$$\left(Bu_k^\pm, u_j^\pm\right) = \pm\delta_{kj}, \qquad \left(Au_k^\pm, u_j^\pm\right) = \mid \lambda_k^\pm \mid^{-1} \delta_{kj}. \tag{4.21}$$

Hence, the form (Au, u) in the union of the systems $\{u_k^+\}_{k=1}^\infty$ and $\{u_k^-\}_{k=1}^\infty$ is reduced to a sum of squares, and the form (Bu, u) is reduced to a sum and difference of squares.

Now, assume that $\lambda = 0$ is an isolated point in the spectrum of operator B, and the operator $P_0 A P_0$, where P_0 is the orthogonal projector onto the kernel E_0 of operator B, has a bounded inverse operator. Then, instead of (4.14) one should consider equation (4.17) in the space $\tilde{E} = E \ominus E_0$. For (4.17) the same conclusions as for (4.18) are valid, that is, there exist two systems of elements $\left\{\tilde{u}_k^\pm\right\}_{k=1}^\infty$ in $\tilde{E}$, which satisfy

$$\begin{aligned} \tilde{B}\tilde{u}_k^\pm &= \tilde{\lambda}_k^\pm \tilde{A}\tilde{u}_k^\pm, \\ \left(\tilde{B}\tilde{u}_k^\pm, \tilde{u}_j^\pm\right) &= \pm\delta_{kj}, \\ \left(\tilde{A}\tilde{u}_k^\pm, \tilde{u}_j^\pm\right) &= |\tilde{\lambda}_k^\pm|^{-1}\delta_{kj}. \end{aligned} \tag{4.22}$$

Noticing that $g_k^\pm := -A_{11}^{-1} A_{12} \tilde{u}_k^\pm \in E_0$, one gets

$$\begin{aligned} 0 &= \tilde{\lambda}_k^\pm \left(A_{11} g_k^\pm + A_{12}\tilde{u}_k^\pm\right), \\ \tilde{B}\tilde{u}_k^\pm &= \tilde{\lambda}_k^\pm \left(A_{21} g_k^\pm + A_{22}\tilde{u}_k^\pm\right). \end{aligned} \tag{4.23}$$

Consequently, the elements $u_k^\pm := g_k^\pm + \tilde{u}_k^\pm$ satisfy the equations (4.1) and correspond to the solutions with nonzero eigenvalues.

1.5 Evolution Equations in Hilbert Spaces

The problems of dynamics of solids can be reduced to solving systems of ordinary differential equations or, which is the same, to finding the solution of some differential equations in finite-dimensional spaces; the dimension of these spaces depends on the total number of degrees of freedom of that system. In contrast to this, the problems of dynamics of a continuous medium and, in particular, the hydrodynamics, lead to differential equations in infinite-dimensional spaces. Moreover, as a rule, these equations involve unbounded operator coefficients. The theory of linear and nonlinear differential equations in Banach spaces is an extensively studied field with many important results presented in a number of monographs.

This section deals solely with linear differential equations in Hilbert spaces; only those facts which are used in the following chapters are discussed.

1.5.1 First Order Linear Differential Equations with Bounded Operator Coefficient

Let us consider a differential equation of the form

$$\frac{\mathrm{d}x}{\mathrm{d}t} = Ax, \tag{5.1}$$

where $x(t)$ is the unknown function with values in a complex Hilbert space E, and A is a bounded linear operator acting on that space. Here and in the sequel, the derivative $\mathrm{d}x/\mathrm{d}t$ is to be understood as the limit of the ratio $\Delta x/\Delta t$ as $\Delta t \to 0$ in the norm of space E.

If one states the *Cauchy problem* for equation (5.1)

$$x(0) = x_0, \tag{5.2}$$

then this initial value problem is always univalently solvable and its solution can be written down as

$$x(t) = \mathrm{e}^{tA} x_0. \tag{5.3}$$

The operator function e^{tA} is to be understood as the sum of the series

$$\mathrm{e}^{tA} := \sum_{n=0}^{\infty} \frac{t^n}{n!} A^n.$$

This series converges everywhere in the complex plane relatively to the operator norm and that is why all the solutions of Cauchy problems admit analytic extensions to holomorphic functions in the whole complex plane.

For the solution (5.3) of the Cauchy problem (5.1), (5.2) and for any $\varepsilon > 0$ one has the estimate

$$\|x(t)\| \leq M_\varepsilon e^{(\omega+\varepsilon)t}\|x_0\|,$$

where $\omega = \max \operatorname{Re}\{\sigma(A)\}$ with $\sigma(A)$, as usual, the spectrum of operator A, and M_ε a constant that does not depend on $x_0 \in E$.

The family of operators e^{tA} form a group relatively to multiplication,

$$e^{tA}e^{\tau A} = e^{(t+\tau)A}.$$

1.5.2 The Cauchy Problem for Equations with Unbounded Operators

Let us now assume that the linear operator A is a closed unbounded operator with the domain of definition $\mathcal{D}(A)$ dense in E. First of all, one should clarify the concept of a solution of equation (5.1). A function $x(t)$ with values in $\mathcal{D}(A)$ for all $t \in [0,T]$ is a solution to equation (5.1) on the interval $[0,T]$, provided this function $x(t)$ has a derivative and satisfies equation (5.1) for every $t \in [0,T]$.

The problem of finding the solution of equation (5.1) satisfying the initial condition

$$x(0) = x_0 \in \mathcal{D}(A),$$

is called a *Cauchy problem on the interval* $[0,T]$.

It is said that the Cauchy problem is *well posed* on the interval $[0,T]$, provided:

1) for any $x_0 \in \mathcal{D}(A)$ there exists a unique solution of the problem;

2) this solution depends continuously on the initial data in the following sense: from $x_n(0) \to 0\,(x_n(0) \in \mathcal{D}(A))$ it follows that for the corresponding solutions $x_n(t) \to 0$ for every $t \in [0,T]$.

It turns out that if the Cauchy problem is well posed on some interval $[0,T](T > 0)$, then it is well posed on any interval $[0,T_1](T_1 > 0)$.

The operator $U(t)$, which assigns to every element $x_0 \in \mathcal{D}(A)$ the value at the point t of the solution of the Cauchy problem (5.1), (5.2), that is, $U(t)x_0 = x(t)$, is bounded and can be continuously extended to a bounded operator on the whole space E (the extension is denoted in the same way). Operators $U(t)$, $t > 0$, form a strongly continuous semigroup, that is, they possess the properties:

1) the function $U(t)x$ is continuous in t for every $x \in E$;

2) $U(t+\tau) = U(t)U(\tau)$, $t, \tau > 0$.

Thus, the solution of a well posed Cauchy problem is given by the formula

$$x(t) = U(t)x_0, \qquad x_0 \in \mathcal{D}(A),$$

where $U(t)$ is a strongly continuous semigroup of operators for $t > 0$.

If x_0 does not belong to $\mathcal{D}(A)$, then, generally speaking, the function $x(t) = U(t)x_0$ does not possess the properties of a solution to equation (5.1). Such functions are called *generalized solutions* of equation (5.1). Each generalized solution is a continuous function on $(0, \infty)$. At infinity, the norm of a solution or the norm of a generalized solution grows not faster than the exponent, and the exponential types of all solutions are bounded from above.

A well posed Cauchy problem is called *uniformly well posed*, if from $x_n(0) \to 0$, $x_n(0) \in \mathcal{D}(A)$, it follows that $x_n(t) \to 0$ uniformly on each finite interval of the semiaxis $[0, +\infty)$. The semigroup $U(t)$ corresponding to a uniformly well posed Cauchy problem has the following additional property: $U(t)x \to x$ as $t \to 0$ for any $x \in E$.

The class of such semigroups is called class C_0. For every semigroup in the class C_0 there exist two numbers M and ω such that

$$\|U(t)\| \le M e^{\omega t}. \tag{5.4}$$

Any semigroup $U(t)$ in the class C_0 corresponds to the uniformly well posed Cauchy problem with the operator $A = U'(0)$. This operator is defined on those elements x in E for which there exists the limit

$$\lim_{t \to 0} \frac{U(t)x - x}{t} =: U'(0)x;$$

$U'(0)$ is called the *generating operator* of the semigroup.

For further consideration, the concept of weak solutions of equation (5.1) is important. Every such solution is a continuous function on $[0, T]$, continuously differentiable on $(0, T]$, and satisfies equation (5.1) on this half-open interval. Thus, a weak solution may not be differentiable at $t = 0$, and its initial value $x(0)$ may not belong to $\mathcal{D}(A)$.

Any solution of equation (5.1) is also a weak solution.

The problem of finding the weak solution satisfying the initial condition $x(0) = x_0$ is called a *weak Cauchy problem.*

1.5.3 Equations With a Negative Self-Adjoint Operator

Let us assume that $A = -B$, where B is a positive-definite self-adjoint operator. Suppose

$$(Bx, x) \ge b(x, x), \qquad b > 0.$$

The Cauchy problem for equation (5.1) with such an operator A is uniformly well posed. The corresponding semigroup of operators $U(t)$, which yields the solution of the Cauchy problem, can be build by means of spectral expansion as a function

$$U(t) = \mathrm{e}^{-tB}$$

of operator B. For this function one has the estimate

$$\|\mathrm{e}^{-tB}\| \leq \mathrm{e}^{-bt}. \tag{5.5}$$

It turns out that the functions $\mathrm{e}^{-tB}x_0$ are weak solutions of equation (5.1) for any $x_0 \in E$. These functions admit analytic extensions to holomorphic functions in the right half-plane of the complex plane.

If operator A in equation (5.1) is self-adjoint and semibounded from above,

$$(Ax, x) \leq a(x, x),$$

then, using the substitution $x = \mathrm{e}^{\omega t}y$ in (5.1), one obtains the equation

$$\frac{\mathrm{d}y}{\mathrm{d}t} = (A - \omega I)y. \tag{5.6}$$

If one chooses $\omega > a$, then the operator $A - \omega I$ is negative-definite and self-adjoint. Hence, it appears that even in this case the Cauchy problem (5.1), (5.2) is uniformly well posed. For every $x_0 \in E$ there exists a unique weak solution of the Cauchy problem, which can be analytically extended on the right half-plane. For all these solutions one has

$$\|x(t)\| \leq \mathrm{e}^{at}.$$

The following estimates are of great importance:

$$\|A^\alpha \mathrm{e}^{tA}\| \leq \frac{C_\alpha}{t^\alpha}\mathrm{e}^{at}, \qquad t > 0. \tag{5.7}$$

A semigroup $U(t)$ which admits an analytic extension on a sector containing the semiaxis $(0, \infty)$ is called an *analytic semigroup*. The semigroup e^{tA} is analytic.

1.5.4 Equations With a Dissipative Operator

As it follows from the inequality (5.5), $\|U(t)\| \leq 1$ for any negative-definite self-adjoint operator A, that is, the semigroup consists of contractive operators.

A linear operator A with a domain of definition dense in E is called *dissipative*, provided the following inequality holds:

$$\mathrm{Re}(Ax, x) \leq 0, \qquad x \in \mathcal{D}(A). \tag{5.8}$$

A dissipative operator A is called *maximal dissipative*, if it does not admit dissipative extensions. The latter situation occurs if and only if operator A is closed and, besides (5.8), the following inequality is valid:

$$\mathrm{Re}(A^*x,x) \le 0, \qquad x \in \mathcal{D}(A^*). \tag{5.9}$$

The Cauchy problem (5.1), (5.2) is uniformly well posed and the corresponding semigroup is contractive if and only if the operator A is maximal dissipative, that is, the inequalities (5.8), (5.9) are true.

Let us notice that in virtue of (5.8) and (5.9) a nonpositive self-adjoint operator is maximal dissipative. The same holds for the operator $\mathrm{i}B$, where B is an arbitrary self-adjoint operator. For the latter operator one gets the following important property: If $A = \mathrm{i}B$, where B is a self-adjoint operator, then the Cauchy problem is uniformly well posed and the corresponding semigroup consists of unitary operators.

A semigroup of unitary operators $U(t)$, $t \ge 0$, can be extended to a group by setting $U(t) := U^{-1}(-t)$ for $t < 0$.

If an operator A satisfies the inequalities

$$\begin{aligned} \mathrm{Re}(Ax,x) &\le \omega(x,x), \qquad x \in \mathcal{D}(A), \\ \mathrm{Re}(A^*x,x) &\le \omega(x,x), \qquad x \in \mathcal{D}(A^*), \end{aligned}$$

for some $\omega > 0$, then the substitution used in Section 1.5.3 reduces equation (5.1) to the equation (5.6) with a maximal dissipative operator. For the corresponding semigroup one has the estimate

$$\|U(t)\| \le \mathrm{e}^{\omega t}.$$

1.5.5 Equations With Perturbed Operators

Let us now consider an equation of the form

$$\frac{\mathrm{d}x}{\mathrm{d}t} = Ax + Qx, \tag{5.10}$$

where Q is a linear and, generally speaking, unbounded operator on E.

If the Cauchy problem (5.1), (5.2) is uniformly well posed and Q is a bounded operator on E, then the Cauchy problem for equation (5.10) is uniformly well posed.

If the estimate (5.4) holds true for the semigroup of the nonperturbed equation (5.1), then the semigroup of the perturbed equation (5.10) satisfies the estimate

$$\|U(t)\| \le M\mathrm{e}^{(\omega+M\|Q\|)t} \tag{5.11}$$

One should mention that stronger assertions are true for the equations with dissipation considered in Section 1.5.3.

One says that an operator Q is *subordinated* to an operator S, provided

$\mathcal{D}(Q) \supset \mathcal{D}(S)$ and there exists a constant c, such that

$$\|Qx\| \leq c\|Sx\|, \qquad x \in \mathcal{D}(S).$$

An operator Q is said to be *quite subordinated* to an operator S, if Q is subordinated to S and for every sufficiently small η one has

$$\|Qx\| \leq \Phi_\eta(x) + \eta\|Sx\|,$$

where $\Phi_\eta(x)$ is a convex continuous functional for $x \in E$.

If S is a self-adjoint positive-definite operator, then an operator Q is quite subordinated to S in the case when Q admits the representation $Q = TS$, where T is a compact operator.

It turns out that whenever $A = -B$, where B is a positive definite self-adjoint operator, and the operator Q is subordinated to a fractional power B^α of the operator B, $0 \leq \alpha < 1$, or quite subordinated to the operator B (the second condition follows from the first one), then the Cauchy problem for equation (5.10) preserves the main properties of the Cauchy problem for equation (5.1) (see Section 1.5.3): the Cauchy problem for equation (5.10) is uniformly well posed; there exists a unique weak solution of the Cauchy problem for any $x \in E$; all weak solutions admit analytic extensions on some sector of the complex plane which contains the positive semiaxis, and the corresponding semigroup is analytic.

1.5.6 Stability

It is said that equation (5.1) possesses the property of *exponential stability* (resp., *stability*), provided the corresponding semigroup belongs to the class C_0 and $\omega < 0$ (resp., $\omega = 0$) in the inequality (5.4). From Section 1.5.3 it follows that in the case $A = -B$, where B is a positive-definite (resp., positive) self-adjoint operator, equation (5.1) has the property of exponential stability (stability). In Section 1.5.4 it was shown that whenever the following inequalities

$$\operatorname{Re}(Ax, x) \leq -\varepsilon(x, x), \qquad x \in \mathcal{D}(A),$$
$$\operatorname{Re}(A^*x, x) \leq -\varepsilon(x, x), \qquad x \in \mathcal{D}(A^*),$$

are satisfied for $\varepsilon > 0$ (resp., $\varepsilon = 0$), then equation (5.1) possesses the property of exponential stability (resp., stability).

The stability for equation (5.1) with a bounded operator A is completely defined by the location of the spectrum of operator A: if $\operatorname{Re}\sigma(A) \leq -\varepsilon$ ($\varepsilon > 0$), then one has exponential stability; if $\operatorname{Re}\sigma(A) \leq 0$, then one gets stability.

For equations with an unbounded operator A the location of the spectrum of operator A does not affect the estimate of the norm of the corresponding semigroup and, therefore, it does not affect stability.

Besides the sufficient conditions of stability mentioned above, connected with the dissipativity of an operator, one can obtain a number of conditions expressed in terms of the system of eigen- and associated elements of operator A. Let us assume that the spectrum of the operator A is located in the half-plane $\operatorname{Re}\lambda \le -\varepsilon$, and its eigenelements f_n corresponding to the eigenvalues λ_n form a Riesz basis for the space E. Let $x_0 \in E$ and assume that its expansion relatively to the basis $\{f_n\}$ has the form

$$x_0 = \sum_{n=1}^{\infty} c_n f_n.$$

Then

$$U(t)x_0 = \sum_{n=1}^{\infty} c_n U(t) f_n = \sum_{n=1}^{\infty} c_n \mathrm{e}^{\lambda_n t} f_n,$$

and, in virtue of (1.3),

$$\begin{aligned} \|U(t)x_0\| &\le \|Q\| \left\{ \sum_{n=1}^{\infty} \left| c_n \mathrm{e}^{\lambda_n t} \right|^2 \right\}^{1/2} \\ &\le \|Q\| \mathrm{e}^{-\varepsilon t} \left\{ \sum_{n=1}^{\infty} |c_n|^2 \right\}^{1/2} \\ &\le \|Q\| \cdot \|Q^{-1}\| \mathrm{e}^{-\varepsilon t} \|x_0\|. \end{aligned}$$

Therefore, the corresponding equation possesses the property of exponential stability for $\varepsilon > 0$ and the property of stability for $\varepsilon = 0$.

A similar situation takes place in the case when there exists a Riesz basis consisting of eigen- and associated elements, but there is only a finite number of associated elements. In this case the space can be decomposed into a direct sum of two invariant subspaces of A. One of them contains a Riesz basis of eigenelements and the above mentioned estimate is true on it; as for the other subspace, it is finite-dimensional, and the semigroup on it is estimated by the magnitude $C\mathrm{e}^{-\varepsilon_1 t}$, where $\varepsilon_1 \in (0, \varepsilon)$ (see Section 1.5.1). Therefore, the property of exponential stability is still valid.

1.5.7 Nonhomogeneous Equations

Let us next consider the Cauchy problem

$$\frac{\mathrm{d}x}{\mathrm{d}t} = Ax + f(t), \qquad 0 \le t \le T,\ x(0) = x_0, \tag{5.12}$$

where $f(t)$ is a continuous function with values in E.

If operator A is bounded, then by the standard method of variation of parameters one obtains the formula

$$x(t) = \mathrm{e}^{tA}x_0 + \int_0^t \mathrm{e}^{(t-\tau)A} f(\tau)\mathrm{d}\tau,$$

which gives the solution of the problem (5.12) for any $x_0 \in E$.

For an equation with an unbounded operator A, to which there corresponds a semigroup $U(t)$, it is natural to seek a solution of problem (5.12) in a similar way, namely,

$$x(t) = U(t)x_0 + \int_0^t U(t-\tau) f(\tau)\mathrm{d}\tau. \tag{5.13}$$

However, here the situation becomes more complicated due, for instance, to the fact that the second term in (5.13) does not belong, generally speaking, to the domain of definition of operator A.

If the Cauchy problem (5.1), (5.2) is uniformly well posed and $x_0 \in \mathcal{D}(A)$, then formula (5.13) gives the solution to problem (5.12) under one of the two additional assumptions:

1) the function $f(t)$ is continuously differentiable;
2) the values $f(t) \in \mathcal{D}(A)$ for $t \in [0, T]$ and the function $Af(t)$ is continuous.

If $x_0 \in E$ and the function $Af(t)$ continuous, then function (5.13) is called the *generalized solution* of the Cauchy problem (5.12).

If $A = -B$ and the operator Q possesses the property of subordination described in Section 1.5.5, then the Cauchy problem

$$\frac{\mathrm{d}x}{\mathrm{d}t} = Ax + Qx + f(t), \qquad x(0) = x_0,$$

has a unique generalized solution of the form (5.13) for any $x_0 \in E$ and for a function $f(t)$ satisfying the Hölder condition

$$\|f(t) - f(\tau)\| \le c|t-\tau|^\gamma$$

with some exponent $\gamma > 0$.

1.5.8 Linear Differential Equations of the Second Order

For differential equations of the second order with unbounded operator coefficients it makes sense to define the concepts of equations of elliptic, hyperbolic and parabolic type. Here one will consider the simplest equations of hyperbolic type in a Hilbert space, that is, equations of the form

$$\frac{\mathrm{d}^2 x}{\mathrm{d}t^2} = A^2 x, \qquad 0 \le t \le T. \tag{5.14}$$

A two times continuously differentiable function $x(t)$ with values in $\mathcal{D}(A^2)$ is considered to be a solution of this equation provided that function satisfies equation (5.14) for $t \in [0, T]$ and has the following property: the derivative $\mathrm{d}x/\mathrm{d}t$ takes values in $\mathcal{D}(A)$ and the function $A\mathrm{d}x/\mathrm{d}t$ is continuous on $[0, T]$.

The Cauchy problem for equation (5.14) involves two initial conditions,

$$x(0) = x_0, \quad \frac{\mathrm{d}x(0)}{\mathrm{d}t} = x_0'. \tag{5.15}$$

The Cauchy problem (5.14), (5.15) has a unique solution for every $x_0 \in \mathcal{D}(A^2)$ and $x_0' \in \mathcal{D}(A) \cap \mathcal{R}(A)$ if and only if A is the generating operator of a strongly continuous group of operators $U(t)$, $-\infty < t < \infty$. Under this assumption, the solution of the Cauchy problem is given by the formula

$$x(t) = (1/2)\,[U(t) + U(-t)]\,x_0 + (1/2)\,[U(t) - U(-t)]\,y_0, \tag{5.16}$$

where y_0 is some solution of the equation $Ay_0 = x_0'$. If A has a bounded inverse operator, then $y_0 = A^{-1}x_0'$.

It was already mentioned in Section 1.5.5 that each operator A of the form $\mathrm{i}B$, where B is a self-adjoint operator is the generating operator of the group of unitary operators $U(t) = \mathrm{e}^{\mathrm{i}tB}$. That is why formula (5.16) holds true for the solutions of equation (5.14) with such an operator. Moreover, if operator B^{-1} is bounded, then the solution of the Cauchy problem for the equation

$$\frac{\mathrm{d}^2 x}{\mathrm{d}t^2} + B^2 x = 0 \tag{5.17}$$

can be written down in the form

$$\begin{aligned} x(t) &= (1/2)\left[\mathrm{e}^{\mathrm{i}tB} + \mathrm{e}^{-\mathrm{i}tB}\right] x_0 + (1/2)\left[\mathrm{e}^{\mathrm{i}tB} - \mathrm{e}^{-\mathrm{i}tB}\right](\mathrm{i}B)^{-1}x_0' \\ &= [\cos(tB)]\,x_0 + [\sin(tB)]\,B^{-1}x_0'. \end{aligned} \tag{5.18}$$

Let us notice that in the considered case B^2 is an arbitrary positive definite operator; in (5.18) one can assume that B is positive definite and equal to the square root of B^2.

Consider now the nonhomogeneous equation

$$\frac{\mathrm{d}^2 x}{\mathrm{d}t^2} + B^2 x = f(t); \tag{5.19}$$

by the method of variation of parameters, one obtains the formula

$$\begin{aligned} x(t) =&[\cos(tB)]x_0 + [\sin(tB)]B^{-1}x_0' \\ &+ \int_0^t (\sin[(t-\tau)B])\, B^{-1}f(\tau)\mathrm{d}\tau. \end{aligned} \tag{5.20}$$

If $f(t)$ is a continuously differentiable function, then this formula gives the solution of the problem (5.19), (5.15) for $x_0 \in \mathcal{D}(B^2)$, $x_0' \in \mathcal{D}(B)$. This formula too makes sense for any $x_0, x_0' \in E$ and for a continuous function $f(t)$. In such cases, (5.20) is called the *generalized solution* of the problem (5.19), (5.15).

We would also need a narrower class of generalized solutions, defined by formula (5.20) for $x_0 \in \mathcal{D}(B)$, $x_0' \in E$ and a continuous function $f(t)$. For every such generalized solution, the continuous functions $Bx(t)$ and $\mathrm{d}x/\mathrm{d}t$ are well defined. Each generalized solution like this is the limit of a sequence of actual solutions $x_n(t)$ in the sense that the functions $x_n(t)$, $Bx_n(t)$ and $\mathrm{d}x_n/dt$ converge uniformly on $[0,T]$ to the functions $x(t)$, $Bx(t)$, dx/dt, respectively.

If $x(t)$ is a solution of the problem (5.19), (5.15), then scalar multiplication of both sides of equation (5.19) by dx/dt and an evaluation of the real part yield

$$(1/2)\frac{\mathrm{d}}{\mathrm{d}t}\left[\left(\frac{\mathrm{d}x}{\mathrm{d}t}, \frac{\mathrm{d}x}{\mathrm{d}t}\right) + (Bx, Bx)\right] = \mathrm{Re}\left(f, \frac{\mathrm{d}x}{\mathrm{d}t}\right).$$

Integrating with respect to t, one obtains the energy indentity

$$\|\mathrm{d}x/\mathrm{d}t\|^2 + \|Bx\|^2 = \|x_0'\|^2 + \|Bx_0^2\|^2 + 2\mathrm{Re}\int_0^t (f(\tau), \mathrm{d}x/\mathrm{d}\tau)\mathrm{d}\tau. \tag{5.21}$$

Taking the limit, one concludes that this identity holds even for the generalized solutions mentioned above. In this regard, such solutions are called *generalized solutions with continuous full energy*.

In many problems of continuous medium mechanics, one meets equations of the form

$$A\frac{\mathrm{d}^2u}{\mathrm{d}t^2} + Bu = \varphi(t) \tag{5.22}$$

Here, operator B deals with the potential energy and is positive definite whenever the nonperturbed state of the system is stable; operator A, connected with the kinetic energy, is bounded and positive.

Let us consider the operator $A^{-1/2}BA^{-1/2}$, which is inverse to the positive bounded operator $A^{1/2}B^{-1}A^{1/2}$ and, therefore, is self-adjoint and positive definite. Let $R := (A^{-1/2}BA^{-1/2})^{1/2}$ be the positive definite square root of this operator. Assume now that $\varphi(t) \in \mathcal{D}\left(A^{-1/2}\right)$ and the function $A^{-1/2}\varphi(t)$ is continuous. Then according to the above mentioned facts, equation

$$\frac{\mathrm{d}^2x}{\mathrm{d}t^2} + R^2x = A^{-1/2}\varphi(t) \tag{5.23}$$

has a generalized solution with continuous energy, which satisfies the initial conditions $x(0) = x_0 \in \mathcal{D}(R), x'(0) = x'_0 \in E$; for this solution $x(t)$, $Rx(t)$, dx/dt are continuous functions.

Let us show that the operator $A^{-1/2}R^{-1}$ is bounded on the whole space E. Let $x \in \mathcal{D}\left(A^{-1/2}\right)$ and $w \in E$. One first gets the estimate

$$\left|\left(A^{-1/2}x, R^{-1}w\right)\right| = \left|\left(R^{-1}A^{-1/2}x, w\right)\right| \leq \|R^{-1}A^{-1/2}x\| \cdot \|w\|.$$

Further,

$$\begin{aligned}\|R^{-1}A^{-1/2}x\|^2 &= \left(R^{-1}A^{-1/2}x, R^{-1}A^{-1/2}x\right) = \left(R^{-2}A^{-1/2}x, A^{-1/2}x\right) \\ &= \left(A^{1/2}B^{-1}x, A^{-1/2}x\right) = (B^{-1}x, x) \leq \|B^{-1}\| \cdot \|x\|^2.\end{aligned}$$

Then, from the latter inequality it follows that

$$\left|\left(A^{-1/2}x, R^{-1}w\right)\right| \leq \|B^{-1}\|^{1/2}\|x\| \cdot \|w\|$$

From the definition of a self-adjoint operator, it follows that $R^{-1}w \in \mathcal{D}\left(A^{-1/2}\right)$ and

$$\|A^{-1/2}R^{-1}w\| \leq \|B^{-1}\|^{1/2} \cdot \|w\|.$$

Now, let $x(t)$ be a solution of equation (5.23) considered above. Then, $x(t) = R^{-1}Rx(t)$; since $Rx(t)$ is continuous, then $x(t) \in \mathcal{D}\left(A^{-1/2}\right)$ and the function

$$u(t) = A^{-1/2}x(t) = \left(A^{-1/2}R^{-1}\right)Rx(t) \tag{5.24}$$

is continuous as well.

If $x(t)$ is an actual solution of (5.23), then by substituting $x(t) = A^{1/2}u(t)$ in (5.22) one obtains

$$\frac{d^2(A^{1/2}u(t))}{dt^2} + A^{-1/2}Bu(t) = A^{-1/2}\varphi(t).$$

Applying the bounded operator A to both sides, one concludes that $u(t)$ is a solution of the equation

$$A^{1/2}\frac{\mathrm{d}^2\left(A^{1/2}u(t)\right)}{\mathrm{d}t^2} + Bu = \varphi(t).$$

If $x(t)$ is just a generalized solution of (5.23) with continuous energy, then $u(t)$ defined by formula (5.24) should be naturally called a generalized solution of equation (5.22) with continuous full (kinetic plus potential) energy. This solution satisfies the identity

$$\left\|\frac{\mathrm{d}(A^{1/2}u)}{\mathrm{d}t}\right\|^2 + \|RA^{1/2}u\|^2$$
$$= \left\|\frac{\mathrm{d}(A^{1/2}u(0))}{\mathrm{d}t}\right\|^2 + \|RA^{1/2}u(0)\|^2 + 2\mathrm{Re}\int_0^t \left(A^{-1/2}\varphi, \mathrm{d}(A^{1/2}u(\tau))/\mathrm{d}\tau\right)\mathrm{d}\tau. \tag{5.25}$$

Let $x \in \mathcal{D}(R^2)$. Then

$$\|Rx\|^2 = (R^2x, x) = \left(A^{1/2}BA^{-1/2}x, x\right) = \left(B^{1/2}A^{-1/2}x, B^{1/2}A^{-1/2}x\right).$$

It can be proved by taking the limit that this equation holds true for $x \in \mathcal{D}(R)$. Then, (5.25) can be written down in the form

$$\left\|\frac{\mathrm{d}(A^{1/2}u)}{\mathrm{d}t}\right\|^2 + \|B^{1/2}u\|^2$$
$$= \left\|\frac{\mathrm{d}\left(A^{1/2}u(0)\right)}{\mathrm{d}t}\right\|^2 + \|B^{1/2}u(0)\|^2 + 2\mathrm{Re}\int_0^t \left(A^{-1/2}\varphi, \frac{\mathrm{d}(A^{1/2}u)}{\mathrm{d}\tau}\right)(\tau)\mathrm{d}\tau. \tag{5.26}$$

If the generalized solution $u(t)$ is continuously differentiable on the interval $[0, T]$, then instead of (5.26) we can obtain the equation of the full energy balance in the more usual form

$$\left\|A^{1/2}\frac{\mathrm{d}u}{\mathrm{d}t}\right\|^2 + \|B^{1/2}u\|^2$$
$$= \|A^{1/2}u_0'\|^2 + \|B^{1/2}u_0\|^2 + 2\mathrm{Re}\int_0^t \left(\varphi(\tau), \frac{\mathrm{d}u}{\mathrm{d}\tau}\right)\mathrm{d}\tau. \tag{5.27}$$

1.5.9 Volterra Integral Equations

Formula (5.13) for a solution of the nonhomogeneous equation (5.12) allows us to reduce the solution of a more complicated differential equation that (5.12) to the solution of an integral equation. It is said that $K(t, s)$ is an operator kernel with weak singularity, provided the operator-valued function $K(t, s)$ is defined in the triangle $0 \le s \le t \le T$, is continuous in t, s in that triangle, and its norm satisfies the inequality

$$\|K(t, s)\| \le \frac{c}{(t-s)^\mu}, \tag{5.28}$$

where $\mu \in [0,1)$ and the constant c does not depend on t and s.

Let us consider the integral equation

$$w(t) = \varphi(t) + \int_0^t K(t,s)w(s)\mathrm{d}s, \tag{5.29}$$

where $\varphi(t)$ is a given function with values in space E continuous on the interval $[0,T]$. It turns out that equation (5.29) has a unique solution which is continuous on the interval $[0,T]$. This solution can be found by the method of successive approximations.

If the function $\varphi(t)$ is continuous only in the half-open interval $(0,T]$ and has a weak singularity at the point $t=0$, that is,

$$\|\varphi(t)\| \le ct^{-v}, \qquad 0 \le v < 1, \tag{5.30}$$

then the solution $w(t)$ possesses the same property.

1.6 Spectral Theory of Operator Pencils

Numerous spectral problems of hydrodynamics can be reduced to the investigation of the spectrum of operator-valued functions of complex variables with values in $\mathcal{L}(E)$. Usually, a function that takes operator values in $\mathcal{L}(E)$ is called an operator pencil. There are linear, polynomial, holomorphic, and continuous pencils. In this section all necessary facts from the theory of such operator-valued functions will be considered.

1.6.1 Eigen- and Associated Elements of an Operator Pencil

Let G be an open region in the complex plane. The spectrum of a holomorphic operator-valued function $A(\lambda)$ on G is defined as the set $\sigma(A(\lambda))$ of all $\lambda \in G$ for which the operator $A(\lambda)$ does not have a bounded inverse operator. The spectrum is a closed set in G. The numbers $\lambda \in G \setminus \sigma(A(\lambda))$ are called regular values of $A(\lambda)$. The operator-valued function $A^{-1}(\lambda)$ is holomorphic on the set $G \setminus \sigma(A(\lambda))$, and is called the resolvent of the function $A(\lambda)$.

A number $\lambda_0 \in G$ is called an *eigenvalue* of the operator-valued function $A(\lambda)$, provided the equation

$$A(\lambda_0)\varphi = 0$$

has a nonzero solution $\varphi = \varphi_0 \in E$; every such element φ_0 is called an *eigenelement* of $A(\lambda)$ corresponding to the eigenvalue λ_0.

The elements $\varphi_1, \dots, \varphi_{m-1}$ are called *associated elements* to the eigenelement

φ_0, provided

$$\sum_{k=0}^{j} \frac{1}{k!} A^{(k)}(\lambda_0)\varphi_{j-k} = 0, \qquad 1 \le j \le m-1,\ A^{(0)} = A.$$

The number m is called the length of the chain formed by an eigenelement φ_0 and the $m-1$ associated elements corresponding to it; the maximal length of such chains is called the *multiplicity* of the eigenelement φ_0 and is denoted by $m(\varphi_0)$.

If the eigensubspace $\ker A(\lambda_0)$ of the operator $A(\lambda_0)$ is finite-dimensional and all the elements of this subspace have finite multiplicities, then in this subspace a canonical basis

$$\varphi_0^{(1)}, \varphi_0^{(2)}, \dots, \varphi_0^{(r)}$$

can be chosen in the following way. First, one needs to find the eigenelement $\varphi_0^{(1)}$ that has the maximal multiplicity m_1. Next, one takes a direct complement M_1 in $\ker A(\lambda_0)$ to the one-dimensional subspace formed by that element, and in this complement there is an eigenelement $\varphi_0^{(2)}$ which has a maximal multiplicity m_2, and so on. Thus, one obtains the basis $\{\varphi_0^{(k)}\}_{k=1}^{r}$ in $\ker A(\lambda_0)$ of eigenelements with multiplicities

$$m_k = \max_{0 \ne \varphi \in M_{k-1}} m(\varphi),$$

where $m_1 \ge m_2 \ge \cdots \ge m_r$. These numbers do not depend on the arbitrary construction of the canonical basis (i.e., the choice of the complements M_k).

The number

$$m(\lambda_0, A(\lambda_0)) := \sum_{k=1}^{r} m_k$$

is called the *algebraic multiplicity* of the eigenvalue λ_0.

It is easy to see that the chain of eigen- and associated elements of the linear pencil $A(\lambda) := A - \lambda I$ coincides with the Jordan chain of operator A; this Jordan chain corresponds to the same eigenvalue $\lambda = \lambda_0$. That is why the algebraic multiplicity of the eigenvalue λ_0 for the pencil $A - \lambda I$ coincides with the algebraic multiplicity of the eigenvalue λ_0 of operator A.

The subspace formed by the linear hull of the eigen- and associated elements is called the *root subspace* of the operator pencil $A(\lambda)$ corresponding to the eigenvalue λ_0.

1.6.2 Root Functions

A function $\varphi(\lambda)$ with values in E is called a *root function* of the operator pencil $A(\lambda)$ relatively to the point $\lambda_0 \in \mathbb{C}$ if:

1) this function is holomorphic in a neighborhood of the point λ_0;
2) $\varphi(\lambda_0) \neq 0$;
3) $A(\lambda_0)\varphi(\lambda_0) = 0$.

If the function $A(\lambda)\varphi(\lambda)$ has a zero of order $m+1$ at the point λ_0, then m is called the *order* of that root function.

It is easy to check that in the Taylor series expansion

$$\varphi(\lambda) = \varphi_0 + (\lambda - \lambda_0)\varphi_1 + \cdots + (\lambda - \lambda_0)^m \varphi_m + \cdots \tag{6.1}$$

by powers of $\lambda - \lambda_0$ of a root function of the m-th order, the element φ_0 is an eigenelement; the elements

$$\varphi_k = \frac{1}{k!} \frac{d^k \varphi(\lambda)}{\mathrm{d}\lambda^k}\bigg|_{\lambda=\lambda_0} \qquad 1 \leq k \leq m$$

are associated elements to that eigenelement corresponding to the eigenvalue λ_0. Conversely, if in the expansion (6.1) φ_0 is an eigenelement and φ_k, $1 \leq k \leq m$, are the corresponding associated elements, then $\varphi(\lambda)$ is a root function with an order not less than m.

If an operator-valued function $B(\lambda)$ is holomorphic in a neighborhood of λ_0 and $\varphi(\lambda)$ is a root function of the pencil $A(\lambda)$ relatively to λ_0 of order m, then $\varphi(\lambda)$ is also a root function of the pencil $B(\lambda)A(\lambda)$ relatively to λ_0 of an order not less than m. If $B(\lambda)$ is invertible at the point λ_0, then the order is preserved.

If $D(\lambda)$ is a holomorphic operator-valued function in a neighborhood of λ_0 that commutes with $A(\lambda)$ and $\varphi(\lambda)$ is a root function of the pencil $A(\lambda)$ relatively to λ_0 such that $D(\lambda_0)\varphi(\lambda_0) \neq 0$, then in virtue of the equation

$$A(\lambda)D(\lambda)\varphi(\lambda) = D(\lambda)A(\lambda)\varphi(\lambda)$$

one concludes that the function $D(\lambda)\varphi(\lambda)$ is also a root function of the pencil $A(\lambda)$ relatively to λ_0, of an order not less than the order of the function $\varphi(\lambda)$. If $\varphi_0, \ldots, \varphi_m$ are the eigenelement and associated elements corresponding to the function $\varphi(\lambda)$, then the eigenelement and the first m associated elements corresponding to the root function $D(\lambda)\varphi(\lambda)$ can be found by the formulas

$$\tilde{\varphi}_0 = D(\lambda_0)\varphi_0,$$

$$\tilde{\varphi}_k = \frac{1}{k!} \frac{\mathrm{d}^k (D(\lambda)\varphi(\lambda))}{\mathrm{d}\lambda^k}\bigg|_{\lambda=\lambda_0} = \sum_{j=0}^{k} \frac{1}{j!} \frac{\mathrm{d}^j D(\lambda_0)}{\mathrm{d}\lambda^j} \varphi_{k-j}.$$

These formulas are useful for various transformations of operator-valued functions.

Similarly, one could consider changes of eigen- and associated elements correpsonding to changes of the parameter. Let $\lambda = \lambda(\mu)$ be a scalar function, holomorphic in a region G_1 and suppose that $\lambda_0 = \lambda(\mu_0)$ is an eigenvalue of the pencil $A(\lambda)$ and $\varphi(\lambda)$ is a root function of this pencil relatively to λ_0 of order m. The function $\varphi(\lambda(\mu))$ is a root function of the pencil $A(\lambda(\mu))$ relatively to the point μ_0. The element $\varphi(\lambda(\mu_0)) = \varphi(\lambda_0)$ is an eigenelement. Using the formula for the derivative of a composite function, one can obtain the associated elements $\tilde{\varphi}_k$ corresponding to this eigenvalue relatively to the pencil $A(\lambda(\mu))$ by

$$\begin{aligned}\tilde{\varphi}_k &= \frac{1}{k!}\frac{\mathrm{d}^k\varphi(\lambda(\mu))}{\mathrm{d}\mu^k}\bigg|_{\mu=\mu_0} \\ &= \sum_{l_1+2l_2+\cdots+jl_j=k} \frac{(l_1+l_2+\cdots+l_j)!}{l_1!l_2!\cdots l_j!}\varphi_{l_1+l_2+\cdots+l_j}\left(\frac{\lambda'(\mu_0)}{1!}\right)^{l_1}\cdots\left(\frac{\lambda^{(j)}(\mu_0)}{j!}\right)^{l_j}.\end{aligned}$$

1.6.3 Fredholm Holomorphic Operator-Valued Functions

A function of the form $A(\lambda) := I - L(\lambda)$, where the values of $L(\lambda)$ are compact operators on E, is called a Fredholm operator-valued function or *Fredholm pencil.* If λ_0 is an eigenvalue of the operator $A(\lambda)$—usually called a *characteristic number* of the pencil $L(\lambda)$—then the number 1 is an eigenvalue of the compact operator $L(\lambda_0)$; in virtue of some properties of compact operators, the geometric multiplicity of the eigenvalue 1 is finite, that is, $\dim\ker A(\lambda_0) < \infty$.

A Fredholm pencil $A(\lambda)$ is called regular in a region $G \subset \mathbb{C}$, provided the operator-valued function $L(\lambda)$ is holomorphic in G, and the operator $A(\lambda)$ has a bounded inverse operator at least at one point in G. The spectrum of a regular pencil consists of isolated points in G and its accumulation points may be located only on the boundary of G. All points of the spectrum are eigenvalues of the operator-valued function $A(\lambda)$.

The eigenelements corresponding to these eigenvalues have finite multiplicities. The resolvent $A^{-1}(\lambda)$ is a meromorphic function in G, which has poles at points coinciding with the eigenvalues. The multiplicity of a pole equals the maximal multiplicity of the eigenelements corresponding to the given eigenvalue.

1.6.4 Linear Pencils. Theorems on Completeness of the System of Eigen- and Associated Elements

If A is a self-adjoint compact operator, then according to Hilbert-Schmidt Theorem its eigenelements form a complete orthogonal system in E. The spectrum of A is real.

Now, let us examine the situation when operator A is weakly perturbed. Namely, let us consider the operator $B = A(I+S)$, where S is a compact operator.

The following result, referred to as the **First Keldysh Theorem** is valid:

If $\ker B = \{0\}$ *and* $A \in \mathfrak{S}_p$ *for some* $p \in (0,\infty)$, *then the eigen- and associated elements of operator* B *form a complete system in* E.

For every small $\varepsilon > 0$ *all the eigenvalues of operator* B, *except perhaps finitely many, are located in the angles*

$$-\varepsilon < \arg\lambda < \varepsilon, \quad \pi - \varepsilon < \arg\lambda < \pi + \varepsilon$$

with aperture 2ε. *If operator* A *has only a finite number of negative (positive) eigenvalues, then* $B - \lambda I$ *has no more than a finite number of eigenvalues within the angle*

$$\pi - \varepsilon < \arg\lambda < \pi + \varepsilon, \qquad (-\varepsilon < \arg\lambda < \varepsilon).$$

This theorem holds true as well in the case when B is obtained by using invertible multipliers of the form $I + S$ not only to the right, but also to the left, that is,

$$B = (I+S_1)A(I+S_2), \qquad S_i \in \mathfrak{S}_\infty, \ \ker B = \{0\}.$$

A similar result, called the **Second Keldysh Theorem**, is true for linear pencils of the form

$$L(\lambda) = I - T - \lambda A. \tag{6.2}$$

It can be formulated as follows:

If $A = A^* \in \mathfrak{S}_p$, $\ker A = \{0\}$, *and* $T \in \mathfrak{S}_\infty$, *then the system of eigen- and associated elements of the pencil* $L(\lambda)$ *is complete in* E.

Usually, we deal with situations when the condition $\ker A = \{0\}$ is met. In this case it is said that the Keldysh pencil is degenerated. We can introduce the projections Π_0 onto the kernel of operator A and $\Pi_1 = I - \Pi_0$ onto the closure of the range of operator A. Each eigenelement φ_0 of the pencil (6.2) admits the expansion $\varphi_0 = \Pi_0\varphi_0 + \Pi_1\varphi_0 =: \varphi_0^0 + \varphi_0^1$. Then, the equation $L(\lambda)\varphi_0 + (I - T - \lambda A)\varphi_0 = 0$ turns into the equation

$$\varphi_0^0 + \varphi_0^1 - T\left(\varphi_0^0 + \varphi_0^1\right) - \lambda A\varphi_0^1 = 0.$$

Consequently, applying the operators Π_0 and Π_1, one obtains

$$\begin{aligned} \varphi_0^0 - \Pi_0 T \Pi_0 \varphi_0^0 &= \Pi_0 T \Pi_1 \varphi_0^1, \\ \varphi_0^1 - \Pi_1 T \Pi_0 \varphi_0^0 - \Pi_1 T \Pi_1 \varphi_0^1 - \lambda A \varphi_0^1 &= 0. \end{aligned}$$

If we assume that the operator $I - \Pi_0 T \Pi_0$ is invertible on the subspace $\ker A$, then from the first equation one gets $\varphi_0^0 = (I - \Pi_0 T \Pi_0)^{-1} \Pi_0 T \Pi_1 \varphi_0^1$. A substitution into the second equation leads to

$$(I - \Pi_1 T \Pi_0 \left(I - \Pi_0 T \Pi_0\right)^{-1} \Pi_0 T \Pi_1 - \Pi_1 T \Pi_1 - \lambda A)\varphi_0^1 = 0.$$

Now, let us consider an element φ_1 associated to the eigenelement φ_0. It satisfies the equation

$$L(\lambda)\varphi_1 + \frac{\mathrm{d}L(\lambda)}{\mathrm{d}\lambda}\varphi_0 = \varphi_1 - T\varphi_1 - \lambda A \varphi_1 - A\varphi_0 = 0.$$

Assuming $\varphi_1 = \Pi_0 \varphi_1 + \Pi_1 \varphi_1 =: \varphi_1^0 + \varphi_1^1$ and carrying out the operations mentioned above, we have the equation

$$\left(I - \Pi_1 T \Pi_0 (I - \Pi_0 T \Pi_0)^{-1} \Pi_0 T \Pi_1 - \Pi_1 T \Pi_1 - \lambda A\right) \varphi_1^1 - A\varphi_0^1 = 0.$$

Therefore, the element φ_1^1 is associated to the eigenelement φ_0^1 of the pencil

$$L_1(\lambda) := I - \Pi_1 T \Pi_0 (I - \Pi_0 T \Pi_0)^{-1} \Pi_0 T \Pi_1 - \Pi_1 T \Pi_1 - \lambda A,$$

acting on the space $E_1 = \overline{\mathcal{R}(A)}$.

The similar situation takes place for all the subsequent associated elements.

The resulting pencil $L_1(\lambda)$ satisfies all the conditions of the Second Keldysh Theorem. From this theorem we get the next result:

If $A = A^* \in \mathfrak{S}_p, T \in \mathfrak{S}_\infty$, and the operator $I - \Pi_0 T \Pi_0$ is invertible on $\ker A$, then the projections of eigen- and associated elements of the pencil (6.2) onto the closure of the range of operator A form a complete system in this subspace.

1.6.5 Keldysh-Type n-Multiple Completeness

Let us consider the Cauchy problem for the differential equation of the n-th order

$$A\left(\frac{\mathrm{d}}{\mathrm{d}t}\right) u := \sum_{k=0}^{n} A_k \frac{\mathrm{d}^k u}{\mathrm{d}t^k} = 0, \quad u^{(j)}(0) = f_j, \qquad 0 \le j \le n-1, \tag{6.3}$$

with bounded operator coefficients A_k.

To every chain $\{\varphi_k\}_{k=0}^{m-1}$ of length m consisting of an eigenelement and associated elements relatively to $A(\lambda)$ there correspond m elementary solutions of equation (6.3) given by

$$u_k(t) = \mathrm{e}^{\lambda_0 t} \sum_{j=0}^{k} \frac{t^j}{j!} \varphi_{k-j}, \qquad 0 \le k \le m-1. \tag{6.4}$$

To every such solution one associates the vector-column of the initial data

$$\left\{u_k^{(j)}(0)\right\}_{j=0}^{n-1} \in E^n.$$

It is easy to show that

$$u_k^{(j)}(0) = \sum_{s=0}^{j} C_j^s \lambda_0^{j-s} \varphi_{k-s}. \tag{6.5}$$

where for $l < 0$ we assume that $\varphi_l = 0$. If the vector-column $\{f_j\}_{j=0}^{n-1}$ of the initial data of the problem (6.3) is a linear combination of the vectors (6.5) (for all possible eigenvalues and the corresponding chains), then the Cauchy problem (6.3) has a solution, which is a linear combination of the corresponding elementary solutions of the form (6.4). Therefore, if the set of all vectors of the form (6.5) is complete in E^n, then the Cauchy problem (6.3) has a solution for a set of initial vectors $\{f_j\}_{j=0}^{n-1}$ dense in E^n.

Following Keldysh, we say that the system of eigen- and associated elements of the pencil $A(\lambda)$ is *n-multiply complete* in E, if the set of all elements of the form (6.5) (for all possible eigenvalues and the corresponding chains of the pencil $A(\lambda)$) is complete in the space E^n. If this set forms a basis in E^n, then the pencil $A(\lambda)$ is *n-multiply basic* in E.

1.6.6 Spectral Factorization of an Operator Pencil

Under some conditions, an operator-valued function $A(\lambda)$ holomorphic in a region $G \subset \mathbb{C}$ can be factorized as

$$A(\lambda) = B(\lambda)(\lambda I - Z),$$

where $B(\lambda)$ is an operator-valued functions holomorphic in G and Z is a bounded operator, called a *root* of $A(\lambda)$. If the spectrum of operator Z does not intersect the spectrum of the pencil $B(\lambda)$, then the multiplier $\lambda I - Z$ is called a *spectral divisor* of the pencil $A(\lambda)$, and Z is called a *spectral root*.

As it was mentioned in Section 1.6.2, each root function of the operator $\lambda I - Z$ is a root function of $A(\lambda)$. Therefore, the eigen- and associated elements of operator Z are eigen- and associated elements of $A(\lambda)$. If the function $B(\lambda)$ is invertible for every $\lambda \in G$, then the spectra, eigen- and associated elements of Z and $A(\lambda)$ coincide.

Thus, the spectral problem for $A(\lambda)$ can be reduced to the investigation of a spectral problem for a single bounded operator.

Let us point out one of the most important cases of possible factorizations.

Suppose that the operator-valued function $A(\lambda)$ is of the form

$$A(\lambda) = A - \lambda I + \sum_{k=1}^{\infty} \lambda^k B_k, \qquad A, B_k \in \mathcal{L}(E),$$

and for some $r > 0$ the following condition holds

$$\|A\| r^{-1} + \sum_{k=1}^{\infty} r^{k-1} \|B_k\| < 1.$$

Then $A(\lambda)$ admits the spectral factorization

$$A(\lambda) = A_+(\lambda)(\lambda I - Z),$$

where $A_+(\lambda)$ is holomorphic and holomorphically invertible for $|\lambda| \leq r$, and the spectrum $\sigma(Z)$ is located within the circle $|\lambda| < r$.

1.6.7 Completeness With Finite Defect of a System of Eigen- and Associated Elements of an Operator-Valued Function

In some problems it is quite natural to consider not all the eigen- and associated elements of a pencil, but only those which correspond to some part of its spectrum. Thus, if the system of eigen- and associated elements of the given pencil is n-multiply complete in E, then it can be expected that the eigen- and associated elements corresponding to the eigenvalues from some sector of the complex plane form an "almost complete" system, that is, the orthogonal complement in E to these elements is finite-dimensional.

Further we shall need the following theorem on completeness of the system of eigen- and associated elements with accuracy up to a finite defect.

Let an operator pencil be of the form

$$L(\lambda) = T - \lambda I - \lambda B(\lambda),$$

where $T = A(I + S)$, $A = A^* \in \mathfrak{S}_p$, $0 < p < \infty$, *with* $\ker A = \{0\}$, $S \in \mathfrak{S}_\infty$, *and* $B(\lambda)$ *is an operator-valued function that takes values in* $\mathcal{L}(E)$ *and is holomorphic in the circle* $|\lambda| < r$. *Then the system of eigen- and associated elements of the operator-valued function* $L(\lambda)$ *corresponding to eigenvalues from the circle* $|\lambda| < r$ *has no more than a finite defect in* E.

1.6.8 Asymptotic Behavior of Branches of Eigenvalues

Let the operator-valued function $L(\mu)$ be of the form

$$L(\mu) = I + T - \mu A + B(\mu),$$
$$T \in \mathfrak{S}_\infty,\ A = A^* \in \mathfrak{S}_\infty,\ \ker A = \{0\}, \tag{6.6}$$

where $B(\mu)$ is an analytic operator-valued function outside of the circle $|\mu| \le R$ that takes values in $\mathcal{L}(E)$ and turns into zero at infinity. Along with (6.6) let us consider the "shortened" linear pencil

$$L_0(\mu) := I - \mu A.$$

If $A = A_+ - A_-, A_\pm > 0$ and the operators $A_\pm$ are infinite-dimensional, then $L_0(\mu)$ has two branches consisting of characteristic numbers of operator A as its eigenvalues:

$$\mu = \mu_j^\pm(A), \quad j = 1, 2, \dots, \qquad \mu_j^+(A) > 0,\ \mu_j^-(A) < 0.$$

Let us denote by

$$n^\pm(r; A) := \sum_{|\mu_n^\pm|<r} 1$$

the distribution functions—with regard to multiplicities—of the eigenvalues of the pencil $L_0(\mu)$ (or, what is the same, of the characteristic values of the operator A). For the operators $A \in \mathfrak{S}_\infty$ usually appearing in problems of mathematical physics, the asymptotic behavior of the functions $n^\pm(r; A)$ for $r \to \infty$ is like a power function, as a rule.

Further, let us denote the similar distribution functions of the eigenvalues of the pencil $L(\mu)$ by $n^\pm(r; L(\mu))$; these eigenvalues are located within the sectors $|\arg \lambda| < \varepsilon$ and $|\arg \lambda - \pi| < \varepsilon$, $\varepsilon > 0$, respectively. The following theorem on spectral asymptotics of eigenvalues of the operator pencil $L(\mu)$ is valid.

If the condition

$$n^+(r; A) = ar^\alpha[1 + \mathrm{o}(1)], \quad r \to \infty, \qquad a > 0,\ \alpha > 0, \tag{6.7}$$

holds true, then under the above formulated assumptions (6.6) the pencil $L(\mu)$ has a countable set of eigenvalues $\mu_n^+(L(\mu))$ in the sector $|\arg \lambda| < \varepsilon$ and

$$n^+(r; L(\mu)) = ar^\alpha[1 + \mathrm{o}(1)], \quad r \to \infty. \tag{6.8}$$

The similar assertion is valid for the countable set of eigenvalues $\mu_n^-(L(\mu))$ within the sector $|\arg \lambda - \pi| < \varepsilon$ and for the asymptotic behavior of the function $n^-(r; L(\mu))$.

We notice that (6.7) is equivalent to equation

$$\mu_n^+(A) = a^{-1/\alpha} n^{1/\alpha}[1 + o(1)], \quad n \to \infty$$

and (6.8) is equivalent to

$$\mu_n^+(L(\mu)) = a^{-1/\alpha} n^{1/\alpha}[1 + o(1)], \quad n \to \infty.$$

Therefore, from the theorem on spectral asymptotics it follows that for each branch of eigenvalues of the shortened pencil $L_0(\mu)$ and of the original pencil $L(\mu)$ the asymptotics components coincide:

$$\mu_n^\pm(L(\mu)) = \mu_n^\pm(A)[1 + o(1)], \quad n \to \infty.$$

1.6.9 Self-Adjoint Operator Pencils

An operator-valued function $A(\lambda)$ with values in $\mathcal{L}(E)$ and defined on some set $G \subset \mathbb{C}$ symmetric relatively to the real axis $\mathbb{R}$ is called self-adjoint, provided $(A(\lambda))^* = A(\bar{\lambda})$. In particular, the polynomial pencil $A(\lambda) = \sum_{k=0}^n \lambda^k A_k$ is self-adjoint if $A_k = A_k^*$, $0 \le k \le n$.

If $A(\lambda)$ is a self-adjoint and holomorphic operator-valued function, then the spectrum of $A(\lambda)$ is symmetric relatively to $\mathbb{R}$. This property follows from the fact that the operators $B \in \mathcal{L}(E)$ and its adjoint B^* are invertible at the same time. Let us notice that in virtue of the symmetry principle, an operator-valued function $A(\lambda)$ that takes self-adjoint values on an open interval $(a, b) \subset \mathbb{R}$ and is holomorphic in an open neighborhood $G \subset \mathbb{C}$ is self-adjoint; moreover, the region of holomorphicity for $A(\lambda)$ is symmetric relatively to $\mathbb{R}$.

If a self-adjoint operator-valued function admits (spectral) factorization, then its (spectral) root Z can be self-adjoint only in special cases (for example, if any two operator coefficients are commuting). However, under some natural conditions Z is similar to a self-adjoint operator.

It is said that the operator $Z \in \mathcal{L}(E)$ is symmetrized from the right by a self-adjoint operator $F \in \mathcal{L}(E)$, provided the operator ZF is self-adjoint. If the symmetrizator $F \gg 0$, then Z is similar to a self-adjoint operator. Indeed, as far as $(ZF)^* = ZF$ and $F \gg 0$, there exists the operators $F^{1/2}$ and $F^{-1/2}$ from $\mathcal{L}(E)$ and

$$Z = F^{1/2}\left(F^{-1/2}(ZF)F^{-1/2}\right)F^{-1/2},$$

where the operator $F^{-1/2}(ZF)F^{-1/2}$ is self-adjoint.

For self-adjoint operator-valued functions the following assertions hold true:

1° *Let $[a,b] \subset \mathbb{R}$, let $U \subset \mathbb{C}$ be a connected neighborhood of the interval $[a,b]$ symmetric relatively to $\mathbb{R}$, and assume that $A(\lambda)$ is a self-adjoint operator-valued function holomorphic in U. If $A(a) \ll 0$, $A(b) \gg 0$, the function $(A(\lambda)\varphi,\varphi)$ for every $\varphi \neq 0$ has exactly one root $p(\varphi)$ in (a,b) and*

$$\inf_{\|\varphi\|=1} (A'(p(\varphi))\varphi,\varphi) > 0,$$

then $A(\lambda)$ admits the factorization $A(\lambda) = A_+(\lambda)(\lambda I - Z)$, where the operator-valued function $A_+(\lambda)$ is holomorphic and invertible in U, $Z \in \mathcal{L}(E)$, and $\sigma(Z) \subset (a,b)$; here Z has a positive definite symmetrizator and, therefore, is similar to a self-adjoint operator.

2° *The result in 1° is true for any self-adjoint operator-valued function $A(\lambda)$ holomorphic in a neighborhood of $[a,b] \subset \mathbb{R}$ provided*

$$A(a) \ll 0, \quad A(b) \gg 0, \quad A'(\lambda) \gg 0, \qquad a \leq \lambda \leq b. \tag{6.9}$$

1.6.10 On Riesz Basicity of the System of Eigenelements of a Self-Adjoint Operator-Valued Function

Let $A(\lambda)$ be a self-adjoint operator-valued function, let $\lambda_0 \in \mathbb{R}$ be one of its eigenvalues and assume that the corresponding eigenelement φ_0 satisfies the following condition

$$(A'(\lambda_0)\varphi_0, \varphi_0) \neq 0.$$

Then, there are no associated elements corresponding to such an eigenelement $\varphi_0 \in E$. Indeed, in the opposite case we would have

$$A(\lambda_0)\varphi_1 + A'(\lambda_0)\varphi_0 = 0.$$

Multiplying this equation by φ_0 we obtain

$$(A'(\lambda_0)\varphi_0, \varphi_0) = -(A(\lambda_0)\varphi_1, \varphi_0) = -(\varphi_1, A(\lambda_0)\varphi_0) = 0,$$

which contradicts our assumption.

Let us point out the following simple properties of an operator $Z \in \mathcal{L}(E)$ similar to a self-adjoint operator: all the root elements of such an operator are eigenelements, and they form a Riesz basis for the closure of their linear hull. Besides that, if $Z \in \mathfrak{S}_\infty$, then the collection of all the eigenelements of operator Z form a Riesz basis in the whole space E.

The following statements can be proved using the properties 1° and 2° from Section 1.6.9 and the above mentioned results.

1° *If the conditions in 1° of Section 1.6.9 hold true and the operator $A(\gamma)$ is compact for some $\gamma \in (a,b)$, then the spectrum of the operator-valued function $A(\lambda)$ in the region U consists of the point γ and no more than a countable set of eigenvalues with finite multiplicities located in (a,b). If this set is infinite, then it converges to the point γ. The eigenelements of $A(\lambda)$ corresponding to the eigenvalues from the interval (a,b) form a Riesz basis for the space E.*

2° *In particular, the conclusions formulated in 1° are valid, if instead of the requirements stated in 1° of Section* 1.6.9 *one uses conditions* (6.9).

Now suppose that conditions (6.9) of uniform positivity at the endpoints of the interval $[a,b] \subset \mathbb{R}$ fail for $A(\lambda)$. Then we could still speak about basicity with finite defect.

3° *Let $A(\lambda)$ be a self-adjoint operator-valued function holomorphic in a neighborhood of zero. If $A(0) \in \mathfrak{S}_\infty$ and $A'(0) \gg 0$, then for a sufficiently small $\varepsilon > 0$ the eigenelements corresponding to eigenvalues from the interval $(-\varepsilon, \varepsilon)$ form a Riesz basis for some subspace with finite defect (i.e., finite co-dimension) in E. In particular, the conditions above hold true if*

$$A(\lambda) = \lambda I - A - \lambda^2 B(\lambda),$$

where $A = A^* \in \mathfrak{S}_\infty$, and $B(\lambda)$ is a self-adjoint operator-valued function holomorphic in a neighborhood of zero.

If the coefficients of the self-adjoint operator-valued function $A(\lambda) := \lambda I + \sum_{k=0}^{\infty} \lambda^k A_k$ belong to the classes $\mathfrak{S}_p$, then its eigenvalues possess not only the property of Riesz basicity but also the property of p-basicity. Let us formulate some results that will be helpful in our further considerations.

4° *Suppose $A(\lambda)$ is an operator-valued function holomorphic in some neighborhood $U \supset [a,b]$, $a < 0 < b$, that satisfies conditions* (6.9) *and the following conditions*

$$A_k \in \mathfrak{S}_{p_k}, \qquad 0 \le k \le m,$$
$$A_k \in \mathcal{L}(E), \qquad k \ge m+1, \; m \ge 1.$$

Then the system of eigenelements of $A(\lambda)$ that correspond to eigenvalues from the interval (a,b) forms a p-basis in the space E for

$$p \ge \tilde{p}_1 := \left[\min\left\{\frac{1}{p_1}, \frac{1}{p_0} + \frac{1}{p_2}, \frac{2}{p_0} + \frac{1}{p_3}, \dots, \frac{(m-1)}{p_0} + \frac{1}{p_m}, \frac{m}{p_0}\right\}\right]^{-1}.$$

If only the third condition (6.9) holds true, then the mentioned system of eigenelements forms a p-basis for $p \geq \tilde{p}_1$ in a subspace which has finite codimension in E, that is, this system is a p-basis in E with finite defect.

5° *If $A_k = 0$ for $k \geq m+1$, that is, $A(\lambda)$ is a polynomial of degree m, and the previous conditions for the coefficients A_k, $0 \leq k \leq m$, are satisfied, then the property of p-basicity, and therefore, the property of p-basicity with a finite defect in E take place for*

$$p \geq \tilde{p}_2 := \left[\min\left\{\frac{1}{p_1}, \frac{1}{p_0} + \frac{1}{p_2}, \frac{2}{p_0} + \frac{1}{p_3}, \dots, \frac{(m-1)}{p_0} + \frac{1}{p_m}\right\}\right]^{-1}.$$

In particular, let us notice that if $A'(0) = I + A_1 \gg 0$, then there exists $\varepsilon > 0$ such that the system of eigenelements of the operator-valued function $A(\lambda)$ corresponding to eigenvalues from the interval $(-\varepsilon, \varepsilon)$ forms a p-basis in a subspace with a finite defect for $p \geq \tilde{p}_1$ (conditions as in 4°) or for $p \geq \tilde{p}_2$ (conditions as in 5°); additionally, if $A(-\varepsilon) \ll 0$, and $A(\varepsilon) \gg 0$, then we have a p-basis in the whole space E for the same p.

1.6.11 Variational Methods for Investigating Continuous Operator-Valued Functions

Let $L(t)$ be a self-adjoint continuous operator-valued function with values in $\mathcal{L}(E)$, defined on some interval $(a,b) \subset \mathbb{R}$. Let us assume that the following conditions hold true:

1° For any $\varphi \neq 0$ from E the function $(L(t)\varphi, \varphi)$ has exactly one root $p(\varphi) \in (a,b)$.

2° The function $g(t) = (L(t)\varphi, \varphi)$ increases at the point $t_0 = p(\varphi)$, that is, there exists $\delta > 0$ such that $(g(t') < g(t_0) < g(t'')(t_0 - \delta < t' < t_0 < t'' < t_0 + \delta)$. In particular, if $L(t)$ is a continuously differentiable operator-valued function, then the condition $(L'(p(\varphi), \varphi) > 0$ for $\varphi \in E \setminus \{0\}$ is considered to be valid.

3° The functional $p(\varphi)$ is continuous on $E \setminus \{0\}$.

4° If $\alpha = \inf p(\varphi)$, $\beta = \sup p(\varphi)$, then $\alpha > a$, $\beta < b$.

The functional $p(\varphi)$ is called the *Rayleigh functional* of the operator-valued function $L(t)$. The functional $p(\varphi) = (A\varphi, \varphi)/(\varphi, \varphi)$ for the linear pencil $L(t) = A - tI$ with $A = A^* \in \mathfrak{S}_\infty$ is the most important already mentioned example of a Rayleigh functional.

For such a continuous operator-valued function $L(t)$ the concepts of spectrum, eigenvalues and eigenelements also make sense. The dimension of the subspace $\ker L(t_0)$ is called the multiplicity of the eigenvalue t_0.

We proceed by stating the minimax properties of eigenvalues of the operator-valued function $L(t)$, which generalize the classical theorems of Fisher-Courant-Weyl and Poincare-Ritz.

Let us assume that besides 1°–4° above the following additional conditions hold true:

5° In the interval (a, b) the function $L(t)$ has a sequence of eigenvalues of finite multiplicity which converges to a number γ (γ can be an eigenvalue of any multiplicity).

6° The system of eigenelements of $L(t)$ corresponding to its eigenvalues from the interval (a, b) is complete in E.

We next arrange the eigenvalues of $L(t)$ located within the interval (γ, b) in nonincreasing order into a sequence $\{\lambda_n^+\}$ (with regard to multiplicities); the eigenvalues of $L(t)$ located within the interval (a, γ) are also labeled in nondecreasing order into the sequence $\{\lambda_n^-\}$.

As consequences of all these assumptions one gets the next results.

7° *The following equations are true (**Fisher-Courant-Weyl Principle**)*

$$\begin{aligned} \lambda_n^+ &= \min_{\dim N^\perp = n-1} \max_{0 \neq \varphi \in N} p(\varphi), \\ \lambda_n^- &= \max_{\dim N^\perp = n-1} \min_{0 \neq \varphi \in N} p(\varphi). \end{aligned} \tag{6.10}$$

8° *The following equations are true (**Poincare-Ritz Principle**)*

$$\begin{aligned} \lambda_n^+ &= \max_{\dim M = n} \min_{0 \neq \varphi \in M} p(\varphi), \\ \lambda_n^- &= \min_{\dim M = n} \max_{0 \neq \varphi \in M} p(\varphi). \end{aligned} \tag{6.11}$$

In equations (6.11), M stands for an arbitrary n-dimensional subspace in E; in (6.10), N is an arbitrary subspace in E with codimension $n - 1$.

As an application of the mentioned principles for self-adjoint operator-valued functions, let us point out the following important property.

9° *If a holomorphic self-adjoint operator pencil $A(\lambda)$ satisfies the conditions stated in* 1° *of Section* 1.6.9, *and besides that, $A(\gamma) \in \mathfrak{S}_\infty$ for some $\gamma \in (a, b)$, then the variational principles* (6.10), (6.11) *hold true for its eigenvalues.*

1.7 Asymptotic Methods for Solving Evolution Equations With a Small Parameter Attached to the Derivative

In the chapters of this book devoted to the investigation of problems connected with the movements of viscous fluids, we will study the case when the viscosity of a fluid is large (more exactly, when the Reynolds number is small). Together with this assumption, asymptotic expansions of solutions of the corresponding problems will be built. In this section one examines the asymptotic method for solving abstract evolution equations of a special form; later on, this method will be applied to more concrete problems. It turns out that the asymptotic method is also applicable to finding eigenvalues and eigenelements of operators depending on a small parameter.

1.7.1 Equations With a Small Parameter Attached to the Derivative

Let A be the generating operator of a semigroup $U(t)$, for which the next estimate is true:

$$\|U(t)\| \le M e^{-at}, \qquad a \ge 0. \tag{7.1}$$

For example, such an estimate with $M = 1$ holds true for a self-adjoint nonpositive operator A.

The operator $(1/\varepsilon)A$, where $\varepsilon > 0$, also generates a semigroup $U_\varepsilon(t)$, for which

$$\|U_\varepsilon(t)\| \le M e^{(-a/\varepsilon)t}. \tag{7.2}$$

Now, let $S(\varepsilon)$ be a linear bounded operator, which depends continuously on $\varepsilon \in [0, \varepsilon_0]$. Then, as it was mentioned in Section 1.5, the operator $(1/\varepsilon)A + S(\varepsilon)$ generates a semigroup $V_\varepsilon(t)$, for which

$$\|V_\varepsilon(t)\| \le M e^{-(a/\varepsilon - M\|S(\varepsilon)\|)t} \tag{7.3}$$

Let us consider the following equation in a Hilbert space E,

$$\varepsilon \frac{dv}{dt} = Av + \varepsilon S(\varepsilon) v + \varepsilon f(t, \varepsilon), \qquad 0 \le t \le T, \tag{7.4}$$

where $f(t, \varepsilon)$ is a continuous function of t and ε. The generalized solution of this equation has the form

$$v(t) = V_\varepsilon(t) v(0) + \int_0^t V_\varepsilon(t - \tau) f(\tau, \varepsilon) d\tau.$$

If $a > 0$, then for a sufficiently small ε (for example, if $\varepsilon\|S(\varepsilon)\| < a/(2M)$) one has the estimate

$$\begin{aligned} \|v(t)\| &\le M e^{-(a/\varepsilon - M\|S(\varepsilon)\|)t} \|v(0)\| \\ &\quad + \int_0^t M e^{-(a/\varepsilon - M\|S(\varepsilon)\|)(t-\tau)} d\tau \cdot \max_{0 \le \tau \le t} \|f(\tau, \varepsilon)\| \\ &\le M e^{-(a/2\varepsilon)t} \|v(0)\| + \frac{2\varepsilon}{a} M \max_{0 \le \tau \le t} \|f(\tau, \varepsilon)\|. \end{aligned} \tag{7.5}$$

If $a = 0$, then

$$\begin{aligned}\|v(t)\| &\leq M\mathrm{e}^{M\|S(\varepsilon)\|t}\|v(0)\| + M\int_0^t \mathrm{e}^{M\|S(\varepsilon)\|\,|t-\tau|}\mathrm{d}\tau \cdot \max_{0\leq\tau\leq t}\|f(\tau,\varepsilon)\| \\ &\leq M\mathrm{e}^{M\|S(\varepsilon)\|T}\|v(0)\| + M\mathrm{e}^{M\|S(\varepsilon)\|T}T\max_{0\leq\tau\leq T}\|f(\tau,\varepsilon)\|. \end{aligned} \tag{7.6}$$

1.7.2 Splitting of a Homogeneous Equation

Let us now consider a homogeneous differential equation in a Hilbert space E of the form

$$\varepsilon\frac{\mathrm{d}x}{\mathrm{d}t} = Ax + \varepsilon Lx, \tag{7.7}$$

where A is the generating operator of a semigroup from the class C_0 and L is a bounded operator.

Let us assume that the whole space E can be decomposed into a direct sum of two subspaces E_1 and E_2, which are invariant relatively to A; besides that, we assume that the spectra of the restrictions A_1 and A_2 of the operator A on these subspaces do not have any common points in the extended complex plane. Since A is closed, from these assumptions it follows that at least one of the operators A_1 and A_2 is bounded. To be specific, let us assume that the operator A_2 is bounded.

Any linear operator C acting on the space E can be represented in a matrix form $C = (C_{ij})_{i,j=1}^2$, where C_{ij} acts from E_j to E_i. Let Π_1 and Π_2 be the projection operators onto the subspaces E_1 and E_2, corresponding to the direct sum decomposition $E = E_1 \dot{+} E_2$:

$$\Pi_1 = \mathrm{diag}(I_1; 0), \quad \Pi_2 = \mathrm{diag}(0; I_2);$$

then, for instance,

$$\Pi_2 C \Pi_1 = \begin{pmatrix} 0 & 0 \\ C_{21} & 0 \end{pmatrix}.$$

Further, the operator $\Pi_i C \Pi_j$ acting on E will be denoted by C_{ij}, which is also the notation of its restriction to E_j.

We now seek a solution $x(t)$ of equation (7.7) in the form

$$x(t) = (I + \varepsilon Y)v(t), \tag{7.8}$$

where the bounded linear operator $Y = Y(\varepsilon)$ does not depend on t and "mixes" spaces, that is, Y maps E_1 to E_2 and E_2 to E_1. Further on, this operator will be called an *interlacing operator.*

The function $v(t)$ from (7.8), in turn, is sought as a solution of the equation

$$\varepsilon\frac{\mathrm{d}v}{\mathrm{d}t} = Av + \varepsilon S(\varepsilon)v, \tag{7.9}$$

where the linear operator $S = S(\varepsilon)$ keeps the subspaces E_1 and E_2 invariant. Substituting (7.8) into equation (7.7), taking into account (7.9), and equating the operator coefficients from the right and from the left sides one gets the next equation that links the operators Y and S,

$$AY - YA = S - L + \varepsilon YS - \varepsilon LY. \tag{7.10}$$

This equation is equivalent to four equations, obtained from it by multiplying both sides by the projections Π_j, $j = 1, 2$. Multiplication by Π_i from left and right leads to the the following equation

$$\Pi_i S\Pi_i = \Pi_i L\Pi_i + \varepsilon\Pi_i LY\Pi_i, \tag{7.11}$$

which was deduced by taking into account the relations

$$A\Pi_i = \Pi_i A, \quad S\Pi_i = \Pi_i S, \quad \Pi_i Y\Pi_i = 0. \tag{7.12}$$

Multiplication by Π_j from the right and by Π_i, $i \neq j$, from the left gives

$$\Pi_j AY\Pi_i - \Pi_j YA\Pi_i = -\Pi_j L\Pi_i + \varepsilon\Pi_j YS\Pi_i - \varepsilon\Pi_j LY\Pi_i \tag{7.13}$$

With the help of (7.11), S can be eliminated from this equation. Indeed, from (7.12) we obtain

$$\Pi_j YS\Pi_i = \Pi_j Y(\Pi_i + \Pi_j)S\Pi_i = \Pi_j Y\Pi_i S\Pi_i,$$

then

$$\Pi_j AY\Pi_i - \Pi_j YA\Pi_i = -\Pi_j L\Pi_i + \varepsilon\Pi_j Y\Pi_i L\Pi_i - \varepsilon\Pi_j LY\Pi_i + \varepsilon^2\Pi_j Y\Pi_i LY\Pi_i.$$

Hence it follows that

$$\Pi_j AY_{ji} - Y_{ji}A\Pi_i = -L_{ji} + \varepsilon Y_{ji}L_{ii} - \varepsilon L_{jj}Y_{ji} + \varepsilon^2 Y_{ji}L_{ij}Y_{ji}. \tag{7.14}$$

1.7.3 Solvability of Commutator Equations

In order to prove the solvability of equations (7.14), let us consider the more general problem of solving the equation

$$A_1 Z - ZA_2 = Q, \tag{7.15}$$

where $A_i (i = 1, 2)$ is an operator acting on a Hilbert space E_i and Q is a bounded operator from E_2 and E_1. The unknown operator Z is also assumed to be a bounded operator from E_2 to E_1.

If the operators A_1 and A_2 are bounded and their spectra are disjoint, then equation (7.15) has a unique solution defined through resolvents of the operators A_1 and A_2 by the formula

$$Z = -\frac{1}{4\pi^2} \oint_{\Gamma_2} \oint_{\Gamma_1} \frac{R_{A_1}(\lambda) Q R_{A_2}(\lambda)}{\lambda - \mu} d\lambda d\mu, \tag{7.16}$$

where the two disjoint closed contours Γ_1 and Γ_2 contain the spectra of A_1 and A_2, respectively.

Now, let us assume that the operators A_1 and A_2 are closed, and their spectra do not intersect each other in the extended complex plane. It means that for one of these operators, A_2 for example, the point at infinity does not belong to its spectrum and, therefore, this operator is bounded. Further, since the extended complex plane cannot be represented as a union of two closed disjoint sets—spectra of A_1 and A_2—the operators A_1 and A_2 have at least one common regular point λ_0.

Let us form the equation

$$R_{A_1}(\lambda_0) Z - Z R_{A_2}(\lambda_0) = -R_{A_1}(\lambda_0) Q R_{A_2}(\lambda_0) \tag{7.17}$$

that involves the bounded operators $R_{A_1}(\lambda_0)$ and $R_{A_2}(\lambda_0)$. In virtue of the above mentioned facts, this equation has a unique solution. Let x be an arbitrary element from E_2. We introduce the element $y = (A_2 - \lambda_0 I)x$ and apply the operators from the previous equation to this element. We get

$$R_{A_1}(\lambda_0) Z (A_2 - \lambda_0 I)x - Zx = -R_{A_1}(\lambda_0) Qx.$$

The first element in the left side and the element in the right side belong to $\mathcal{D}(A_1)$, therefore, $Zx \in \mathcal{D}(A_1)$. Applying the operator $A_1 - \lambda_0 I$ to both sides of the last equation, we obtain

$$Z(A_2 - \lambda_0 I)x - (A_1 - \lambda_0 I)Zx = -Qx,$$

or

$$A_1 Zx - Z A_2 x = Qx.$$

Thus, the bounded operator Z, which was found by solving equation (7.17), is a solution of equation (7.15). Moreover, Z maps the space E_2 into the domain of definition $\mathcal{D}(A_1) \subset E_1$. It is easy to check, reversing the previous reasoning, that such a solution of equation (7.15) is unique.

If we estimate the norm of the operator Z defined by formula (7.16), then we get the inequality

$$\|Z\|_{E_2 \to E_1} \le c\|Q\|_{E_2 \to E_1}, \tag{7.18}$$

where the constant c depends on estimates for the resolvents of the operators A_1 and A_2 on the contours Γ_1, Γ_2, on the lengths of those contours, and the distance between them; c does not depend on operator Q. A similar reasoning shows that the inequality (7.18) also holds for the solution of equation (7.17). Thus, to each operator $Q \in \mathcal{L}(E_2, E_1)$ we associate the uniquely defined operator $Z = TQ$; the latter operator is the only solution of equation (7.15), and

$$\|TQ\|_{E_2 \to E_1} \le c\|Q\|_{E_2 \to E_1}.$$

Sometimes it is said that T is a *transformer* on the space of operators $\mathcal{L}(E_2, E_1)$.

Now we return to equation (7.14). Let us temporarily denote $A_1 = \Pi_j A \Pi_j$, $A_2 = \Pi_i A \Pi_i$, and let Q stand for the right side of the equation. Then, (7.14) has the form

$$A_1 Y_{ji} - Y_{ji} A_2 = Q. \tag{7.19}$$

Since A_1 and A_2 are supposed to have disjoint spectra in the extended complex plane, then there is a transformer T_{ji} that corresponds to equation (7.19) and acts boundedly from the space E_i to the space E_j.

Applying this transformer to both sides of the equation (7.14) we obtain

$$Y_{ji} = -T_{ji} L_{ji} + \varepsilon T_{ji}(Y_{ji} - L_{jj} Y_{ji} + \varepsilon Y_{ji} L_{ij} Y_{ji}). \tag{7.20}$$

With the help of the principle of contractive mappings it is easy to find out that for any sufficiently small ε, $0 \le \varepsilon \le \varepsilon_0$, the quadratic operator equation (7.20) has a unique solution $Y_{ji}(\varepsilon)$, which is uniformly bounded in ε for $\varepsilon \in [0, \varepsilon_0]$.

Thus, for a sufficiently small ε there exist two bounded operators $Y_{12}(\varepsilon)$, $Y_{21}(\varepsilon)$ and the interlacing operator

$$Y(\varepsilon) = \begin{pmatrix} 0 & Y_{12}(\varepsilon) \\ Y_{21}(\varepsilon) & 0 \end{pmatrix}. \tag{7.21}$$

With these operators at hand and using (7.11) it is easy to define the operator $S(\varepsilon) = \operatorname{diag}(S_{11}, S_{22})$ with the diagonal entries

$$S_{ii} = S_{ii}(\varepsilon) = L_{ii}(\varepsilon) + \varepsilon L_{ij} Y_{ji}(\varepsilon), \qquad j \ne i, \ i = 1, 2. \tag{7.22}$$

Let us now go back to equation (7.9) and notice that this equation is much simpler than the initial equation (7.7), because all the operator coefficients in (7.9) leave the subspaces E_1 and E_2 invariant. That is exactly why this equation can be split into two independent equations in the subspaces E_1 and E_2:

$$\varepsilon\frac{dv_1}{dt} = A_1 v_1 + \varepsilon S_{11}(\varepsilon)v_1,$$
$$\varepsilon\frac{dv_2}{dt} = A_2 v_2 + \varepsilon S_{22}(\varepsilon)v_2, \tag{7.23}$$

where $v_i = \Pi_i v$. The derivation of this splitting was the purpose of the previous reasoning.

Let us assume next that for equation (7.7) it is necessary to solve the Cauchy problem with the initial condition $x(0) = x_0$. Then, from (7.8) we obtain an initial condition for the function $v(t)$ in the form

$$(I + \varepsilon Y(\varepsilon))v(0) = x_0. \tag{7.24}$$

For any sufficiently small ε the operator $I + \varepsilon Y(\varepsilon)$ is invertible and, therefore,

$$v(0) = (I + \varepsilon Y(\varepsilon))^{-1}\, x_0 = \left(\sum_{k=0}^{\infty}(-1)^k \varepsilon^k Y^k(\varepsilon)\right) x_0. \tag{7.25}$$

We introduce the notation $x_{0,i} := \Pi_i x_0$; using the basic property of an interlacing operator we have

$$Y^{2m}x_{0,i} \in E_i, \quad Y^{2m+1}x_{0,i} \in E_j, \qquad m = 0, 1, 2, \ldots.$$

Therefore,

$$\begin{aligned} v_1(0) &= \sum_{m=0}^{\infty} \varepsilon^{2m}Y^{2m}x_{0,1} - \sum_{m=0}^{\infty} \varepsilon^{2m+1}Y^{2m+1}x_{0,2} \\ &= (I - \varepsilon^2 Y^2)^{-1}x_{0,1} - \varepsilon Y(I - \varepsilon^2 Y^2)^{-1}x_{0,2}; \\ v_2(0) &= -\sum_{m=0}^{\infty} \varepsilon^{2m+1}Y^{2m+1}x_{0,1} + \sum_{m=0}^{\infty} \varepsilon^{2m}Y^{2m}x_{0,2} \\ &= -\varepsilon Y(I - \varepsilon^2 Y^2)^{-1}x_{0,1} + (I - \varepsilon^2 Y^2)^{-1}x_{0,2}. \end{aligned} \tag{7.26}$$

1.7.4 Asymptotic Expansions of Solutions

Let us assume that the operator L in equation (7.7) depends on ε and admits the asymptotic expansion

$$L(\varepsilon) \sim \sum_{k=0}^{\infty} \varepsilon^k L^{(k)}, \tag{7.27}$$

where $L^{(k)}$ are bounded linear operators. Then in (7.8), (7.9) it is natural to look for operators Y and S in the form of asymptotic expansions

$$Y(\varepsilon) \sim \sum_{k=0}^{\infty} \varepsilon^k Y^{(k)}, \quad S(\varepsilon) \sim \sum_{k=0}^{\infty} \varepsilon^k S^{(k)}. \tag{7.28}$$

We let a_N denote the N-th partial sum of the power series $a(\varepsilon) = \sum_{k=0}^{\infty} \varepsilon^k a^{(k)}$, that is, $a_N(\varepsilon) := \sum_{k=0}^{N} \varepsilon^k a^{(k)}$.

Substituting the expansions (7.27), (7.28) into equation (7.7) and equating the operator coefficients of like powers of ε, we obtain for $j \neq i$

$$\begin{aligned}
\Pi_j A Y_{ji}^{(0)} - Y_{ji}^{(0)} A \Pi_i &= -L_{ji}^{(0)}, \\
\Pi_j A Y_{ji}^{(1)} - Y_{ji}^{(1)} A \Pi_i &= -L_{ji}^{(1)} + Y_{ji}^{(0)} L_{ii}^{(0)} - L_{ji}^{(0)} Y_{ji}^{(0)} \\
\Pi_j A Y_{ji}^{(k)} - Y_{ji}^{(k)} A \Pi_i &= -L_{ji}^{(k)} + \sum_{l=0}^{k-1} Y_{ji}^{(l)} L_{ii}^{(k-1-l)} - \sum_{l=0}^{k-1} L_{jj}^{(l)} Y_{ji}^{(k-1-l)} \\
&\quad + \sum_{l=0}^{k-2} Y_{ji}^{(l)} \sum_{m=0}^{k-2-l} L_{ij}^{(m)} Y_{ji}^{(k-2-l-m)}, \qquad k = 2, 3, \ldots.
\end{aligned} \tag{7.29}$$

As soon as we find $Y_{ji}^{(k)}$ from these recurrent relations, then from the equations obtained by substituting expansions (7.27), (7.28) into (7.11) we get

$$S_{ii}^{(0)} = L_{ii}^{(0)}, \quad S_{ij}^{(k)} = L_{ii}^{(k)} + \sum_{l=0}^{k-1} L_{ij}^{(l)} Y_{ji}^{(k-1-l)}, \qquad k = 1, 2, \ldots. \tag{7.30}$$

Suppose we managed to solve the first N equations in (7.29) and to find $Y^{(k)}$, $k = 1, 2, \ldots, N-1$. Then an approximate solution of equation (7.7) can be built in the following way. We form the operators $S^{(k)}$, $k = 0, 1, \ldots, N-1$ by formulas (7.30), and set up two equations similar to (7.23):

$$\varepsilon \frac{\mathrm{d}v_i^{(N)}}{\mathrm{d}t} = A_i v_i^{(N)} + \varepsilon S_{ii,N-1} v_i^{(N)}, \qquad i = 1, 2 \tag{7.31}$$

In order to get some initial data for solving these equations, we notice that

$$\begin{aligned}
&(I - \varepsilon^2 Y^2)^{-1} - (I - \varepsilon^2 Y_m^2)^{-1} \\
&\quad = \varepsilon^2 (I - \varepsilon^2 Y^2)^{-1} \left[Y(Y - Y_m) + (Y - Y_m) Y_m \right] (I - \varepsilon^2 Y_m^2)^{-1},
\end{aligned}$$

hence, taking into account the estimate $\|Y - Y_m\| = \mathrm{O}(\varepsilon^{m+1})$, $\varepsilon \to 0$, we have

$$\|(I - \varepsilon^2 Y^2)^{-1} - (I - \varepsilon^2 Y_m^2)^{-1}\| = \mathrm{O}(\varepsilon^{m+3}), \qquad \varepsilon \to 0.$$

Similarly, it can be proved that

$$\|\varepsilon Y(I-\varepsilon^2Y^2)^{-1}-\varepsilon Y_m(I-\varepsilon^2Y_m^2)^{-1}\| = \mathrm{O}\,(\varepsilon^{m+2}), \qquad \varepsilon\to 0.$$

Hence, let us denote

$$v_i^{(N)}(0) = (I-\varepsilon^2Y_{N-2}^2)^{-1}x_{0,i}-\varepsilon Y_{N-1}(I-\varepsilon^2Y_{N-1}^2)^{-1}x_{0,j}, \qquad Y_m=0, \text{ if } m<0.$$

Then

$$\|v_i(0)-v_i^{(N)}(0)\| = \mathrm{O}(\varepsilon^{N+1}), \qquad \varepsilon\to 0. \tag{7.32}$$

This relation is valid provided in the expansion of $v_i^{(N)}(0)$ as a power series we keep only those components containing powers of ε, which do not exceed N. Finally, we obtain the following formula for the approximate solution $x^{(N)}$:

$$x^{(N)}(t) = (I+\varepsilon Y_{N-1})v^{(N)}(t). \tag{7.33}$$

Let us now analyse the asymptotic behavior of this equation. From equations (7.14) and (7.29) we obtain

$$\begin{aligned}&\Pi_jA(Y_{ji}-Y_{ji}^{(0)})-(Y_{ji}-Y_{ji}^{(0)})A\Pi_i\\ &\qquad = -(L_{ji}-L_{ji}^{(0)})+\varepsilon Y_{ji}L_{ii}-\varepsilon L_{jj}Y_{ji}+\varepsilon^2Y_{ji}L_{ij}L_{ji}.\end{aligned}$$

Here, in virtue of (7.27), the magnitude of the right side is of an order not less than ε. Therefore, from the boundedness of the transformer T_{ij} it follows that

$$\|Y_{ji}-Y_{ji}^{(0)}\| = \mathrm{O}(\varepsilon), \qquad \varepsilon\to 0.$$

Applying the method of mathematical induction, we may assume that the following relation is already proved,

$$\|Y_{ji}-Y_{ji,m-1}\| = \mathrm{O}(\varepsilon^m), \qquad \varepsilon\to 0; \tag{7.34}$$

let us write down (7.14) in the form

$$\Pi_jAY_{ji}-Y_{ji}A\Pi_i = -L_{ji}+\varepsilon\Phi(L,Y_{ji}). \tag{7.35}$$

If the operators $Y_{ji}^{(k)}$, $k=0,1,\ldots,m$, are found from the equations (7.29), then it means that

$$\Pi_jAY_{ji,m}-Y_{ji,m}A\Pi_i = -L_{ji,m}+\varepsilon\Phi(L_{m-1},Y_{ji,m-1})+\varepsilon^{m+1}Q, \tag{7.36}$$

where Q is a polynomial of the operators $L_{lr}^{(k)}$, $Y_{ji}^{(k)}$, $k = 0, 1, \ldots, m-1$, $l, r = 1, 2$.

Subtracting (7.36) from (7.35), we have

$$\Pi_j A(Y_{ji} - Y_{ji,m}) - (Y_{ji} - Y_{ji,m})A\Pi_i$$
$$= -(L_{ji} - L_{ji,m}) + \varepsilon \left[\sum_{l,r=1}^{2} P_{l,r}(L_{lr} - L_{lr,m-1}) + R(Y_{ji} - Y_{ji,m-1}) - \varepsilon^{m+1} Q \right],$$

where $P_{l,r}$ and R are polynomials of L_{lr}, Y, $L_{lr}^{(k)}$, and $Y_{ji}^{(k)}$. Hence, from the assumption (7.34) and the uniform boundedness in ε of $Y = Y(\varepsilon)$ it follows that

$$\|Y_{ji} - Y_{ji,m}\| = \mathrm{O}(\varepsilon^{m+1}). \tag{7.37}$$

Further, from the equations (7.30) it follows that

$$S_{ii,m} = L_{ii,m} + \varepsilon L_{ij,m} Y_{ji,m} + \varepsilon^{m+1} Q_1$$

and, therefore,

$$S_{ii} - S_{ii,m} = L_{ii} - L_{ii,m} + \varepsilon \left[(L_{ij} - L_{ij,m}) Y_{ji} + L_{ji,m}(Y_{ji} - Y_{ji,m}) \right] - \varepsilon^{m+1} Q_1.$$

Hence, from (7.37), (7.27), (7.28) we obtain

$$\|S - S_m\| = \mathrm{O}(\varepsilon^{m-1}). \tag{7.38}$$

For an estimate of the difference between $x(t)$ and the approximate solution, let us introduce an auxiliary function $\tilde{v}^{(N)}(t)$, which is a solution of equation (7.9) with the initial condition $\tilde{v}^{(N)}(0) = v^{(N)}(0)$. From the inequalities (7.5) and (7.6) for $f(t) \equiv 0$, and from (7.32) it follows that

$$\|v(t) - \tilde{v}^{(N)}(t)\| \le M_1 \|v(0) - v^{(N)}(0)\| = \mathrm{O}(\varepsilon^{m+1}).$$

Then

$$\begin{aligned} \|x(t) - x^{(N)}(t)\| &\le \|(I + \varepsilon Y)v(t) - (I + \varepsilon Y_{N-1})v^{(N)}(t)\| \\ &\le \|(I + \varepsilon Y)v(t) - \tilde{v}^{(N)}(t)\| + \|(I + \varepsilon Y)(\tilde{v}^{(N)}(t) - v^{(N)}(t))\| \\ &\quad + \varepsilon \|(Y - Y_{n-1})v^{(N)}(t)\| \end{aligned} \tag{7.39}$$

The function $w(t) := \tilde{v}^{(N)}(t) - v^{(N)}(t)$ is a solution of the differential equation

$$\varepsilon \frac{\mathrm{d}w}{\mathrm{d}t} = Aw + \varepsilon S(\varepsilon) w + \varepsilon (S(\varepsilon) - S_{N-1}(\varepsilon)) v^{(N)},$$

which satisfies the initial condition $w(0) = 0$. In virtue of the inequalities (7.5), (7.6) and the estimate (7.38), the magnitude of $\|w(t)\|$ has an order which is not less than ε^{N+1} for $a > 0$, and equals ε^N for $a = 0$. The first and third summands in the right side of (7.39) have an order not less than ε^{N+1}.

Finally, we come to the following conclusion: the approximate solution $x^{(N)}(t)$ differs from the exact solution $x(t)$ in a magnitude of order not less than ε^{N+1} for $a > 0$, and of order ε^N for $a = 0$.

1.7.5 The Special Case of a Splitable Operator Kernel

We now consider the special case when operator A equals zero on the subspace E_2 and its restriction A_1 to the subspace E_1 generates a semigroup of negative type, therefore, this restriction has a bounded inverse operator A_1^{-1}. Under such assumptions equations (7.29) can be simplified to

$$\begin{aligned} A_1 Y_{12}^{(0)} &= -L_{12}^{(0)}, A_1 Y_{12}^{(1)} = -L_{12}^{(1)} + Y_{12}^{(0)} L_{22}^{(0)} - L_{11}^{(0)} Y_{12}^{(0)}, \\ A_1 Y_{12}^{(k)} &= -L_{12}^{(k)} + \sum_{l=0}^{k-1} Y_{12}^{(l)} L_{22}^{(k-1-l)} - \sum_{l=0}^{k-1} L_{11}^{(l)} Y_{12}^{(k-1-l)} \\ &\quad + \sum_{l=0}^{k-2} Y_{12}^{(l)} \sum_{m=0}^{k-2-l} L_{21}^{(m)} Y_{12}^{(k-2-l-m)}, \qquad k = 2, 3, \ldots . \end{aligned}$$

Hence

$$\begin{aligned} Y_{12}^{(0)} &= -A_1^{-1} L_{12}^{(0)}, Y_{12}^{(1)} = -A_1^{-1} \left(L_{12}^{(1)} - Y_{12}^{(0)} L_{22}^{(0)} + L_{11}^{(0)} Y_{12}^{(0)} \right), \\ Y_{12}^{(k)} &= -A_1^{-1} \left(L_{12}^{(k)} - \sum_{l=0}^{k-1} Y_{12}^{(l)} L_{22}^{(k-1-l)} + \sum_{l=0}^{k-1} L_{11}^{(l)} Y_{12}^{(k-1-l)} \right. \\ &\quad \left. - \sum_{l=0}^{k-2} Y_{12}^{(l)} \sum_{m=0}^{k-2-l} L_{21}^{(m)} Y_{12}^{(k-2-l-m)} \right), \qquad k = 2, 3, \ldots . \end{aligned} \tag{7.40}$$

Similarly we obtain

$$\begin{aligned} Y_{21}^{(0)} &= L_{21}^{(0)} A_1^{-1}, Y_{21}^{(1)} = \left(L_{21}^{(1)} - Y_{21}^{(0)} L_{11}^{(0)} + L_{22}^{(0)} Y_{21}^{(0)} \right) A_1^{-1}, \\ Y_{21}^{(k)} &= \left(L_{21}^{(k)} - \sum_{l=0}^{k-1} Y_{21}^{(l)} L_{11}^{(k-1-l)} + \sum_{l=0}^{k-1} L_{22}^{(l)} Y_{21}^{(k-1-l)} \right. \\ &\quad \left. - \sum_{l=0}^{k-2} Y_{21}^{(l)} \sum_{m=0}^{k-2-l} L_{21}^{(m)} Y_{21}^{(k-2-l-m)} \right) A_1^{-1}, \qquad k = 2, 3, \ldots . \end{aligned} \tag{7.41}$$

According to our assumptions, the operator A_1 generates a semigroup for which the estimate (7.1) with $a > 0$ is true. From this estimate and based on the conclusions of the previous section we deduce that in order to build the approximate solution $x^{(N)}(t)$ with an error $O(\varepsilon^{N+1})$, one should use the component $S_{11,N-1}(\varepsilon)$ of the operator $S_{N-1}(\varepsilon)$ from the first equation (7.31) and the component $S_{22,N}(\varepsilon)$ from the second one.

1.7.6 The Case of a Nonstationary Perturbation

If one considers the equation

$$\varepsilon \frac{\mathrm{d}x}{\mathrm{d}t} = Ax + \varepsilon L(t, \varepsilon)x \tag{7.42}$$

where the operator $L(t, \varepsilon)$ depends smoothly on t, then, by using the methods developed in Sections 1.7.1–1.7.3, one gets that the operators Y and S also depend on t. Let us assume that

$$x(t) = (I + \varepsilon Y(t))v(t) \tag{7.43}$$

and suppose that $v(t)$ satisfies the equation

$$\varepsilon \frac{\mathrm{d}v}{\mathrm{d}t} = Av + \varepsilon S(t)v. \tag{7.44}$$

Here, the operators $S(t)$ leave the subspaces E_1 and E_2 invariant and the operators $Y(t)$ map E_i into E_j, $i \neq j$.

The substitution of (7.43) into (7.42) in conjunction with (7.44) leads to the equation

$$AY - YA = \varepsilon \frac{\mathrm{d}Y}{\mathrm{d}t} + S - L + \varepsilon YS - \varepsilon LY.$$

Hence, we again obtain the equations (7.11), but instead of (7.14) we have

$$\Pi_j AY_{ji} - Y_{ji}A\Pi_i = \varepsilon \frac{\mathrm{d}Y_{ji}}{\mathrm{d}t} - L_{ji} + \varepsilon Y_{ji}L_{ii} - \varepsilon L_{jj}Y_{ji} + \varepsilon^2 Y_{ji}L_{ij}Y_{ji}. \tag{7.45}$$

Equation (7.45) is a Ricatti differential equation relatively to the operator $Y_{ji}(t)$. This equation appears in numerous problems and the theory of such equations is rapidly developing. Here we will only make some comments connected with the construction of a formal asymptotic solution for the special case which was considered in the previous section. Specifically, formulas (7.40) and (7.41) need to be changed beginning with $k = 1$ in the following way: the component $A_1^{-1}(dY_{12}^{(k-1)}/dt)$ is added to the right sides of (7.40), and the component $(\mathrm{d}Y_{21}^{(k-1)}/\mathrm{d}t)A_1^{-1}$ is subtracted from the right sides of (7.41). To make all these changes possible, it is natural to include a requirement on the differentiability of the operators $L^{(k)}(t)$ for a certain number of times.

1.7.7 Nonhomogeneous Equations

Let us next consider the equation

$$\varepsilon \frac{\mathrm{d}x}{\mathrm{d}t} = Ax + \varepsilon L(\varepsilon)x + \varepsilon f(t, \varepsilon), \tag{7.46}$$

where $f(t,\varepsilon)$ admits the asymptotic expansion

$$f(t,\varepsilon) \sim \sum_{k=0}^{\infty} \varepsilon^k f^{(k)}(t). \tag{7.47}$$

We construct a particular solution for (7.46). First, let us assume that the operator A has a bounded inverse, A^{-1}. Then the solution $x(t)$ can be sought in the asymptotic series form

$$x(t) \sim \sum_{k=0}^{\infty} \varepsilon^k h^{(k)}(t). \tag{7.48}$$

By substituting (7.27), (7.47), (7.48) into (7.46) and then equating the coefficients of the same powers of ε, one gets the system of equations

$$\begin{aligned} Ah^{(0)} &= 0, \\ \frac{\mathrm{d}h^{(0)}}{\mathrm{d}t} &= Ah^{(1)} + L^{(0)}h^{(0)} + f^{(0)}, \\ \frac{\mathrm{d}h^{(k-1)}}{\mathrm{d}t} &= Ah^{(k)} + \sum_{l=0}^{k-1} L^{(l)} H^{(k-1-l)} + f^{(k-1)}, \qquad k = 2,3,\ldots; \end{aligned}$$

the coefficients in the series (7.48) can be found from this system as follows:

$$\begin{aligned} h^{(0)} &= 0, \\ h^{(1)} &= A^{-1} f^{(0)}, \\ &\ldots, \\ h^{(k)} &= A^{-1}\left(\frac{\mathrm{d}h^{(k-1)}}{\mathrm{d}t} - \sum_{l=0}^{k-1} L^{(l)} h^{(k-1-l)} - f^{(k-1)}\right). \end{aligned} \tag{7.49}$$

If we denote $h_{N-1}(t) := \sum_{k=0}^{N-1} \varepsilon^k h^{(k)}$, then this function satisfies equation (7.46) with an accuracy of order $\mathrm{O}(\varepsilon^N)$. In the case when operator A satisfies condition (7.1) for $a > 0$, then from inequality (7.5) it follows that $h_{N-1}(t)$ differs from the solution of equation (7.46) satisfying the same initial condition in a magnitude of order $\mathrm{O}(\varepsilon^{N+1})$.

In the more complicated case when operator A does not have a bounded inverse operator, let us assume that A possesses the properties from Section 1.7.2, and A_1 has a bounded inverse operator. Here, the particular solution to equation (7.46) can be sought in the form

$$x(t) = (I + \varepsilon Y)v + h, \tag{7.50}$$

where the function $v(t)$ is a solution of the equation

$$\varepsilon\frac{\mathrm{d}v}{\mathrm{d}t} = Av + \varepsilon S(\varepsilon)v + \varepsilon g, \tag{7.51}$$

with a function $g = g(t,\varepsilon)$ that will be defined later on. We substitute (7.50) into (7.46), and, taking into account (7.51), equate the coefficients of v in both sides and all the components which do not contain v. Then, for Y and S we obtain equation (7.10), which was studied above. The second equation is of the form

$$\varepsilon\frac{\mathrm{d}h}{\mathrm{d}t} = Ah + \varepsilon Lh - \varepsilon(I+\varepsilon Y)g + \varepsilon f. \tag{7.52}$$

Now we choose g in such a way that this equation can be split into two equations in the subspaces E_1 and E_2. Specifically, we assume that

$$(I+\varepsilon Y)g = Lh. \tag{7.53}$$

Then, equation (7.52) splits into two equations,

$$\varepsilon\frac{\mathrm{d}h_1}{\mathrm{d}t} = A_1 h_1 + \varepsilon\Pi_1 f, \tag{7.54}$$

$$\varepsilon\frac{\mathrm{d}h_2}{\mathrm{d}t} = A_2 h_2 + \varepsilon\Pi_2 f, \tag{7.55}$$

where $h_i = \Pi_i h$. The second equation is an equation with a bounded operator, and consequently it is solvable for any initial data from E_2. In virtue of the properties indicated above of operator A_1, equation (7.54) also has a (possibly generalized) solution for any initial data from E_1. If $h_i(t)$ and $h(t) = h_1(t) + h_2(t)$ are already found, then the function $g(t)$ can be determined from equation (7.53), which is solvable at least for a sufficiently small ε. Finally, the solution of equation (7.51) for a certain function $g(t)$ and operators Y, S makes it possible to build the particular solution (7.50) of the nonhomogeneous equation (7.46).

The operator A_1 in equation (7.54) has a bounded inverse operator, therefore, the asymptotic expansion of a solution can be found in the way illustrated at the beginning of this section, namely, $h_1 \sim \sum_{k=0}^{\infty}\varepsilon^k h_1^{(k)}$, with

$$h_1^{(0)} = 0, h_1^{(1)} = -A_1^{-1}\Pi_1 f^{(0)}, \ldots, h_1^{(k)} = A_1^{-1}\left(\frac{\mathrm{d}h_1^{(k-1)}}{\mathrm{d}t} - \Pi_1 f^{(k-1)}\right).$$

The operator A_2 does not have a bounded inverse operator, hence the asymptotic expansion for h_2 may include components of the so-called "boundary layer" type. This difficulty does not occur if the operator A satisfies the conditions in Section 1.7.4, that is, when $A_2 = 0$. Then, equation (7.55) takes the form $dh_2/dt = \Pi_2 f$, and, therefore,

$$h_2^{(k)} = \int_0^t \Pi_2 f^{(k)}(s)\mathrm{d}s + h_2^{(k)}(0). \tag{7.56}$$

If an initial condition for $h_2(t)$ is given, it is natural to consider $h_2^{(0)} = h_2(0)$ and $h_2^{(k)}(0) = 0$ for $k = 1, 2, \ldots$.

Now we proceed by finding the asymptotic expansion of the function $g \sim \sum_{k=0}^{\infty} \varepsilon^k g^{(k)}$. From equation (7.53) we obtain

$$\begin{aligned} g^{(0)} &= L^{(0)}h^{(0)}, \\ g^{(1)} &= -Y^{(0)}g^{(0)} + L^{(0)}h^{(1)} + L^{(1)}h^{(0)}, \ldots, \\ g^{(k)} &= -\sum_{l=0}^{k-1} Y^{(l)}g^{(k-1-l)} + \sum_{l=0}^{k} L^{(l)}h^{(k-l)}. \end{aligned} \tag{7.57}$$

As soon as the functions $g^{(k)}(t)$ are found, we can set up and solve a differential equation similar to (7.46) for the approximate solution $v^{(N)}(t)$, namely,

$$\varepsilon \frac{\mathrm{d}v^{(N)}}{\mathrm{d}t} = Av^{(N)} + \varepsilon S_{N-1}(\varepsilon)v^{(N)} + \varepsilon g_{N-1}. \tag{7.58}$$

Then, the approximate particular solution of equation (7.46) is of the form

$$x^{(N)}(t) = (I + \varepsilon Y_{N-1})\, v^{(N)}(t) + h^{(N)}(t). \tag{7.59}$$

1.7.8 Eigenvalue Problems

Let us consider the spectral problem

$$(A + \varepsilon L(\varepsilon))x = -\varepsilon\lambda x, \qquad x \in E, \tag{7.60}$$

associated to the evolution problem (7.7) with the operators A and $L = L(\varepsilon)$, whose properties were illustrated in Sections 1.7.2–1.7.3. Problem (7.7) reduces to equation (7.60) provided the dependence of its solutions on t can be expressed by the multiplier $\exp(-\lambda t)$.

We seek a solution to (7.60) of a form similar to the solution produced by our previous reasoning, that is,

$$x = (I + \varepsilon Y(\varepsilon))v, \tag{7.61}$$

where $Y(\varepsilon)$ is an interlacing operator and v satisfies the condition

$$Av + \varepsilon S(\varepsilon)v = -\varepsilon\lambda v, \tag{7.62}$$

with $S(\varepsilon)$ leaving the subspaces E_1 and E_2 invariant. The substitution of (7.61) into (7.60) in conjunction with (7.62) leads again to the above investigated equation (7.10) that connects the operators Y and S.

It is obvious that equation (7.62) splits into two independent equations on the subspaces E_1 and E_2:

$$A_1 v_1 + \varepsilon S_{11}(\varepsilon) v_1 = -\varepsilon \lambda v_1, \tag{7.63}$$

$$A_2 v_2 + \varepsilon S_{22}(\varepsilon) v_2 = -\varepsilon \lambda v_2. \tag{7.64}$$

Hence, it follows that for a sufficiently small ε the spectrum of the perturbed operator $A + \varepsilon L(\varepsilon)$ splits into two sets σ_1 and σ_2, which are possibly intersecting each other.

In the particular case when $A_2 = 0$, equation (7.64) has the form

$$-\lambda v_2 = S_{22}(\varepsilon) v_2.$$

Since the operator $S_{22}(\varepsilon)$ is bounded, σ_2 is a bounded set. If in addition to that $S_{22}(\varepsilon)$ is a compact operator, then its nonzero spectrum consists of a sequence of eigenvalues $\{\lambda_k\}_{k=1}^{\infty}$, $\lambda_k = \lambda_k(S_{22}(\varepsilon))$ with finite multiplicities; this sequence may have only one limiting point, $\lambda = 0$. Let $\lambda = \lambda_k$ be one of these eigenvalues and $v_2 = v_{2k}$ be the corresponding eigenelement. Then, according to (7.61), to this eigenvalue λ_k there corresponds an eigenelement $x = x_{2k}$ of the problem (7.64), which is located "near" the space E_2:

$$x_{2k} = (I + \varepsilon Y_{12}(\varepsilon)) v_{2k}. \tag{7.65}$$

If the operator A_1 in equation (7.63) is unbounded and, for instance, has a compact inverse operator, then the numbers $-\varepsilon\lambda$, for which this equation has nontrivial solutions, are close to the eigenvalues $\lambda_k(A_1)$ of the operator A_1, and the corresponding eigenelements

$$x_{1k} = (I + \varepsilon Y_{21}(\varepsilon)) v_{1k} \tag{7.66}$$

are close to the eigenelements of the operator A_1, which are located in the space E_1.

Equations of the form

$$A_1 v_1 + \varepsilon S_{11}(\varepsilon) v_1 = -\mu v_1 \tag{7.67}$$

are considerably investigated in the classical theory of perturbations. We apply this theory to the special case when A_1 is a negative definite self-adjoint operator that has a compact inverse operator. Let $\mu^{(0)}$ be a simple eigenvalue of operator A_1 and $v_1^{(0)}$ is the corresponding normalized eigenelement. We look for the solutions $v_1 = v_1(\varepsilon)$ and $\mu = \mu(\varepsilon)$ in the form of the series

$$v_1 \sim \sum_{k=0}^{\infty} \varepsilon^k v_1^{(k)}, \quad \mu \sim \sum_{k=0}^{\infty} \varepsilon^k \mu^{(k)}. \tag{7.68}$$

Substituting (7.68) into (7.67) and equating the coefficients of like powers of ε we obtain the following system of equations

$$A_1 v_1^{(0)} = -\mu^{(0)} v_1^{(0)}, \quad A_1 v_1^{(k)} + \sum_{l=0}^{k-1} S_{11}^{(l)} v_1^{(k-1-l)} = -\sum_{l=0}^{k} \mu^{(l)} v_1^{(k-l)}. \tag{7.69}$$

From the first equation it follows that $v_1^{(0)}$ is considered to be a normalized eigenelement of operator A_1. Let us write down equation (7.69) for $k = 1$:

$$A_1 v_1^{(1)} + \mu^{(0)} v_1^{(1)} = -\mu^{(1)} v_1^{(0)} - S_{11}^{(0)} v_1^{(0)}. \tag{7.70}$$

This equation is solvable provided the right side is orthogonal to the element $v_1^{(0)}$. Hence

$$\mu^{(1)} = -\left(S_{11}^{(0)} v_1^{(0)}, v_1^{(0)}\right).$$

For this specific value of $\mu^{(1)}$, from equation (7.70) we can determine the element $v_1^{(1)}$ up to a summand of the form $cv_1^{(0)}$; by choosing the coefficient c properly we can assume that $v_1^{(1)}$ is orthogonal to $v_1^{(0)}$.

For $k = 2$ we obtain

$$A_1 v_1^{(2)} + \mu^{(0)} v_1^{(2)} = -\mu^{(1)} v_1^{(1)} - \mu^{(2)} v_1^{(0)} - S_{11}^{(0)} v_1^{(1)} - S_{11}^{(1)} v_1^{(0)}.$$

The condition of solvability for this equation yields

$$\mu^{(2)} = -\left(S_{11}^{(0)} v_1^{(1)}, v_1^{(0)}\right) - \left(S_{11}^{(1)} v_1^{(0)}, v_1^{(0)}\right).$$

Now, an approximate expression for the eigenvalue $\lambda_0 = \lambda_0(\varepsilon)$ can be written down in the form

$$\lambda_0 \approx \left\{\mu^{(0)} - \varepsilon\left(S_{11}^{(0)} v_1^{(0)}, v_1^{(0)}\right) - \varepsilon^2\left(S_{11}^{(0)} v_1^{(1)}, v_1^{(0)}\right) - \varepsilon^2\left(S_{11}^{(1)} v_1^{(0)}, v_1^{(0)}\right)\right\} / \varepsilon. \tag{7.71}$$

Using (7.71) we get an approximate expression for the eigenelement x_1 in the form

$$x_1 \approx \left(I + \varepsilon\left(Y_{21}^{(0)} + \varepsilon Y_{21}^{(1)}\right)\right)\left(v_1^{(0)} + \varepsilon v_1^{(1)} + \varepsilon^2 v_1^{(2)}\right), \tag{7.72}$$

which satisfies equation (7.60) with an accuracy of order not less that ε^3.

1.8 A General Scheme for Solving Boundary Value Problems

This section deals with a general scheme based on functional analysis methods used in finding generalized and weak solutions of some linear boundary value problems that appear in mathematical physics. The concepts of a Hilbert space pair, generating operator, abstract boundary value problems, and several specific operators related to such problems are introduced and widely used. The abstract analogue of Green Formula is also presented. The general constructions are illustrated by examples of classical boundary value problems for the Laplace operator.

Further on this scheme will be used in other chapters of the book where the motion of ideal and viscous fluids is investigated.

1.8.1 Hilbert Pairs. Generating Operators

Let F be a Hilbert space densely embedded in a Hilbert space E. This property means that F is a dense linear subset in E and there exists $a > 0$ such that for any $x \in F$

$$\|x\|_E \leq a\|x\|_F. \tag{8.1}$$

Under such assumptions one says that the spaces E and F form a *Hilbert pair* $(F; E)$.

Let us fix $y \in E$ and consider the linear functional $(x, y)_E$ on $F (x \in F)$. This functional is bounded in F. Indeed,

$$|(x,y)_E| \leq \|x\|_E \cdot \|y\|_E \leq a\|x\|_F\|y\|_E.$$

According to Riesz Theorem (Section 1.1.5) this functional can be represented by means of the scalar product in F, that is, there exists a unique element $z \in F$ such that

$$(x,y)_E = (x,z)_F;$$

from the last equation it follows that

$$\|z\|_F \leq a\|y\|_E. \tag{8.2}$$

Thus we get a mapping from E to F defined by $Vy = z$. The operator V is linear. According to inequality (8.2) this operator acts boundedly from E to F and, therefore,

$$a^{-1}\|Vy\|_E \leq \|Vy\|_F \leq a\|y\|_E. \tag{8.3}$$

By the definition of operator V we have

$$(x,y)_E = (x, Vy)_F \tag{8.4}$$

for all $x \in F$. If in this equation we set $x = Vy_1, y_1 \in E$, then

$$(Vy_1, y)_E = (Vy_1, Vy)_F = \overline{(Vy, Vy_1)}_F = (y_1, Vy)_E.$$

Hence, V is a self-adjoint operator on E. Moreover, if $y_1 = y$, then

$$(Vy, y)_E = (Vy, Vy)_F \geq 0.$$

Further, if $Vy = 0$ then from (8.4) we get $(x, y)_E = 0$ for all $x \in F$. Since F is dense in E it follows that $y = 0$. Hence, $(Vy, y) > 0$ when $y \neq 0$, that is, operator V is positive.

Therefore, the scalar product in E can be obtained by formula (8.4) from the scalar product in F by means of a bounded positive operator V.

If F is compactly embedded in E, then operator V is compact. Indeed, if a set $\{y\}$ is bounded in E, then, by inequality (8.3), the set $\{Vy\}$ is bounded in E; hence, the set $\{Vy\}$ is relatively compact in E.

We can introduce the inverse operator $A := V^{-1}$. It is an unbounded, self-adjoint, positive definite operator with the domain of definition $\mathcal{D}(A)$ dense in E; this domain coincides with the range of operator V, that is, $\mathcal{D}(A) = \mathcal{R}(V)$. From (8.4) we obtain the identity

$$(x, Az)_E = (x, z)_F \tag{8.5}$$

for all $x \in F$, $z \in \mathcal{D}(A)$. If x also belongs to $\mathcal{D}(A)$, then this identity can be written down in the form

$$(A^{1/2}x, A^{1/2}z)_E = (x, z)_F, \qquad x, z \in \mathcal{D}(A). \tag{8.6}$$

Assuming that $z = x$ we obtain

$$\|A^{1/2}x\|_E = \|x\|_F, \qquad x \in \mathcal{D}(A).$$

The domain of definition of the square root of operator A is given by the closure in the norm $\|A^{1/2}x\|_E$ of the domain of operator A, and in this case the domain of definition $\mathcal{D}(A^{1/2})$ coincides with the closure of $\mathcal{D}(A)$ in the norm of the space F, that is, it is a subspace of F. It turns out that $\mathcal{D}(A^{1/2}) = F$. In the opposite case there would exist a nonzero element $u_0 \in F$ such that $(u_0, z)_F = 0$ for all $z \in \mathcal{D}(A)$, and hence all $z \in \mathcal{D}(A^{1/2})$. Then, in virtue of (8.5),

$$0 = (u_0, z)_F = (u_0, Az)_E,$$

and because $\mathcal{R}(A) = E$ we conclude that $u_0 = 0$, a contradiction.

Hence, the following identity is true,

$$\|x\|_F = \|A^{1/2}x\|_E \ (x \in F). \tag{8.7}$$

Therefore, space F is the domain of definition of the positive definite self-adjoint operator $A^{1/2}$; the norm in F is defined by formula (8.7).

If F is compactly embedded in E, then operator $A^{-1/2}$ which is the inverse to operator $A^{1/2}$ is compact.

Operator A is called the *generating operator* of the Hilbert pair $(F; E)$. It is easy to show that this operator is uniquely defined by the pair $(F; E)$.

Let us mention here that sometimes a space F may consist of special elements, which do not have the same origin as the elements of a space E, but we can indicate a one-to-one correspondence between F and a dense set in E, and that correspondence preserves all the algebraic operations (it is a so-called embedding operator of F into E); then F can be identified with its image. Provided inequality (8.1) holds true, the above mentioned construction can be repeated once again.

1.8.2 Hilbert Pairs Connected With the Spaces $H^1(\Omega)$ and $L_2(\Omega)$

The spaces we just mentioned above provide the most important examples of Hilbert pairs.

1° The space $H^1(\Omega)$ is densely embedded in the space $L_2(\Omega)$, thus there exists a generating operator that corresponds to the pair $(H^1(\Omega); L_2(\Omega))$, that is, a self-adjoint positive definite operator A, which, in virtue of (8.5), satisfies the following identity

$$(u, Av)_{L_2(\Omega)} := \int_\Omega uAv\mathrm{d}\Omega = (u, v)_{H^1(\Omega)} := \int_\Omega uv\mathrm{d}\Omega + \int_\Omega \nabla u \cdot \nabla v\mathrm{d}\Omega$$

for all $u \in H^1(\Omega), v \in \mathcal{D}(A)$.

In order to determine the analytic nature of operator A, let us assume that the boundary of Ω is piecewise-smooth, and the function $v(x)$ is two times continuously differentiable. Then, integrating by parts, we get

$$\int_\Omega uAv\mathrm{d}\Omega = \int_\Omega uv\mathrm{d}\Omega - \int_\Omega u\Delta v\mathrm{d}\Omega + \int_{\partial\Omega} u\frac{\partial v}{\partial n}\mathrm{d}S. \tag{8.8}$$

Assuming that $u(x)$ is an arbitrary finite function we get the equation

$$\int_\Omega u(v - \Delta v - Av)\mathrm{d}\Omega = 0,$$

whence $Av = v - \Delta v$. Moreover, in virtue of (8.8),

$$\int_{\partial\Omega} u \frac{\partial v}{\partial n} \mathrm{d}S = 0$$

for all $u \in H^1(\Omega)$ and, especially, for all continuously differentiable functions in $\bar{\Omega}$. The collection of traces of continuously differentiable functions in $\bar{\Omega}$ is dense in $L_2(\partial\Omega)$, therefore, from the last equation it follows that

$$\frac{\partial v}{\partial n} = 0 \qquad \text{on } \partial\Omega. \tag{8.9}$$

Thus, the boundary condition (8.9) is true for all the functions from $\mathcal{D}(A)$ which are twice continuously differentiable, and operator A coincides with the operator $I - \Delta$ on these functions.

The inverse operator $V = A^{-1}$ provides the generalized solution to the boundary value problem

$$v - \Delta v = f \quad \text{in } \Omega, \qquad \frac{\partial v}{\partial n} = 0 \quad \text{on } \partial\Omega. \tag{8.10}$$

Let us recall that the classical solution of this problem is a two times continuously differentiable function $v(x)$ which satisfies equation (8.10). When $f \in L_2(\Omega)$, the solution $v = A^{-1}f$ may not have any continuous second derivatives, that is why this solution is called generalized. For the classical solution of problem (8.10) we should perform the computations in (8.8) in reversed order; we get

$$(u, v)_{H^1(\Omega)} = (u, f)_{L_2(\Omega)}$$

for any $u \in H^1(\Omega)$. Now, by the definition of the generating operator (8.5), we have $v = A^{-1}f$. Thus, any classical solution of problem (8.10) is a generalized solution, too.

Let us notice that whenever the embedding $H^1(\Omega) \hookrightarrow L_2(\Omega)$ is compact (i.e., Ω is a region of the second type), the operator A^{-1} is compact.

2° Let us consider the subspace H^1_Ω in the space $H^1(\Omega)$, which consists of all functions from $H^1(\Omega)$ with zero mean value. This space is densely embedded in the subspace $L_{2,\Omega}$ of the space $L_2(\Omega)$; the space $L_{2,\Omega}$ consists of all functions in $L_2(\Omega)$ orthogonal to constant functions. In H^1_Ω we introduce the Dirichlet norm (2.4), assuming that Ω is a region of the first type.

For the generating operator A of the pair $(H^1_\Omega; L_{2,\Omega})$ the following identity is satisfied

$$\int_\Omega uAv \mathrm{d}\Omega = \int_\Omega \nabla u \cdot \nabla v \mathrm{d}\Omega, \qquad u \in H^1_\Omega,\ v \in \mathcal{D}(A). \tag{8.11}$$

If function $v(x)$ is two times continuously differentiable, then, as it was pointed out above, using integration by parts we get

$$Av = -\Delta v \quad \text{in } \Omega, \qquad \frac{\partial v}{\partial n} = 0 \quad \text{on } \partial\Omega.$$

Therefore, the self-adjoint positive operator $V = A^{-1}$ yields the generalized solution to the homogeneous Neumann problem for the Poisson equation

$$-\Delta v = f \quad \text{in } \Omega, \qquad \frac{\partial v}{\partial n} = 0 \quad \text{on } \partial\Omega. \tag{8.12}$$

Here, we assume that $f \in L_{2,\Omega}$, that is, $f \in L_2(\Omega)$ and $\int_\Omega f(x)\mathrm{d}\Omega = 0$, and the solution $v(x)$ is to be founded in H^1_Ω.

3° As a third example we can consider the subspace $H^1_{0,\Gamma}(\Omega)$ in the space $H^1(\Omega)$; this subspace consists of all the functions from $H^1(\Omega)$, which are equal to zero on a part Γ of the boundary $\partial\Omega$ with $\mathrm{mes}\Gamma > 0$. We will denote the complement of the set $\bar{\Gamma}$ in $\partial\Omega$ by S. The norm, equivalent to the norm in $H^1(\Omega)$, is introduced like the Dirichlet norm in $H^1_{0,\Gamma}(\Omega)$. Identity (8.11) holds true for the operator A which generates the pair $(H^1_{0,\Gamma}(\Omega); L_2(\Omega))$ whenever $u \in H^1_{0,\Gamma}(\Omega)$.

For a two times continuously differentiable function $v(x)$ and $u(x) \in H^1_{0,\Gamma}(\Omega)$ we get the following

$$\int_\Omega uAv\mathrm{d}\Omega = -\int_\Omega u\Delta v\mathrm{d}\Omega + \int_{\partial\Omega} u\frac{\partial v}{\partial n}\mathrm{d}S = -\int_\Omega u\Delta v\mathrm{d}\Omega + \int_S u\frac{\partial v}{\partial n}\mathrm{d}S.$$

Hence,

$$Av := -\Delta v \quad \text{in } \Omega, \qquad \frac{\partial v}{\partial n} = 0 \quad \text{on } S.$$

Therefore, operator V, which is an inverse operator to the generating operator A, yields the generalized solution of Zaremba problem for the Poisson equation

$$-\Delta v = f \quad \text{in } \Omega, \qquad v = 0 \quad \text{on } \Gamma, \qquad \frac{\partial v}{\partial n} = 0 \quad \text{on } S. \tag{8.13}$$

In the particular case when $\Gamma = \partial\Omega$ we obtain the homogeneous Dirichlet problem for the Poisson equation

$$-\Delta v = f \quad \text{in } \Omega, \qquad v = 0 \quad \text{on } \partial\Omega. \tag{8.14}$$

Its solution $v(x)$ belongs to $H^1_{0,\Gamma}(\Omega)$ when $f \in L_2(\Omega)$.

1.8.3 Hilbert Scale of Spaces. Space $E^{-1/2}$

Let $(F; E)$ be a Hilbert pair and A the corresponding generating operator. Using operator A, a whole series of new Hilbert spaces can be introduced in the following way. We let E^α, $\alpha > 0$ be the domain of definition of the power A^α of operator A. This linear set turns into a Hilbert space relatively to the scalar product

$$(x,y)_{E^\alpha} := (A^\alpha x, A^\alpha y)_E, \qquad x, y \in \mathcal{D}(A^\alpha),$$

and the norm

$$\|x\|_{E^\alpha} = \|A^\alpha x\|_E.$$

For consistency it is convenient to introduce the alternative notation $F = E^{1/2}$. Then, in virtue of (8.7),

$$\|x\|_F = \|x\|_{E^{1/2}}.$$

The situation is just a bit more complicated for spaces that have negative exponent. We first introduce in E the new norm

$$\|x\|_{E^{-\alpha}} := \|A^{-\alpha}x\|_E,$$

and then the space E is completed in this norm. As a result we get a Hilbert space which is denoted by $E^{-\alpha}$.

Therefore, we obtain the family of Hilbert spaces E^α, $-\infty < \alpha < \infty$, called a *Hilbert scale* of spaces. When $\alpha < \beta$, the space E^β is densely embedded in the space E^α. When $\alpha > 1$, the restriction of operator A to E^α univalently and continuously maps the space E^α onto $E^{\alpha-1}$. When $\alpha \leq 1$, the extension by continuity of the operator A maps E^α to $E^{\alpha-1}$ (in particular, the space $E^{1/2} = F$ to $E^{-1/2}$). The operator $A^{1/2}$ acts in a similar way; in particular, it maps $E^{1/2} = F$ onto E, and E onto $E^{-1/2}$.

Let us illustrate the details for the space $E^{-1/2}$. This space can be obtained as the completion of $E^0 = E$ in the norm $\|x\|_{E^{-1/2}} = \|A^{-1/2}x\|_{E^0}$. Let us fix an element $x \in E_0$. In $E^{1/2} = F$, we define the bounded linear functional $l_x(y) := (x,y)_{E^0}$ that corresponds to this element. Indeed,

$$\begin{aligned} |l_x(y)| &= |(x,y)_{E^0}| = |(A^{-1/2}x, A^{1/2}y)_{E^0}| \\ &\leq \|A^{-1/2}x\|_{E^0} \cdot \|A^{1/2}y\|_{E^0} = \|A^{-1/2}x\|_{E^0} \cdot \|y\|_F. \end{aligned}$$

Hence,

$$\|l_x\|_{F^*} \leq \|A^{-1/2}x\|_{E^0}.$$

Actually it turns out that we have an equality. Indeed, when $z = A^{-1}x$,

$$l_x(z) = (x, A^{-1}x)_{E^0} = (A^{-1/2}x, A^{-1/2}x)_{E^0} = \|A^{-1/2}x\|_{E^0} \cdot \|A^{-1/2}x\|_{E^0}$$
$$= \|A^{-1/2}x\|_{E^0} \cdot \|A^{1/2}A^{-1}x\|_{E^0} = \|A^{-1/2}x\|_{E^0} \cdot \|z\|_F,$$

therefore,

$$\|l_x\|_{F^*} = \|A^{-1/2}x\|_{E^0}. \tag{8.15}$$

We have already shown that the space E can be embedded in the space F^*, which is the space of all bounded linear functionals on F. In this case equation (8.15) holds true. Hence, the completion $E^{-1/2}$ of the space $E^0 = E$ in the norm $\|A^{-1/2}x\|_{E^0}$ can be isometrically identified with a subspace in F^*. Let us show that this subspace coincides with the whole F^*. In the opposite case there would exist a nonzero element y_0 in the space F such that all functionals from this subspace equal zero on this element. But in this case

$$l_x(y_0) = (x, y_0)_{E^0} = 0$$

for any $x \in E^0$, hence $y_0 = 0$.

Thus the space $E^{-1/2}$ is isometrically identified with the space F^*, which is the dual space to the space F, that is,

$$E^{-1/2} = F^* = (E^{1/2})^*.$$

1.8.4 Self-Adjoint Extensions of Positive Definite Symmetric Operators. Generalized and Weak Solutions of Equations

Let A_0 be a linear symmetric positive definite operator acting in a Hilbert space E and defined on a linear set E_0 dense in E. These properties amount to

$$(A_0x, y)_E = (x, A_0y)_E, (A_0x, x)_E \geq \frac{1}{a^2}\|x\|_E^2, \tag{8.16}$$

whenever $x, y \in E_0$. In the set E_0 we introduce a new scalar product by means of the formula

$$(x, y)_F := (A_0x, y)_E. \tag{8.17}$$

The completion of E_0 in the corresponding norm will be denoted by F.

Inequality (8.16) can be now written down in the form

$$\|x\|_E \leq a\|x\|_F,$$

for each $x \in E_0$. From this inequality it follows that every fundamental sequence in the norm of space F is also fundamental in the norm of space E; therefore, such sequences converge to elements from E. Hence, each element from F corresponds to an element from E. We claim that this correspondence is one-to-one. In the opposite case there would exist a sequence $\{u_n\}$ from E_0 such that $u_n \to \tilde{u} \neq 0$ in F and $u_n \to 0$ in E. As $n \to \infty$ we would have

$$(u_n, y)_F \to (\tilde{u}, y)_F$$

for any $y \in E_0$. On the other hand,

$$(u_n, y)_F = (A_0 u_n, y)_E \to 0.$$

Hence, $(\tilde{u}, y)_F = 0$ for all $y \in E_0$ and, therefore, $\tilde{u} = 0$, a contradiction.

Thus, space F is densely embedded in the space E, that is, these spaces form a Hilbert pair. Consequently, there exists a generating operator A for which

$$(Ax, y)_E = (x, y)_F$$

for all $y \in F$ and $x \in \mathcal{D}(A)$. When $y \in E_0$, a comparison of this identity with (8.17) shows that the equality $Ax = A_0 x$ is valid for all $x \in E_0$, that is, the self-adjoint operator A is an extension of the operator A_0. This extension is called the *Friederichs extension*, or *rigid extension*, of a symmetric positive definite operator.

Apparently operator A_0 is connected with the set of solutions of the equation

$$A_0 u = f \quad (f \in E). \tag{8.18}$$

Solutions of the related equation

$$Au = f \tag{8.19}$$

are called *generalized solutions* of equation (8.18). If operator A has a bounded inverse operator, a unique generalized solution u exists for every $f \in E$ and

$$\|u\|_F \leq c\|f\|_E.$$

Generalized solutions fill the whole domain of definition of operator A. However, if we inspect the construction of a generating operator, we could see that $\mathcal{D}(A)$ does not have an explicit description. On the other hand, the domain of definition of the operator $A^{1/2}$ coincides with the whole space F. That is why a further extension of operator A can be quite useful. We already know that the extension of A to the subspace $E^{1/2} = F$ univalently maps space $E^{1/2}$ onto space $E^{-1/2}$.

Therefore, equation (8.19) can be considered for any $f \in E^{-1/2}$. It has a unique solution from $E^{1/2} = F$. Such solutions are called *weak solutions* of equation (8.18).

Let us notice that in the problems (8.10), (8.12)–(8.14) we have been using the operator A^{-1} in order to prove the existence of generalized solutions, which belong to some specific subspaces of the space $H^1(\Omega)$. But those solutions do not fill that subspace completely. The description of the set of generalized solutions is a really complicated problem, which can be solved by imposing additional conditions on the boundary of the region Ω and on the parts S and Γ of that boundary.

For weak solutions, the similar problem is quite evident. In problem (8.10), for every function $f \in (H^1(\Omega))^*$ a unique weak solution exists in $H^1(\Omega)$; moreover, any function $u \in H^1(\Omega)$ is a weak solution of problem (8.10) for a certain $f \in (H^1(\Omega))^*$.

In problem (8.12), a unique weak solution exists in H^1_Ω for every $f \in H^{-1}_\Omega := (H^1_\Omega)^*$. Such weak solutions fill the whole space H^1_Ω.

Similarly, in problem (8.13) a unique weak solution exists in $H^1_{0,\Gamma}(\Omega)$ for each function $f \in (H^1_{0,\Gamma}(\Omega))^*$. Weak solutions fill the whole space $H^1_{0,\Gamma}(\Omega)$.

For the Dirichlet problem (8.14) a unique weak solution exists in $H^1_0(\Omega)$ for every $f \in H^{-1}(\Omega) = (H^1_0(\Omega))^*$; the whole space $H^1_0(\Omega)$ consists of such weak solutions.

1.8.5 Nonhomogeneous Boundary Value Problems

At the end of Section 1.8.1 we pointed out that the construction described there can be extended in the case when one of the spaces is univalently, linearly and continuously mapped onto a dense set in the other space. Here the more often encountered case of a nonunivalent mapping is considered.

Let γ be a linear operator defined on space F and acting boundedly into a Hilbert space G, that is,

$$\|\gamma u\|_G \leq b\|u\|_F, \qquad u \in F. \tag{8.20}$$

We let N denote the kernel of operator γ, that is, the collection of elements $u \in F$ such that $\gamma u = 0$. In virtue of (8.20), N is a subspace in F. We next let M denote the orthogonal complement to N in F, so we have $F = N \oplus M$.

Let us temporarily denote the range of operator γ by G_+. Operator γ provides a univalent mapping from the space M onto G_+. This makes it possible to introduce the structure of a Hilbert space in G_+ by setting

$$(\varphi, \psi)_{G_+} := (u, v)_F,$$

where $u, v \in M$, $\gamma u = \varphi$, $\gamma v = \psi$. Therefore,

$$\|\varphi\|_{G_+} = \|u\|_F, \qquad \gamma u = \varphi,\ u \in M.$$

Assume $u \in F$ and $\gamma u = \varphi$. Then there is an element $v \in M$ such that $\gamma v = \varphi$. Hence $\gamma(u - v) = 0$, that is, $u - v \in N$. Further,

$$\|u\|_F^2 = \|v + (u-v)\|_F^2 = \|v\|_F^2 + \|u - v\|_F^2.$$

Hence it follows that

$$\|\varphi\|_{G_+}^2 = \|v\|_F^2 \leq \|u\|_F^2$$

for any $u \in F$ with $\gamma u = \varphi$. In other words,

$$\|\varphi\|_{G_+} = \min_{\gamma u = \varphi} \|u\|_F. \tag{8.21}$$

Let us assume that G_+ is densely embedded in space G. From (8.20) we get

$$\|\varphi\|_G = \|\gamma u\|_G \leq b\|u\|_F = b\|\varphi\|_{G_+},$$

where $\varphi \in G_+$, $u \in M$. It means that the spaces G and G_+ form a Hilbert pair; there is a Hilbert scale G^α associated with this pair, constructed in such a way that $G_+ = G^{1/2}$. The space $G^{-1/2}$ can be identified with the space $(G_+)^*$.

Let T be the adjoint of operator γ with respect to the scalar product in G. Since the operator γ maps isometrically the space M onto the space $G_+ = G^{1/2}$, the operator T maps isometrically the space $(G_+)^* = G^{-1/2}$ onto the space M. By definition, when $\psi \in G^{-1/2}$,

$$(T\psi, v)_F = (\psi, \gamma v)_G \tag{8.22}$$

for any $v \in M$. Let ∂ be the inverse operator to T, that is, $\partial := T^{-1}$.

From the properties of the two operators γ and T it follows that the operator $C := \gamma T$ maps isometrically the space $G^{-1/2}$ onto the space $G^{1/2}$. The restriction of operator C to G is a bounded operator on $G^{1/2}$. Indeed, in virtue of (8.20) and taking into account the inequality

$$\|\varphi\|_{G^{-1/2}} \leq b\|\varphi\|_G, \qquad \varphi \in G,$$

which can be derived from (8.20) and from the properties of the scale G^α, we have that

$$\|\gamma T\varphi\|_G \leq b\|\gamma T\varphi\|_{G^{1/2}} = b\|\varphi\|_{G^{-1/2}} \leq b^2\|\varphi\|_G, \qquad \varphi \in G.$$

If the embedding $G_+ \hookrightarrow G$ is compact, then the restriction of the operator C to G is also compact.

From (8.22) for $v = T\varphi$ ($\varphi, \psi \in G$) we obtain

$$(T\psi, T\varphi)_F = (\psi, \gamma T\varphi)_G = (\psi, C\varphi)_G.$$

Hence it follows that operator C is self-adjoint and positive. In view of these properties we can state that operator $C^{-1} = \partial \cdot \gamma^{-1}$ is the generating operator of the Hilbert pair $(G_+; G)$.

Let us next consider the equation $\partial v + \sigma\gamma v = \varphi \in G$ that involves the combination of operators $\partial + \sigma\gamma$, where σ is a bounded linear operator on G. Employing the substitution $\partial v = \psi$ we obtain the equation

$$\psi + \sigma C\psi = \varphi.$$

Operator C and hence σC are compact on G, therefore, the just obtained equation is a Fredholm equation. If ψ is a solution of this equation, then $v = T\psi \in M$ is a solution of the initial equation; consequently, the Fredholm theorems are true for this equation.

1.8.6 Spaces of Harmonic Functions

1° Let us use the scheme developed in the previous section for the case when $F = H^1(\Omega)$, γ is the trace operator acting on functions $v \in H^1(\Omega)$ on the surface $\partial\Omega$, and Ω is a region of the third type. The subspace N coincides with space $H_0^1(\Omega)$. We need first to clarify what is the orthogonal complement M to the subspace $H_0^1(\Omega)$ in the space $\tilde{H}^1(\Omega)$ with the scalar product corresponding to the norm (2.8), that is,

$$(u, v)_{\tilde{H}^1(\Omega)} := \int_\Omega \nabla u \cdot \nabla v d\Omega + \int_{\partial\Omega} \gamma u \cdot \gamma v dS. \tag{8.23}$$

If u is a finite function in Ω, then it belongs to $H_0^1(\Omega)$ and, therefore, when $p \in M$ we have

$$0 = \int_\Omega \nabla u \cdot \nabla p d\Omega. \tag{8.24}$$

If function $p(x)$ has second order generalized derivatives, then this equation leads to $\Delta p = 0$. Conversely, if $p \in \tilde{H}^1(\Omega)$ satisfies Laplace equation in the sense of generalized functions, then (8.24) holds true, and because the set of finite functions is dense in $H_0^1(\Omega)$, function $p(x)$ is orthogonal to $H_0^1(\Omega)$ in $\tilde{H}^1(\Omega)$.

The resulting subspace can be naturally called the space of *harmonic functions* and we denote it by $H_h^1(\Omega)$. Thus,

$$\tilde{H}^1(\Omega) = H_0^1(\Omega) \oplus H_h^1(\Omega),$$

where orthogonality is understood in the sense of the scalar product (8.23).

The operator γ maps isometrically the space $M = H_h^1(\Omega)$ onto the space $G_+ = H^{1/2}(\partial\Omega)$; in the latter space the norm is introduced according to the formula (in virtue of (8.21))

$$\|\varphi\|_{H^{1/2}(\partial\Omega)} = \min_{\gamma u = \varphi} \|u\|_{\tilde{H}^1(\Omega)} \qquad u \in \tilde{H}^1(\Omega).$$

Here the minimum is reached on a harmonic function $p(x)$, which satisfies $\gamma p = \varphi$.

From the above mentioned isometry between the spaces M and G_+ it follows that the Dirichlet problem

$$\Delta p = 0 \quad \text{in } \Omega, \qquad p = \varphi \quad \text{on } \partial\Omega, \tag{8.25}$$

has a unique solution $p \in \tilde{H}^1(\Omega)$ for each function $\varphi \in H^{1/2}(\partial\Omega)$.

Let us next consider the operator T, which is the adjoint to operator γ, and its inverse operator $\partial = T^{-1}$. By definition

$$\int_\Omega \nabla p \cdot \nabla q d\Omega + \int_{\partial\Omega} \gamma p \cdot \gamma q dS = \int_{\partial\Omega} \partial p \cdot \gamma q dS, \qquad p, q \in H_h^1(\Omega). \tag{8.26}$$

For every smooth function $p(x)$, in virtue of the Green Formula, we get

$$\int_\Omega \nabla p \cdot \nabla q d\Omega = -\int_\Omega \Delta p \cdot q d\Omega + \int_{\partial\Omega} \frac{\partial p}{\partial n} \gamma q dS = \int_{\partial\Omega} \frac{\partial p}{\partial n} \gamma q dS. \tag{8.27}$$

(8.26) and (8.27) lead to the following

$$\int_{\partial\Omega} \partial p \cdot \gamma q dS = \int_{\partial\Omega} \left(\frac{\partial p}{\partial n} + \gamma p \right) \gamma q dS.$$

Hence, it follows, that $\partial p = \partial p / \partial n + \gamma p$. Further, the operator ∂ defined by formula (8.26) will be denoted by $\partial / \partial n + I$. The isometry implemented by this operator makes it possible to derive the following conclusion: The problem

$$\Delta p = 0 \quad \text{in } \Omega, \qquad \left(\frac{\partial}{\partial n} + I \right) p = \psi \quad \text{on } \partial\Omega,$$

has a unique solution $p \in \tilde{H}^1(\Omega)$ for each

$$\psi \in H^{-1/2}(\partial\Omega) := (H^{1/2}(\partial\Omega))^*.$$

Moreover, we have

$$\|p\|_{\tilde{H}^1(\Omega)} = \|\psi\|_{H^{-1/2}(\partial\Omega)}.$$

2° Let us now consider the case when $F = H^1_{\partial\Omega}(\Omega)$, the space consisting of all functions $u \in H^1(\Omega)$ such that

$$\int_{\partial\Omega} \gamma u \mathrm{d}S = 0. \tag{8.28}$$

In this space we introduce the Dirichlet norm and the corresponding scalar product, which is different from the scalar product (8.23), namely,

$$(u, v)_{H^1_{\partial\Omega}(\Omega)} := \int_\Omega \nabla u \cdot \nabla v \mathrm{d}\Omega.$$

Here, γ is the trace operator and the subspace N coincides with $H^1_0(\Omega)$. The orthogonal complement $M := H^1_{h,\partial\Omega}(\Omega)$ to N consists of harmonic functions $u \in H^1(\Omega)$ that satisfy (8.28). The operator γ maps isometrically the subspace $H^1_{h,\partial\Omega}(\Omega)$ onto the subset of the space $H^{1/2}(\partial\Omega)$ consisting of functions orthogonal to the constants in $L_2(\partial\Omega)$. We will denote this subspace by $H^{1/2}_{\partial\Omega}$. Its norm can be obtained by the formula

$$\|\varphi\|_{H^{1/2}_{\partial\Omega}} := \min_{\gamma u = \varphi} \|u\|_{H^1_{\partial\Omega}(\Omega)}, \qquad u \in H^1_{\partial\Omega}(\Omega).$$

The minimum here is reached on a harmonic function $p \in H^1_{h,\partial\Omega}(\Omega)$ such that $\gamma p = \varphi$.

Now, let us consider the operator ∂. By definition, when $p, q \in M$ we get

$$\int_\Omega \nabla p \cdot \nabla q \mathrm{d}\Omega = \int_{\partial\Omega} \partial p \cdot \gamma q \mathrm{d}S. \tag{8.29}$$

If $p(x)$ is smooth, then in virtue of Green Formula we have

$$\int_\Omega \nabla p \cdot \nabla q \mathrm{d}\Omega = \int_{\partial\Omega} \frac{\partial p}{\partial n} \gamma q \mathrm{d}S, \qquad \Delta p = 0, \tag{8.30}$$

and, therefore, $\partial = \partial/\partial n$. The isometry implemented by the operator ∂ makes it possible to show that the Neumann problem

$$\Delta p = 0 \quad \text{in } \Omega, \qquad \frac{\partial p}{\partial n} = \psi \quad \text{on } \partial\Omega, \tag{8.31}$$

has a unique solution $p \in H^1_{\partial\Omega}(\Omega)$ for each $\psi \in H^{-1/2}_{\partial\Omega} := \left(H^{1/2}_{\partial\Omega}\right)^*$. Moreover, we have

$$\|p\|_{H^1_{\partial\Omega}(\Omega)} = \|\psi\|_{H^{-1/2}_{\partial\Omega}}. \tag{8.32}$$

In some cases, instead of $F = H^1_{\partial\Omega}(\Omega)$ it is necessary to consider the space $\breve{H}^1(\Omega)$ with the scalar product

$$(u, v)_{\breve{H}^1(\Omega)} := \int_\Omega \nabla u \cdot \nabla v \mathrm{d}\Omega + \int_{\partial\Omega} \gamma u \mathrm{d}S \cdot \int_{\partial\Omega} \gamma v \mathrm{d}S. \tag{8.33}$$

The corresponding norm is equivalent to the usual norm in $H^1(\Omega)$; the two spaces $\breve{H}^1(\Omega)$ and $H^1(\Omega)$ coincide. If we consider the trace operator γ on $\partial\Omega$, then again $N = H^1_0(\Omega)$, and the orthogonal complement M to N in $\breve{H}^1(\Omega)$ is the set of harmonic functions $H^1_h(\Omega)$:

$$\breve{H}^1(\Omega) = H^1_0(\Omega) \oplus H^1_h(\Omega). \tag{8.34}$$

The orthogonality of the subspace in (8.39) is understood in the sense of (8.33); condition (8.28) is not necessary.

The operator γ maps isometrically $H^1_h(\Omega)$ onto $H^{1/2}(\partial\Omega)$, which has the norm

$$\|\varphi\|_{H^{1/2}(\partial\Omega)} := \min_{\gamma u = \varphi} \|u\|_{\breve{H}^1(\Omega)} \qquad u \in \breve{H}^1(\Omega);$$

the minimum here is reached on a function $p \in H^1_h(\Omega)$ such that $\gamma p = \varphi$.

3° Let us assume now, as in Example 3° of Section 1.8.2, that the boundary of the region Ω consists of two disjoint parts, $\partial\Omega = \bar{S} \cup \Gamma$, where $\operatorname{mes}\Gamma > 0$. We use the scheme of the previous section for the space $F = H^1_\Gamma(\Omega)$ with the Dirichlet norm by taking $G = L_{2,\Gamma} := L_2(\Gamma) \ominus \{1\}$ and by letting γ_Γ be the composition of the trace operator and the restriction operator from $\partial\Omega$ to Γ.

The subspace N consists of all functions from $H^1_\Gamma(\Omega)$ whose traces equal zero on Γ. In order to find the orthogonal complement to N in $H^1_\Gamma(\Omega)$ we use the Green formula (8.27). When $q \in N$ and $p \in M$ is smooth, we have

$$0 = \int_\Omega \nabla q \cdot \nabla p \mathrm{d}\Omega = -\int_\Omega q \Delta p \mathrm{d}\Omega + \int_{\partial\Omega} \gamma q \frac{\partial p}{\partial n} \mathrm{d}S = -\int_\Omega q \Delta p \mathrm{d}\Omega + \int_S \gamma q \frac{\partial p}{\partial n} \mathrm{d}S.$$

Assuming that q is finite in Ω we get $\Delta p = 0$. Taking into account the fact that the set of traces of continuous functions in $\bar{\Omega}$ which are equal to zero on Γ is dense in $L_2(S)$, we obtain that $\partial p/\partial n = 0$ on S. Hence, the subspace M consists of harmonic functions in Ω such that $\partial p/\partial n = 0$ on S. Let us denote $N := H^1_{0,\Gamma}(\Omega)$ and $M := H^1_{h,S}(\Omega)$. Then

$$H^1_\Gamma(\Omega) = H^1_{0,\Gamma}(\Omega) \oplus H^1_{h,S}(\Omega), \tag{8.35}$$

where the orthogonality is understood in the sense of the scalar product generated by the Dirichlet norm.

The operator γ_Γ induces an isometrical mapping from the space $H^1_{h,S}(\Omega)$ onto a space of functions $H^{1/2}_\Gamma$, which is dense in the space $L_{2,\Gamma}$, and is equipped with the norm

$$\|\varphi\|_{H^{1/2}_\Gamma} := \min_{\gamma q=\varphi} \|q\|_{H^1_\Gamma(\Omega)}, \qquad q \in H^1_\Gamma(\Omega). \tag{8.36}$$

This fact can be reformulated in the following way: For every function $\varphi \in H^{1/2}_\Gamma$ with $\int_\Gamma \varphi d\Gamma = 0$ the Zaremba problem for the Laplace equation in $H^1_\Gamma(\Omega)$,

$$\Delta p = 0 \quad \text{in } \Omega, \qquad p = \varphi \quad \text{on } \Gamma, \qquad \frac{\partial p}{\partial n} = 0 \quad \text{on } S, \tag{8.37}$$

has a unique solution p in $H^1_\Gamma(\Omega)$; moreover we have

$$\|p\|_{H^1_\Gamma(\Omega)} = \|\varphi\|_{H^{1/2}_\Gamma}. \tag{8.38}$$

Let us notice that since problem (8.37) has the solution $p = p_0 = 1$ for $\varphi = \varphi_0 = 1$, then that problem can be solved as well for any $\varphi \in H^{1/2}_\Gamma \dot{+} \mathbb{R} = H^{1/2}(\Gamma)$.

From identity (8.29) it follows again that the operator ∂, which maps isometrically the space $H^1_{h,S}(\Omega)$ onto the dual space $H^{-1/2}_\Gamma := (H^{1/2}_\Gamma)^*$ to $H^{1/2}_\Gamma$, coincides with the operator $\partial/\partial n$. Hence, it appears that for every $\psi \in H^{-1/2}_\Gamma$ with $\int_\Gamma \psi \mathrm{d}\Gamma = 0$ there exists a unique solution p in $H^1_\Gamma(\Omega)$ of the Neumann problem for the Laplace equation,

$$\Delta p = 0 \quad \text{in } \Omega, \qquad \frac{\partial p}{\partial n} = \psi \quad \text{on } \Gamma, \qquad \frac{\partial p}{\partial n} = 0 \quad \text{on } S. \tag{8.39}$$

Here again

$$\|p\|_{H^1_\Gamma(\Omega)} = \|\psi\|_{H^{-1/2}_\Gamma}. \tag{8.40}$$

1.8.7 Embedding and Mapping. The Abstract Green Formula

We now examine a situation that unites the two ones considered in the previous section. We assume that a space F is densely embedded in the space E and let γ be a bounded linear operator defined on F that maps F onto a dense set G_+ in another space G. Let N be the kernel of operator γ; we will assume that this kernel is dense in E. Then there is a generating operator corresponding to the Hilbert pair $(N; E)$, and we denote it by L. The inverse operator of operator L maps the whole space E into N and, therefore, yields the generalized solution of the problem

$$Lu = f, \quad \gamma u = 0,$$

for every $f \in E$. As for the weak solutions, they exist for any $f \in E^{-1/2}$ and fill the space $E^{1/2} = N$ completely.

Let M be the orthogonal complement to subspace N in the space F. The operator L can be extended to $\mathcal{D}(L) \oplus M$ by setting it equal to zero on M.

We proceed by introducing the operator ∂. As indicated in Section 1.8.5 above, the operator ∂ is defined on the subspace M as the inverse to the adjoint operator of γ. We next extend ∂ to $\mathcal{D}(L)$ in the following way. First, we consider the inequality

$$|(Lu, v)_E| \leq \|Lu\|_E \cdot \|v\|_E \leq a\|Lu\|_E\|v\|_F = a\|Lu\|_E\|\gamma v\|_{G^{1/2}},$$

which holds true for any $u \in \mathcal{D}(L)$ and $v \in M$. This inequality shows that $(Lu, v)_E$ can be considered as a bounded linear functional on $G^{1/2}$. Let us denote the element from $G^{-1/2}$ that represents this functional in terms of the scalar product on G by $-\partial u$. Then

$$(Lu, v)_E = -(\partial u, \gamma v)_G, \qquad u \in \mathcal{D}(L),\ v \in F,$$

and

$$\|\partial u\|_{G^{-1/2}} \leq a\|Lu\|_E.$$

Further, ∂ can be linearly extended on $\mathcal{D}(L) \oplus M$.

The reasoning outlined above leads to the following abstract Green formula:

$$(Lu, v)_E = (u, v)_F - (\partial u, \gamma v)_G \tag{8.41}$$

for any $v \in F$ and $u \in \mathcal{D}(L)$.

Indeed, according to the direct sum decomposition $F = N \oplus M$ we have $u = u_1 + u_2$, $v = v_1 + v_2$, where $u_1 \in \mathcal{D}(L)$, $v_1 \in N$, $u_2, v_2 \in M$. Then, using the definitions of operators L and ∂ on $\mathcal{D}(L)$ we get

$$\begin{aligned}(Lu, v)_E &= (L(u_1 + u_2), v_1 + v_2)_E = (Lu_1, v_1 + v_2)_E \\ &= (Lu_1, v_1)_E + (Lu_1, v_2)_E = (u_1, v_1)_F - (\partial u_1, \gamma v_2)_G.\end{aligned}$$

Moreover,

$$(u, v)_F = (u_1, v_1)_F + (u_2, v_2)_F.$$

Finally, in virtue of (8.22) and using the equation $\partial = T^{-1}$, we have

$$\begin{aligned}(\partial u, \gamma v)_G &= (\partial(u_1 + u_2), \gamma v_2)_G \\ &= (\partial u_1, \gamma v_2)_G + (\partial u_2, \gamma v_2)_G \\ &= (\partial u_1, \gamma v_2)_G + (u_2, v_2)_F.\end{aligned}$$

Formula (8.41) follows now easily by combining all these equations.

Chapter 2

Fundamental Spaces and Operators of Linear Hydrodynamics

In this chapter we investigate the fundamental operators of vector analysis and their applications to the study of vector fields from $\boldsymbol{L}_2(\Omega)$. Since such fields are not required to be smooth, the Riesz theorem on the representation of linear functionals is going to play a significant role. We begin by introducing the spaces of velocity fields of an ideal fluid in a closed or open container. Our goal will be to investigate the velocity fields that describe the movement of a viscous incompressible fluid. We introduce the form that defines the velocity of energy dissipation and derive the corresponding fundamental Korn inequality. The Stokes operator, which plays a major role in all the investigations of the movement of a viscous fluid, will also be defined. In the last part of this chapter we discuss and solve the basic boundary value problems pertaining to the movement of a viscous fluid in an open container.

2.1 Fundamental Spaces and Hydrodynamics Operators for an Ideal Fluid

In many problems of hydrodynamics, we deal with regions that are filled with fluids and have a nonsmooth boundary. A cylindrical container, entirely or partially filled with fluid, is just the simplest example of such a region. In this regard, it becomes necessary to consider the fundamental operators of vector analysis and hydrodynamics on spaces of vector fields defined in nonsmooth regions.

In this section, we investigate properties of the field divergence operator, the normal component of a field on the boundary, the gradient, and the Laplace operator. As a rule, the ranges of all these operators belong to some spaces of generalized vector functions. We will present an analogue of the well-known Weyl decomposition and its development specifically adapted to the problems of movement of a fluid with free boundary.

2.1.1 Fields with Finite Kinetic Energy

Let Ω be a bounded simply-connected region filled with a moving fluid. The velocity vector $\boldsymbol{u} = (u_1, u_2, u_3)$ of this fluid can be defined at every point x in Ω and, therefore, we get a velocity field $\boldsymbol{u}(x)$ on Ω. It is just natural to assume that the corresponding mass of the fluid has a finite kinetic energy T. If the fluid is homogeneous and has density ρ, then

$$2T = \rho \int_\Omega |\boldsymbol{u}|^2 \mathrm{d}\Omega < \infty.$$

Hence, all the fields that yield a finite kinetic energy form the Hilbert space $\boldsymbol{L}_2(\Omega)$, with the norm

$$\|\boldsymbol{u}\|_{\boldsymbol{L}_2(\Omega)} = \left(\int_\Omega |\boldsymbol{u}|^2 \mathrm{d}\Omega\right)^{1/2}.$$

This norm corresponds to the scalar product

$$(\boldsymbol{u}, \boldsymbol{v})_{\boldsymbol{L}_2(\Omega)} := \int_\Omega \boldsymbol{u} \cdot \boldsymbol{v} \mathrm{d}\Omega,$$

where $\cdot$ denotes the scalar product of vectors in $\mathbb{R}^3$.

2.1.2 Potential Fields

Potential fields of the form $\boldsymbol{v} = \nabla p$ play a special role. The fact that the kinetic energy for such a field is finite can be expressed as

$$\int_\Omega |\nabla p|^2 \mathrm{d}\Omega < \infty. \tag{1.1}$$

Hence, the potential fields with finite kinetic energy are gradients of functions from the space $L_2^1(\Omega)$. Further, we will assume that the region Ω is of the first type, whence we get that the space $L_2^1(\Omega)$ coincides with the space $H^1(\Omega)$ and, therefore, $\boldsymbol{v} = \nabla p$, $p \in H^1(\Omega)$.

Taking into account that the generalized derivatives are defined by the equations

$$\int_\Omega \frac{\partial p}{\partial x_i} u \mathrm{d}\Omega = -\int_\Omega p \frac{\partial u}{\partial x_i} \mathrm{d}\Omega,$$

where u is an arbitrary finite function on Ω, we conclude that a field $\boldsymbol{v} \in \boldsymbol{L}_2(\Omega)$ is a potential field if and only if there exists a function $p \in H^1(\Omega)$ such that

$$\int_\Omega \boldsymbol{v} \cdot \boldsymbol{u} d\Omega = -\int_\Omega p \sum_{i=1}^{3} \frac{\partial u_i}{\partial x_i} d\Omega = -\int_\Omega p \operatorname{div} \boldsymbol{u} d\Omega, \tag{1.2}$$

where $\boldsymbol{u}$ is an arbitrary finite field in Ω.

The potential p of a field $\boldsymbol{v} = \nabla p$ is defined up to a constant. Any subspace of the space $H^1(\Omega)$ that intersects the one-dimensional subspace of constants only at zero and gives the entire space $H^1(\Omega)$ as the result of a direct sum with the above mentined subspace of constants will be called a *space of potentials.* By Subsection 1.2.7, the norm of the space $H^1(\Omega)$ is equivalent, in any space of potentials, to the Dirichlet norm,

$$\|p\|_{H^1(\Omega)} = \left(\int_\Omega |\nabla p|^2 d\Omega\right)^{1/2}, \tag{1.3}$$

the constants of equivalence depending on the choice of a specific space of potentials. The spaces H^1_Ω and $H^1_\Gamma(\Omega)$ introduced in Section 1.2 are examples of such spaces of potentials.

The norm of a potential field $\boldsymbol{u} = \nabla p$ in $\boldsymbol{L}_2(\Omega)$ coincides with the norm (1.3); consequently, all such potential fields form a closed subset of the space $\boldsymbol{L}_2(\Omega)$ and therefore, we get a subspace of $\boldsymbol{L}_2(\Omega)$ that will be denoted by $\boldsymbol{G}(\Omega)$.

It is also usefull to consider the set $\boldsymbol{G}_0(\Omega)$ of all potential fields in $\boldsymbol{L}_2(\Omega)$ with potentials from $H^1_0(\Omega)$. Since the norm on $H^1(\Omega)$ is equivalent on the subspace $H^1_0(\Omega)$ to the Dirichlet norm (1.3), $\boldsymbol{G}_0(\Omega)$ forms a subspace of both the spaces $\boldsymbol{L}_2(\Omega)$ and $\boldsymbol{G}(\Omega)$.

2.1.3 Divergence of Fields with Finite Kinetic Energy

Since the layers of an ideal fluid can easily slide along each other, it is natural to assume that the velocity field $\boldsymbol{u}$ is not smooth. However, the incompressibility condition for a fluid can be written down as $\operatorname{div} \boldsymbol{u} = 0$, an equations that requires all the components of the field to be differentiable. Actually, the incompressibility condition has an integral character, and the velocity field may not necessarily be differentiable.

We now define the divergence of a field in $\boldsymbol{L}_2(\Omega)$. Let $\boldsymbol{u} \in \boldsymbol{L}_2(\Omega)$. We first consider the functional

$$\Phi(p) = -\int_\Omega \nabla p \cdot \boldsymbol{u} d\Omega$$

defined on $H^1_0(\Omega)$. The following inequality holds true,

$$|\Phi(p)| = \left|\int_\Omega \nabla p \cdot \boldsymbol{u} d\Omega\right| \leq \|\boldsymbol{u}\|_{\boldsymbol{L}_2(\Omega)} \cdot \|\nabla p\|_{\boldsymbol{L}_2(\Omega)} = \|\boldsymbol{u}\|_{\boldsymbol{L}_2(\Omega)} \cdot \|p\|_{H^1_0(\Omega)}. \tag{1.4}$$

It means that $\Phi(p)$ is a bounded linear functional on the space $H_0^1(\Omega)$ and, therefore, $\Phi(p)$ is an element in the space $(H_0^1(\Omega))^* := H^{-1}(\Omega)$. Next, let us denote the function that generates this functional by means of the scalar product on $L_2(\Omega)$ by $\operatorname{div} \boldsymbol{u}$. Then

$$\Phi(p) = \int_\Omega p \operatorname{div} \boldsymbol{u} d\Omega = -\int_\Omega \nabla p \cdot \boldsymbol{u} d\Omega, \qquad p \in H_0^1(\Omega),\ \boldsymbol{u} \in \boldsymbol{L}_2(\Omega). \tag{1.5}$$

According to this definition, we get that $\operatorname{div} \boldsymbol{u} \in H^{-1}(\Omega)$ for any field $\boldsymbol{u} \in \boldsymbol{L}_2(\Omega)$. Therefore, the operator div maps the space $\boldsymbol{L}_2(\Omega)$ into the space $H^{-1}(\Omega)$, and by inequality (1.4) we have

$$\|\operatorname{div} \boldsymbol{u}\|_{H^{-1}(\Omega)} \leq \|\boldsymbol{u}\|_{\boldsymbol{L}_2(\Omega)}. \tag{1.6}$$

2.1.4 The Space of Solenoidal Fields

A field $\boldsymbol{u} \in \boldsymbol{L}_2(\Omega)$ is called *solenoidal* provided $\operatorname{div} \boldsymbol{u} = 0$. By inequality (1.6) it follows that the set of all solenoidal fields is a subspace $\boldsymbol{J}(\Omega)$ of the space $\boldsymbol{L}_2(\Omega)$. It consists of all fields $\boldsymbol{u} \in \boldsymbol{L}_2(\Omega)$ for which

$$\int_\Omega \nabla p \cdot \boldsymbol{u} d\Omega = 0$$

for all $p \in H_0^1(\Omega)$. In other words, $\boldsymbol{J}(\Omega)$ is an orthogonal complement to the subspace $\boldsymbol{G}_0(\Omega)$ and, therefore, the following orthogonal decomposition is true,

$$\boldsymbol{L}_2(\Omega) = \boldsymbol{J}(\Omega) \oplus \boldsymbol{G}_0(\Omega). \tag{1.7}$$

2.1.5 Laplace Operator on the Space $H^1(\Omega)$

The Δ operator defined by the formula

$$\Delta p := \operatorname{div} \nabla p, \qquad p \in H^1(\Omega),$$

will be called the *Laplace operator*.

By definition, Δ acts from the space $H^1(\Omega)$ into the space $H^{-1}(\Omega) = (H_0^1(\Omega))^*$. Moreover,

$$\|\Delta p\|_{H^{-1}(\Omega)} = \|\operatorname{div} \nabla p\|_{H^{-1}(\Omega)} \leq \|\nabla p\|_{\boldsymbol{L}_2(\Omega)} \leq \|p\|_{H^1(\Omega)}.$$

The weak extension $\tilde{\Delta}$ of the Laplace operator Δ that produces a one-to-one continuous mapping from the space $H_0^1(\Omega)$ onto the space $H^{-1}(\Omega)$ was considered in Subsection 1.8.2. The operators Δ and $\tilde{\Delta}$ coincide on finite functions; moreover, since Δ and $\tilde{\Delta}$ are bounded, they coincide on the entire space $H_0^1(\Omega)$. Hence, the restriction of the Laplace operator to the space $H_0^1(\Omega)$ maps univalently this space onto the space $H^{-1}(\Omega)$. It turns out that this map is an isometry.

The following corollary follows from the above mentioned facts:

The operator div *maps isometrically the space* $\boldsymbol{G}_0(\Omega)$ *onto the entire space* $H^{-1}(\Omega)$.

2.1.6 Normal Component of a Field on the Boundary

Let the field $\boldsymbol{u} \in \boldsymbol{L}_2(\Omega)$ be such that

$$\operatorname{div} \boldsymbol{u} \in \boldsymbol{L}_2(\Omega). \tag{1.8}$$

We consider the functional

$$F(p) := \int_\Omega \nabla p \cdot \boldsymbol{u} \mathrm{d}\Omega + \int_\Omega p \operatorname{div} \boldsymbol{u} \mathrm{d}\Omega \tag{1.9}$$

defined on the space $H^1(\Omega)$, and notice the estimates

$$\begin{aligned} |F(p)| &\le \|\boldsymbol{u}\|_{\boldsymbol{L}_2(\Omega)} \cdot \|\nabla p\|_{\boldsymbol{L}_2(\Omega)} + \|\operatorname{div} \boldsymbol{u}\|_{L_2(\Omega)} \cdot \|p\|_{L_2(\Omega)} \\ &\le (\|\boldsymbol{u}\|_{\boldsymbol{L}_2(\Omega)} + \|\operatorname{div} \boldsymbol{u}\|_{L_2(\Omega)})\|p\|_{H^1(\Omega)}. \end{aligned} \tag{1.10}$$

Therefore, $F(p)$ is a bounded linear functional on the space $H^1(\Omega)$. Further, if $p \in H_0^1(\Omega)$, then in virtue of (1.5) we get $F(p) = 0$. Hence, F is a bounded linear functional on the orthogonal complement to the subspace $H_0^1(\Omega)$ in the space $H^1(\Omega)$. According to Subsection 1.8.6, we know that this orthogonal complement is the space $H_{\mathrm{h}}^1(\Omega)$ of all the harmonic functions in $H^1(\Omega)$. Since the space $H_{\mathrm{h}}^1(\Omega)$ is isometrically isomorphic to the space $H^{1/2}(\partial\Omega)$, then $F(p)$ can be considered as a bounded linear functional on $H^{1/2}(\partial\Omega)$. We represent this functional by means of the scalar product on $L_2(\partial\Omega)$ and denote the corresponding element in $H^{-1/2}(\partial\Omega) = (H^{1/2}(\partial\Omega))^*$ by $\boldsymbol{u}_n$. Thus we obtain the formula

$$F(p) = \int_\Omega \nabla p \cdot \boldsymbol{u} \mathrm{d}\Omega + \int_\Omega p \operatorname{div} \boldsymbol{u} \mathrm{d}\Omega = \int_{\partial\Omega} \gamma p \cdot \boldsymbol{u}_n \mathrm{d}S. \tag{1.11}$$

We call the element $\boldsymbol{u}_n \in H^{-1/2}(\partial\Omega)$ the *normal component* of the field $\boldsymbol{u}$ on the boundary $\partial\Omega$. From (1.10) we have the following inequality,

$$\|\boldsymbol{u}_n\|_{H^{-1/2}(\partial\Omega)} \le \|\boldsymbol{u}\|_{\boldsymbol{L}_2(\Omega)} + \|\operatorname{div} \boldsymbol{u}\|_{L_2(\Omega)}. \tag{1.12}$$

Assuming in (1.11) that $p(x) \equiv 1$, we obtain the Ostrogradski formula,

$$\int_\Omega \operatorname{div} \boldsymbol{u} \mathrm{d}\Omega = \int_{\partial\Omega} \boldsymbol{u}_n \mathrm{d}S, \qquad \operatorname{div} \boldsymbol{u} \in \boldsymbol{L}_2(\Omega),\ \boldsymbol{u}_n \in H^{-1/2}(\partial\Omega). \tag{1.13}$$

For a solenoidal field, from (1.12) we have

$$\|\boldsymbol{u}_n\|_{H^{-1/2}(\partial\Omega)} \leq \|\boldsymbol{u}\|_{\boldsymbol{L}_2(\Omega)}, \tag{1.14}$$

and formula (1.13) gives

$$\int_{\partial\Omega} u_n \mathrm{d}S = 0. \tag{1.15}$$

REMARK. A significant roughness was performed in inequality (1.10) while substituting $\|p\|_{L_2(\Omega)}$ for $\|p\|_{H^1(\Omega)}$. Condition (1.8) can be weakened, if we require $\operatorname{div}\boldsymbol{u}$ to belong to the dual space to a space into which the spaces $H_0^1(\Omega)$ and $H^1(\Omega)$ are densely embedded. For example, the following would be sufficient,

$$\operatorname{div}\boldsymbol{u} \in L_q(\Omega), \qquad q > 6/5.$$

2.1.7 Green Formula for the Laplace Operators. Harmonic Fields

Let us assume that $q \in H^1(\Omega)$ and $\Delta q \in L_2(\Omega)$. We use formula (1.11) with $\boldsymbol{u} = \nabla q$. Then

$$\int_\Omega \nabla p \cdot \nabla q \mathrm{d}\Omega + \int_\Omega p\Delta q \mathrm{d}\Omega = \int_{\partial\Omega} \gamma p (\nabla q)_n \mathrm{d}S.$$

We next denote $(\nabla q)_n =: \partial q/\partial n$. Thus we obtain the *Green formula* for the Laplace operator,

$$\int_\Omega p\Delta q \mathrm{d}\Omega = -\int_\Omega \nabla p \cdot \nabla q \mathrm{d}\Omega + \int_{\partial\Omega} \gamma p \frac{\partial q}{\partial n} \mathrm{d}S, \qquad p \in H^1(\Omega). \tag{1.16}$$

In addition, we have $\partial q/\partial n \in H^{-1/2}(\partial\Omega)$ and

$$\left\|\frac{\partial q}{\partial n}\right\|_{H^{-1/2}(\partial\Omega)} \leq \|\nabla q\|_{L_2(\Omega)} + \|\Delta q\|_{L_2(\Omega)} \leq \|q\|_{H^1(\Omega)} + \|\Delta q\|_{L_2(\Omega)}.$$

We now consider a field $\boldsymbol{u} \in \boldsymbol{L}_2(\Omega)$ that is solenoidal and potential at the same time, i.e., one has $\operatorname{div}\boldsymbol{u} = 0$ and $\boldsymbol{u} = \nabla q$ $(q \in H^1(\Omega))$. Hence, $\operatorname{div}\nabla q = \Delta q = 0$. Such fields are naturally called *harmonic*. The set of all harmonic fields is a subspace $\boldsymbol{G}_{\mathrm{h}}(\Omega)$ of the space of all solenoidal fields:

$$\boldsymbol{G}_{\mathrm{h}}(\Omega) = \boldsymbol{J}(\Omega) \cap \boldsymbol{G}(\Omega).$$

From Green formula (1.16) it follows that

$$\int_\Omega \nabla p \cdot \nabla q \mathrm{d}\Omega = \int_{\partial\Omega} \gamma p \frac{\partial q}{\partial n} \mathrm{d}S$$

for any $p \in H^1(\Omega)$.

Comparing this formula with equation (1.8.29), we conclude that the above introduced operator $\partial/\partial n$ coincides with the operator ∂ on harmonic functions from (1.8.29) (in (1.8.29) ∂ was denoted by $\partial/\partial n$, too). From the results proved in Subsection 1.8.6 it follows that the operator $\partial/\partial n$ maps isometrically the space $H^1_{\mathrm{h},\partial\Omega}$ of all harmonic functions with zero mean value on the boundary of region Ω to the space $H^{-1/2}_{\partial\Omega} = (H^{1/2}_{\partial\Omega})^*$.

2.1.8 Weyl Decomposition

Let $\boldsymbol{u}$ be a solenoidal field. We construct $\boldsymbol{u}_n$ and find, by using the result mentioned above, a harmonic function $q \in H^1_{\mathrm{h},\partial\Omega}(\Omega)$ with zero mean value on $\partial\Omega$, for which $\partial q/\partial n = \boldsymbol{u}_n$. In other words, this function is a solution of the Neumann problem:

$$\Delta q = 0 \quad \text{in } \Omega, \qquad \frac{\partial q}{\partial n} = u_n \quad \text{on } \Omega, \qquad \int_{\partial\Omega} q \mathrm{d}S = 0.$$

Let us denote $\boldsymbol{v} = \boldsymbol{u} - \nabla q$. Then $\operatorname{div} \boldsymbol{v} = -\Delta q = 0$ and $\boldsymbol{v}_n = \boldsymbol{u}_n - \partial q/\partial n = 0$. The fields $\boldsymbol{v}$ and ∇q are orthogonal. Indeed, by (1.11) we have

$$\int_\Omega \boldsymbol{v} \cdot \nabla q \mathrm{d}\Omega = \int_{\partial\Omega} \boldsymbol{v}_n \gamma q \mathrm{d}S = 0.$$

Hence, we obtain a decomposition of the space $\boldsymbol{J}(\Omega)$ of solenoidal fields into the orthogonal sum of the space of potential harmonic fields $\boldsymbol{G}_{\mathrm{h}}(\Omega)$ and the subspace $\boldsymbol{J}_0(\Omega)$ of solenoidal fields with zero normal component on the boundary:

$$\boldsymbol{J}(\Omega) = \boldsymbol{J}_0(\Omega) \oplus \boldsymbol{G}_{\mathrm{h}}(\Omega). \tag{1.17}$$

Hence

$$\boldsymbol{L}_2(\Omega) = \boldsymbol{J}_0(\Omega) \oplus \boldsymbol{G}_{\mathrm{h}}(\Omega) \oplus \boldsymbol{G}_0(\Omega). \tag{1.18}$$

This decomposition is a slight variation of the decomposition discovered by H. Weyl and, in what follows, will be referred to as the *Weyl decomposition*. In this subsection we obtained the Weyl decomposition for any region of the third type and, in particular, for a Lipschitz region.

If the boundary $\partial\Omega$ of a region Ω is a solid boundary S, then, for a moving fluid, the nonleaking condition $\boldsymbol{u}_n = 0$ on $\partial\Omega$ holds true. Hence, the space $\boldsymbol{J}_0(\Omega)$ consists of the velocity fields of movement of an ideal (incompressible) fluid in a closed container.

Let us notice that the space $\boldsymbol{G}_{\mathrm{h}}(\Omega) \oplus \boldsymbol{G}_0(\Omega)$ contains all potential fields. Indeed, if $\boldsymbol{u} = \nabla p$, where $p \in H^1_{\partial\Omega}(\Omega)$, then $\nabla p = \boldsymbol{v} + \nabla p_1 + \nabla p_2$, where $\boldsymbol{v} \in \boldsymbol{J}_0(\Omega)$, $\nabla p_1 \in \boldsymbol{G}_{\mathrm{h}}(\Omega)$, $\nabla p_2 \in \boldsymbol{G}_0(\Omega)$. Since $\operatorname{div} \boldsymbol{v} = 0$ and $\boldsymbol{v}_n = 0$, then, by virtue of (1.11), the field $\boldsymbol{v}$ is orthogonal to all potential fields. Taking the scalar product of the components of ∇p and $\boldsymbol{v}$ we have

$$\int_\Omega |\boldsymbol{v}|^2 \mathrm{d}\Omega = 0,$$

i.e., $\boldsymbol{v} = 0$ and, therefore, $\nabla p = \nabla p_2 + \nabla p_2$.

2.1.9 Approximation by Smooth Fields

According to the fact that every function from the space $H_0^1(\Omega)$ is a limit of finite functions, it follows that the set of potential fields with finite potentials is dense in $\boldsymbol{G}_0(\Omega)$.

Let us now consider the approximation of fields from $\boldsymbol{J}_0(\Omega)$ by smooth fields. First, let the region Ω be the interior of a sphere B with radius R and centered at the origin. The given field $\boldsymbol{u} \in \boldsymbol{J}_0(B)$ is considered to be extended by zero outside the sphere B. Let $R_1 < R$ and let $\boldsymbol{a}$ be a vector with $|\boldsymbol{a}| < R - R_1$.

We define the field $\boldsymbol{u}_1(x) := \boldsymbol{u}[(R/R_1)(x - a)]$, which equals zero when $|\boldsymbol{x} - \boldsymbol{a}| > R_1$. We take an arbitrary potential field ∇q on the sphere B and consider the integral

$$\begin{aligned} I &:= \int_B \boldsymbol{u}_1(x) \cdot \nabla q(x) \mathrm{d}B = \int_B \boldsymbol{u}\left(\frac{R}{R_1}(x-a)\right) \cdot \nabla q(x) \mathrm{d}B \\ &= \int_{B_1} \boldsymbol{u}\left(\frac{R}{R_1}(x-a)\right) \cdot \nabla q(x) \mathrm{d}B, \end{aligned}$$

where $B_1 = \boldsymbol{a} + (R/R_1)B$. If we perform the substitution $(R/R_1)(x-a) = y$, then

$$I = \left(\frac{R_1}{R}\right)^3 \int_B \boldsymbol{u}(y) \cdot \nabla_x q\left(a + \frac{R_1}{R} y\right) \mathrm{d}B = \left(\frac{R_1}{R}\right)^2 \int \boldsymbol{u}(y) \cdot \nabla_y q_1(y) \mathrm{d}B,$$

where $q_1(y) = q[a + (R_1/R)y]$. Since $\boldsymbol{u} \in \boldsymbol{J}_0(B)$, this field is orthogonal to all potential fields in B and, therefore, $I = 0$ for any $q \in H_{\partial B}^1(B)$. That means that the field $\boldsymbol{u}_1(x)$ is solenoidal and, according to its construction, it equals zero on a neighborhood of the boundary of the sphere B (for $|\boldsymbol{x}| > |\boldsymbol{a}| + R$).

Let us now introduce an averaging function $\rho(x)$ with its support in the sphere $|\boldsymbol{x}| < (R - R_1)/2$ and construct the averaged field

$$\boldsymbol{u}_{R_1}(x) = \int \rho(z) \boldsymbol{u}\left(\frac{R}{R_1}(x-z)\right) \mathrm{d}\Omega_z = \int \rho(x-y) \boldsymbol{u}\left(\frac{R_1}{R} y\right) \mathrm{d}\Omega_y.$$

This field is solenoidal on B as a parameter integral of solenoidal fields on B, equals zero on $|\boldsymbol{x}| > (R_1 + R)/2$, and it is smooth by the smoothness of the averaging function $\rho(x)$. Similar to the scalar case, it can be proved that $\boldsymbol{u}_{R_1} \to \boldsymbol{u}$ in the space $\boldsymbol{L}_2(B)$ as $R_1 \to R$. Finally, it is proved that for every sphere $B \subset \Omega$, the set of solenoidal finite fields is dense in the space $\boldsymbol{J}_0(B)$.

Hence we have an important corollary:

If a field $\boldsymbol{v} \in \boldsymbol{L}_2(\Omega)$ is orthogonal to all finite solenoidal fields, then this field is potential.

Indeed, we note first that the field $\boldsymbol{v}$ is orthogonal to the solenoidal fields on any sphere $B \subset \Omega$ and that the set of all such fields is dense in $\boldsymbol{J}_0(B)$, hence, the field $\boldsymbol{v}$ is orthogonal to $\boldsymbol{J}_0(B)$. Then, as it was shown in Subsection 2.1.8, we get that the field $\boldsymbol{v}$ is potential in B. Further, from the fact that the field $\boldsymbol{v}$ is locally potential in the simply-connected region Ω, it follows that it is potential in Ω, too. This important result is a special case of the well-known DeRham Theorem and is stated here without a proof.

Finally, it can be proved that the set of all finite solenoidal fields on Ω is dense in the space $\boldsymbol{J}_0(\Omega)$. Indeed, if a field $\boldsymbol{v}$ from $\boldsymbol{J}_0(\Omega)$ is orthogonal to all the finite solenoidal fields, then it is potential by the above mentioned observations. Since the intersection of $\boldsymbol{J}_0(\Omega)$ and $\boldsymbol{G}(\Omega)$ consists only of zero, it follows that $\boldsymbol{v} = 0$. Hence the required assertion follows.

2.1.10 The Space of Velocity Fields for an Ideal Fluid in an Open Container

If an ideal fluid is placed into a closed container, then the boundary of the region filled with fluid consists of two parts: a solid boundary S and a free surface whose interior is denoted by Γ. The nonleaking condition $\boldsymbol{u}_n = 0$ holds true on the solid boundary S, but it does not hold true again on the free surface. Consequently, the space of all the velocity fields of the fluid movement in an open container is wider than $\boldsymbol{J}_0(\Omega)$. The needed extension of the set $\boldsymbol{J}_0(\Omega)$ can be obtained by adding those potential harmonic fields for which the normal component of the field—i.e., the normal derivative of the potential—equals zero on S.

Let us recall that the space $H^1_{\mathrm{h},S}(\Omega)$ of all such harmonic potentials was introduced in Example 3 of Subsection 1.8.6. The orthogonal decomposition

$$H^1_\Gamma(\Omega) = H^1_{0,\Gamma}(\Omega) \oplus H^1_{\mathrm{h},S}(\Omega) \tag{1.19}$$

stated there as formula (1.8.35) involves the space $H^1_{0,\Gamma}(\Omega)$ of potentials in $H^1_\Gamma(\Omega)$ with zero mean value on Γ and equipped with the Dirichlet norm.

If $\Gamma \subset \partial\Omega$ has positive measure, then the potential fields in $\boldsymbol{G}(\Omega)$ can be normed by making the mean of their potentials equal to zero on Γ and not on $\partial\Omega$. The orthogonal decomposition of potential fields in $\boldsymbol{L}_2(\Omega)$ that corresponds to the decomposition (1.19) is

$$\boldsymbol{G}(\Omega) = \boldsymbol{G}_{0,\Gamma}(\Omega) \oplus \boldsymbol{G}_{\mathrm{h},S}(\Omega). \tag{1.20}$$

Here, $\boldsymbol{G}_{0,\Gamma}(\Omega)$ is the subspace of all potential fields with a potential that equals zero on Γ. It is easy to see that

$$\boldsymbol{G}_{0,\Gamma}(\Omega) = \boldsymbol{G}_0(\Omega) \oplus \boldsymbol{G}_{0,\Gamma,\mathrm{h}}(\Omega), \tag{1.21}$$

where $\boldsymbol{G}_{0,\Gamma,\mathrm{h}}(\Omega)$ is the set of all harmonic fields with a potential that equals zero on Γ.

The subspace $\boldsymbol{G}_{\mathrm{h},S}(\Omega)$ in (1.20) consists of all harmonic fields for which the normal component of the field equals zero on S. It is exactly this particular subspace that needs to be added to $\boldsymbol{J}_0(\Omega)$ in order to obtain the whole set $\boldsymbol{J}_{0,S}(\Omega)$ of all solenoidal fields with zero normal component on S, that is,

$$\boldsymbol{J}_{0,S}(\Omega) = \boldsymbol{J}_0(\Omega) \oplus \boldsymbol{G}_{\mathrm{h},S}(\Omega). \tag{1.22}$$

The Weyl decomposition (1.18) and the decompositions (1.20)–(1.22) generate the decomposition

$$\boldsymbol{L}_2(\Omega) = \boldsymbol{J}_0(\Omega) \oplus \boldsymbol{G}_{\mathrm{h},S}(\Omega) \oplus \boldsymbol{G}_{0,\Gamma,\mathrm{h}}(\Omega) \oplus \boldsymbol{G}_0(\Omega). \tag{1.23}$$

Further, in addressing problems of the fluid movement in an open container, it will be more convenient to use the decomposition

$$\boldsymbol{L}_2(\Omega) = \boldsymbol{J}_0(\Omega) \oplus \boldsymbol{G}_{\mathrm{h},S}(\Omega) \oplus \boldsymbol{G}_{0,\Gamma}(\Omega), \tag{1.24}$$

where $\boldsymbol{G}_{0,\Gamma}(\Omega)$ is either defined as in (1.21), or even by the decomposition

$$\boldsymbol{L}_2(\Omega) = \boldsymbol{J}_{0,S}(\Omega) \oplus \boldsymbol{G}_{0,\Gamma}(\Omega). \tag{1.25}$$

2.1.11 Systems of Nonmixing Fluids

Let Ω be a closed immovable container filled with a system of m nonmixing fluids placed one on top of another—similar to a puffpastry—and such that the fluid with a bigger density ρ_1 is on the bottom (with respect to the gravitational acceleration $\boldsymbol{g}$) and the fluid with the smallest density ρ_m is on top. In other words, the inequalities $\rho_1 > \cdots > \rho_k > \cdots > \rho_m > 0$ hold true. We denote the region occupied by the fluid with density ρ_k in a stable position by Ω_k. The boundary of the container Ω will be denote as before by S and the parts of the boundary attached to the region Ω_k by S_k, $k = 1, \ldots, m$. The normals $\boldsymbol{n}_i$ to the boundaries Γ_i between the ith and the $(i+1)$th fluid are oriented inside the $(i+1)$th fluid ($i = 1, \ldots, m-1$).

The kinetic energy T corresponding to the velocity field $\hat{\boldsymbol{u}} := \{\boldsymbol{u}^1, \ldots, \boldsymbol{u}^m\}$ of the movement of the whole system of fluids has obviously the form

$$2T = \sum_{k=1}^{m} \rho_k \int_{\Omega_k} |\boldsymbol{u}^k|^2 \mathrm{d}\Omega_k, \tag{1.26}$$

where $\boldsymbol{u}^k = \sum_{j=1}^{3} u_j^k(x)\boldsymbol{e}_j$ is the velocity field on the region Ω_k. If the kinetic energy is finite, the right side of (1.26) can be looked upon as a squared norm on a Hilbert space and we will denote that space by $\hat{\boldsymbol{L}}_2(\Omega)$ (similarly to the case of one fluid). The sets $\hat{\boldsymbol{u}} = \{\boldsymbol{u}^k\}_{k=1}^m$ of the vector fields for which the right side of (1.26) is finite are elements in this space. Obviously, $\hat{\boldsymbol{L}}_2(\Omega)$ is a direct sum of the Hilbert spaces $\boldsymbol{L}_2(\Omega_k)$,

$$\hat{\boldsymbol{L}}_2(\Omega) := \bigoplus_{k=1}^{m} \boldsymbol{L}_2(\Omega_k),$$

for which the Weyl decomposition (1.18) is preserved, where

$$\hat{\boldsymbol{J}}_0(\Omega) = \bigoplus_{k=1}^{m} \boldsymbol{J}_0(\Omega_k), \qquad \hat{\boldsymbol{G}}_0(\Omega) = \bigoplus_{k=1}^{m} \boldsymbol{G}_0(\Omega_k), \qquad \hat{\boldsymbol{G}}_{\mathrm{h}}(\Omega) = \bigoplus_{k=1}^{m} \boldsymbol{G}_{\mathrm{h}}(\Omega_k).$$

Here the orthogonality is understood in the sense of the scalar product

$$(\hat{\boldsymbol{u}}, \hat{\boldsymbol{v}})_{\hat{\boldsymbol{L}}_2(\Omega)} := \sum_{k=1}^{m} \rho_k \int_{\Omega_k} \boldsymbol{u}^k(x) \cdot \boldsymbol{v}^k(x) \mathrm{d}\Omega_k,$$

generated by the norm (1.26).

Let us consider now sets of potential fields $\{\boldsymbol{v}^k\}_{k=1}^m = \widehat{\nabla p} := \{\nabla p_k\}_{k=1}^m$. For regions of the first type, every function p_k is an element in the space $H^1(\Omega_k)$ and the set $\hat{p} := \{p_k\}_{k=1}^m$ of all these functions is an element in the space $\bigoplus_{k=1}^{m} H^1(\Omega_k) =: \hat{H}^1(\Omega)$. The Dirichlet norm generates the following seminorm

$$\left(\sum_{k=1}^{m} \rho_k \int_{\Omega_k} |\nabla p_k|^2 \mathrm{d}\Omega_k \right)^{1/2} \tag{1.27}$$

on the space $\hat{H}^1(\Omega)$. This seminorm equals zero on the m-dimensional subspace consisting of the sets of constants $\{c_k\}_{k=1}^m$.

If we consider a direct complement to this subspace, then on that complement the norm of the space $\hat{H}^1(\Omega)$ is equivalent to the norm (1.27). Such a direct complement can be obtained by selecting a space of potentials in every space $H^1(\Omega_k)$ and then taking their orthogonal sum.

However, we would need yet another construction that "bonds" the spaces $H^1(\Omega_k)$ to each other. The subspace $H^1_\Gamma(\Omega)$ consisting of all the sets $\hat{p} \in \hat{H}^1(\Omega)$ for which the following equalities hold true,

$$\begin{aligned} \int_{\Gamma_i} (\rho_i p_i - \rho_{i+1} p_{i+1}) \mathrm{d}\Gamma_i = 0, \quad i = 1, \ldots, m-1, \\ \int_{\Gamma_{m-1}} \rho_m p_m \mathrm{d}\Gamma_{m-1} = 0, \end{aligned} \tag{1.28}$$

is an important example of such a construction.

If a set $\{c_k\}_{k=1}^m$ belongs to this subspace, then

$$\begin{aligned} \rho_i c_i - \rho_{i+1} c_{i+1} = 0, \quad i = 1, \ldots, m-1, \\ \rho_m c_m = 0, \end{aligned}$$

hence $c_k = 0$ for every $k = 1, \ldots, m$.

If $\hat{p} = \{p_k\}_{k=1}^m$ is an arbitrary set in $\hat{H}^1(\Omega)$, then let us denote

$$\begin{aligned} \int_{\Gamma_i} (\rho_i p_i - \rho_{i+1} p_{i+1}) \mathrm{d}\Gamma_i =: \beta_i, \quad i = 1, \ldots, m-1, \\ \int_{\Gamma_{m-1}} \rho_m p_m \mathrm{d}\Gamma_{m-1} =: \beta_m, \end{aligned}$$

and form the next system of equations with respect to the constants α_k:

$$\begin{aligned} \int_{\Gamma_i} (\rho_i \alpha_i - \rho_{i+1} \alpha_{i+1}) \mathrm{d}\Gamma_i = \beta_i, \\ \int_{\Gamma_{m-1}} \rho_m \alpha_m \mathrm{d}\Gamma_{m-1} = \beta_m. \end{aligned}$$

Since the corresponding homogeneous system has the zero solution only, the nonhomogeneous system has a unique solution $\{\alpha_k\}_{k=1}^m$. Subtracting the constants α_k from the functions p_k we obtain that the set $\{p_k - \alpha_k\}_{k=1}^m \in \hat{H}^1_\Gamma(\Omega)$.

If we consider the seminorm

$$\|\hat{p}\|^2 := \sum_{i=1}^{m-1} \left| \int_{\Gamma_i} (\rho_i p_i - \rho_{i+1} p_{i+1}) \mathrm{d}\Gamma_i \right|^2 + \left| \int_{\Gamma_m} \rho_m p_m \mathrm{d}\Gamma_{m-1} \right|^2,$$

then it turns out that this seminorm does not equal zero on nonzero sets of constants and, by the trace theorem, we get $\|\hat{p}\| \leq c \|\hat{p}\|_{\hat{H}^1(\Omega)}$. Therefore, we can introduce an equivalent norm on the space $\hat{H}_\Gamma(\Omega)$:

$$\begin{aligned} \|\hat{p}\|^2_{\hat{H}^1_\Gamma(\Omega)} := \sum_{k=1}^m \rho_k \int_{\Omega_k} |\nabla p_k|^2 \mathrm{d}\Omega_k + \sum_{i=1}^{m-1} \left| \int_{\Gamma_i} (\rho_i p_i - \rho_{i+1} p_{i+1} \mathrm{d}\Gamma_i \right|^2 \\ + \left| \int_{\Gamma_m} \rho_m p_m \mathrm{d}\Gamma_{m-1} \right|^2. \end{aligned}$$

This norm is equivalent to the norm generated by the Dirichlet form (1.27) on the subspace $\hat{H}^1_\Gamma(\Omega)$.

We use now the abstract scheme developed in Subsection 1.8.5. We take the space $\hat{H}^1_\Gamma(\Omega)$ as F, the space $\hat{L}_{2,\Gamma} := \bigoplus_{i=1}^{m-1} L_{2,\Gamma_i}(\Gamma_i)$ as G, and let the operator γ be defined by the equation

$$\gamma\hat{p} := \{(\rho_i p_i - \rho_{i+1}p_{i+1})|_{\Gamma_i}\}_{i=1}^{m-1}, \qquad \hat{p} \in \hat{H}^1_\Gamma(\Omega).$$

According to the trace theorem, this operator acts boundedly from the space $\hat{H}^1_\Gamma(\Omega)$ to the space $\hat{L}^1_{2,\Gamma}(\Omega)$. The kernel of the operator γ is the subspace $\hat{H}^1_{0,\Gamma}(\Omega)$ of the space $\hat{H}^1_\Gamma(\Omega)$ consisting of all sets $\hat{p} = \{p_k\}_{k=1}^m \in \hat{H}^1_\Gamma(\Omega)$ for which

$$\rho_i p_i - \rho_{i+1}p_{i+1} = 0, \quad \text{on } \Gamma_i, \qquad i = 1, \dots, m-1. \tag{1.29}$$

Let us next find out how the orthogonal complement to the subspace $\hat{H}^1_{0,\Gamma}(\Omega)$ in the space $\hat{H}^1_\Gamma(\Omega)$ looks like. We note that $\hat{H}^1_{0,\Gamma}(\Omega)$ includes the subspace $\hat{H}^1_0(\Omega) := \bigoplus_{k=1}^{m} \hat{H}^1_0(\Omega_k)$ and that is why the orthogonal complement to $\hat{H}^1_{0,\Gamma}(\Omega)$ is contained into the orthogonal complement to $\hat{H}^1_0(\Omega)$ that consists of sets of harmonic potentials. Now, let $\hat{q} = \{q_k\}_{k=1}^m$ be a set orthogonal to $\hat{H}^1_{0,\Gamma}(\Omega)$. Then, for any $\hat{p} \in \hat{H}^1_{0,\Gamma}(\Omega)$ we have

$$\begin{aligned} 0 &= \sum_{k=1}^{m} \rho_k \int_{\Omega_k} \nabla p_k \cdot \nabla q_k \mathrm{d}\Omega_k \\ &= \sum_{k=1}^{m} \rho_k \int_{S_k} p_k \frac{\partial q_k}{\partial n} \mathrm{d}S_k + \sum_{i=1}^{m-1} \int_{\Gamma_i} \left(\rho_i p_i \frac{\partial q_i}{\partial n_i} - \rho_{i+1}p_{i+1}\frac{\partial q_{i+1}}{\partial n_i} \right) \mathrm{d}\Gamma_i, \end{aligned} \tag{1.30}$$

by Green formula for the regions Ω_k and by taking into account that $\Delta q_k = 0$. Here we used the fact that the normal $\boldsymbol{n}_i$ is directed outward the region Ω_i on the boundary Γ_i and inward on the boundary Γ_{i-1}.

Let now all the functions p_k for $k \neq k_0$ be equal to zero and let the function p_{k_0} be finite with support in $\overline{\Omega}_{k_0}$ and equal to zero on Γ_{k_0-1} and Γ_{k_0} (we assume that $\Gamma_0 = \emptyset$). Then by (1.30) it follows that $\partial q_{k_0}/\partial n = 0$ on S_{k_0}. Similarly, we obtain that

$$\frac{\partial q_k}{\partial n} = 0 \quad \text{on } S_k, \qquad k = 1, \dots, m. \tag{1.31}$$

Therefore, equation (1.30) can be transformed into

$$\sum_{i=1}^{m-1} \int_{\Gamma_i} \left(\rho_i p_i \frac{\partial q_i}{\partial n_i} - \rho_{i+1}p_{i+1}\frac{\partial q_{i+1}}{\partial n_i} \right) \mathrm{d}\Gamma_i = 0. \tag{1.32}$$

Further, we choose a finite function φ on the surface Γ_{i_0}, set $p_{i_0+1} = \rho_{i_0}\varphi$ and $p_{i_0} = \rho_{i_0+1}\varphi$ on Γ_{i_0} ($\rho_0 = 0$ and $p_0 = 0$) and extend these functions to finite functions with supports in $\overline{\Omega}_{i_0+1}$ and $\overline{\Omega}_{i_0}$, respectively, and, finally, take $p_i = 0$ for $i \neq i_0 + 1$ and $i \neq i_0$. Thus, we get a set $\hat{p} = \{p_k\}_{k=1}^m \in \hat{H}^1_{0,\Gamma}(\Omega)$. Using this specific $\hat{p}$, (1.32) is reduced to but one term with the index $i = i_0$ and since the finite function φ is arbitrary we have that

$$\frac{\partial q_{i_0}}{\partial n_{i_0}} - \frac{\partial q_{i_0+1}}{\partial n_{i_0}} = 0.$$

Similarly, we can prove that

$$\frac{\partial q_i}{\partial n_i} = \frac{\partial q_{i+1}}{\partial n_i} \quad \text{on } \Gamma_i, \qquad i = 1, \ldots, m-1. \tag{1.33}$$

Conversely, if a set of harmonic functions $\{q_k\}_{k=1}^m$ satisfies conditions (1.31) and (1.33), then by Green formula and by equations (1.33) we have that

$$\sum_{k=1}^m \rho_k \int_{\Omega_k} \nabla p_k \cdot \nabla q_k \mathrm{d}\Omega_k = 0,$$

i.e., the sets $\{p_k\}_{k=1}^m \in \hat{H}^1_{0,\Gamma}(\Omega)$ and $\{q_k\}_{k=1}^m$ are orthogonal to each other.

To conclude, the space $\hat{H}^1_{\mathrm{h},S}(\Omega)$ defined as the orthogonal complement of $\hat{H}^1_{0,\Gamma}(\Omega)$ in $\hat{H}^1_\Gamma(\Omega)$ consists of all the sets of harmonic functions $\{q_k\}_{k=1}^m$ satisfying conditions (1.31) and (1.33):

$$\hat{H}^1_\Gamma(\Omega) = \hat{H}^1_{0,\Gamma}(\Omega) \bigoplus \hat{H}^1_{\mathrm{h},S}(\Omega). \tag{1.34}$$

From the results in Section 1.8, it follows that the operator γ yields a one-to-one mapping from the subspace $\hat{H}^1_{\mathrm{h},S}(\Omega)$ onto its range. Since for the elements $(\gamma\hat{p})_i$ satisfy the conditions

$$\int_{\Gamma_i} (\gamma\hat{p})_i \mathrm{d}\Gamma_i = 0, \qquad i = 1, \ldots, m-1$$

for any $\hat{p} \in \hat{H}^1_\Gamma(\Omega)$, the range of the operator γ is contained into the space $\bigoplus_{i=1}^{m-1} H^{1/2}_{\Gamma_i}(\Gamma_i)$. Further, if a set $\{\varphi_i\}_{i=1}^{m-1}$ belongs to this space, then by choosing an arbitrary function p_m in the space $H^{1/2}_{\Gamma_{m-1}}(\Gamma_{m-1})$ one can find all the functions p_i and p_{i+1} on Γ_i that appear in equations (1.29). They belong to the subspace $H^{1/2}_{\Gamma_i}(\Gamma_i)$. Then we can extend these functions to the regions Ω_k so that the extensions belong to the spaces $H^1(\Omega_k)$, respectively. The resulting set $\hat{p} = \{p_k\}_{k=1}^m$ has the property $\gamma\hat{p} = \{\varphi_i\}_{i=1}^{m-1}$.

Since operator γ is an isometry, we conclude that the following problem

$$\begin{cases} \Delta q_k = 0 \quad \text{on } \Omega_k, \quad \dfrac{\partial q_k}{\partial n} = 0 \quad \text{on } S_k, & k = 1, \dots, m, \\ \rho_i q_i - \rho_{i+1} q_{i+1} = \varphi_i, \quad \dfrac{\partial q_i}{\partial n_i} = \dfrac{\partial q_{i+1}}{\partial n_i} \quad \text{on } \Gamma_i, & i = 1, \dots, m-1, \\ \displaystyle\int_{\Gamma_i} (\rho_i q_i - \rho_{i+1} q_{i+1}) d\Gamma_i = 0, & i = 1, \dots, m-1, \\ \displaystyle\int_{\Gamma_{m-1}} \rho_m q_m d\Gamma_{m-1} = 0, \end{cases} \tag{1.35}$$

is solvable and has a unique solution for any set $\{\varphi_i\}_{i=1}^{m-1} \in \hat{H}_\Gamma^{1/2}(\Gamma) := \bigoplus_{i=1}^{m-1} H_{\Gamma_i}^{1/2}(\Gamma_i)$, i.e., for any set $\{\varphi_i\}_{i=1}^{m-1} \in \hat{H}^{1/2}(\Gamma)$ with

$$\int_{\Gamma_i} \varphi d\Gamma_i = 0, \qquad i = 1, \dots, m-1.$$

Now let us consider the operator ∂. By Green formula and (1.31) we obtain

$$(\hat{p}, \hat{q})_{\hat{H}^1(\Omega)} = \sum_{k=1}^{m} \rho_k \int_{\Omega_k} \nabla p_k \cdot \nabla q_k d\Omega_k = \sum_{i=1}^{m-1} \int_{\Gamma_i} \left(\rho_i q_i \frac{\partial p_i}{\partial n_i} - \rho_{i+1} q_{i+1} \frac{\partial p_{i+1}}{\partial n_i} \right) d\Gamma_i$$

for any sets $\{p_k\}_{k=1}^m$ and $\{q_k\}_{k=1}^m$ in the space $\hat{H}^1_{\text{h},S}(\Omega)$.

Further, by (1.33) we have

$$(\hat{p}, \hat{q})_{\hat{H}^1(\Omega)} = \sum_{i=1}^{m-1} \int_{\Gamma_i} \frac{\partial p_i}{\partial n_i} (\rho_i q_i - \rho_{i+1} q_{i+1}) d\Gamma_i. \tag{1.36}$$

According to the definition of operator ∂ indicated in (1.8.22) and using the operator γ introduced earlier we get

$$(\hat{p}, \hat{q})_{\hat{H}^1(\Omega)} = \sum_{i=1}^{m-1} \int_{\Gamma_i} (\partial \hat{p})_i (\gamma \hat{q})_i d\Gamma_i = \sum_{i=1}^{m-1} \int_{\Gamma_i} (\partial \hat{p})_i (\rho_i q_i - \rho_{i+1} q_{i+1}) d\Gamma_i \tag{1.37}$$

Comparing (1.36) and (1.37) we conclude that

$$\partial \hat{p} = \partial \{p_k\}_{k=1}^m = \left\{ \frac{\partial p_i}{\partial n_i}\bigg|_{\Gamma_i} \right\}_{i=1}^{m-1}. \tag{1.38}$$

Since operator ∂ is an isomorphism, we get that the following problem

$$\begin{cases} \Delta p_k = 0 \quad \text{on } \Omega_k, \quad \dfrac{\partial p_k}{\partial n} = 0 \quad \text{on } S_k & k = 1, \dots, m, \\ \dfrac{\partial p_i}{\partial n_i} = \dfrac{\partial p_{i+1}}{\partial n_i} = \psi_i \quad \text{on } \Gamma_i & i = 1, \dots, m-1, \\ \displaystyle\int_{\Gamma_i} (\rho_i p_i - \rho_{i+1} p_{i+1}) \mathrm{d}\Gamma_i = 0, & i = 1, \dots, m-1, \\ \displaystyle\int_{\Gamma_{m-1}} \rho_m p_m \mathrm{d}\Gamma_{m-1} = 0, \end{cases} \tag{1.39}$$

is solvable and has a unique solution for any set $\{\psi_i\}_{i=1}^{m-1} \in \bigoplus_{i=1}^{m-1} (H_{\Gamma_i}^{1/2}(\Gamma_i))^*$.

The physical connotation of problem (1.39) is that the potentials of the displacements p_k of fluid particles in every region Ω_k involved in this problem can be restored by means of the displacements ψ_i of the boundaries Γ_i between the fluids. In regard with (1.35), that problem can be considered as the problem of finding the acceleration potentials q_k in the regions Ω_k by means of the functions φ_i that equal the difference of pressures $\rho_i q_i - \rho_{i+1} q_{i+1}$ on the boundaries Γ_i. We recall that by the linearized Euler equations for an ideal fluid, the acceleration potential $q(t,x)$ and the pressure in the fluid $p(t,x)$ are connected by the relation $p = -\rho q$.

2.1.12 Spaces of Velocity Fields for Systems of Nonmixing Ideal Fluids

We next describe the subspace in $\hat{\boldsymbol{L}}_2(\Omega)$ corresponding to sets of velocity fields $\{\boldsymbol{u}^k\}_{k=1}^m$ of the movement of a system of nonmixing incompressible ideal fluids. For such a set, we have the solenoidal conditions $\operatorname{div} \boldsymbol{u}^k = 0$ in every region Ω_k, the nonleaking conditions $(\boldsymbol{u}^k)_n = 0$ on each part of the boundary S_k, and the nonleaking conditions of one fluid into another that can be stated as

$$(\boldsymbol{u}^{i+1} - \boldsymbol{u}^i) \cdot \boldsymbol{n}_i = 0 \quad \text{on } \Gamma_i, \qquad i = 1, \dots, m-1. \tag{1.40}$$

The sets of fields from the subspace $\hat{\boldsymbol{J}}_0(\Omega) := \bigoplus_{k=1}^m \boldsymbol{J}_0(\Omega_k)$ satisfy all these conditions. For such fields the boundaries between different fluids are the solid walls. It seems appropriate to introduce some more sets of potential fields $\hat{\boldsymbol{v}} = \{\boldsymbol{v}^k\}_{k=1}^m = \{\nabla p_k\}_{k=1}^m$. By the solenoidal conditions it follows that the functions p_k are harmonic. The nonleaking conditions will then become

$$\frac{\partial p_k}{\partial n} = 0 \quad \text{on } S_k, \qquad k = 1, \dots, m,$$

while the inseparability conditions will take the form

$$\frac{\partial p_{i+1}}{\partial n_i} - \frac{\partial p_i}{\partial n_i} = 0 \quad \text{on } \Gamma_i, \qquad i = 1, \dots, m-1.$$

Since the potentials p_k are determined up to a constant, by the above-mentioned facts it follows that all the sets $\{p_k\}_{k=1}^m$ with a certain choice of constants form the space $\hat{H}^1_{\mathrm{h},S}(\Omega)$.

Let us denote

$$\hat{\boldsymbol{G}}_{\mathrm{h},S}(\Omega) := \{\hat{\boldsymbol{v}} = \{\nabla q_k\}_{k=1}^m : \{q_k\}_{k=1}^m \in \hat{H}^1_{\mathrm{h},S}(\Omega)\}. \tag{1.41}$$

By the Weyl decomposition, the subspaces $\hat{\boldsymbol{J}}_0(\Omega)$ and $\hat{\boldsymbol{G}}_{\mathrm{h},S}(\Omega)$ are mutually orthogonal in $\hat{\boldsymbol{L}}_2(\Omega)$. It is quite natural to assume that the space

$$\hat{\boldsymbol{J}}_{0,S}(\Omega) := \hat{\boldsymbol{J}}_0(\Omega) \oplus \hat{\boldsymbol{G}}_{\mathrm{h},S}(\Omega) \tag{1.42}$$

consists of all the sets of velocity fields for the movement of a system of nonmixing ideal fluids.

From the orthogonal decompositions (1.34) and (1.42) we finally obtain

$$\hat{\boldsymbol{L}}_2(\Omega) = \hat{\boldsymbol{G}}_{0,\Gamma}(\Omega) \oplus \hat{\boldsymbol{G}}_{\mathrm{h},S}(\Omega) \oplus \hat{\boldsymbol{J}}_0(\Omega), \tag{1.43}$$

where

$$\hat{\boldsymbol{G}}_{0,\Gamma}(\Omega) := \{\hat{\boldsymbol{u}} = \{\nabla p_k\}_{k=1}^m : \{p_k\}_{k=1}^m \in \hat{H}^1_{0,\Gamma}(\Omega)\}. \tag{1.44}$$

2.2 Spaces and Hydrodynamics Operators for a Viscous Fluid

In this section we introduce the basic Hilbert spaces of fields pertaining to the movement of a viscous incompressible fluid in closed or open containers. The norm in these spaces is derived from the form representing the velocity of energy dissipation. We will also consider the vector Laplace operator and deduce the Green formula for it. Further, we will investigate the Korn identity and the Korn inequality that play an important role in our study. With the aid of the abstract scheme explained in Section 1.3, we will next introduce the fundamental operator of a viscous fluid hydrodynamics, the Stokes operator, for which we will subsequently investigate the main boundary value problems.

2.2.1 Forces of Internal Friction. Energy Dissipation

When a viscous fluid is moving, the surface forces of friction between the layers of the moving fluid are proportional to the gradients of velocities. Therefore, it is natural to assume that the velocity field of a viscous fluid has differentiable components. In this regard, let us consider for such a field $\boldsymbol{u}(x)$ that $u_i(x) \in H^1(\Omega)$, $i = 1, 2, 3$. The space of all such fields is denoted by $\boldsymbol{H}^1(\Omega)$. The norm on this space is introduced by the formula

$$\|\boldsymbol{u}\|^2_{\boldsymbol{H}^1(\Omega)} = \sum_{i=1}^{3} \|u_i\|^2_{H^1(\Omega)}.$$

Equivalent norms on $H^1(\Omega)$ generate equivalent norms on $\boldsymbol{H}^1(\Omega)$.

When a homogeneous viscous fluid is moving, the tensor of tensions due to the viscous forces has the components

$$\mu\left(\frac{\partial u_i}{\partial x_j} + \frac{\partial u_j}{\partial x_i}\right) =: \mu\tau_{ij},$$

where μ is a positive constant called the *coefficient of viscousity*. The presence of viscous forces leads to energy dissipation and its velocity is defined by the formula

$$\mu E(\boldsymbol{u}, \boldsymbol{u}) := \frac{1}{2}\mu \int_\Omega \sum_{i,j=1}^{3} |\tau_{ij}|^2 \mathrm{d}\Omega. \tag{2.1}$$

It is obvious that

$$E(\boldsymbol{u}, \boldsymbol{u}) \le 2 \int_\Omega \sum_{i,j=1}^{3} \left|\frac{\partial u_i}{\partial x_j}\right|^2 \mathrm{d}\Omega \le 2\|\boldsymbol{u}\|^2_{\boldsymbol{H}^1(\Omega)}. \tag{2.2}$$

The bilinear form that corresponds to the quadratic form $E(\boldsymbol{u}, \boldsymbol{u})$ is given by

$$E(\boldsymbol{u}, \boldsymbol{v}) = \frac{1}{2} \int_\Omega \sum_{i,j=1}^{3} \left(\frac{\partial u_i}{\partial x_j} + \frac{\partial u_j}{\partial x_i}\right)\left(\frac{\partial v_i}{\partial x_j} + \frac{\partial v_j}{\partial x_i}\right) \mathrm{d}\Omega, \tag{2.3}$$

and it can be defined for any pair $\boldsymbol{u}$, $\boldsymbol{v}$ in $\boldsymbol{H}^1(\Omega)$. The quantity $\mu E(\boldsymbol{u}, \boldsymbol{v})$ has the following physical meaning: it is the work done in a unit of time by the viscous forces that correspond to the velocity field $\boldsymbol{u}$ on the possible displacement determined by the velocity field $\boldsymbol{v}$. This fact is evident from the representation

$$\mu E(\boldsymbol{u}, \boldsymbol{v}) = \mu \int_\Omega \sum_{i,j=1}^{3} \left(\frac{\partial u_i}{\partial x_j} + \frac{\partial u_j}{\partial x_i}\right) \frac{\partial v_i}{\partial x_j} \mathrm{d}\Omega. \tag{2.4}$$

2.2.2 Divergence Operator. Solenoidal Fields

The operator div in the space $\boldsymbol{H}^1(\Omega)$ can be found using the simple formula $\operatorname{div}\boldsymbol{u} = \sum_{i=1}^{3} \partial u_i/\partial x_i$, that follows from definition (1.15) and Ostrogradsky formula. This operator acts boundedly from the space $\boldsymbol{H}^1(\Omega)$ into the space $L_2(\Omega)$.

Let us denote the subspace of $\boldsymbol{H}^1(\Omega)$ that consists of all its solenoidal fields by $\boldsymbol{J}^1(\Omega)$.

Let $\boldsymbol{u} \in \boldsymbol{J}^1(\Omega)$ be given and let us consider the field $\partial\boldsymbol{u}/\partial x_i$ that belongs to $\boldsymbol{L}_2(\Omega)$. For any finite function $q(x)$ on Ω we have

$$\begin{aligned}\int_\Omega q\operatorname{div}\frac{\partial\boldsymbol{u}}{\partial x_i}\mathrm{d}\Omega &= -\int_\Omega \nabla q\cdot\frac{\partial\boldsymbol{u}}{\partial x_i}\mathrm{d}\Omega = \int_\Omega \nabla\frac{\partial q}{\partial x_i}\cdot\boldsymbol{u}\mathrm{d}\Omega\\ &= -\int_\Omega \frac{\partial q}{\partial x_i}\operatorname{div}\boldsymbol{u}\mathrm{d}\Omega. \end{aligned}\tag{2.5}$$

Here the definition of div was used twice; we also used integration by parts and the finiteness of the function $\partial q/\partial x_i$. Finally, we get

$$\int_\Omega q\operatorname{div}\frac{\partial\boldsymbol{u}}{\partial x_i}\mathrm{d}\Omega = 0.$$

Since the finite functions are dense in $H_0^1(\Omega)$ and $\operatorname{div}(\partial\boldsymbol{u}/\partial x_i) \in H^{-1}(\Omega)$, the last equality amounts to

$$\operatorname{div}\frac{\partial\boldsymbol{u}}{\partial x_i} = 0. \tag{2.6}$$

Hence, the operator $\partial/\partial x_i$ maps the space $\boldsymbol{J}^1(\Omega)$ to the space $\boldsymbol{J}(\Omega)$.

For any field $\boldsymbol{u}$ in $\boldsymbol{H}^1(\Omega)$ we can define a vector $\operatorname{rot}\boldsymbol{u} = \nabla\times\boldsymbol{u}$, that has the following projections onto the coordinate axes:

$$\begin{aligned}(\operatorname{rot}\boldsymbol{u})_1 &= \frac{\partial u_3}{\partial x_2} - \frac{\partial u_2}{\partial x_3},\\ (\operatorname{rot}\boldsymbol{u})_2 &= \frac{\partial u_1}{\partial x_3} - \frac{\partial u_3}{\partial x_1},\\ (\operatorname{rot}\boldsymbol{u})_3 &= \frac{\partial u_2}{\partial x_1} - \frac{\partial u_1}{\partial x_2}.\end{aligned}$$

The field $\operatorname{rot}\boldsymbol{u}$ is solenoidal. Indeed, for any finite function q we have

$$\int_\Omega q\operatorname{div}(\operatorname{rot}\boldsymbol{u})\mathrm{d}\Omega = -\int_\Omega \nabla q\cdot\operatorname{rot}\boldsymbol{u}\mathrm{d}\Omega = \int_\Omega \operatorname{rot}\nabla q\cdot\boldsymbol{u}\mathrm{d}\Omega = 0.$$

This shows that

$$\operatorname{div}\operatorname{rot}\boldsymbol{u} = 0. \tag{2.7}$$

2.2.3 Vector Laplace Operator. Green Formula

The components u_i of every vector field $\boldsymbol{u} \in \boldsymbol{H}^1(\Omega)$ have partial derivatives belonging to $L_2(\Omega)$ hence we can apply the operator div to the field ∇u_i and get $\operatorname{div}\nabla u_i = \Delta u_i$. The field with the components Δu_i is denoted by $\Delta \boldsymbol{u}$. The operator $\Delta \boldsymbol{u}$ defined on $\boldsymbol{H}^1(\Omega)$ is called the *vector Laplace operator.* We next deduce an important identity for this operator. For a fixed i let us consider the field $\boldsymbol{\tau}_i(\boldsymbol{u})$ with the components $\tau_{ij} = \partial u_i/\partial x_j + \partial u_j/\partial x_i$, $j = 1, 2, 3$. If $q \in H_0^1(\Omega)$, then

$$\int_\Omega (\operatorname{div} \boldsymbol{\tau}_i) q \mathrm{d}\Omega = -\int_\Omega \boldsymbol{\tau}_i \cdot \nabla q \mathrm{d}\Omega = -\int_\Omega \sum_{j=1}^{3} \left(\frac{\partial u_i}{\partial x_j} + \frac{\partial u_j}{\partial x_i} \right) \frac{\partial q}{\partial x_j} \mathrm{d}\Omega$$

$$= -\int_\Omega \nabla u_i \cdot \nabla q \mathrm{d}\Omega - \int_\Omega \frac{\partial \boldsymbol{u}}{\partial x_i} \cdot \nabla q \mathrm{d}\Omega = \int_\Omega \Delta u_i \cdot q \mathrm{d}\Omega + \int_\Omega \operatorname{div} \frac{\partial \boldsymbol{u}}{\partial x_i} q \mathrm{d}\Omega.$$

Hence, we obtain the identity

$$\Delta u_i = \operatorname{div} \boldsymbol{\tau}_i - \operatorname{div} \frac{\partial \boldsymbol{u}}{\partial x_i}, \tag{2.8}$$

which is valid for any field $\boldsymbol{u}$ in $\boldsymbol{H}^1(\Omega)$. Let us note that all the components of this identity belong to the space $H^{-1}(\Omega)$.

In the particular case when $\operatorname{div} \boldsymbol{u} = 0$, by (2.6) the last term in (2.8) can be omitted, and we get

$$\Delta u_i = \operatorname{div} \boldsymbol{\tau}_i, \qquad u \in \boldsymbol{J}^1(\Omega),\ i = 1, 2, 3. \tag{2.9}$$

Now, let Ω be a region of the third type, $\boldsymbol{u} \in \boldsymbol{H}^2(\Omega) \cap \boldsymbol{J}(\Omega)$, and $\boldsymbol{v} \in \boldsymbol{H}^1(\Omega)$. Then $\operatorname{div} \boldsymbol{\tau}_i(\boldsymbol{u}) \in L_2(\Omega)$, hence we can use formula (1.11) and the identity (2.8) for the field $\boldsymbol{\tau}_i$. We have

$$\int_\Omega \boldsymbol{\tau}_i \cdot \nabla v_i \mathrm{d}\Omega = -\int_\Omega \operatorname{div} \boldsymbol{\tau}_i v_i \mathrm{d}\Omega + \int_{\partial\Omega} (\boldsymbol{\tau}_i)_n v_i \mathrm{d}S = -\int_\Omega \Delta u_i v_i \mathrm{d}\Omega + \int_{\partial\Omega} (\boldsymbol{\tau}_i)_n v_i \mathrm{d}S.$$

Adding by i we obtain the equality

$$\int_\Omega \sum_{i,j=1}^{3} \left(\frac{\partial u_i}{\partial x_j} + \frac{\partial u_j}{\partial x_i} \right) \frac{\partial v_i}{\partial x_j} \mathrm{d}\Omega = -\int_\Omega \Delta \boldsymbol{u} \cdot \boldsymbol{v} \mathrm{d}\Omega + \int_{\partial\Omega} \sum_{i=1}^{3} (\boldsymbol{\tau}_i)_n v_i \mathrm{d}S,$$

or, in short, by (2.4),

$$E(\boldsymbol{u}, \boldsymbol{v}) = -\int_\Omega \Delta \boldsymbol{u} \cdot \boldsymbol{v} \mathrm{d}\Omega + \int_{\partial\Omega} \sum_{i=1}^{3} (\boldsymbol{\tau}_i(\boldsymbol{u}))_n v_i \mathrm{d}S, \tag{2.10}$$

for any $\boldsymbol{u} \in \boldsymbol{H}^2(\Omega) \cap \boldsymbol{J}(\Omega)$, $\boldsymbol{v} \in \boldsymbol{H}^1(\Omega)$. This equality is called the *Green formula* for the vector Laplace operator.

In the particular case when $\Delta \boldsymbol{u} = \boldsymbol{0}$ we have

$$E(\boldsymbol{u}, \boldsymbol{v}) = \int_{\partial\Omega} \sum_{i=1}^{3} (\boldsymbol{\tau}_i(\boldsymbol{u}))_n v_i \mathrm{d}S. \tag{2.11}$$

2.2.4 Movement of a Viscous Fluid in a Closed Container. Korn Identity and Korn Inequality

If a viscous fluid wholly fills the immovable container Ω, then the stickness condition $\boldsymbol{u}|_S = 0$ holds true on the walls of the container $\partial\Omega =: S$.

For $\boldsymbol{u} \in \boldsymbol{H}^1(\Omega)$, let us consider the identities

$$\frac{1}{4}\left(\frac{\partial u_i}{\partial x_j} + \frac{\partial u_j}{\partial x_i}\right)^2 - \frac{1}{4}\left(\frac{\partial u_i}{\partial x_j} - \frac{\partial u_j}{\partial x_i}\right)^2 = \frac{\partial u_i}{\partial x_j} \cdot \frac{\partial u_j}{\partial x_i}, \qquad i,j = 1,2,3.$$

Introducing the notations $r_{ij} := \partial u_i/\partial x_j - \partial u_j/\partial x_i$ and adding up by j we obtain

$$\frac{1}{4}\sum_{j=1}^{3}|\tau_{ij}|^2 - \frac{1}{4}\sum_{j=1}^{3}|r_{ij}|^2 = \nabla u_i \cdot \frac{\partial \boldsymbol{u}}{\partial x_i}.$$

We now integrate this identity on the region Ω. Then

$$\frac{1}{4}\int_\Omega \sum_{j=1}^{3}|\tau_{ij}|^2 d\Omega - \frac{1}{4}\int_\Omega \sum_{j=1}^{3}|r_{ij}|^2 d\Omega = \int_\Omega \nabla u_i \cdot \frac{\partial \boldsymbol{u}}{\partial x_i} d\Omega.$$

If $\boldsymbol{u}|_S = 0$, then the right side of the above equality becomes

$$\int_\Omega \nabla u_i \cdot \frac{\partial \boldsymbol{u}}{\partial x_i} d\Omega = -\int_\Omega u_i \operatorname{div}\frac{\partial \boldsymbol{u}}{\partial x_i} d\Omega.$$

Finally, if $\operatorname{div}\boldsymbol{u} = 0$, then in virtue of (2.6), $\operatorname{div}(\partial\boldsymbol{u}/\partial x_i) = 0$ and the right side of (2.12) equals zero. Hence, we have

$$\int_\Omega \sum_{j=1}^{3}|\tau_{ij}|^2 d\Omega = \int_\Omega \sum_{j=1}^{3}|r_{ij}|^2 d\Omega. \tag{2.13}$$

Further, it is obvious that

$$\frac{1}{4}|\tau_{ij}|^2 + \frac{1}{4}|r_{ij}|^2 = \frac{1}{2}\left|\frac{\partial u_i}{\partial x_j}\right|^2 + \frac{1}{2}\left|\frac{\partial u_j}{\partial x_i}\right|^2.$$

Adding up by i and j and integrating, we obtain the identity

$$\begin{aligned}\int_\Omega \sum_{i,j=1}^{3}\left|\frac{\partial u_i}{\partial x_j}\right|^2 d\Omega &= \frac{1}{4}\int_\Omega \sum_{i,j=1}^{3}|\tau_{ij}|^2 d\Omega + \frac{1}{4}\int_\Omega \sum_{i,j=1}^{3}|r_{ij}|^2 d\Omega \\ &= \frac{1}{2}\int_\Omega \sum_{i,j=1}^{3}|\tau_{ij}|^2 d\Omega = E(\boldsymbol{u},\boldsymbol{u}).\end{aligned} \tag{2.14}$$

This identity is called the *Korn identity.* If we drop the condition $\operatorname{div} \boldsymbol{u} = 0$, then this identity should be replaced by an inequality. Let us assume that the field $\boldsymbol{u}$ is finite. Then, by performing the same operations as in (2.5), we obtain

$$\int_\Omega \nabla u_i \cdot \frac{\partial \boldsymbol{u}}{\partial x_i} \mathrm{d}\Omega = -\int_\Omega \nabla \frac{\partial u_i}{\partial x_j} \cdot \boldsymbol{u} \mathrm{d}\Omega = \int_\Omega \frac{\partial u_i}{\partial x_j} \operatorname{div} \boldsymbol{u} \mathrm{d}\Omega.$$

Adding up by i the identities in (2.12), we have

$$\frac{1}{4}\int_\Omega \sum_{i,j=1}^{3} |\tau_{ij}|^2 \mathrm{d}\Omega - \frac{1}{4}\int_\Omega \sum_{i,j=1}^{3} |r_{ij}|^2 \mathrm{d}\Omega = \int_\Omega |\operatorname{div} \boldsymbol{u}|^2 \mathrm{d}\Omega \geq 0,$$

hence,

$$\int_\Omega \sum_{i,j=1}^{3} |\tau_{ij}|^2 \mathrm{d}\Omega \geq \int_\Omega \sum_{i,j=1}^{3} |r_{ij}|^2 \mathrm{d}\Omega.$$

Repeating the steps that led us to (2.14) we obtain the important Korn inequality,

$$\int_\Omega \sum_{i,j=1}^{3} \left| \frac{\partial u_i}{\partial x_j} \right|^2 \mathrm{d}\Omega \leq E(\boldsymbol{u}, \boldsymbol{u}). \tag{2.15}$$

If the norm on space $\boldsymbol{H}^1(\Omega)$ is given by the formula

$$\|\boldsymbol{u}\|^2_{\boldsymbol{H}^1(\Omega)} = \int_\Omega \sum_{i,j=1}^{3} \left| \frac{\partial u_i}{\partial x_j} \right|^2 \mathrm{d}\Omega + \left| \int_{\partial\Omega} \boldsymbol{u} \mathrm{d}S \right|^2,$$

then inequality (2.15) can be written down as

$$\|\boldsymbol{u}\|^2_{\boldsymbol{H}^1(\Omega)} \leq E(\boldsymbol{u}, \boldsymbol{u}).$$

Putting together this inequality and (2.2), we obtain

$$\|\boldsymbol{u}\|^2_{\boldsymbol{H}^1(\Omega)} \leq E(\boldsymbol{u}, \boldsymbol{u}) \leq 2\|\boldsymbol{u}\|^2_{\boldsymbol{H}^1(\Omega)}. \tag{2.16}$$

According to the fact that the set of all the finite fields is dense in $\boldsymbol{H}^1_0(\Omega)$ we can prove, by passing to the limit, that inequality (2.16) holds true for any field in $\boldsymbol{H}^1_0(\Omega)$.

Inequality (2.16) shows that $E(\boldsymbol{u}, \boldsymbol{u})$ can be viewed upon as a square norm that is equivalent to the main norm on the space $\boldsymbol{H}^1_0(\Omega)$.

Let us note that $\operatorname{rot} \boldsymbol{u} = 0$ for all potential fields $\boldsymbol{u}$ in $\boldsymbol{H}^1(\Omega)$ and, therefore (see (2.14)),

$$\int_\Omega \sum_{i,j=1}^{3} \left| \frac{\partial u_i}{\partial x_j} \right|^2 \mathrm{d}\Omega = E(\boldsymbol{u}, \boldsymbol{u}).$$

Further, the closure of the set of all finite solenoidal fields in $\boldsymbol{H}^1_0(\Omega)$ will be denoted by $\boldsymbol{J}^1_0(\Omega)$.

2.2.5 Stokes Operator

We consider the pair of Hilbert spaces $E = \boldsymbol{J}_0(\Omega)$ and $F = \boldsymbol{J}_0^1(\Omega)$. A scalar product on the space $\boldsymbol{J}_0^1(\Omega)$ can be introduced by means of the form $E(\boldsymbol{u}, \boldsymbol{v})$. Let us find out how does the generating operator $A = A_0$ of this pair look like. Identity (1.8.5) can be used as the definition of this operator. In our case, that identity has the following form,

$$\int_\Omega \boldsymbol{u} \cdot A_0 \boldsymbol{v} \mathrm{d}\Omega = E(\boldsymbol{u}, \boldsymbol{v}), \qquad \boldsymbol{u} \in \boldsymbol{J}_0^1(\Omega),\ \boldsymbol{v} \in \mathcal{D}(A_0) \subset \boldsymbol{J}_0^1(\Omega). \tag{2.17}$$

Let us assume that the field $\boldsymbol{v}$ is two times differentiable. Then the Green formula leads to

$$E(\boldsymbol{u}, \boldsymbol{v}) = -\int_\Omega \boldsymbol{u} \cdot \Delta \boldsymbol{v} \mathrm{d}\Omega, \tag{2.18}$$

where the boundary integral can be omitted in virtue of the fact that $\boldsymbol{u}$ belongs to the space $\boldsymbol{J}_0^1(\Omega)$. From (2.17) and (2.18) it follows that

$$\int_\Omega \boldsymbol{u} \cdot (A_0 \boldsymbol{v} + \Delta \boldsymbol{v}) \mathrm{d}\Omega = 0.$$

Since $\boldsymbol{J}_0^1(\Omega)$ is dense in $\boldsymbol{J}_0(\Omega)$, then, from the last equality, it follows that the field $A_0 \boldsymbol{v} + \Delta \boldsymbol{v}$ is orthogonal to $\boldsymbol{J}_0(\Omega)$ and, therefore, this field is a potential. Let us denote it by ∇q. Then

$$A_0 v = -\Delta \boldsymbol{v} + \nabla q.$$

From the general theory developed in Section 1.8 it is known that the equation $A_0 \boldsymbol{v} = \boldsymbol{f}$ is solvable for any $\boldsymbol{f} \in \boldsymbol{J}_0(\Omega)$ and the solution $\boldsymbol{v}$ belongs to $\boldsymbol{J}_0^1(\Omega)$. This solution is the generalized solution to the problem

$$-\Delta \boldsymbol{v} + \nabla q = \boldsymbol{f}, \quad \operatorname{div} \boldsymbol{v} = 0 \quad \text{on } \Omega, \quad \boldsymbol{v} = \boldsymbol{0} \quad \text{on } S. \tag{2.19}$$

If we apply the projection P_0 of the space $\boldsymbol{L}_2(\Omega)$ onto the subspace $\boldsymbol{J}_0(\Omega)$ to both sides of this equation, we obtain the following equation,

$$-P_0 \Delta \boldsymbol{v} = \boldsymbol{f}, \qquad \boldsymbol{f} \in \boldsymbol{J}_0(\Omega).$$

Hence, the operator A_0 is a self-adjoint extension of the operator $-P_0\Delta$. In what follows this operator will be called the *Stokes operator*.

According to the general scheme, operator A_0 can be extended further if we consider the weak solutions of equation (2.19). Thus, to any field $\boldsymbol{f}$ in $(\boldsymbol{J}_0^1(\Omega))^*$ it corresponds a unique weak solution of the problem (2.19), and every field $\boldsymbol{v} \in \boldsymbol{J}_0^1(\Omega)$ is a weak solution of the problem (2.19) for some $\boldsymbol{f} \in (\boldsymbol{J}_0^1(\Omega))^*$.

2.2.6 Spaces of Velocity Fields for a Viscous Incompressible Fluid in an Open Container

Let us assume that Ω is a Lipshitz region located in the lower halfspace $x_3 < 0$. The boundary of this region consists of a part of the plane $x_3 = 0$ and the surface S. The region Ω is filled with a viscous incompressible fluid. The surface S consists of the immovable solid walls of the container. The free surface of the fluid, Γ, is located in the plane part of the boundary $\partial\Omega$ of the region Ω; Γ is a connected region bounded by a finite number of disjoint, closed, ordinary Lipshitz curves.

Let us consider the linear set $\boldsymbol{J}_{00,S}(\Omega)$ of fields in $\boldsymbol{J}^1(\Omega)$ that equal zero in a neighborhood (each field in its own neighborhood) of the surface S. We denote the closure of this set in the space $\boldsymbol{L}_2(\Omega)$ by $\boldsymbol{J}_{0,S}(\Omega)$. Finite solenoidal fields in Ω belong to $\boldsymbol{J}_{00,S}(\Omega)$, and since the set of such fields is dense in $\boldsymbol{J}_0(\Omega)$, then

$$\boldsymbol{J}_0(\Omega) \subset \boldsymbol{J}_{0,S}(\Omega) \subset \boldsymbol{J}(\Omega). \tag{2.20}$$

We will show now that the set $\boldsymbol{J}_{00,S}(\Omega)$ contains a sufficiently wide set of fields of a special kind. Let $\varphi(x_1, x_2)$ be an arbitrary finite function on Γ, and let Γ_0 be an open set containing the support of function φ, which, together with its closure, are located inside Γ. The corresponding cylinder located in region Ω for any sufficiently small h is defined as $C_h = \Gamma_0 \times [-h, 0]$. Let $\psi(x_3)$ be a smooth function on the interval $[-h, 0]$ with $\psi(-h) = 0$. We define the field $\boldsymbol{v} = (v_1, v_2, v_3)$ for $x_3 < 0$ as equal to zero outside C_h and determined within C_h by the equations

$$\begin{aligned} v_1 &= \psi(x_3)\varphi(x_1, x_2), \\ v_2 &= 0, \\ v_3 &= -\Psi(x_3)\frac{\partial\varphi(x_1, x_2)}{\partial x_1}, \end{aligned} \tag{2.21}$$

where $\Psi(x_3)$ is an antiderivative of $\psi(x_3)$. The field $\boldsymbol{v}$ with $\Psi(-h) = 0$ belongs to the space $\boldsymbol{J}^1(\Omega)$, and equals zero in a neighborhood of the surface S, i.e., $\boldsymbol{v} \in \boldsymbol{J}_{00,S}(\Omega)$. Hence, the set $\boldsymbol{J}_{00,S}(\Omega)$ contains fields for which the first or second component, v_1 or v_2, coincides on Γ with an arbitrary finite function on Γ, and the third component, v_3, is proportional to the derivative of that function with respect to x_1 or x_2, respectively.

Taking into account the embedding (2.20), let us describe the orthogonal complement to the subspace $\boldsymbol{J}_{0,S}(\Omega)$ in the space $\boldsymbol{J}(\Omega)$. If the potential p of a harmonic field in $\boldsymbol{J}_{00,S}(\Omega)$ equals zero on Γ, then for any field $\boldsymbol{v}$ in $\boldsymbol{J}_{00,S}(\Omega)$ we have by (1.11) that

$$\int_\Omega \boldsymbol{v} \cdot \nabla p d\Omega = \int_{\partial\Omega} v_n \gamma p dS = \int_\Gamma v_3 \gamma p d\Gamma = 0, \tag{2.22}$$

i.e., the field ∇p is orthogonal to $\boldsymbol{J}_{00,S}(\Omega)$ and, therefore, ∇p is orthogonal to $\boldsymbol{J}_{0,S}(\Omega)$. Further, by the Weyl decomposition and the inclusion (2.20), each element that is orthogonal to $\boldsymbol{J}_{0,S}(\Omega)$ in $\boldsymbol{J}(\Omega)$ is a harmonic potential field ∇p. Now let us assume that the field ∇p is orthogonal to the fields described by (2.21). Then

$$0 = \int_\Omega \boldsymbol{v} \cdot \nabla p \mathrm{d}\Omega = \int_{\partial\Omega} v_n \gamma p \mathrm{d}\Omega = \int_\Gamma v_3 \gamma p \mathrm{d}\Gamma = -\Psi(0) \int_\Gamma \frac{\partial \varphi}{\partial x_1} \gamma p \mathrm{d}\Gamma. \qquad (2.23)$$

If the function $\psi(x_3)$ is such that

$$\Psi(0) = \int_{-h}^{0} \psi(x_3) \mathrm{d}x_3 \neq 0,$$

then from (2.23) it follows that

$$\int_\Gamma \frac{\partial \varphi}{\partial x_1} \gamma p \mathrm{d}\Gamma = 0 \qquad (2.24)$$

for any finite function φ in Γ. Hence, $(\partial p/\partial x_1)_\Gamma = 0$. Similarly, $(\partial p/\partial x_2)_\Gamma = 0$ and, therefore, $p|_\Gamma = \text{const}$. Subtracting this constant from p we obtain that the orthogonal complement to $\boldsymbol{J}_{0,S}(\Omega)$ in $\boldsymbol{J}(\Omega)$ consists of harmonic potential fields that equal zero on Γ. In Section 2.1, this space was denoted by $\boldsymbol{G}_{0,\Gamma,h}(\Omega)$. From the decompositions (1.22)–(1.25) it follows that the above introduced space $\boldsymbol{J}_{0,S}(\Omega)$ coincides with the space of velocity fields of an ideal fluid in an open container that was introduced in Section 2.1.

Let us denote the closure of the set $\boldsymbol{J}_{00,S}(\Omega)$ in the norm topology of the space $\boldsymbol{J}^1(\Omega)$ by $\boldsymbol{J}^1_{0,S}(\Omega)$. This space plays the main role in our further investigation of the movements of a viscous fluid in an open container.

From the Korn inequality introduced in Subsection 2.2.4, it follows that the Dirichlet norm is equivalent to the norm $(E(\boldsymbol{u}, \boldsymbol{u}))^{1/2}$ on the space $\boldsymbol{J}^1_0(\Omega)$. The Korn inequality can be generalized even further, so that the above mentioned equivalence takes place on the space $\boldsymbol{J}^1_{0,S}(\Omega)$, too.

Let us next consider the operator γ_n of the normal trace on the free surface Γ that, for every field $\boldsymbol{v} \in \boldsymbol{H}^1(\Omega)$, puts into correspondence the function $v_3(x_1, x_2, 0)$ on Γ. If $\boldsymbol{v} \in \boldsymbol{J}_{0,S}(\Omega)$, then in virtue of (1.13) we have that

$$\int_\Gamma \gamma_n \boldsymbol{v} \mathrm{d}\Gamma = 0.$$

Hence, the operator γ_n maps $\boldsymbol{J}^1_{0,S}(\Omega)$ onto the space $H^{1/2}_\Gamma$.

As the construction (2.21) of the field $\boldsymbol{v}$ in $\boldsymbol{J}_{00,S}(\Omega)$ shows, the derivatives of any finite functions on Γ can be considered as values of the operator γ_n. Hence, it follows that the range of this operator is dense in the space $L_{2,\Gamma}$. Indeed, if $q(x_1, x_2)$

in $L_{2,\Gamma}$ is orthogonal to all the normal traces of functions in $\boldsymbol{J}_{00,S}(\Omega)$, then it is orthogonal to the derivatives of all the finite functions in Γ and, therefore, its generalized derivatives equal zero. The function q is a constant, and since its mean value equals zero, q is the zero function.

If we denote the intersection of region Ω and the plane $x_3 = -\alpha$ with a sufficiently small $\alpha > 0$ by Γ_α, then by reasoning as above we get

$$\int_{\Gamma_\alpha} v_3 \mathrm{d}\Gamma_\alpha = 0,$$

and, since

$$\int_\Gamma v_3 \mathrm{d}\Gamma = \lim_{\alpha \to 0} \int_{\Gamma_\alpha} v_3 \mathrm{d}\Gamma_\alpha,$$

for any smooth v_3 we have

$$\frac{\partial}{\partial \alpha} \int_{\Gamma_\alpha} v_3 \mathrm{d}\Gamma_\alpha = \int_\Gamma \frac{\partial v_3}{\partial x_3} \mathrm{d}\Gamma = 0. \tag{2.25}$$

2.2.7 Main Boundary Value Problems for the Fluid Movement in an Open Container

In the previous subsection we introduced the embedding of $\boldsymbol{J}^1_{0,S}(\Omega)$ in $\boldsymbol{J}_{0,S}(\Omega)$ and the operator γ_n that maps the space $\boldsymbol{J}^1_{0,S}(\Omega)$ into the space $L_{2,\Gamma}$ by taking the normal trace of a field, $\gamma_n \boldsymbol{u} := u_3|_\Gamma$. Thus, we find ourselves in the same situation as in Section 1.8.

In this context, let us consider the pair of Hilbert spaces $F = \boldsymbol{J}^1_{0,S}(\Omega)$ and $E = \boldsymbol{J}_{0,S}(\Omega)$, and try to find its generating operator A. It is defined by the identity

$$\int_\Omega \boldsymbol{u} \cdot A\boldsymbol{v} \mathrm{d}\Omega = E(\boldsymbol{u}, \boldsymbol{v}), \qquad \boldsymbol{u} \in \boldsymbol{J}^1_{0,S}(\Omega),\ \boldsymbol{v} \in \mathcal{D}(A) \subset \boldsymbol{J}^1_{0,S}(\Omega). \tag{2.26}$$

For every two times differentiable field $\boldsymbol{v}$, the Green formula can be written down as

$$E(\boldsymbol{u}, \boldsymbol{v}) = -\int_\Omega \boldsymbol{u} \cdot \Delta \boldsymbol{v} \mathrm{d}\Omega + \int_\Gamma \sum_{i=1}^3 u_i \tau_{i3}(\boldsymbol{v}) \mathrm{d}\Gamma.$$

Then

$$\int_\Omega \boldsymbol{u} \cdot A\boldsymbol{v} \mathrm{d}\Omega = -\int_\Omega \boldsymbol{u} \cdot \Delta \boldsymbol{v} \mathrm{d}\Omega + \int_\Gamma \sum_{i=1}^3 u_i \tau_{i3}(\boldsymbol{v}) \mathrm{d}\Gamma. \tag{2.27}$$

For a finite field $\boldsymbol{u}$ we obtain

$$\int_\Omega \boldsymbol{u} \cdot (A\boldsymbol{v} + \Delta \boldsymbol{v}) \mathrm{d}\Omega = 0,$$

hence,

$$Av = -\Delta v + \nabla q.$$

Substituting this expression for $A\boldsymbol{v}$ into (2.27), we obtain the equation

$$\int_{\Omega} \boldsymbol{u} \cdot \nabla q \mathrm{d}\Omega = \int_{\Gamma} \sum_{i=1}^{3} u_i \tau_{i3}(\boldsymbol{v}) \mathrm{d}\Gamma,$$

or

$$\int_{\Gamma} \left[\sum_{i=1}^{2} u_i \tau_{i3}(\boldsymbol{v}) + u_3(-q + \tau_{33}(\boldsymbol{v})) \right] \mathrm{d}\Gamma = 0. \tag{2.28}$$

Let us take now $\boldsymbol{u}$ as the field in (2.21) with $\psi(x_3)$ satisfying the additional conditions $\psi(0) = 0$ and $\Psi(0) \neq 0$ (for example, $\psi(x_3) = 4x_3(h^2 - x_3^2)$). Then from (2.28) it follows that

$$\int_{\Gamma} (-q + \tau_{33}(\boldsymbol{v})) \frac{\partial \varphi}{\partial x_1} \mathrm{d}\Gamma = 0. \tag{2.29}$$

This means that the generalized derivative with respect to x_1 of the function $-q + \tau_{33}(\boldsymbol{v})$ equals zero on Γ. Similarly, it can be checked that the derivative with respect x_2 equals zero, too. Consequently, $-q + \tau_{33}(v)$ is constant on Γ. Subtracting this constant from q we obtain

$$-q + \tau_{33}(\boldsymbol{v}) = 0 \quad \text{on } \Gamma.$$

Let us now choose a field $\boldsymbol{u}$ as in (2.21) with $\psi(0) = 1$. Then, by taking into account (2.29), we get

$$\int_{\Gamma} \tau_{13}(\boldsymbol{v}) \varphi \mathrm{d}\Gamma = 0.$$

Hence, $\tau_{13}(\boldsymbol{v}) = 0$ on Γ. Similarly, it can be checked that $\tau_{23}(\boldsymbol{v}) = 0$ on Γ.

Finally, we conclude that any two times differentiable solution $\boldsymbol{v}$ of the equation $A\boldsymbol{v} = \boldsymbol{f}$ is the solution of the problem

$$\begin{cases} -\Delta \boldsymbol{v} + \nabla q = \boldsymbol{f}, \quad \operatorname{div} \boldsymbol{v} = 0 \quad \text{in } \Omega, \quad \boldsymbol{v} = \boldsymbol{0} \text{ on } S, \\ \tau_{13}(\boldsymbol{v}) = \tau_{23}(\boldsymbol{v}) = -q + \tau_{33}(\boldsymbol{v}) = 0 \quad \text{on } \Gamma. \end{cases} \tag{2.30}$$

This problem has the unique generalized solution $A^{-1}\boldsymbol{f}$ for any field $\boldsymbol{f} \in \boldsymbol{J}_{0,S}(\Omega)$. A unique weak solution of the problem (2.30) exists for any field $\boldsymbol{f} \in \boldsymbol{J}_{0,S}^{-1}(\Omega) = (\boldsymbol{J}_{0,S}^{1}(\Omega))^*$. Each field $\boldsymbol{v} \in \boldsymbol{J}_{0,S}^{1}(\Omega)$ is a weak solution to the problem (2.30) for some field $\boldsymbol{f} \in \boldsymbol{J}_{0,S}^{-1}(\Omega)$.

Now, let us consider the normal trace operator γ_n on the space $\boldsymbol{J}_{0,S}^{1}(\Omega)$. As it was mentioned in the previous section, this operator maps the space $\boldsymbol{J}_{0,S}^{1}(\Omega)$ into a dense subset of the space $L_{2,\Gamma}$.

Obviously, the kernel $\boldsymbol{N}(\Omega)$ of operator γ_n consists of all fields $\boldsymbol{v}$ from $\boldsymbol{J}^1_{0,S}(\Omega)$ for which $v_3|_\Gamma = 0$. From the density of the set of finite solenoidal fields in the subspace $\boldsymbol{J}_0(\Omega)$ it follows that $\boldsymbol{N}(\Omega)$ is dense in $\boldsymbol{J}_0(\Omega)$ and, therefore, the spaces $\boldsymbol{N}(\Omega)$ and $\boldsymbol{J}_0(\Omega)$ form a Hilbert pair. As in Subsection 1.8.7, let us denote the generating operator of this pair by L. It is defined by the following identity:

$$E(\boldsymbol{u},\boldsymbol{v}) = \int_\Omega \boldsymbol{u}\cdot L\boldsymbol{v}\mathrm{d}\Omega, \qquad \boldsymbol{u}\in\boldsymbol{N}(\Omega),\ \boldsymbol{v}\in\mathcal{D}(L)\subset\boldsymbol{N}(\Omega).$$

For any two times differentiable fields $\boldsymbol{v},\boldsymbol{u}\in\boldsymbol{N}(\Omega)$, the Green formula can be written down as

$$E(\boldsymbol{u},\boldsymbol{v}) = -\int_\Omega \boldsymbol{u}\cdot\Delta\boldsymbol{v}\mathrm{d}\Omega + \int_\Gamma\sum_{i=1}^2 u_i\tau_{i3}(\boldsymbol{v})\mathrm{d}\Gamma;$$

hence

$$\int_\Omega \boldsymbol{u}\cdot L\boldsymbol{v}\mathrm{d}\Omega = -\int_\Omega \boldsymbol{u}\cdot\Delta\boldsymbol{v}\mathrm{d}\Omega + \int_\Gamma\sum_{i=1}^2 u_i\tau_{i3}(\boldsymbol{v})\mathrm{d}\Gamma. \tag{2.31}$$

Writing down this equality for finite solenoidal fields $\boldsymbol{u}$ we obtain, as above, that the field $L\boldsymbol{v}$ differs from the field $-\Delta\boldsymbol{v}$ by some potential field ∇q:

$$L\boldsymbol{v} = -\Delta\boldsymbol{v} + \nabla q, \qquad \nabla q\in\boldsymbol{G}(\Omega). \tag{2.32}$$

Then, equality (2.31) becomes

$$\int_\Omega \boldsymbol{u}\cdot\nabla q\mathrm{d}\Omega = \int_\Gamma\sum_{i=1}^2 u_i\tau_{i3}(\boldsymbol{v})\mathrm{d}\Gamma.$$

Since the elements $\boldsymbol{u}\in\boldsymbol{N}(\Omega)$ are orthogonal to the potential fields in $\boldsymbol{L}_2(\Omega)$ we have

$$\int_\Gamma\sum_{i=1}^2 u_i\tau_{i3}(\boldsymbol{v})\mathrm{d}\Gamma = 0.$$

Let us choose the field $\boldsymbol{u}\in\boldsymbol{N}(\Omega)$ as in (2.21) with the function $\psi(x_3)$ such that $\psi(0) = 1$, $\psi(-h) = 0$, $\Psi(0) = \Psi(-h) = 0$ (for example, one can take $\Psi(x_3) := x_3(x_3+h)^2/h^2$). Then, we get that the function $\tau_{13}(\boldsymbol{v})$ is orthogonal to all the finite functions on Γ and, therefore, $\tau_{13}(\boldsymbol{v}) = 0$. Similarly, it can be shown that $\tau_{23}(\boldsymbol{v}) = 0$.

Hence, the self-adjoint operator L^{-1} in $\boldsymbol{J}_0(\Omega)$ provides the solution to the boundary value problem

$$\begin{cases} -\Delta\boldsymbol{v} + \nabla q = \boldsymbol{f}, \quad \operatorname{div}\boldsymbol{v} = 0 \quad \text{on } \Omega, \quad \boldsymbol{v} = \boldsymbol{0} \quad \text{on } S, \\ \tau_{13}(\boldsymbol{v}) = \tau_{23}(\boldsymbol{v}) = 0, \quad v_3 = 0, \quad q = 0 \quad \text{on } \Gamma. \end{cases} \tag{2.33}$$

For any field $\boldsymbol{f} \in \boldsymbol{J}_0(\Omega)$ the problem (2.33) has a unique generalized solution $\boldsymbol{v}$ that belongs to the space $\boldsymbol{N}(\Omega) \subset \boldsymbol{J}^1_{0,S}(\Omega)$. For any field $\boldsymbol{f} \in (\boldsymbol{N}(\Omega))^*$ there exists only one weak solution for the problem (2.33); this solution belongs to $\boldsymbol{N}(\Omega)$, too. Every field $\boldsymbol{v} \in \boldsymbol{N}(\Omega)$ is a weak solution of the problem (2.33) for some field $\boldsymbol{f} \in (\boldsymbol{N}(\Omega))^*$.

Let us investigate now the orthogonal complement $\boldsymbol{M}(\Omega)$ to the space $\boldsymbol{N}(\Omega)$ in $\boldsymbol{J}^1_{0,S}(\Omega)$; this complement consists of fields $\boldsymbol{u}$ for which $E(\boldsymbol{u},\boldsymbol{v}) = 0$ for any $\boldsymbol{v} \in \boldsymbol{N}(\Omega)$. If $\boldsymbol{u}$ is two times differentiable, then the Green formula becomes

$$0 = E(\boldsymbol{u},\boldsymbol{v}) = -\int_\Omega \boldsymbol{v}\cdot\Delta\boldsymbol{u}\,\mathrm{d}\Omega + \int_\Gamma \sum_{i=1}^{2} v_i\tau_{i3}(\boldsymbol{u})\mathrm{d}\Gamma. \tag{2.34}$$

Choosing $\boldsymbol{v}$ to be a finite field, we conclude that there is a function $p \in H^1(\Omega)$ such that

$$-\Delta\boldsymbol{u} + \nabla p = \boldsymbol{0} \qquad \text{in } \Omega.$$

Then, from (2.34) we have

$$0 = -\int_\Omega \boldsymbol{v}\cdot\nabla p\,\mathrm{d}\Omega + \int_\Gamma \sum_{i=1}^{2} v_i\tau_{i3}(\boldsymbol{u})\mathrm{d}\Gamma,$$

or, since $\operatorname{div}\boldsymbol{v} = 0$ and $\boldsymbol{v}_n|_{\partial\Omega} = 0$,

$$\int_\Gamma \sum_{i=1}^{2} v_i\tau_{i3}(\boldsymbol{u})\mathrm{d}\Gamma = 0.$$

Reasoning as above, we conclude that the subspace $\boldsymbol{M}(\Omega)$ consists of the fields $\boldsymbol{u}$ for which

$$\begin{cases} -\Delta\boldsymbol{u} + \nabla p = 0, \quad \operatorname{div}\boldsymbol{u} = 0 \quad \text{on } \Omega, \quad \boldsymbol{u} = \boldsymbol{0} \quad \text{on } S, \\ \tau_{13}(\boldsymbol{u}) = \tau_{23}(\boldsymbol{u}) = 0 \quad \text{on } \Gamma. \end{cases} \tag{2.35}$$

The operator γ_n implements a one-to-one correspondence between the subspace $\boldsymbol{M}(\Omega)$ and its range $H^{1/2}_\Gamma$. The inverse operator to γ_n yields the generalized solution to the problem

$$\begin{cases} -\Delta\boldsymbol{u} + \nabla p = 0, \quad \operatorname{div}\boldsymbol{u} = 0 \quad \text{on } \Omega, \quad \boldsymbol{u} = \boldsymbol{0} \quad \text{on } S, \\ \tau_{13}(\boldsymbol{u}) = \tau_{23}(\boldsymbol{u}) = 0, \quad u_3 = \varphi, \quad \text{on } \Gamma, \end{cases} \tag{2.36}$$

for any $\varphi \in H^{1/2}_\Gamma$. Any field in $\boldsymbol{M}(\Omega)$ is a generalized solution of the problem (2.36) for some $\varphi \in H^{1/2}_\Gamma$.

Now, let us construct an inverse operator to the adjoint of γ_n. In our general scheme, this inverse operator was denoted by ∂ and it was given by identity (1.8.22), that in this case turns out to be

$$E(\boldsymbol{u},\boldsymbol{v}) = \int_\Gamma \partial\boldsymbol{u}\cdot\gamma_n\boldsymbol{v}\,d\Gamma, \qquad \boldsymbol{u},\boldsymbol{v}\in \boldsymbol{M}(\Omega)\subset \boldsymbol{J}^1_{0,S}(\Omega), \tag{2.37}$$

or,

$$E(\boldsymbol{u},\boldsymbol{v}) = \int_\Gamma \partial\boldsymbol{u}\cdot v_3\,d\Gamma.$$

Using Green formula and the conditions in (2.35) we obtain

$$-\int_\Omega \nabla p\cdot\boldsymbol{v}\,d\Omega + \int_\Gamma \tau_{33}(\boldsymbol{u})v_3\,d\Gamma = \int_\Gamma \partial\boldsymbol{u}v_3\,d\Gamma,$$

or,

$$\int_\Gamma (-p+\tau_{33}(\boldsymbol{u}))v_3\,d\Gamma = \int_\Gamma \partial\boldsymbol{u}v_3\,d\Gamma.$$

Hence,

$$\partial\boldsymbol{u} = -p+\tau_{33}(\boldsymbol{u}).$$

Therefore, the inverse operator to ∂ gives the generalized solution of the problem

$$\begin{cases} -\Delta\boldsymbol{u}+\nabla p = 0, \quad \operatorname{div}\boldsymbol{u} = 0 \quad \text{on } \Omega, \quad \boldsymbol{u}=\boldsymbol{0} \quad \text{on } S, \\ \tau_{13}(\boldsymbol{u}) = \tau_{23}(\boldsymbol{u}) = 0, \quad -p+\tau_{33}(\boldsymbol{u}) = \psi \quad \text{on } \Gamma, \end{cases} \tag{2.33}$$

for any $\psi\in (H^{1/2}_\Gamma)^* = H^{-1/2}_\Gamma$. As in the general scheme, this operator will be denoted by T.

2.2.8 Spaces of Velocity Fields for a System of Viscous Fluids

As in Subsection 2.1.11, we assume that the container Ω is filled with a system of heavy fluids with densities ρ_k, where $\rho_1 > \cdots > \rho_k > \cdots > \rho_m > 0$, but now the fluids are supposed to be viscous with the coefficients of dynamic viscousity $\mu_k > 0$, $k=1,\ldots,m$. The kinetic energy for the field $\hat{\boldsymbol{u}} = \{\boldsymbol{u}^k\}_{k=1}^m$ of the system velocities still has the form (1.26) and defines the scalar product on the space $\hat{\boldsymbol{L}}_2(\Omega)$. The following bilinear functional is connected with the velocity of energy dissipation,

$$\hat{E}(\hat{\boldsymbol{u}},\hat{\boldsymbol{v}}) := \sum_{k=1}^m \mu_k E_k(\boldsymbol{u}^k,\boldsymbol{v}^k), \tag{2.39}$$

where E_k is determined by formula (2.3) for the region Ω_k.

Since the considered system consists of viscous fluids, the stickness conditions hold true on the solid boundary S, that is, $\boldsymbol{u}^k = 0$ on S_k, and the condition of inseparability of one fluid from another is valid on the boundaries Γ_i between fluids, i.e., $\boldsymbol{u}^i = \boldsymbol{u}^{i+1}$ on Γ_i. Based on this, we introduce the linear set $\hat{\boldsymbol{J}}_{00,S}(\Omega)$ of the fields

$\hat{\boldsymbol{u}} = \{\boldsymbol{u}^k\}_{k=1}^m$ from $\bigoplus_{k=1}^m \boldsymbol{J}^1(\Omega_k) =: \hat{\boldsymbol{J}}^1(\Omega)$ for which the components $\boldsymbol{u}^k$ belong to the sets $\boldsymbol{J}_{00,S}(\Omega_k)$ defined as in Subsection 2.2.6. Each of these components equals zero in a neighborhood of S_k, and the following conditions hold true on the boundaries Γ_i:

$$\boldsymbol{u}^i = \boldsymbol{u}^{i+1} \quad \text{on } \Gamma_i, \qquad i = 1, \dots, m-1. \tag{2.40}$$

We denote the closure of the linear set $\hat{\boldsymbol{J}}_{00,S}(\Omega)$ in the space $\hat{\boldsymbol{L}}_2(\Omega)$ by $\hat{\boldsymbol{J}}_{0,S}(\Omega)$, and the closure of the same set in the space $\hat{\boldsymbol{J}}^1(\Omega) = \bigoplus_{k=1}^m \boldsymbol{J}^1(\Omega_k)$ with the scalar product (2.39) by $\hat{\boldsymbol{J}}^1_{0,S}(\Omega)$. Obviously, the latter space contains all the fields $\hat{\boldsymbol{u}} = \{\boldsymbol{u}^k\}_{k=1}^m$ of velocities of a system of incompressible viscous fluids with a finite velocity of energy dissipation.

The following assertions are stated without proof, because their proofs are just a little bit different from the corresponding ones carried out in Section 2.1 and Subsections 2.2.6 and 2.2.7.

The space $\hat{\boldsymbol{J}}_{0,S}(\Omega)$ coincides with the space denoted in the same way in Subsection 2.1.12 and the orthogonal decomposition (1.42) holds true for this space. Furthermore, the space $\hat{\boldsymbol{J}}^1_{0,S}(\Omega)$ is the set

$$\begin{aligned}\hat{\boldsymbol{J}}_{0,S}(\Omega) = \{\hat{\boldsymbol{u}} = \{\boldsymbol{u}^k\}_{k=1}^m : \boldsymbol{u}^k \in \boldsymbol{J}^1(\Omega_k),\ \boldsymbol{u}^k = 0 \text{ on } S_k,\ k = 1, \dots, m;\\ \boldsymbol{u}^i = \boldsymbol{u}^{i+1} \text{ on } \Gamma_i,\ i = 1, \dots, m-1\},\end{aligned} \tag{2.41}$$

and $\hat{\boldsymbol{J}}^1_{0,S}(\Omega)$ is a dense set in $\hat{\boldsymbol{J}}_{0,S}(\Omega)$.

Let us consider the Hilbert pair $(\hat{\boldsymbol{J}}^1_{0,S}(\Omega); \hat{\boldsymbol{J}}_{0,S}(\Omega))$. The generating operator A of this pair can be determined by using the identity

$$\sum_{k=1}^m \rho_k \int_{\Omega_k} \boldsymbol{u}^k \cdot (A\hat{\boldsymbol{v}})_k d\Omega_k = \sum_{k=1}^m \mu_k E_k(\boldsymbol{u}^k, \boldsymbol{v}^k), \tag{2.42}$$

where

$$\hat{\boldsymbol{u}} = \{\boldsymbol{u}^k\}_{k=1}^m \in \hat{\boldsymbol{J}}^1_{0,S}(\Omega), \quad \hat{\boldsymbol{v}} = \{\boldsymbol{v}^k\}_{k=1}^m \in \mathcal{D}(A) \subset \hat{\boldsymbol{J}}^1_{0,S}(\Omega).$$

Using Green formula (2.10) for each component Ω_k of the decomposition $\Omega = \bigcup_{k=1}^m \Omega_k$, we can easily obtain the Green formula for Ω, that is,

$$\begin{aligned}\sum_{k=1}^m \mu_k E_k(\boldsymbol{u}^k, \boldsymbol{v}^k) = &-\sum_{k=1}^m \mu_k \int_{\Omega_k} \boldsymbol{u}^k \cdot \Delta \boldsymbol{v}^k d\Omega_k\\ &+ \sum_{i=1}^{m-1} \int_{\Gamma_i} \sum_{j=1}^3 u_j^i [\mu_i \tau_{j3}(\boldsymbol{v}^i) - \mu_{i+1}\tau_{j3}(\boldsymbol{v}^{i+1})] d\Gamma_i.\end{aligned} \tag{2.43}$$

This formula helps to verify that every two times differentiable solution of the equation $A\hat{\boldsymbol{v}} = \hat{\boldsymbol{f}}$ is a solution to the problem

$$\begin{cases} \rho_k(A\hat{\boldsymbol{v}})_k := -\mu_k\Delta\boldsymbol{v}^k + \nabla p_k = \rho_k\boldsymbol{f}_k, & \\ \operatorname{div}\boldsymbol{v}^k = 0 \quad \text{in } \Omega_k, \quad \boldsymbol{v}^k = 0 \quad \text{on } S_k, & k = 1,\dots,m \\ \mu_i\tau_{j3}(\boldsymbol{v}^i) - \mu_{i+1}\tau_{j3}(\boldsymbol{v}^{i+1}) = 0 \quad \text{on } \Gamma_i, & j = 1,2 \\ -p_i + \mu_i\tau_{33}(\boldsymbol{v}^i) = -p_{i+1} + \mu_{i+1}\tau_{33}(\boldsymbol{v}^{i+1}) \quad \text{on } \Gamma_i, & i = 1,\dots,m-1. \end{cases} \tag{2.44}$$

This problem has the only generalized solution $\hat{\boldsymbol{v}} = A^{-1}\hat{\boldsymbol{f}}$ for any field $\hat{\boldsymbol{f}} \in \hat{\boldsymbol{J}}_{0,S}(\Omega)$; for each field $\hat{\boldsymbol{f}} \in \hat{\boldsymbol{J}}_{0,S}^{-1}(\Omega) := (\hat{\boldsymbol{J}}_{0,S}^1(\Omega))^*$ there exists a unique weak solution $\hat{\boldsymbol{v}} \in \hat{\boldsymbol{J}}_{0,S}^1(\Omega)$; every field $\hat{\boldsymbol{v}} \in \hat{\boldsymbol{J}}_{0,S}^1(\Omega)$ is a weak solution of the problem (2.44) for a field $\hat{\boldsymbol{f}} \in \hat{\boldsymbol{J}}_{0,S}^{-1}(\Omega)$.

Let us introduce the normal trace operator $\hat{\gamma}_n\hat{\boldsymbol{v}} := \{v_3^i|_{\Gamma_i}\}_{i=1}^{m-1}$ on the space $\hat{\boldsymbol{J}}_{0,S}^1(\Omega)$. The kernel $\hat{\boldsymbol{N}}(\Omega)$ of the operator $\hat{\gamma}_n$ consists of all fields $\hat{\boldsymbol{v}} \in \hat{\boldsymbol{J}}_{0,S}^1(\Omega)$ for which

$$\hat{\gamma}_n\hat{v} = \{v_3^i|_{\Gamma_i}\}_{i=1}^{m-1} = 0.$$

It can be proved that a set of fields $\hat{\boldsymbol{u}} = \{\boldsymbol{u}^k\}_{k=1}^m$ lies in the orthogonal complement $\hat{\boldsymbol{M}}(\Omega)$ of the kernel $\hat{\boldsymbol{N}}(\Omega)$ in $\hat{\boldsymbol{J}}_{0,S}^1(\Omega)$ if

$$\begin{cases} -\mu_k\Delta\boldsymbol{u}^k + \nabla q_k = 0, & \\ \operatorname{div}\boldsymbol{u}^k = 0 \quad \text{in } \Omega_k, \quad \boldsymbol{u}^k = 0 \quad \text{on } S_k, & k = 1,\dots,m; \\ \mu_i\tau_{j3}(\boldsymbol{u}^i) - \mu_{i+1}\tau_{j3}(\boldsymbol{u}^{i+1}) = 0 \quad \text{on } \Gamma_i, & i = 1,\dots,m-1,\ j = 1,2. \end{cases} \tag{2.45}$$

The operator $\hat{\gamma}_n$ yields a one-to-one correspondence between the space $\hat{\boldsymbol{M}}(\Omega)$ and its range $\hat{H}_\Gamma^{1/2} = \bigoplus_{i=1}^{m-1} H_{\Gamma_i}^{1/2}(\Gamma_i)$ (see Section 2.1). The inverse operator to $\hat{\gamma}_n$ gives the generalized solution of the problem

$$\begin{cases} -\mu_k\Delta\boldsymbol{u}^k + \nabla q_k = 0, & \\ \operatorname{div}\boldsymbol{u}^k = 0 \quad \text{in } \Omega_k, \quad \boldsymbol{u}^k = 0 \quad \text{on } S_k, & k = 1,\dots,m \\ \mu_i\tau_{j3}(\boldsymbol{u}^i) - \mu_{i+1}\tau_{j3}(\boldsymbol{u}^{i+1}) = 0 \quad \text{on } \Gamma_i, & i = 1,\dots,m-1,\ j = 1,2 \\ v_3^i = \varphi_i \quad \text{on } \Gamma_i, & i = 1,\dots,m-1 \end{cases} \tag{2.46}$$

for any $\varphi := \{\varphi_i\}_{i=1}^{m-1} \in \hat{H}_\Gamma^{1/2}$. Any field $\hat{\boldsymbol{u}} \in \hat{\boldsymbol{M}}(\Omega)$ is the generalized solution of problem (2.46) for some $\varphi \in \hat{H}_\Gamma^{1/2}$.

Further, the operator ∂ defined as the inverse operator of the adjoint operator to $\hat{\gamma}_n$ is determined by the identity

$$\hat{E}(\hat{\boldsymbol{u}}, \hat{\boldsymbol{v}}) = \sum_{i=1}^{m-1} \int_{\Gamma_i} (\partial\hat{\boldsymbol{u}})_i(\hat{\gamma}_n\hat{\boldsymbol{v}})_i d\Gamma_i$$

for any $\hat{\boldsymbol{u}}, \hat{\boldsymbol{v}} \in \hat{M}(\Omega) \subset \hat{\boldsymbol{J}}^1_{0,S}(\Omega)$. Using the Green formula (2.43) it can be proved that

$$(\partial \hat{\boldsymbol{u}})_i = \{[-q_i + \mu_i \tau_{33}(\boldsymbol{u}^i)] - [-q_{i+1} + \mu_{i+1}\tau_{33}(\boldsymbol{u}^{i+1})]\}|_{\Gamma_i}; \tag{2.47}$$

the operator T defined as the inverse operator of ∂ gives the generalized solution of the problem

$$\begin{cases} -\mu_k \Delta \boldsymbol{u}^k + \nabla q_k = 0, & \\ \operatorname{div} \boldsymbol{u}^k = 0 \quad \text{in } \Omega_k, \quad \boldsymbol{u}^k = 0 \quad \text{on } S_k, & k = 1, \dots, m \\ \mu_i \tau_{j3}(\boldsymbol{u}^i) - \mu_{i+1}\tau_{j3}(\boldsymbol{u}^{i+1}) = 0 \quad \text{on } \Gamma_i, & i = 1, \dots, m-1 \\ & j = 1, 2 \\ [-q_i + \mu_i \tau_{33}(\boldsymbol{u}^i)] - [-q_{i+1} + \mu_{i+1}\tau_{33}(\boldsymbol{u}^{i+1})] & \\ \qquad = \psi_i \quad \text{on } \Gamma_i, & i = 1, \dots, m-1 \end{cases} \tag{2.48}$$

for any $\psi = \{\psi_i\}_{i=1}^{m-1} \in (\hat{H}^{1/2}_\Gamma)^* =: \hat{H}^{-1/2}_\Gamma$.

Summing up the main results of this section, let us stress one more time that for a single viscous fluid in an open container the space $\boldsymbol{J}^1_{0,S}(\Omega)$ with the scalar product (2.40) is the space of velocity fields with a finite velocity of energy dissipation; for a system of fluids, the space $\hat{\boldsymbol{J}}^1_{0,S}(\Omega)$ (see (2.41)) with the scalar product (2.39) is the space of velocity fields.

Appendix A

Remarks and Reference Comments to Part I

This Appendix highlights some references consulted by the authors while they were redacting the auxiliary mathematical part that supplies a basis for the investigation of boundary value and spectral problems in all the subsequent parts of the book. The list of references below is by no means exhaustive and can be substantially extended.

A.1 Chapter 1

1.1. One of the best presentations of the theory of linear operators on Hilbert spaces is given in the book by N. I. Akhiezer and I. M. Glazman [AG] (Chapters 1–6). Parts of the monographs of L. A. Lusternik and V. I. Sobolev [1] (Chapter 8), A. V. Balakrishan [1] (Chapters 1, 3), F. Riesz and B. Sz.-Nagy [1] (Chapters 5–9), M. S. Birman and M. Z. Solomjak [BS], and I. Gohberg and S. Goldberg [GG] (Chapters 1–8) offer some insights of this subject, too.

To the readers who are not particularly interested in proofs, but would like to get acquainted with the basic information, we recommend the reference book in functional analysis edited by S. Krein [F].

Results on s-numbers of compact operators and the classes $\mathfrak{S}_p$ are discussed in detail in the monographs of I. Gohberg and M. Krein [GK] (Chapters 2–3), and I. Gohberg, S. Goldberg, and M. A. Kashook [GGK]. The book by I. Gohberg and M. Krein [GK] (Chapter 6) addresses in great detail questions on the theory of bases. The concept of p-basis was introduced by V. A. Prigorskii [1].

1.2. The notion of generalized derivatives was introduced by S. L. Sobolev in 1935 in [1]. He also introduced the spaces $W_p^l(\Omega)$ that have been eventually called after his name and play an important role in many problems raised by mathematical physics. The theory of Sobolev spaces is presented to a large extent in S. L. Sobolev's monograph [6]. This theory has been also presented and generalized in the books by S. M. Nikolskii [1], S. L. Sobolev [5], V. G. Mazja [1], S. V. Uspenskii, G. V. Demidenko, and V. G. Perepyolkin [1]. An accessible presentation of the basic principles in this subject can be found in the book by V. A. Solonnikov and N. N. Uraltseva [1].

The spaces $L_2^1(\Omega)$ and properties of regions of the first type have been investigated in the book by J. Denie and J.-L. Lionse [1]. We should also point out that in O. Nikodim's earlier work [1] an example of a region Ω for which the space $L_2^1(\Omega)$ is not included in $L_2(\Omega)$ was constructed. More detailed answers to questions on these topics can be found in V. G. Mazja's monograph [1].

1.3. A thorough investigation of self-adjoint operators on spaces with indefinite metrics has been carried out in the basic work by L. S. Pontryagin [1], where he answered a question raised by S. L. Sobolev, who encountered such operators in problems related to the motion of a rotating top with a cavity filled with an ideal fluid. Later on, this important topic was developed by M. G. Krein, I. S. Iokhvidov, G. Langer, and some of their collaborators and students.

A detailed discussion of the theory of linear operators on spaces with indefinite metrics and the corresponding bibliography can be found in the book by T. Y. Azizov and I. S. Iokhvidov [AI]. The theorem in Subsection 1.3.7 on completeness and basicity of the system of root elements of a J-self-adjoint (generally speaking, unbounded) operator in the class $\mathcal{K}(\mathcal{H})$ is due to T. Y. Azizov [1] and was stated for the first time in his habilitation thesis [2].

1.4. The problem of determining eigenvalues and eigenvectors of the linear pencil of operators $B - \lambda A$ is a classic one. As for the minimax properties and asymtotics of eigenvalues, we refer to the surveys by M. S. Birman and M. Z. Solomyak [1], [2].

1.5. The monograph by Y. L. Daletskii and M. G. Krein [1] contains interesting information on evolution equations with a bounded operator coefficient. The book by S. G. Krein [K] (see also the survey by S. G. Krein and M. I. Khazan [1], and D. Henry's monograph [1]) is devoted to the theory of linear equations with an unbounded operator coefficient. The stability property for equations with a bounded operator coefficient was investigated by Y. L. Daletskii and M. G. Krein in their book [1].

In his work [1], J. Zabchik proved that for any real numbers $a \leq b$ one can construct an unboubded operator A, the coefficient of an equation in a Hilbert space, for which the corresponding semigroup belongs to the class C_0 on that space, operator

A has type $\omega = b$, and $\max\{\mathrm{Re}\,\sigma(A)\} = a$. Hence, the location of the spectrum itself does not influence the behavior of solutions at infinity. There are several works devoted to determine necessary conditions for a and b to coincide (see, for example, S. G. Krein and M. I. Khazan [1], pp. 161–165). The connection described in Subsection 1.5.6 between stability properties for the solutions of an equation and basicity properties of the associated eigenvector system of the corresponding spectral problem was pointed out in the works by Greenlee [1], [2], and A. I. Miloslavskii [1].

An investigation of the problem on the existence of solutions with continuous full energy for abstract evolution equations of the second order has been made (more likely, for the first time) in the the book by N. D. Kopachevsky, S. G. Krein, and Ngo Zuy Kan [KKN]. In the mixed problem for hyperbolic equations, these issues were considered by O. A. Ladyzhenkaya [1].

1.6. The spectral theory of operator pencils has been initiated by M. V. Keldysh in [1] (see also his article [2], and M. V. Keldysh and V. B. Lidsky [1]). He introduced the basic concepts of this theory, proved the theorems on completeness of the associated eigenvector system for the important classes of linear and polynomial pencils (Subsections 1.6.4–1.6.5), and found the asymptotics of the eigenvalues.

A new stage in the development of spectral theory of operator pencils has been set fourth by M. G. Krein and G. Langer in [1], [2], where quadratric pencils were investigated in detail and the method of pencil factorization was aplied for the first time. Later on, a considerable number of mathematicians worked on these problems. The monograph of A. S. Markus [M] and G. V. Radzievskii's survey [3] contain in extended form the main part of the results considered in Section 1.6. In particular, A. S. Markus [M] includes M. V. Keldysh's results and some of their generalizations, as well as results obtained by A. S. Markus and V. I. Matsaev [1].

The results in Subsection 1.6.10 on p-basicity of the eigenvector system of a holomorphic operator-valued function were obtained by N. D. Kopachevsky [20],V. A. Grinstein and N. D. Kopachevsky [1], [2], V. A. Grinstein [1], [2], and N. D. Kopachevsky and O. I. Nemirskaya [1].

Let us point out that the concept of root functions was independently introduced by S. G. Krein and V. P. Trofimov [1] (see also V. P. Trofimov [1]) and by V. I. Matsaev and Y. A. Palant [1].

For properties of Fredholm pencils, we refer to the book by I. T. Gohberg and M. G. Krein [GK], V. P. Trofimov's work [1], and the monograph of I. Gohberg, S. Goldberg and M. A. Kaashoek [GGK].

Variational methods of investigating continuous operator-valued functions were developed by Y. S. Abramov [1]; in Subsection 1.6.11 such results are stated in the same way as they were presented in A. S. Markus's book [M].

1.7. The asymptotic method based on equation splitting was initiated by N. N. Bogolyubov and N. M. Krylov. In S. G. Krein's monograph this method is discussed in a general setting. In our book, we considered it in a simplified form, when the operator A is constant. Therefore, we were able to state it with full substantiation. For our purposes, a more particular case is required, i.e., the kernel of operator A is splitting (Subsection 1.7.5). This is exactly the case to be considered for problems of oscillations of a body containing strongly viscous fluids (Part II).

In the matrix case, equation (7.15) was investigated by A. M. Lyapunov. The solution (7.16) of this and some more general equations can be found in the books by Y. L. Daletskii and M. G. Krein [1], S. G. Krein [K], and I. Gohberg, S. Goldberg, and M. A. Kaashoek [GGK].

A. G. Baskakov and his students devoted their works [1], [2] to applications of asymptotic splitting in eigenvalue problems.

1.8. The investigation of operators generated by forms and their connections with extensions of symmetric operators was initiated in the works of K. O. Friedrichs. The scales of Hilbert spaces were introduced by J.-L. Lions [1] and S. G. Krein [4] (see also the article by S. G. Krein and Y. I. Petunin [1], and the books by S. G. Krein, Y. I. Petunin, and E. M. Semyonov [1], and J.-L. Lions and E. Magenes [1]).

The abstract scheme for handling boundary value problems and the Green formula (8.41) are due to S. G. Krein and were stated for the first time in the monograph by N. D. Kopachevsky, S. G. Krein, and Ngo Zuy Kan [KKN]. In a different form, the abstract Green formula was introduced in the book by V. E. Lyantse and O. G. Storozh [1].

A.2 Chapter 2

2.1. The authors believe that the velocity fields with finite kinetic energy, i.e., the elements of the space $\boldsymbol{L}_2(\Omega)$, are the main objects to be employed in the theory of motion of an ideal fluid in a bounded region Ω. Besides, various subspaces of this space (potential fields, solenoidal fields, etc.) appear more naturally in this framework. It was H. Weyl [1] who highlighted the necessity of considering the fundamental operators of vector analysis for nonsmooth fields. For example, the definition of a solenoidal field stated in Subsection 2.1.4 is due to him.

In Subsection 2.1.3, where the concept of divergence of a field in $\boldsymbol{L}_2(\Omega)$ was introduced, it was essential to use Sobolev spaces with "negative smoothness." Such spaces appeared in the works by J. Lerey [1] and P. Lax [1] devoted to the theory of partial differential equations. A more detailed and an extended investigation of the properties of spaces with negative indexes is given in the article [1] and the monograph [2] by Y. M. Berezanskii. The properties of Sobolev spaces in regions Ω with nonsmooth boundaries were considered in detail in V. G. Mazja's monograph [1].

The normal component of a field on the boundary cannot be defined for any field in $\boldsymbol{L}_2(\Omega)$. Some additional restrictions should be taken into account. Subsection 2.1.6 deals with matters similar to those considered in R. Temam's book [1].

H. Weyl [1] showed that the components of a harmonic field are infinitely differentiable. He also obtained in this work the decompositions (1.17) and (1.18).

We should point out that other statements of the above mentioned basic results are contained in O. A. Ladyuzhenskaya's book [3] (see also E. B. Bykhovskii and N. V. Smirnov [1]).

The De Rham theorem mentioned in Subsection 2.1.9 can be found in his monograph [1].

Generalizations of the Weyl decomposition, either in the case of velocity fields for the motion of an ideal fluid in an open container (Subsection 2.1.10) or, in the more general case of the motion of a system of nonmixing ideal fluids with different constant densities (Subsection 2.1.11), were studied by N. D. Kopachevsky [2], [3], [9], [13].

2.2. In regard with the study of the motion of a viscous incompressible fluid in a bounded region, the authors' point of view is that the space of velocity fields with finite kinetic energy and finite velocity of energy dissipation is the main object within this theory. The formula for the velocity of energy can be found in the book by N. E. Kochin, I. A. Kibel, and N. V. Roze [1]. By Korn inequality, it follows that this space of vector fields coincides with the space $\boldsymbol{H}^1(\Omega)$, up to an equivalent norm. The proof of Korn inequality in Subsection 2.2.4 is due to K. Friedrichs [1], who considered this inequality in a simpler case. In our book, Korn inequality is used in a more general setting (see Subsection 2.2.6). Its validity in this case was proved by J. Gobert [1] (see also V. P. Mikhailov [1]).

There are a lot of studies devoted to Korn inequality, where upper estimates of the square norm for functions in Sobolev spaces are expressed as an integral of squares of some differential polynomials of those functions (see, for example, the books by O. V. Besov, V. P. Ilyin, and S. M. Nikolskii [1], and V. M. Goldstein and Y. G. Reshetnyak [1]).

The Stokes operator A_0 was introduced as a Friedrichs extension of the operator $-P_0\Delta$ by S. G. Krein in [1].

The spaces of velocity fields for a viscous incompressible fluid in an open container were considered for the first time by S. G. Krein [6], [7], and later on by S. G. Krein and G. I. Laptev [1].

The two main boundary value problems (2.30) and (2.38) of the motion of a fluid in an open container were introduced and investigated by S. G. Krein [6]. In our book these problems are addressed based on the abstract scheme developed in Section 1.8 and the Green formula (2.10) for the vector Laplace operator.

The study of the corresponding main boundary value problems of the motion of a system of nonmixing viscous incompressible fluids was carried out by N. D. Kopachevsky [21], [22]. The investigation of the two similar main boundary value problems turns out to be quite useful while addressing several different questions in the mechanics of continuous media.

PART II

MOTION OF BODIES WITH CAVITIES CONTAINING IDEAL FLUIDS

In this part of the book, we address several linear problems of hydrodynamics connected with the movements and oscillations of an ideal fluid.

Chapter 3 is devoted to the study of oscillations of a heavy fluid in wholly or partially filled containers. We will assume that the containers are immovable or perform a motion that is close to the nonperturbed state.

Chapter 4 deals with problems on oscillations of a capillary fluid and some problems of hydroelasticity that are closely related in a mathematical sense. Specific problems on fluid oscillations in a container with one or two elastic ends are considered.

In Chapter 5 we discuss methods for investigating spectral problems in the theory of oscillations for an ideal homogeneous fluid that differ from the general scheme presented in Section 1.8. Namely, we will consider applications of the stream function to plane problems, applications of the theory of shallow water, and some general methods for solving elliptic boundary value problems containing a spectral parameter in the equation and boundary condition.

In Chapter 6 we study problems on oscillations of an ideal rotating fluid, that are very important in applications and are quite interesting from a mathematical point of view. In these problems, internal waves caused by the Coriolis forces appear side by side with ordinary surface waves.

Chapter 3

Oscillations of a Heavy Ideal Fluid in Stationary and Nonstationary Containers

The problems considered in this chapter are relatively simple compared with some other problems in linear hydrodynamics. However, some of them require not only methods of the ordinary theory of self-adjoint operators and the corresponding evolution equations, but also methods of the theory of operators in spaces with indefinite metrics.

In Section 3.1 we will set up the equations for the motion of a rigid body with a cavity filled wholly or partially with an incompressible fluid. Further, according to their complexity, some other problems are considered, such as the problem on small movements of a fluid in a wholly filled immovable container (Section 3.2), the same problem for a partially filled immovable container (Section 3.3), and the problem on joint movements of a fluid and the affixed wholly (Section 3.4) or partially (Section 3.5) filled container.

3.1 Equations of the Motion of a Rigid Body with a Cavity Filled with an Incompressible Fluid

While investigating problems of the dynamics of the system "body + fluid", one should consider the equations of the motion of a rigid body with regard to the influence on the body dynamics of the moving fluid in the cavity. In this section, we will present the necessary information related to the mathematical statement of this complicated problem. The kinematic interdependences between the main magnitudes in stationary, nonstationary, and auxiliary coordinate systems, the equations of the motion of the body, and the equations of the fluid motion in the cavity are some of the issues discussed in the sequel. The main linear problems of the hydrodynamics for ideal and viscous fluids will be formulated as well.

3.1.1 Basic Concepts of Kinematics

In the three-dimensional space $\mathbb{R}^3$, let us introduce a stationary Cartesian coordinate system $O'x_1'x_2'x_3'$, with origin at the point O' and axial vectors $\boldsymbol{e}_k'$ $(k = 1, 2, 3)$. The location of an arbitrary point in space is described by its radius vector $\boldsymbol{r}' = \sum_{k=1}^{3} x_k' \boldsymbol{e}_k'$.

Besides the stationary (absolute) coordinate system, it is convenient to consider a nonstationary (relative) coordinate system $Ox_1x_2x_3$, with origin at a point O called the *pole*. If we denote the axial vectors of the nonstationary system by $\boldsymbol{e}_k$ $(k = 1, 2, 3)$, then the radius vector $\boldsymbol{r}'$ of any given point and the radius vector $\boldsymbol{r} = \sum_{k=1}^{3} x_k \boldsymbol{e}_k$ of the same point in the nonstationary coordinate system are connected by the relation

$$\boldsymbol{r}' = \boldsymbol{R}_0 + \boldsymbol{r}, \tag{1.1}$$

where $\boldsymbol{R}_0 = \sum_{k=1}^{3} x_{0k}' \boldsymbol{e}_k'$ is the radius vector of the pole O.

If $\boldsymbol{a}(t)$ is a variable vector defined in the nonstationary coordinate system as $\boldsymbol{a}(t) = \sum_{k=1}^{3} a_k(t) \boldsymbol{e}_k$, then the velocity of its changing $\mathrm{d}'\boldsymbol{a}/\mathrm{d}t$ in the stationary coordinate system and the velocity of changing $\mathrm{d}\boldsymbol{a}/\mathrm{d}t$ in the nonstationary system are connected by the relation

$$\frac{\mathrm{d}'\boldsymbol{a}}{\mathrm{d}t} = \frac{\mathrm{d}\boldsymbol{a}}{\mathrm{d}t} + \boldsymbol{\omega} \times \boldsymbol{a}, \tag{1.2}$$

where $\boldsymbol{\omega}(t)$ is the momentary angular velocity of the nonstationary coordinate system.

For a moving point, the vector $\boldsymbol{v}_{\mathrm{abs}} = \mathrm{d}'\boldsymbol{r}'/\mathrm{d}t$ is considered to be its absolute velocity and the vector $\boldsymbol{u} = \mathrm{d}\boldsymbol{r}/\mathrm{d}t$ gives its relative velocity.

By differentiating equation (1.1), we get the following relation

$$\boldsymbol{v}_{\mathrm{abs}} = \boldsymbol{v}_0 + \boldsymbol{\omega} \times \boldsymbol{r} + \boldsymbol{u}, \tag{1.3}$$

where $\boldsymbol{v}_0$ is the absolute velocity of the pole O. The vector $\boldsymbol{v}_0 + \boldsymbol{\omega} \times \boldsymbol{r}$ is called the *displacement velocity* of the point.

The absolute acceleration $\boldsymbol{w}_{\mathrm{abs}} = \mathrm{d}'\boldsymbol{v}_{\mathrm{abs}}/\mathrm{d}t$ can be obtained from (1.3) by differentiation,

$$\boldsymbol{w}_{\mathrm{abs}} = \frac{\mathrm{d}'\boldsymbol{v}_0}{\mathrm{d}t} + \frac{\mathrm{d}'\boldsymbol{\omega}}{\mathrm{d}t} \times \boldsymbol{r} + \boldsymbol{\omega} \times \frac{\mathrm{d}\boldsymbol{r}}{\mathrm{d}t} + \frac{\mathrm{d}'\boldsymbol{u}}{\mathrm{d}t} + \boldsymbol{\omega} \times (\boldsymbol{\omega} \times \boldsymbol{r}), \tag{1.4}$$

or

$$\boldsymbol{w}_{\text{abs}} = \frac{\mathrm{d}'\boldsymbol{v}_0}{\mathrm{d}t} + \boldsymbol{\varepsilon} \times \boldsymbol{r} + \boldsymbol{\omega} \times (\boldsymbol{\omega} \times \boldsymbol{r}) + 2\boldsymbol{\omega} \times \boldsymbol{u} + \frac{\mathrm{d}\boldsymbol{u}}{\mathrm{d}t}, \tag{1.5}$$

where $\boldsymbol{\varepsilon} = \mathrm{d}\boldsymbol{\omega}/\mathrm{d}t$ is the angular acceleration of the nonstationary coordinate system. The first three terms in (1.5) form the *displacement acceleration*, the fourth term is referred to as the *Coriolis acceleration*, and the last term stands for the *relative acceleration*, respectively. Then

$$\boldsymbol{w}_{\text{abs}} = \boldsymbol{w}_{\text{dis}} + \boldsymbol{w}_{\text{cor}} + \boldsymbol{w}_{\text{rel}}. \tag{1.6}$$

We assume now that the nonstationary coordinate system is rigidly connected with a moving rigid body. Then, relatively to the nonstationary system, all the points of the body are in stable state and, therefore, $\boldsymbol{u} = \boldsymbol{0}$ for them. The above mentioned formulas for points of the body become

$$\boldsymbol{v}_{\text{abs}} = \boldsymbol{v}_0 + \boldsymbol{\omega} \times \boldsymbol{r}, \tag{1.7}$$

$$\boldsymbol{w}_{\text{abs}} = \boldsymbol{w}_{\text{dis}} = \boldsymbol{w}_0 + \boldsymbol{\varepsilon} \times \boldsymbol{r} + \boldsymbol{\omega} \times (\boldsymbol{\omega} \times \boldsymbol{r}), \tag{1.8}$$

where $\boldsymbol{\omega}(t)$ and $\boldsymbol{\varepsilon}(t)$ are the angular velocity and the angular acceleration of the body.

3.1.2 Equations of Motion for an Incompressible Fluid

If the moving body has a cavity filled with an incompressible fluid, then the motion of the fluid in Euler variables is described by the velocity vector $\boldsymbol{v}(t, x')$ and the field of pressures $P' = P(t, x')$ for every point $x' = (x_1', x_2', x_3')$ in the cavity. For a viscous fluid with constant density ρ and the kinematic viscousity ν, the equations of the motion are the well-known Navier-Stokes equations,

$$\frac{\mathrm{d}'\boldsymbol{v}}{\mathrm{d}t} = -\frac{1}{\rho}\nabla' P' + \nu\Delta'\boldsymbol{v} + \boldsymbol{F}', \qquad (\mathrm{div})'\boldsymbol{v} = 0, \tag{1.9}$$

where the prime denotes symbols in the stationary system, and $\boldsymbol{F}' = \boldsymbol{F}'(t, x')$ is a prescribed field of mass external forces.

In investigating the dynamics problems of the system "body + fluid", the equations of motion of a fluid in the nonstationary coordinate system are used more often than equations (1.9). In order to get these equations, we first represent the field of absolute velocity of a fluid (according to (1.3)) by

$$\boldsymbol{v} = \boldsymbol{v}_0 + \boldsymbol{\omega} \times \boldsymbol{r} + \boldsymbol{u}(t, x), \tag{1.10}$$

where $\boldsymbol{u}(t, x)$ is the field of relative velocity of the fluid in the cavity, and then use formula (1.5) for the field of absolute acceleration,

$$\boldsymbol{w}(t,x)=\frac{\mathrm{d}'\boldsymbol{v}}{\mathrm{d}t}=\boldsymbol{w}_0+\boldsymbol{\varepsilon}\times\boldsymbol{r}+\boldsymbol{\omega}\times(\boldsymbol{\omega}\times\boldsymbol{r})+2\boldsymbol{\omega}\times\boldsymbol{u}+\frac{\partial\boldsymbol{u}}{\partial t}+(\boldsymbol{u}\cdot\nabla)\boldsymbol{u}. \tag{1.11}$$

In this formula, the last two terms equal the relative acceleration of a fluid particle, i.e., the full derivative $\mathrm{d}\boldsymbol{u}/\mathrm{d}t$ of the field $\boldsymbol{u}(t,x)$.

Using the invariance of the operators div, ∇, and Δ to linear substitutions of variables, from (1.10) and (1.11) we obtain the Navier-Stokes equations in the nonstationary coordinate system $Ox_1x_2x_3$ as

$$\frac{\partial\boldsymbol{u}}{\partial t}+(\boldsymbol{u}\cdot\nabla)\boldsymbol{u}+2\boldsymbol{\omega}\times\boldsymbol{u}+\boldsymbol{\varepsilon}\times\boldsymbol{r}=-\frac{1}{\rho}\nabla p+\nu\Delta\boldsymbol{u}+\boldsymbol{f},\qquad \operatorname{div}\boldsymbol{u}=0. \tag{1.12}$$

Here,

$$p:=P(t,x)-P_0(t,x),\qquad P_0(t,x)=\rho\left(\frac{1}{2}(\boldsymbol{\omega}\times\boldsymbol{r})^2-U-\boldsymbol{r}\cdot\boldsymbol{w}_0\right)+\text{const} \tag{1.13}$$

is the presure caused by the influence of inertial, external, potential, and centrifugal forces, and $U(t,x)$ is the potential of the external field of mass forces coresponding to the decomposition

$$\boldsymbol{F}'=-\nabla U+\boldsymbol{f}(t,x). \tag{1.14}$$

In (1.12), one should assume that the coefficient of kinematic viscousity equals zero, i.e., $\nu=0$, in the case of an ideal (nonviscous) fluid.

3.1.3 Boundary Conditions

On the solid boundary S of the region Ω filled with a viscous fluid, one should take into account the stickness condition, in a way that is mathematically obvious: on that boundary, the particles of the viscous fluid move with the same velocity as the particles of the rigid body. Therefore, the field of relative velocity, $\boldsymbol{u}=\boldsymbol{u}(t,x)$ equals zero there,

$$\boldsymbol{u}(t,x)=0\quad\text{on } S. \tag{1.15}$$

If the moving heavy (noncapillary) fluid has a free surface $\Gamma=\Gamma(t)$, i.e., has contact with a gas of constant pressure p_a, then the tensions in fluid and gas must coincide on $\Gamma(t)$. This dynamic requirement amounts to

$$\sum_{j=1}^{3}\tilde{\tau}'_{ij}(\boldsymbol{v})n'_j:=\sum_{j=1}^{3}\left(-P\delta_{ij}-\rho\nu\left(\frac{\partial v_i}{\partial x'_j}+\frac{\partial v_j}{\partial x'_i}\right)\right)n'_j=-p_a n'_i,\qquad i=1,2,3, \tag{1.16}$$

where $\tilde{\tau}'_{ij}(\boldsymbol{v})$ are the components of the tension tensor that was build taking into account the velocity $\boldsymbol{v}$ (see (1.10)), and $\boldsymbol{n} = \sum_{j=1}^{3} n'_j \boldsymbol{e}'_j$ is the external normal to $\Gamma(t)$.

For a capillary fluid, i.e., a fluid for which the capillary (surface) forces play an essential role on $\Gamma(t)$, the difference between the right and left sides of (1.16) does no longer equal zero, it equals the capillary leap of pressures, $p_{\text{cap}} n'_i$. Condition (1.16) and its linearization will be discussed later.

Let us point out that the well known kinematic condition holds true on $\Gamma(t)$, which means that a fluid particle located on the free surface at some moment will remain on that surface during the whole period of the movement.

If the fluid is ideal, then the boundary conditions (1.15) and (1.16) should be changed. Since in this case the fluid particles can easily move along the solid boundary, the stickness condition (1.15) should be replaced by the nonleaking condition

$$u_n = 0 \quad \text{on } S. \tag{1.17}$$

Further, assuming in (1.16) that $\nu = 0$, we obtain the following dynamic condition in the case of a capillary fluid,

$$P(t, x) - p_a = p_{\text{cap}} \quad \text{on } \Gamma(t). \tag{1.18}$$

3.1.4 Motion Equations for a Gyrostate

We are going to set up the motion equations of a rigid body with a cavity wholly filled with an ideal or viscous fluid.

Since the fluid is incompressible, the mass center C of the considered mechanical system is stationary relatively to the body, and the inertia moments of the system relatively to any of the axes connected to the body are constant and do not depend on the motion of fluid in the cavity. Such a system is called a *gyrostate*. The motion equation of the mass center has the form

$$m\boldsymbol{w}_C = \boldsymbol{G}, \tag{1.19}$$

where $m = m_{\text{b}} + m_{\text{f}}$ is the mass of the entire system which is the sum of the body mass m_{b} and the fluid mass m_{f}, $\boldsymbol{w}_C$ is the acceleration of the mass centre, and $\boldsymbol{G}$ is the main vector of external forces.

For a gyrostate, equation (1.19) does not change for whatever motion laws of the fluid in the cavity one assumes and therefore, this equation should be considered independently of other equations.

Sometimes it is convenient to consider besides the stationary, $O'x'_1x'_2x'_3$, and the nonstationary, $Ox_1x_2x_3$, coordinate systems an auxilliary system $Oy_1y_2y_3$ that translationally moves relatively to $O'x'_1x'_2x'_3$. Its axial vectors coincide with $\boldsymbol{e}'_k$,

$k = 1, 2, 3$ and its origin is the pole O. Formula (1.3) holds true for such a system, i.e., we can consider that $\mathrm{d}'\boldsymbol{a}/\mathrm{d}t$ is a derivative in the auxiliary system.

Assuming that the motion of the pole O is given, let us consider the motion of a body with a fluid in the system $Oy_1y_2y_3$. Obviously, the velocity $\boldsymbol{v}$ of an arbitrary point in the system $Oy_1y_2y_3$ can be represented (according to (1.12)) as

$$\boldsymbol{v} = \boldsymbol{\omega} \times \boldsymbol{r} + \boldsymbol{u}, \tag{1.20}$$

where $\boldsymbol{\omega}$ is the angular velocity of the body, $\boldsymbol{r}$ is a radius vector of the point considered from O, and $\boldsymbol{u}$ is the velocity of that point in the nonstationary system (for points of the rigid body $\boldsymbol{u} = \boldsymbol{0}$).

Let us write down the equation of moments in the auxiliary system as

$$\frac{\mathrm{d}'\boldsymbol{K}}{\mathrm{d}t} = \boldsymbol{M}. \tag{1.21}$$

Here $\boldsymbol{K} = \boldsymbol{K}(t)$ is a kinetic moment of the entire system relatively to the pole O in the system's motion relatively to the auxiliary system, $\boldsymbol{M} = \boldsymbol{M}(t)$ is the main moment relatively to all the external forces acting on the body with the fluid (considered also in the system $Oy_1y_2y_3$). In particular, $\boldsymbol{M}$ contains the moment of inertia forces (that equals $-m\boldsymbol{r}_C \times \boldsymbol{w}_C$) caused by the motion of the auxiliary system. This moment equals zero in each of the following situations:

a) when the auxiliary system is stationary, i.e., the pole O is an immovable point;

b) when the auxiliary system moves rectilinearly and uniformly: $\boldsymbol{w}_C = \boldsymbol{0}$;

c) when the pole O coincides with the mass center ($\boldsymbol{r}_C = \boldsymbol{0}$).

In all these cases $\boldsymbol{M}$ in (1.21) is a moment relatively to the pole O of all external forces influencing the system.

Suppose now that the body and the fluid have densities ρ_0 and ρ and occupy the regions Ω_0 and Ω, respectively. Then, taking (1.20) into account, we have

$$\begin{aligned}\boldsymbol{K} &:= \int_{\Omega\cup\Omega_0} \boldsymbol{r} \times \boldsymbol{v}\mathrm{d}m = \rho_0 \int_{\Omega_0} \boldsymbol{r} \times (\boldsymbol{\omega} \times \boldsymbol{r})\mathrm{d}\Omega + \rho \int_{\Omega} \boldsymbol{r} \times (\boldsymbol{\omega} \times \boldsymbol{r})\mathrm{d}\Omega + \boldsymbol{L} \\ &=: \boldsymbol{J}\boldsymbol{\omega} + \boldsymbol{L}, \\ \boldsymbol{L} &:= \rho \int_{\Omega} \boldsymbol{r} \times \boldsymbol{u}\mathrm{d}\Omega. \end{aligned} \tag{1.22}$$

Here $\boldsymbol{J} = \boldsymbol{J}_{\mathrm{b}} + \boldsymbol{J}_{\mathrm{f}}$ is the tensor of inertia of the whole system relatively to the pole O, and it is equal to the sum of the inertia tensor of the body, $\boldsymbol{J}_{\mathrm{b}}$, and the inertia tensor of the fluid, $\boldsymbol{J}_{\mathrm{f}}$. The vector $\boldsymbol{L}$ is called the *gyrostatical moment*.

Since the amount of motion of the fluid in the nonstationary system, $\rho \int_{\Omega} \boldsymbol{u}\mathrm{d}\Omega$, equals zero (the mass center is stable), $\boldsymbol{L}$ does not depend on the choice of the pole

O and can be computed relatively to any point connected with the cavity. If the body contains several cavities Ω_i $(i = 1, \ldots, m)$ filled with fluid, then we should consider the sum $\sum_{i=1}^{m} \boldsymbol{L}_i$ of gyrostatical moments for every cavity as $\boldsymbol{L}$.

Let us substitute equations (1.22) into the moment equation (1.21) and use formula (1.3) for $\boldsymbol{J\omega}$ and $\boldsymbol{L}$. Then, in the nonstationary coordinate system $Ox_1x_2x_3$ we obtain the equation

$$\boldsymbol{J}\boldsymbol{\varepsilon} + \boldsymbol{\omega} \times (\boldsymbol{J}\boldsymbol{\omega}) + \frac{\mathrm{d}\boldsymbol{L}}{\mathrm{d}t} + \boldsymbol{\omega} \times \boldsymbol{L} = \boldsymbol{M}, \tag{1.23}$$

where $\boldsymbol{\varepsilon} = \mathrm{d}\boldsymbol{\omega}/\mathrm{d}t$ is the angular acceleration of the body.

Finally, while investigating the dynamics of a gyrostate, one should jointly solve the hydrodynamic equations (1.12) with the boundary condition (1.15) and the moment equation (1.23). Actually, as it is well known in the mechanics of a solid body, we can add to what we did so far the kinematic relations for the Euler angles or for the direction cosines, $a_{ij} := \boldsymbol{e}'_j \cdot \boldsymbol{e}_i$.

Let us next write down the boundary value problem of the fluid movement in a linearized form and the moment equation. The corresponding problems with initial data for determining the laws of motion for the mass center and how are the direction cosines changing with time can be easily solved for any known angular velocity of the body.

In the nonperturbed state, let the gyrostate perform a translational movement (along with pole O) according to the known law and uniformly rotate relatively to pole O with the constant angular velocity $\boldsymbol{\omega}_0$. Then, the fluid is stable relatively to the cavity and the nonperturbed pressure $P_0 = P_0(t, x)$ is defined by formula (1.13) as

$$P_0(t, x) = \rho \left(\frac{1}{2} (\boldsymbol{\omega}_0 \times \boldsymbol{r})^2 - U - \boldsymbol{r} \cdot \boldsymbol{w}_0 \right) + \text{const}, \tag{1.24}$$

where $U = U(t)$ is the potential of external forces of nonperturbed motion (see (1.14)), and $\boldsymbol{w}_0 = \boldsymbol{w}_0(t)$ is the given acceleration of the pole O.

Now, let us assume that in the perturbed motion, the velocity field $\boldsymbol{u} = \boldsymbol{u}(t, x)$, the perturbed pressure $p(t, x) = P(t, x) - P_0(t, x)$, and the deviation of the angular velocity of the body from $\boldsymbol{\omega}_0$, denoted by $\boldsymbol{\omega}(t)$, have small magnitudes of the same order (together with their derivatives). Then the linearized equations of the gyrostate motion, together with the boundary and initial conditions, take the form

$$\begin{aligned} &\boldsymbol{J}\boldsymbol{\varepsilon} + \boldsymbol{\omega} \times (\boldsymbol{J}\boldsymbol{\omega}_0) + \boldsymbol{\omega}_0 \times (\boldsymbol{J}\boldsymbol{\omega}) + \frac{\mathrm{d}}{\mathrm{d}t} \left(\rho \int_\Omega \boldsymbol{r} \times \boldsymbol{u} \mathrm{d}\Omega \right) + \boldsymbol{\omega}_0 \times \left(\rho \int_\Omega \boldsymbol{r} \times \boldsymbol{u} \mathrm{d}\Omega \right) \\ &\quad = \boldsymbol{M}, \\ &\boldsymbol{\omega}(0) = \boldsymbol{\omega}^0, \end{aligned} \tag{1.25}$$

$$\frac{\partial \boldsymbol{u}}{\partial t} + 2\omega_0 \times \boldsymbol{u} + \boldsymbol{\varepsilon} \times \boldsymbol{r} = -\frac{1}{\rho}\nabla p + \nu\Delta \boldsymbol{u},$$
$$\operatorname{div} \boldsymbol{u} = 0 \quad \text{in } \Omega,$$
$$\boldsymbol{u} = \boldsymbol{0} \quad \text{on } S,$$
$$\boldsymbol{u}(0, x) = \boldsymbol{u}^0(x). \tag{1.26}$$

If the cavity contains an ideal fluid, then we should assume $\nu = 0$ in equation (1.26) and take the nonleaking condition instead of the stickness condition on the solid boundary S.

3.1.5 Dynamics Equations of the System "Body + Fluid" With a Partially Filled Cavity

Suppose that a rigid body of mass m_b has a cavity containing a viscous incompressible fluid of mass m_f and a gas whose pressure is constant and equals p_a. We do not take into account the influence of the gas motion. We will consider now the simplest situation that can occur by assuming that the translational movement of the body with fluid is nonperturbed, the external force is the gravitation, and the fluid is heavy (noncapillary). Eventually, we will consider more complicated situations after a thorough investigation of the simpler problems.

Finally, let us suppose that in the nonperturbed motion the fluid occupies a region Ω that is bounded by a solid boundary S and by a plane free surface Γ. We choose an arbitrary point as a pole O that is rigidly connected with the body and consider that in the nonperturbed motion the nonstationary system $Ox_1x_2x_3$ and the auxiliary system $Oy_1y_2y_3$ coincide. The surface Γ is decribed by the equation $x_3 = b$ and the points of the region Ω occupied by fluid are such that $x_3 < b$.

In the coordinate system $Oy_1y_2y_3$, the system "body + fluid" is influenced by the mass forces whose acceleration $\boldsymbol{g} = \boldsymbol{g}_{\text{grav}} - \boldsymbol{w}_0$ equals the sum of the gravitational accelaration $\boldsymbol{g}_{\text{grav}}$ and the acceleration of forces of inertia, $-\boldsymbol{w}_0 = -\mathrm{d}^2\boldsymbol{R}_0/\mathrm{d}t^2$. We assume that in the nonperturbed motion the acceleration of mass forces $\boldsymbol{g}$ acts against the axis Ox_3 (and, in particular, has constant direction). As long as $\boldsymbol{g}_{\text{grav}}$ is a constant vector, the latter assumption is met in each of the following situations:

a) the acceleration $\boldsymbol{w}_0$ of the translational movement has a constant value and a constant direction;

b) the acceleration $\boldsymbol{w}_0$ is collinear to the vector $\boldsymbol{g}_{\text{grav}}$ and changes its values arbitrarily;

c) the acceleration $\boldsymbol{w}_0$ has constant direction and $\boldsymbol{g}_{\text{grav}} = 0$ (the low-gravity case).

With regard to the perturbed motion, we assume that the angles between the corresponding coordinate axes of the stationary and the auxiliary systems, the angular velocity $\boldsymbol{\omega}$ of the rigid body, and the projection of the acceleration $\boldsymbol{g}$ onto the axes Ox_1 and Ox_2 have small magnitudes of the first order. We also consider that the relative velocity $\boldsymbol{u}(t,x)$, the dynamic pressure $p(t,x)$, and the function $x_3 - b = \zeta(t, x_1, x_2)$ that defines the deviation of the perturbed moving free surface $\Gamma(t)$ from the nonperturbed surface Γ (with equation $x_3 = b$) have magnitudes of the same order. In what follows, the entire problem is stated as a linear problem.

For a more detailed description of the motion of the considered hydromechanical system, let us introduce the semi-stationary coordinate system $Oz_1z_2z_3$ in addition to the stationary and nonstationary coordinate systems $Oy_1y_2y_3$ and $Ox_1x_2x_3$. In the theory of gyroscopes, the system $Oz_1z_2z_3$ is called a *Rezalle system.* This system has an important role in the nutational and the wobbling motion of the body, but does not take part in its proper rotation (the axis Oz_3 coincides with the axis Ox_3).

It is evident from Fig. 3.1 that in order to match the stationary trihedron $Oy_1y_2y_3$ with the semi-stationary system $Oz_1z_2z_3$ one should perform two rotations:

1) a rotation by the angle δ_2 around the axis Oy_2 until $Oy_1y_2y_3$ matches the trihedron Oz_1z_2w;

2) a rotation by the angle δ_1 around the axis Oz_1 until Oz_1y_2w matches the semi-statinary trihedron $Oz_1z_2z_3$.

Now, in order to match the semi-stationary trihedron $Oz_1z_2z_3$ with the nonstationary trihedron $Ox_1x_2x_3$, we should perform one more rotation by the angle δ_3 around the axis $Oz_3 = Ox_3$ (as a result, the axis Oz_1 is converted into the axis Ox_1, and the axis Oz_2 into Ox_2, respectively).

Let us next introduce the vector $\boldsymbol{\delta} = \sum_{k=1}^{3} \delta_k \boldsymbol{e}_k$, where $\boldsymbol{e}_k$ is the axial vector of Ox_k, and call it the *angular displacement vector* of the body. Such a name is justified by the fact that according to our assumptions, all the angles δ_i are small and, therefore, the unit vectors $\boldsymbol{e}_{z_1}$ and $\boldsymbol{e}_{y_2}$ equal $\boldsymbol{e}_1$ and $\boldsymbol{e}_2$, respectively, with an accuracy of the second order. Then, for the angular velocity of the rotation of the body we approximately obtain

$$\boldsymbol{\omega} = \sum_{k=1}^{3} \frac{\mathrm{d}\delta_k}{\mathrm{d}t} \boldsymbol{e}_k.$$

A similar formula holds true for the angular acceleration,

$$\boldsymbol{\varepsilon} = \frac{\mathrm{d}\boldsymbol{\omega}}{\mathrm{d}t} = \sum_{k=1}^{3} \frac{\mathrm{d}^2\delta_k}{\mathrm{d}t^2} \boldsymbol{e}_k.$$

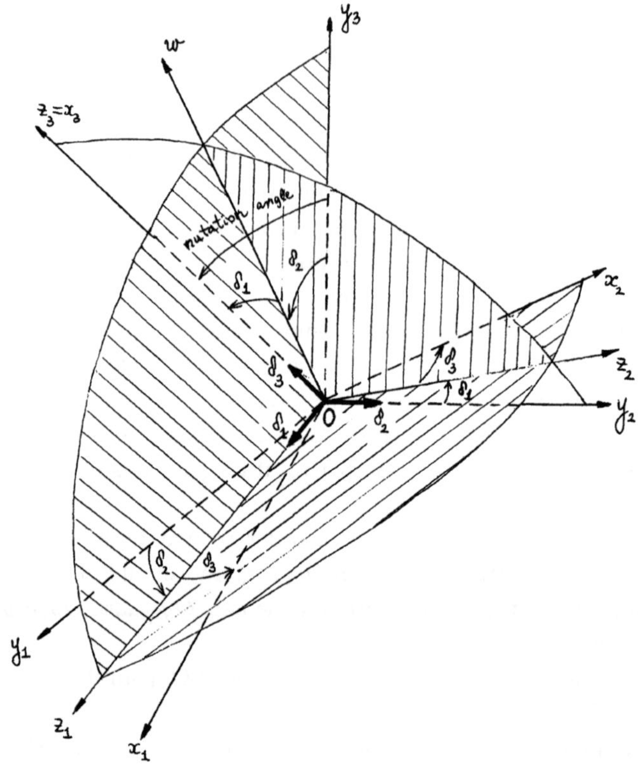

Figure 3.1

Since the given system is not a gyrostate, then in describing its dynamics we should jointly consider the motion amount equation, the moment equation, and the Navier-Stokes equations. The first two equations have the form

$$\frac{\mathrm{d}'\boldsymbol{Q}}{\mathrm{d}t} = \boldsymbol{G} + m\boldsymbol{g}, \qquad m = m_{\mathrm{b}} + m_{\mathrm{f}},$$
$$\frac{\mathrm{d}'\boldsymbol{K}}{\mathrm{d}t} = \boldsymbol{M} + (m_{\mathrm{b}}\boldsymbol{r}_{\mathrm{b}} + m_{\mathrm{f}}\boldsymbol{r}_{\mathrm{f}}) \times \boldsymbol{g},$$

where

$$\begin{aligned} \boldsymbol{Q} &= m_{\mathrm{b}}\frac{\mathrm{d}'\boldsymbol{r}_{\mathrm{b}}}{\mathrm{d}t} + m_{\mathrm{f}}\frac{\mathrm{d}'\boldsymbol{r}_{\mathrm{f}}}{\mathrm{d}t}, \\ \boldsymbol{K} &= \boldsymbol{J}_{\mathrm{b}}\boldsymbol{\omega} + \rho\int_{\Omega} \boldsymbol{r} \times (\boldsymbol{\omega} \times \boldsymbol{r} + \boldsymbol{u})\mathrm{d}\Omega = \boldsymbol{J}\boldsymbol{\omega} + \boldsymbol{L} \\ \boldsymbol{L} &= \rho\int_{\Omega} (\boldsymbol{r} \times \boldsymbol{u})\mathrm{d}\Omega. \end{aligned} \tag{1.27}$$

Here we used the same notations as in Subsection 3.1.4 (see (1.21) and (1.22)). We notice that $\boldsymbol{G}$ and $\boldsymbol{M}$ do not contain the gravitation and inertia forces. The radius vectors that represent the mass center of the body and fluid are denoted by $\boldsymbol{r}_{\mathrm{b}}$ and $\boldsymbol{r}_{\mathrm{f}}$, respectively, and the prime indicates that the derivative with respect to t is calculated in the auxiliary system $Oy_1y_2y_3$. Let us note that, since $\boldsymbol{\omega}$ and $\boldsymbol{u}$ are small, the integration in (1.27) is performed not over the time changing region $\Omega(t)$ occupied by the fluid, but over the nonperturbed region Ω.

Obviously,

$$\begin{aligned} m_{\mathrm{b}}\boldsymbol{r}_{\mathrm{b}} + m_{\mathrm{f}}\boldsymbol{r}_{\mathrm{f}} &= m_{\mathrm{b}}\boldsymbol{r}_{\mathrm{b}} + \rho\int_\Omega \boldsymbol{r}\mathrm{d}\Omega + \rho\int_\Gamma \boldsymbol{r}\zeta\mathrm{d}\Gamma \\ &=: m\boldsymbol{r}_{\mathrm{C}} + \rho\int_\Gamma (x_1\boldsymbol{e}_1 + x_2\boldsymbol{e}_2)\zeta\mathrm{d}\Gamma, \end{aligned} \tag{1.28}$$

with an accuracy of the second order (the radius vector $\boldsymbol{r}_{\mathrm{C}}$ of the mass center C of the system in the nonperturbed state is stationary relatively to the system $Ox_1x_2x_3$). While setting up the last expression in (1.28) we used the fact that for points on Γ the radius vector is given by $\boldsymbol{r} = b\boldsymbol{e}_3 + x_1\boldsymbol{e}_1 + x_2\boldsymbol{e}_2$, and took into account the volume preserving condition

$$\int_\Gamma \zeta\mathrm{d}\Gamma = 0.$$

Let us also note that with an accuracy up to the second order we have

$$\boldsymbol{g} = -g\boldsymbol{e}_3 + \sum_{i=1}^{2} g_i\boldsymbol{e}_i, \quad g_i := \boldsymbol{g}\cdot\boldsymbol{e}_i, \quad g_1 \approx g\delta_2, \quad g_2 \approx -g\delta_1. \tag{1.29}$$

By (1.28) and (1.29), equations (1.27) have the following form after linearization,

$$\begin{aligned} &\frac{\mathrm{d}'\boldsymbol{Q}}{\mathrm{d}t} = \boldsymbol{G} + m\boldsymbol{g}, \qquad \boldsymbol{g} = \boldsymbol{g}_{\mathrm{grav}} - \boldsymbol{w}_0,\ m = m_{\mathrm{b}} + m_{\mathrm{f}}, \\ &\frac{\mathrm{d}'\boldsymbol{K}}{\mathrm{d}t} = \boldsymbol{M} + m\boldsymbol{r}_{\mathrm{C}}\times\boldsymbol{g} + \rho g\int_\Gamma (\mathrm{e}_3\times\boldsymbol{r})\zeta\mathrm{d}\Gamma; \\ &\boldsymbol{Q} = m\boldsymbol{\omega}\times\boldsymbol{r}_{\mathrm{C}} + \rho\int_\Gamma (x_1\mathrm{e}_1 + x_2\mathrm{e}_2)\frac{\partial\zeta}{\partial t}\mathrm{d}\Gamma, \\ &\boldsymbol{K} = \boldsymbol{J}\boldsymbol{\omega} + \boldsymbol{L}, \\ &\boldsymbol{L} = \rho\int_\Omega (\boldsymbol{r}\times\boldsymbol{u})\mathrm{d}\Omega. \end{aligned} \tag{1.30}$$

Now, let us write down the linearized equations and the boundary conditions of the hydrodynamic part of the problem. Since $\boldsymbol{u}(t,x)$ and $\boldsymbol{\omega}(t)$ have small magnitudes of the same order, from (1.12) and (1.15) we obtain

$$\begin{aligned} &\boldsymbol{\varepsilon}\times\boldsymbol{r} + \frac{\partial\boldsymbol{u}}{\partial t} = -\frac{1}{\rho}\nabla p + \nu\Delta\boldsymbol{u}, \\ &\operatorname{div}\boldsymbol{u} = 0 \quad \text{in } \Omega, \qquad \boldsymbol{u} = \boldsymbol{0} \quad \text{on } S, \end{aligned} \tag{1.31}$$

where $p(t,x) = P(t,x) - P_0(t,x)$ and $P_0(t,x)$ is given by

$$P_0(t,x) = \rho \boldsymbol{g} \cdot \boldsymbol{r} + p_a + \rho g b = -\rho g(x_3 - \delta_2 x_1 + \delta_1 x_2) + p_a + \rho g b, \tag{1.32}$$

according to (1.13) and (1.29). In the nonperturbed motion, when $\delta_i = 0$ for $i = 1, 2$, $P_0(t,x)$ coincides with the nonperturbed pressure $P_0(x) = -\rho g(x_3 - b) + p_a$. On Γ (i.e., for $x_3 = b$), $P_0(t,x)$ coincides with the gas pressure p_a.

Since $\boldsymbol{v} = \boldsymbol{\omega} \times \boldsymbol{r} + \boldsymbol{u}$ and all the tensions equal zero for the pure rotation $\boldsymbol{\omega} \times \boldsymbol{r}$, the linearization of the tangent dynamic conditions (1.16) with regard to Γ leads to the expressions

$$\tilde{\tau}_{13}(\boldsymbol{u}) := \rho\nu\left(\frac{\partial u_i}{\partial x_3} + \frac{\partial u_3}{\partial x_i}\right) = 0 \quad \text{on } \Gamma_i \qquad i = 1, 2. \tag{1.33}$$

For the normal tension we have (with an accuracy up to the second order)

$$\tilde{\tau}_{33}(\boldsymbol{u}) := -P + 2\rho\nu\frac{\partial u_3}{\partial x_3} = -p_a \qquad \text{for } x_3 = b + \zeta(t, x_1, x_2).$$

Taking into account that $P = P_0 + p$ and by (1.32), we obtain after linearization that

$$p - 2\rho\nu\frac{\partial u_3}{\partial x_3} = \rho g(\zeta - \delta_2 x_1 + \delta_1 x_2) \quad \text{on } \Gamma. \tag{1.34}$$

The corresponding linearized kinematic condition (see Subsection 3.1.3) for the moving surface $\Gamma(t)$ with equation $x_3 = b + \zeta(t, x_1, x_2)$ has the form

$$\frac{\partial \zeta}{\partial t} = u_3 \qquad \text{on } \Gamma, \tag{1.35}$$

which shows that the third component of the relative velocity $\boldsymbol{u}(t,x)$ coincides with the displacement velocity along the axis Ox_3 of the free surface $\Gamma(t)$.

Let us formulate now in full the linear boundary value problem on the movement of a system "body + fluid" with a partially filled cavity. In order to define the system's evolution, it is necessary to solve jointly the equations (1.30) with boundary conditions for the body (i.e., to define the body's position and velocity) and equations (1.31) with the boundary conditions (1.33)–(1.35) and the initial conditions

$$\boldsymbol{u}(0,x) = \boldsymbol{u}^0(x), \qquad \zeta(o, x_1, x_2) = \zeta^0(x_1, x_2), \tag{1.36}$$

that determine the velocity field of the fluid and the location of its free surface for $t = 0$.

As in the case of a gyrostate, one should add the kinematic relations for the direction cosines to all these equations.

For an ideal fluid, we assume that $\nu = 0$ in (1.33) and (1.34), and take into consideration the boundary nonleaking condition (1.17) instead of the stickness condition (1.31).

3.1.6 Transition to Undimensional Variables

While investigating and numerically solving problems of fluid oscillations, it is convenient to use undimensional variables (this will also be the case in other applied problems). We will illustrate the transition to the undimensional form of equations by using the problem (1.30)–(1.35) on small oscillations of a viscous heavy fluid in a partially filled container.

Let us denote a characteristic size (for example, the diameter) of the container by l, and introduce the characteristic time $t_* = (l/g)^{1/2}$, where $g > 0$ is the gravitation acceleration. Then, assuming that the axial vectors $\boldsymbol{e}_i$ of the Cartesian coordinate system $Ox_1x_2x_3$ are undimensional, we obtain for the velocity field $(\boldsymbol{u}(t,x))_{\text{dim}}$,

$$(\boldsymbol{u})_{\text{dim}} = (\boldsymbol{u})_{\text{undim}} l/t_*,$$

where $(\boldsymbol{u})_{\text{undim}}$ is an undimensional velocity field. Similarly, for the pressure, the angular velocity, and the angular acceleration, we obtain

$$(p)_{\text{dim}} = (p)_{\text{undim}} \rho l^2/t_*^2,$$
$$(\boldsymbol{\omega})_{\text{dim}} = (\boldsymbol{\omega})_{\text{undim}}/t_*,$$
$$(\boldsymbol{\varepsilon})_{\text{dim}} = (\boldsymbol{\varepsilon})_{\text{undim}}/t_*^2.$$

All the other physical quantities can be replaced in (1.30)–(1.35) in the same way.

Mere calculations show that, after the above mentioned transition to undimensional variables, equations (1.30)–(1.35) preserve their form but all the magnitudes are undimensional (they will be denoted by the same symbols). In (1.31)–(1.35) one should formally replace ρ by 1, g by 1, and the kinematic viscosity by the number $(Re)^{-1}$ where $Re := l^2/(\nu t_*)$ is the so-called *Reynolds number.* If in a problem the capillary forces are taken into account (see Chapter 4), then the coefficient of surface tension σ on the boundary Γ of the "fluid-gas" should be replaced—according to the transition to undimensional variables—by the magnitude $(Bo)^{-1}$, where $Bo := \rho g l^2/\sigma$ is the *Bond number.*

Let us note that the undimensional combinations of Re and Bo are called *criteria (numbers) of similarity*; the Reynolds number characterizes the order of the relation between the inertial forces acting in the hydrosystem and the viscous forces, and the Bond number characterizes the order of relation between the gravitation forces and the capillary forces. In particular, for $Re = \infty$ ($\nu = 0$) we obtain the model of an ideal fluid, for $Bo = 0$ we have the conditions of complete low-gravity (the model of a capillary fluid), and for $Bo = \infty$ ($\sigma = 0$) we get the model of a heavy fluid.

Further, we will consider (whenever it is necessary) that the transition to undimensional variables in the considered problem has already been made. However, it is more convenient to use also the dimensional form of the equations and the relations that follow from these equations (for example, the law of system energy balance) in order to make the physical sense of every term in such relations more evident.

3.2 Motion of an Ideal Fluid in a Closed Stationary Container

The main method employed in this section and which will also be used in the sequel is the method of projecting the motion equations onto orthogonal subspaces of the space of fields $\boldsymbol{L}_2(\Omega)$. In our context, this method will help us to "split up" the problem into some independent components, thus essentially simplifying it.

3.2.1 Basic Equations

We start the investigation of linear hydrodynamic problems with the simplest one. Let us consider that an ideal incompressible fluid is located in some immovable container Ω and completely fills it. The fluid is influenced by the field of external forces with the mass density $\boldsymbol{F} = \boldsymbol{F}(t, x)$, where t is the time variable and $x = (x_1, x_2, x_3)$ is a point in Ω.

We denote the velocity field of the fluid at the point x and time t by $\boldsymbol{u} = \boldsymbol{u}(t, x)$, the pressure by $P = P(t, x)$, and the density, that is considered to be constant, by ρ. For the sake of convenience, we single out from the field $\boldsymbol{F}$ the external gravitational field with acceleration $\boldsymbol{g}$, and consider that the latter field is constant and directed against the axis Ox_3. Then, in the stable state, when $\boldsymbol{F} = \boldsymbol{F}_0 = \boldsymbol{g}$ and $\boldsymbol{u}(t, x) \equiv 0$, the equilibrium pressure settles in the fluid,

$$P_0(x_3) = -\rho g x_3 + p_a,$$

where p_a is a constant.

Let us consider that the movements of the fluid in the container are small. The second law of Newton applied to the particles of the ideal fluid leads to the Euler equation

$$\frac{\partial \boldsymbol{u}}{\partial t} + (\boldsymbol{u} \cdot \nabla)\boldsymbol{u} = -\frac{1}{\rho}\nabla P + \boldsymbol{F}.$$

We represent the full pressure $P = P(t, x)$ as a sum, $P(t, x) = P_0(x_3) + p(t, x)$, and call the term $p(t, x)$, added to the static pressure $P_0(x_3)$, the *dynamic pressure*. Next, we similarly represent the field of external forces $\boldsymbol{F} = \boldsymbol{F}(t, x)$ as $\boldsymbol{F} = -\nabla(g x_3) + \boldsymbol{f}(t, x)$ where we omitted the gravitational component. By linearizing the Euler equation, i.e., by assuming $\boldsymbol{u}$, ∇p, and $\boldsymbol{f}$ to have small magnitudes of the same order and by taking linear terms only, we obtain

$$\frac{\partial \boldsymbol{u}}{\partial t} + \frac{1}{\rho}\nabla p = \boldsymbol{f} \qquad \text{in } \Omega. \tag{2.1}$$

We should add to this equation the incompressibility condition for a fluid, that for $\rho = \text{const}$ leads to the solenoidality of the velocity field,

$$\operatorname{div} \boldsymbol{u} = 0 \qquad \text{in } \Omega. \tag{2.2}$$

Since we consider that the fluid is moving in an immovable closed rigid container Ω, the nonleaking condition is true on the boundary $S = \partial\Omega$, that is,

$$\boldsymbol{u} \cdot \boldsymbol{n} = u_n = 0 \qquad \text{on } S. \tag{2.3}$$

For a complete formulation of the considered evolution problem, one should add to (2.1)–(2.3) the initial condition

$$\boldsymbol{u}(0, x) = \boldsymbol{u}^0(x). \tag{2.4}$$

Relations (2.1)–(2.4) provide the full set of equations, boundary and initial conditions that describe the small movements of a fluid in a closed container.

3.2.2 Existence of Solutions

Relations (2.2) and (2.3) mean that the velocity field $\boldsymbol{u}$ (for each fixed t) belongs to the space $\boldsymbol{J}_0(\Omega)$ that was introduced in Section 2.1. The derivative $\partial\boldsymbol{u}/\partial t$ belongs to this space as well. The term $\rho^{-1}\nabla p$ in (2.1) belongs to the space of potential fields $\boldsymbol{G}(\Omega)$. Let us recall the the subspaces $\boldsymbol{J}_0(\Omega)$ and $\boldsymbol{G}(\Omega)$ are orthogonal complements to each other in the space $\boldsymbol{L}_2(\Omega)$. Therefore, equation (2.1) can be understood as the decomposition of the given field $\boldsymbol{f}$ into two mutually orthogonal components.

Let us denote the operator of orthogonal projection of the space $\boldsymbol{L}_2(\Omega)$ onto the subspace $\boldsymbol{J}_0(\Omega)$ by P_0. Then, assuming $\boldsymbol{f} \in \boldsymbol{L}_2(\Omega)$ (for every t), from (2.1) we obtain

$$\frac{\partial \boldsymbol{u}}{\partial t} = P_0 \boldsymbol{f}, \qquad \rho^{-1}\nabla p = (I - P_0)\boldsymbol{f}. \tag{2.5}$$

From the second equation we find p, and from the first equation and (2.4) we obtain

$$\boldsymbol{u}(t) = \boldsymbol{u}^0 + \int_0^t (P_0 \boldsymbol{f})(\tau)\mathrm{d}\tau.$$

Here we assume that the initial field $\boldsymbol{u}^0 \in \boldsymbol{J}_0(\Omega)$, i.e., $\boldsymbol{u}^0 \in \boldsymbol{L}_2(\Omega)$, $\operatorname{div} \boldsymbol{u}^0 = 0$ in Ω, $u_n^0 = 0$ on S.

An important physical conclusion follows from (2.5): the potential component of the field of external mass forces does not affect the velocity of the fluid motion in a wholly filled container, but affects the value of its pressure.

Let us recall (see Section 2.1) that for finding the potential component of a field $\boldsymbol{f}$ one should find the solution q to the inhomogeneous Newmann problem

$$\Delta q = \operatorname{div} \boldsymbol{f} \quad \text{in } \Omega, \qquad \frac{\partial q}{\partial n} = f_n \quad \text{on } S.$$

Then

$$\begin{aligned} \boldsymbol{f} &= \nabla q + \boldsymbol{f}_0, \\ \boldsymbol{f}_0 &= P_0 \boldsymbol{f} \in \boldsymbol{J}_0(\Omega), \\ p &= \rho q, \\ \boldsymbol{u}(t) &= \boldsymbol{u}^0 + \int_0^t \boldsymbol{f}_0(\tau)\mathrm{d}\tau. \end{aligned}$$

Let us also note that if we take the scalar product in $L_2(\Omega)$ of both sides of (2.1) and $\rho\boldsymbol{u}$, then from the following relation

$$\int_\Omega \frac{\partial \boldsymbol{u}}{\partial t} \cdot \boldsymbol{u}\mathrm{d}\Omega = \frac{1}{2}\frac{\mathrm{d}}{\mathrm{d}t}\int_\Omega |\boldsymbol{u}|^2 \mathrm{d}\Omega = \int_\Omega \boldsymbol{f} \cdot \boldsymbol{u}\mathrm{d}\Omega = \int_\Omega \boldsymbol{f}_0 \cdot \boldsymbol{u}\mathrm{d}\Omega$$

we get the equation for kinetic energy balance, namely,

$$\frac{\mathrm{d}}{\mathrm{d}t}T_{\mathrm{kin}} = \frac{\mathrm{d}}{\mathrm{d}t}\left(\frac{1}{2}\rho\int_\Omega |\boldsymbol{u}|^2 \mathrm{d}\Omega\right) = \rho\int_\Omega \boldsymbol{f} \cdot \boldsymbol{u}\mathrm{d}\Omega = \rho\int_\Omega \boldsymbol{f}_0 \cdot \boldsymbol{u}\mathrm{d}\Omega. \tag{2.6}$$

Hence, in particular, when $\boldsymbol{f}_0(t,x) \equiv \mathbf{0}$, it appears that for a potential field of external forces the kinetic energy of a fluid is constant and equals $(\rho/2)\int_\Omega |\boldsymbol{u}^0|^2 \mathrm{d}\Omega$.

3.3 Small movements of an Ideal Fluid in an Open Immovable Container

This problem is related to the classical Cauchy-Poisson problem on waves on the surface of a heavy fluid located in a half-space or a strip region. However, by virtue of the boundedness of the region occupied by fluid, the frequency spectrum of the fluid oscillations is discrete and not continuous as in the classical case.

3.3.1 Statement of the Problem and the Basic Equations

Let us suppose that an immovable container is partially filled with a heavy ideal fluid. The part of the container wall that is in contact with fluid is denoted by S. In this chapter we do not take into account the forces of surface tension and that is why, in the stable state, the free surface Γ of the fluid is plane and horizontal, i.e., it is orthogonal to the direction of the acceleration of the gravitation field $\boldsymbol{g}$. The pressure in the gas located above the fluid is considered to be constant and equal to the external atmospheric pressure p_a; we do not take into account the gas motion. As in Section 3.2, we choose the coordinate system $Ox_1x_2x_3$ such that the axis Ox_3 is vertical and directed against the vector $\boldsymbol{g}$. Then, in the stable state, the fluid pressure $P_0(x_3)$ is given by the relation

$$P_0(x_3) = p_a - \rho g x_3,$$

where we took into account that the equation of Γ has the form $x_3 = 0$.

We consider small movements of the fluid close to the stable state. The velocity field $\boldsymbol{u} = \boldsymbol{u}(t,x)$, $x = (x_1, x_2, x_3)$, the dynamic pressure $p(t,x)$, and the deviation $x_3 = \zeta(t, x_1, x_2)$ of the free moving surface $\Gamma(t)$ from the equilibrium surface Γ are considered to have small magnitudes of the same order (together with their derivatives). The linearized Euler equations have the same form as before,

$$\frac{\partial \boldsymbol{u}}{\partial t} + \frac{1}{\rho}\nabla p = \boldsymbol{f}, \quad \operatorname{div} \boldsymbol{u} = 0 \qquad \text{in } \Omega. \tag{3.1}$$

On the moistened solid boundary S, the nonleaking condition is valid,

$$u_n := \boldsymbol{u} \cdot \boldsymbol{n} = 0 \qquad \text{on } S. \tag{3.2}$$

On the free moving surface $\Gamma(t)$ of the fluid, the pressure equals the atmospheric pressure p_a; the linearization of this condition and the evident kinematic relation lead to the boundary conditions

$$p = \rho g \zeta, \quad u_n = u_3 = \frac{\partial \zeta}{\partial t} \qquad \text{on } \Gamma. \tag{3.3}$$

For a complete formulation of the evolution problem we should add to (3.1)–(3.3) the following initial conditions,

$$\boldsymbol{u}(0,x) = \boldsymbol{u}^0(x), \qquad \zeta(0, x_1, x_2) = \zeta^0(x_1, x_2). \tag{3.4}$$

3.3.2 Projection of Euler Equations

As it was mentioned in Section 2.1, the velocity field $\boldsymbol{u}$ of an ideal fluid located in an open container has vortex and potential components. That is why the solution $\boldsymbol{u}$ of the problem (3.1)–(3.4) will be looked for in the subspace $\boldsymbol{J}_{0,S}(\Omega)$. Projecting equation (3.1) onto $J_0(\Omega)$ we obtain (as previously) the relation

$$\frac{\mathrm{d}}{\mathrm{d}t} P_0 \boldsymbol{u} = P_0 \boldsymbol{f},$$

hence, with regard to (3.4), we have

$$P_0 \boldsymbol{u} = \int_0^t (P_0 \boldsymbol{f})(\tau) \mathrm{d}\tau + P_0 \boldsymbol{u}^0. \tag{3.5}$$

In the given problem, all our considerations are related to the subspace of potential fields $\boldsymbol{G}_{\mathrm{h},S}(\Omega)$. Let us denote $(I - P_0)\boldsymbol{u} = \nabla \Phi$ and $(I - P_0)\boldsymbol{f} = \nabla f$. Then, applying the operator $(I - P_0)$ to (3.1), we obtain the equation

$$\nabla \left(\frac{\partial \Phi}{\partial t} + \frac{1}{\rho} p - f \right) = 0;$$

we hence get the Cauchy-Lagrange integral

$$\frac{\partial \Phi}{\partial t} + \frac{1}{\rho} p = f + c(t) \qquad \text{in } \Omega, \tag{3.6}$$

with an arbitrary function of time $c(t)$. Let us choose this function as

$$c(t) = -\int_{\Gamma} f(t, x_1, x_2, 0) \mathrm{d}\Gamma.$$

Now, by writing down equation (3.6) for points on Γ and by using the first boundary condition in (3.3) we obtain

$$\frac{\partial \Phi}{\partial t} + g\zeta = f_1, \qquad \int_{\Gamma} f_1(t, x_1, x_2, 0) \mathrm{d}\Gamma = 0. \tag{3.7}$$

Since the potential component $\nabla\Phi$ of the velocity field $\boldsymbol{u}$ belongs to the space $\boldsymbol{G}_{\mathrm{h},S}(\Omega)$, we get

$$\begin{aligned} \nabla\Phi &= 0 \quad \text{in } \Omega, \\ \frac{\partial \Phi}{\partial n} &= 0 \quad \text{on } S, \\ \frac{\partial \Phi}{\partial n} &= \frac{\partial \zeta}{\partial t} \quad \text{on } \Gamma; \end{aligned} \tag{3.8}$$

the last boundary condition follows from the kinematic relation (3.3).

3.3.3 Existence of Solutions

Following Example 3° in Subsection 1.8.6, let us consider the Hilbert space $H^1_{\mathrm{h},S}(\Omega)$ and the trace operator γ_Γ on Γ mapping $H^1_{\mathrm{h},S}(\Omega)$ into $L_{2,\Gamma}$. In this case, the canonical operator T associates the solution $q(x)$ of the Neumann problem

$$\begin{aligned} \Delta q &= 0 \quad \text{in } \Omega, \\ \frac{\partial q}{\partial n} &= 0 \quad \text{on } S, \\ \frac{\partial q}{\partial n} &= \psi \quad \text{on } \Gamma. \end{aligned} \tag{3.9}$$

to each function $\psi \in H_\Gamma^{-1/2}$ on the boundary Γ. The self-adjoint positive operator $C = \gamma_\Gamma T$ is related with this problem. Taking into account (3.8) and (3.9) we obtain

$$\Phi|_\Gamma = \gamma_\Gamma T \left(\frac{\partial \zeta}{\partial t} \right) = C \frac{\partial \zeta}{\partial t} \quad \text{on } \Gamma, \tag{3.10}$$

and therefore, with regard to (3.7) we obtain the basic equation

$$C\frac{\mathrm{d}^2\zeta}{\mathrm{d}t^2} + g\zeta = f_1. \tag{3.11}$$

The initial conditions (3.4) for equation (3.11) lead to the following,

$$\zeta(0) = \zeta^0, \qquad \frac{\mathrm{d}\zeta(0)}{\mathrm{d}t} = [(I - P_0)\boldsymbol{u}^0] \cdot \boldsymbol{n} =: \zeta^1. \tag{3.12}$$

Let us now consider the Cauchy problem (3.11), (3.12), in the space $L_{2,\Gamma}$. Problems of this kind were considered in Section 1.5. According to the conclusions in that section, the generalized solution of problem (3.11), (3.12) can be written as

$$\begin{aligned}\zeta(t) = {} & \cos(g^{1/2}tC^{-1/2})\zeta^0 + g^{1/2}C^{1/2}\sin(g^{1/2}tC^{-1/2})\zeta^1 \\ & + \int_0^t (gC)^{-1/2}\sin[g^{1/2}C^{-1/2}(t-\tau)]f_1(\tau)\mathrm{d}\tau. \end{aligned} \tag{3.13}$$

To ensure the finiteness of the kinetic energy of the fluid at every moment of time, it is necessary that $\nabla\Phi \in \boldsymbol{G}_{\mathrm{h},S}(\Omega)$, i.e.,

$$\begin{aligned}\left(\frac{\partial\Phi}{\partial n}\right)_\Gamma = \frac{\partial\zeta}{\partial t} = {} & -g^{1/2}C^{-1/2}\sin(g^{1/2}tC^{-1/2})\zeta^0 + g\cos(g^{1/2}tC^{-1/2})\zeta^1 \\ & + C^{-1}\int_0^t \cos[g^{1/2}(t-\tau)C^{-1/2}]f_1(\tau)\mathrm{d}\tau \in H_\Gamma^{-1/2}. \end{aligned} \tag{3.14}$$

Hence, the requirement (3.14) will be satisfied if

$$\zeta^0 \in H = L_{2,\Gamma}, \qquad \zeta^1 \in H_\Gamma^{-1/2}, \tag{3.15}$$

and $f_1(t)$ is a continuous function with values in $H_\Gamma^{1/2}$. Whenever these conditions are satisfied, the component $\nabla\Phi$ of the velocity field $\boldsymbol{u}$ belongs to $\boldsymbol{G}_{\mathrm{h},S}(\Omega)$ and the kinetic energy corresponding to this field equals

$$\frac{1}{2}\rho\int_\Omega |\nabla\Phi|^2\mathrm{d}\Omega = \frac{1}{2}\rho\left(C\frac{\mathrm{d}\zeta}{\mathrm{d}t}, \frac{\mathrm{d}\zeta}{\mathrm{d}t}\right)_{L_{2,\Gamma}} = \frac{1}{2}\rho\int_\Gamma \left(C\frac{\mathrm{d}\zeta}{\mathrm{d}t}\right)\frac{\mathrm{d}\zeta}{\mathrm{d}t}\mathrm{d}\Gamma.$$

That is why operator C can be naturally called the *operator of kinetic energy* of potential fluid flows.

Summing up the results derived so far for the evolution problem (3.1)–(3.4), we conclude that if at the initial moment of time the kinetic and potential energy of the system are finite, i.e.,

$$\zeta^0 \in H = L_{2,\Gamma}, \qquad \boldsymbol{u}^0 \in \boldsymbol{J}_{0,S}(\Omega), \tag{3.16}$$

and the field of external forces, $\boldsymbol{f}(t)$, is a continuous function with values in $\boldsymbol{L}_2(\Omega)$, then problem (3.1)–(3.4) at any moment of time has a generalized solution $\boldsymbol{u}(t)$ with continuous kinetic and potential energy. In this case, the law of balance for full energy holds true,

$$\frac{1}{2}\rho\|\boldsymbol{u}(t)\|^2_{\boldsymbol{L}_2(\Omega)} + \frac{1}{2}\rho g\int_\Gamma |\zeta|^2 \mathrm{d}\Gamma = \frac{1}{2}\rho\|\boldsymbol{u}^0\|^2_{\boldsymbol{L}_2(\Omega)} + \frac{1}{2}\rho g\int_\Gamma |\zeta^0|^2 \mathrm{d}\Gamma + \int_0^t \left(\int_\Omega \boldsymbol{f}(\tau)\cdot\boldsymbol{u}(\tau)\mathrm{d}\Omega\right)\mathrm{d}\tau. \tag{3.17}$$

3.3.4 Proper Oscillations

If the field of external forces is negligible, $\boldsymbol{f}(t,x) \equiv \boldsymbol{0}$, the fluid movements in a container can be represented as a superposition of the so-called *proper oscillations*, i.e., solutions of the problem (3.1)–(3.4), which depend on time according to the $\exp(\mathrm{i}\omega t)$-law:

$$\begin{aligned} \boldsymbol{u}(t,x) &= \mathrm{e}^{\mathrm{i}\omega t}\boldsymbol{v}(x), \\ p(t,x) &= \mathrm{e}^{\mathrm{i}\omega t}q(x), \\ \zeta(t,x_1,x_2) &= \mathrm{e}^{\mathrm{i}\omega t}h(x_1,x_2). \end{aligned} \tag{3.18}$$

Here, ω is the frequency of proper oscillations and the space multipliers $\boldsymbol{v}(x)$, $q(x)$, and $h(x_1,x_2)$ are called *modes of oscillations.*

In this case, by setting $f_1 \equiv 0$ in equation (3.11) we get the spectral problem

$$h = \lambda Ch, \qquad \lambda = \omega^2/g, \; h \in L_{2,\Gamma}. \tag{3.19}$$

Since the operator $C : L_{2,\Gamma} \to L_{2,\Gamma}$ is positive and compact, then, according to the conclusions in Subsection 1.4.1, problem (3.19) has a discrete spectrum $\{\lambda_k\}_{k=1}^\infty$ which consists of the characteristic numbers of operator C, i.e., $\lambda_k = \mu_k(C) = 1/\lambda_k(C)$ and $0 < \lambda_1 \le \cdots \le \lambda_k \le \cdots$, $\lambda_k \to \infty$ $(k \to \infty)$. The system of eigenelements $\{h_k\}_{k=1}^\infty$ corresponding to the eigenvalues λ_k (that are counted according to their multiplicity, as usual) forms an orthonormal basis in $L_{2,\Gamma}$,

$$(h_k, h_l)_{L_{2,\Gamma}} = \delta_{kl}, \qquad (Ch_k, h_l) = \lambda_k^{-1}\delta_{kl}. \tag{3.20}$$

Hence, the frequencies ω_k of proper oscillations of a heavy fluid in a partially filled container can be calculated by the formulas

$$\omega_k^{\pm} = \pm(g/\lambda_k(C))^{1/2}, \qquad k = 1,2,\ldots \tag{3.21}$$

According to the Minimax Principle stated in Subsection 1.4.1, it follows that the numbers $\lambda_k(C)$ can be found as succesive maximums of the variational quotient

$$\int_\Gamma |\Phi|^2 \mathrm{d}\Gamma \Big/ \int_\Omega |\nabla\Phi|^2 \mathrm{d}\Omega, \tag{3.22}$$

subject to the additional conditions

$$\begin{aligned} \Delta\Phi &= 0 \quad \text{in } \Omega, \\ \frac{\partial\Phi}{\partial n} &= 0 \quad \text{on } S, \\ \int_\Gamma \Phi d\Gamma &= 0. \end{aligned} \tag{3.23}$$

The solutions of the variational problem (3.22) and (3.23) have the asymptotic behavior

$$\lambda_k(C) = \left(\frac{\operatorname{mes}\Gamma}{4\pi}\right)^{1/2} k^{-1/2}[1 + o(1)], \qquad k \to \infty, \tag{3.24}$$

hence we get the next asymptotic formula for oscillation frequencies,

$$\omega_k^{\pm} = \pm g^{1/2} k^{1/4} \left(\frac{\operatorname{mes}\Gamma}{4\pi}\right)^{-1/4} [1 + o(1)], \qquad k \to \infty. \tag{3.25}$$

In conclusion, let us note that the modes of proper oscillations

$$\{\boldsymbol{v}_k(x)\}_{k=1}^{\infty} = \{\nabla\Phi_k(x)\}_{k=1}^{\infty} = \{Th_k(x_1, x_2)\}_{k=1}^{\infty}$$

form an orthogonal basis in the subspace $\boldsymbol{G}_{\mathrm{h},S}(\Omega)$.

3.4 Small Joint Movements of a Fluid and a Container

The problem considered in this section was solved by N. E. Zhukovsky. He has shown for the first time that the motion of a hydrodynamical system with a cavity wholly filled with an ideal fluid is equivalent to the motion of a rigid body with changed characteristics.

3.4.1 Statement of the Problem and the Basic Equations

Suppose that we investigate the motion of a gyrostate, i.e., a rigid body with a cavity Ω wholly filled with an incompressible fluid. The pole O of the nonstationary coordinate system $Ox_1x_2x_3$ is rigidly connected to the body and is chosen at an immovable point, if there is any, or at the mass center C of the whole system.

Assuming that in the nonperturbed state the system is stable (in particular, it does not rotate), the motion equation of the fluid and the boundary conditions can be written down in the nonstationary coordinate system. From (1.26) with $\omega_0 = 0$, these equations become

$$\begin{aligned} \frac{\partial \boldsymbol{u}}{\partial t} + \boldsymbol{\varepsilon} \times \boldsymbol{r} &= -\frac{1}{\rho}\nabla p + \boldsymbol{f}, \\ \operatorname{div} \boldsymbol{u} &= 0 \quad \text{in } \Omega, \end{aligned} \tag{4.1}$$

$$u_n = 0 \quad \text{on } S. \tag{4.2}$$

Here we used the same notations as in Sections 3.1–3.3: $\boldsymbol{u} = \boldsymbol{u}(t,x)$ is the velocity field in the cavity Ω, $p = p(t,x)$ is the dynamic pressure, $\boldsymbol{\varepsilon}$ is the angular acceleration of the body, and $\boldsymbol{r}$ is the radius vector related to the pole O.

The motion of the container filled with fluid is described by the kinetic moment equation of the whole system, that can be obtained from (1.29) for $\omega_0 = 0$, namely,

$$\boldsymbol{J}\boldsymbol{\varepsilon} + \rho\frac{\mathrm{d}}{\mathrm{d}t}\int_\Omega (\boldsymbol{r}\times\boldsymbol{u})\mathrm{d}\Omega = \boldsymbol{M}, \tag{4.3}$$

where $\boldsymbol{J}$ is the inertia tensor of the whole system, i.e., of the body together with the heavy fluid, and $\boldsymbol{M}$ is the main moment relatively to the pole O of all the forces influencing the system.

Let us recall that for a gyrostate the motion law of the mass center is determined from equation (1.19), i.e., it has the same form as the motion equation of a system with mass m and without any fluid. For this reason, this law will not be considered further.

3.4.2 Finding the Velocity Field and the Pressure

Equations (4.1) and (4.2) point out that the velocity field $\boldsymbol{u}$ can be found in the subspace $\boldsymbol{J}_0(\Omega)$. Applying the projection operator P_0 onto this subspace to both sides of (4.1) we obtain

$$\frac{\mathrm{d}\boldsymbol{u}}{\mathrm{d}t} = -P_0(\boldsymbol{\varepsilon}\times\boldsymbol{r}) + P_0\boldsymbol{f}. \tag{4.4}$$

Hence it appears that for finding $\mathrm{d}\boldsymbol{u}/\mathrm{d}t$ and $\boldsymbol{u}$ it is sufficient to obtain the potential components of the fields $\boldsymbol{\varepsilon}\times\boldsymbol{r}$ and $\boldsymbol{f}$. Since $\operatorname{div}(\boldsymbol{\varepsilon}\times\boldsymbol{r}) = 0$, the field $\nabla\psi = (I-P_0)(\boldsymbol{\varepsilon}\times\boldsymbol{r})$ can be found by means of a solution of the problem

$$\begin{aligned} &\Delta\psi = 0 \quad \text{in } \Omega,\\ &\frac{\partial\psi}{\partial n} = (\boldsymbol{\varepsilon}\times\boldsymbol{r})\cdot\boldsymbol{n} \quad \text{on } S \end{aligned} \tag{4.5}$$

Thus, $P_0(\boldsymbol{\varepsilon}\times\boldsymbol{r}) = \boldsymbol{\varepsilon}\times\boldsymbol{r} - \nabla\psi$. We can find the field $P_0\boldsymbol{f}$ in the same way. If the initial velocity field $\boldsymbol{u}^0 \in \boldsymbol{J}_0(\Omega)$ is given, then from (4.4) we get

$$\boldsymbol{u}(t) = \boldsymbol{u}^0 - \int_0^t P_0(\boldsymbol{\varepsilon}\times\boldsymbol{r})(\tau)\mathrm{d}\tau + \int_0^t (P_0\boldsymbol{f})(\tau)\mathrm{d}\tau. \tag{4.6}$$

If we project (4.1) onto the subspace of potential fields, we obtain the equation

$$\psi = f - \frac{1}{\rho}p \quad \text{in } \Omega, \tag{4.7}$$

where $\nabla f := (I-P_0)\boldsymbol{f}$. Thus, we can find the pressure $p = p(t,x)$ at any point $x\in\Omega$.

3.4.3 Defining the Motion Law of the Body

Using (4.4) we can omit $\mathrm{d}\boldsymbol{u}/\mathrm{d}t$ in the kinetic moment equation (4.3) and we thus obtain

$$\boldsymbol{J\varepsilon} - \rho\int_\Omega [\boldsymbol{r} \times P_0(\boldsymbol{\varepsilon} \times \boldsymbol{r})]\mathrm{d}\Omega = \boldsymbol{M}_1, \qquad \boldsymbol{M}_1 := \boldsymbol{M} - \rho\int_\Omega (\boldsymbol{r} \times P_0\boldsymbol{f})\mathrm{d}\Omega. \tag{4.8}$$

The integral in the left side of (4.8) is a linear operator acting on the three-dimensional space of vectors $\boldsymbol{\varepsilon}$, or, in other words, it is a tensor which will be denoted by $\boldsymbol{\Pi}$. Let us find out its properties.

First of all, we notice that this tensor is symmetric. Indeed, using properties of the mixed product and the orthoprojector P_0, we have

$$\begin{aligned}(\boldsymbol{\Pi\varepsilon}) \cdot \boldsymbol{\eta} &= \int_\Omega [\boldsymbol{r} \times P_0(\boldsymbol{\varepsilon} \times \boldsymbol{r})]\mathrm{d}\Omega \cdot \boldsymbol{\eta} = \int_\Omega (\boldsymbol{\eta} \times \boldsymbol{r}) \cdot P_0(\boldsymbol{\varepsilon} \times \boldsymbol{r})\mathrm{d}\Omega \\ &= \int_\Omega P_0(\boldsymbol{\eta} \times \boldsymbol{r}) \cdot P_0(\boldsymbol{\varepsilon} \times \boldsymbol{r})\mathrm{d}\Omega = \boldsymbol{\varepsilon} \cdot (\Pi\boldsymbol{\eta}).\end{aligned}$$

Hence, we get that $\boldsymbol{\Pi}$ is nonnegative, namely, for $\boldsymbol{\eta} = \boldsymbol{\varepsilon}$

$$(\boldsymbol{\Pi\varepsilon}) \cdot \boldsymbol{\varepsilon} = \int_\Omega |P_0(\boldsymbol{\varepsilon} \times \boldsymbol{r})|^2\mathrm{d}\Omega \geq 0.$$

If, in the last inequality, the right side is zero, then $P_0(\boldsymbol{\varepsilon} \times \boldsymbol{r}) = 0$, i.e., the field $\boldsymbol{\varepsilon} \times \boldsymbol{r}$ is a potential. This is true if and only if $\boldsymbol{\varepsilon} = \boldsymbol{0}$. Therefore, the tensor $\boldsymbol{\Pi}$ is positive definite,

$$(\boldsymbol{\Pi\varepsilon}) \cdot \boldsymbol{\varepsilon} \geq \lambda_\Pi|\boldsymbol{\varepsilon}|^2, \qquad \lambda_\Pi > 0.$$

Let us next write down equation (4.8) as

$$(\boldsymbol{J} - \rho\boldsymbol{\Pi})\boldsymbol{\varepsilon} = \boldsymbol{M}_1, \tag{4.9}$$

and prove that the tensor $\boldsymbol{J} - \rho\boldsymbol{\Pi}$ is positive definite, too, and therefore has an inverse. We introduce the tensors of the inertia moments of the rigid body, $\boldsymbol{J}_\mathrm{b}$, and of the hardened mass of the fluid, $\boldsymbol{J}_\mathrm{f}$. Clearly $\boldsymbol{J} = \boldsymbol{J}_\mathrm{b} + \boldsymbol{J}_\mathrm{f}$. Then

$$\boldsymbol{J}_\mathrm{f}\boldsymbol{\varepsilon} = \rho\int_\Omega (\boldsymbol{r} \times \boldsymbol{\varepsilon}) \times \boldsymbol{r}\mathrm{d}\Omega,$$

hence

$$\boldsymbol{J}_\mathrm{f}\boldsymbol{\varepsilon} \cdot \boldsymbol{\varepsilon} = \rho\int_\Omega (\boldsymbol{r} \times \boldsymbol{\varepsilon}) \times \boldsymbol{r} \cdot \boldsymbol{\varepsilon}\mathrm{d}\Omega = \rho\int_\Omega (\boldsymbol{\varepsilon} \times \boldsymbol{r}) \cdot (\boldsymbol{\varepsilon} \times \boldsymbol{r})\mathrm{d}\Omega.$$

Therefore,

$$\begin{aligned}(\boldsymbol{J}_\mathrm{f} - \rho\boldsymbol{\Pi})\boldsymbol{\varepsilon} \cdot \boldsymbol{\varepsilon} &= \rho\int_\Omega [(I - P_0)(\boldsymbol{\varepsilon} \times \boldsymbol{r})] \cdot (\boldsymbol{\varepsilon} \times \boldsymbol{r})\mathrm{d}\Omega \\ &= \rho\int_\Omega |(I - P_0)(\boldsymbol{\varepsilon} \times \boldsymbol{r})|^2\mathrm{d}\Omega \geq 0.\end{aligned}$$

This expression equals zero if $(I - P_0)(\boldsymbol{\varepsilon} \times \boldsymbol{r}) = \boldsymbol{0}$, i.e., $\boldsymbol{\varepsilon} \times \boldsymbol{r} \in \boldsymbol{J}_0(\Omega)$; in particular,

$(\varepsilon \times r) \cdot n = (n \times \varepsilon) \cdot r = 0$ on S. This linear function of the coordinates x_1, x_2, x_3 on the closed surface S can be equal to zero if and only if $\varepsilon \times n = 0$. From this relation, and since n is changing on S, it follows that $\varepsilon = \mathbf{0}$.

Hence, the tensor $J_{\mathrm{as}} = J_{\mathrm{f}} - \rho \Pi$ of the associate moment of inertia is positive definite, and equation (4.9) becomes

$$(J_{\mathrm{b}} + J_{\mathrm{as}})\varepsilon = M_1. \tag{4.10}$$

Thus, we came to the following important conclusion: if the field of mass forces f is a potential (and $M_1 = M$), then a body with a cavity wholly filled with an ideal fluid is moving under the influence of external forces as any other rigid body under the influence of the same forces, with the changed moment of inertia $J_{\mathrm{ch}} = J_{\mathrm{b}} + J_{\mathrm{as}}$.

3.4.4 Zhukovsky Potentials

The problem (4.5) can be divided into three separate problems:

$$\Delta\psi_i = 0 \quad \text{in } \Omega,$$
$$\frac{\partial \psi_i}{\partial n} = e_i \times r \cdot n \quad \text{on } S, \qquad i = 1, 2, 3. \tag{4.11}$$

The solutions of each of them depend only on the form of the cavity Ω filled with fluid. The functions $\psi_i(x)$ are called the *Zhukovsky potentials*. If they are known, then the solution of (4.5) has the form $\psi = \sum_{i=1}^{3} \varepsilon_i \psi_i$, where ε_i are the projections of the angular acceleration ε onto the coordinate axes. Hence we obtain

$$(I - P_0)(\varepsilon \times r) = \sum_{i=1}^{3} \varepsilon_i \nabla \psi_i.$$

Then, for the associated tensor of inertia moments J_{as} we finally have

$$J_{\mathrm{as}}\varepsilon \cdot \varepsilon = \sum_{i,j=1}^{3} \varepsilon_i \varepsilon_j \langle \psi_i, \psi_j \rangle,$$

where, according to Zhukovsky's notations,

$$\langle \psi_i, \psi_j \rangle = \rho \int_\Omega \nabla\psi_i \cdot \nabla\psi_j \, d\Omega.$$

In conclusion, we should notice that for several cavities with typical forms, the matrices containing the entries $\langle \psi_i, \psi_j \rangle$ have been calculated using computer programs.

3.5 Small Joint Movements Around a Fixed Point of a Body and a Fluid Partially Filling the Cavity

The problem considered in this section links in some sense the problems investigated in Sections 3.2 and 3.3 and, therefore, we will use similar mathematical methods for solving it.

3.5.1 Statement of the Problem and the Basic Equations

Let us assume that a rigid body is fixed at a certain point O and performs small movements relatively to that point. The body has a cavity Ω partially filled with an ideal fluid. For the sake of simplicity, we assume that, in the nonperturbed state, there are no external forces acting upon the body except for the gravitation force. Then, in the stable state, the equilibrium surface Γ of the fluid is plane and orthogonal to the gravitation acceleration $\boldsymbol{g}_{\text{grav}} = \boldsymbol{g}$.

In order to describe the small movements of the system, let us introduce the stationary coordinate system $Oy_1y_2y_3$, with the axis Oy_3 directed against the vector $\boldsymbol{g}$, and the nonstationary system $Ox_1x_2x_3$ rigidly connected to the body. We assume that, in the nonperturbed state, the systems $Oy_1y_2y_3$ and $Ox_1x_2x_3$ coincide, the equation of the equilibrium surface Γ has the form $x_3 = b$, and the mass center C of the entire system, which is located on the same vertical line as the pole O, has coordinates $(0, 0, -a)$.

The location of the nonstationary system $Ox_1x_2x_3$ relatively to the stationary system $Oy_1y_2y_3$ is described by the vector of angular displacement $\boldsymbol{\delta}(t)$; for small deviations of the body, the following relations hold true (as it was already mentioned in Section 3.1)

$$\boldsymbol{\delta}(t) = \sum_{k=1}^{3} \delta_k(t)\boldsymbol{e}_k,$$

$$\boldsymbol{\omega}(t) = \frac{\mathrm{d}\boldsymbol{\delta}}{\mathrm{d}t} = \sum_{k=1}^{3} \frac{\mathrm{d}\delta_k(t)}{\mathrm{d}t}\boldsymbol{e}_k = \sum_{k=1}^{3} \omega_k(t)\boldsymbol{e}_k,$$

$$\boldsymbol{\varepsilon}(t) = \frac{\mathrm{d}\boldsymbol{\omega}}{\mathrm{d}t} = \sum_{k=1}^{3} \frac{\mathrm{d}\omega_k}{\mathrm{d}t}\boldsymbol{e}_k = \sum_{k=1}^{3} \varepsilon_k\boldsymbol{e}_k = \sum_{k=1}^{3} \frac{\mathrm{d}^2\delta_k}{\mathrm{d}t^2}\boldsymbol{e}_k,$$

where $\boldsymbol{e}_k$ are the axial vectors of Ox_k, $\boldsymbol{\omega}(t)$ is the angular velocity, and $\boldsymbol{\varepsilon}(t)$ is the angular acceleration of the body.

In the nonstationary coordinate system $Ox_1x_2x_3$ the equations determining the relative velocity $\boldsymbol{u} = \boldsymbol{u}(t,x)$ and the dynamic pressure $p(t,x) = P(t,x) - P_0(t,x)$ can be obtained from (1.31) and (1.32) taking into account the small external field of mass forces $\boldsymbol{f}(t,x)$ for $\nu = 0$, namely,

$$\frac{\partial \boldsymbol{u}}{\partial t} + \boldsymbol{\varepsilon} \times \boldsymbol{r} = -\frac{1}{\rho}\nabla p + \boldsymbol{f},$$

$$\operatorname{div} \boldsymbol{u} = 0 \quad \text{in } \Omega. \tag{5.1}$$

On the solid boundary S of the region Ω filled with fluid, the nonleaking condition should hold true,

$$u_n = \boldsymbol{u} \cdot \boldsymbol{n} = 0 \quad \text{on } S, \tag{5.2}$$

and, on the equilibrium surface Γ given by $x_3 = b$, we have the dynamic and kinematic conditions (1.34) and (1.35) written as

$$p = \rho g(\zeta - \delta_2 x_1 + \delta_1 x_2),$$
$$\frac{\partial \zeta}{\partial t} = u_3 \quad \text{on } \Gamma. \tag{5.3}$$

While deriving the law of change of the kinetic moment of the whole system, we notice that, according to (1.29), for the considered problem we have

$$\boldsymbol{g} = -g(\boldsymbol{e}_3 - \delta_2 \boldsymbol{e}_1 + \delta_1 \boldsymbol{e}_2),$$

and the radius vector $\boldsymbol{r}_C$ of the mass center C equals $-a\boldsymbol{e}_3$. Therefore,

$$m\boldsymbol{r}_C \times \boldsymbol{g} = ma\boldsymbol{e}_3 \times (\boldsymbol{e}_3 - \delta_2 \boldsymbol{e}_1 + \delta_1 \boldsymbol{e}_2) = -mga(\delta_1 \boldsymbol{e}_1 + \delta_2 \boldsymbol{e}_2).$$

From (1.30) we obtain

$$\boldsymbol{J}\boldsymbol{\varepsilon} + \rho \int_\Omega \boldsymbol{r} \times \frac{\partial \boldsymbol{u}}{\partial t} \mathrm{d}\Omega + mga(\delta_1 \boldsymbol{e}_1 + \delta_2 \boldsymbol{e}_2) - \rho g \int_\Omega (\boldsymbol{e}_3 \times \boldsymbol{r}) \zeta \mathrm{d}\Gamma = \boldsymbol{M}, \tag{5.4}$$

where $\boldsymbol{J}$ is the inertia tensor of the system that corresponds to the nonperturbed state, $\boldsymbol{\varepsilon}$ is the angular acceleration of the body, and $\boldsymbol{M}$ is a summarized moment relatively to the pole O of the small forces influencing the system.

For a complete mathematical statement of the problem we should add to equations (5.1)–(5.4) the following initial conditions,

$$\boldsymbol{u}(0, x) = \boldsymbol{u}^0(x),$$
$$\zeta(0, x_1, x_2) = \zeta^0(x_1, x_2),$$
$$\boldsymbol{\delta}(0) = \boldsymbol{\delta}^0,$$
$$\boldsymbol{\delta}'(0) = \boldsymbol{\omega}^0. \tag{5.5}$$

3.5.2 The Law of Full Energy Balance

Before investigating the problem (5.1)–(5.5), let us find the law of full energy balance of the system for the classical solutions of this problem. First, we multiply equation (5.1) by $\rho\boldsymbol{u}(t, x)$ and integrate it on Ω. Taking into account the boundary conditions (5.2) and (5.3) and the incompressibility condition, we obtain

$$\rho \int_\Omega \frac{\partial \boldsymbol{u}}{\partial t} \cdot \boldsymbol{u} \mathrm{d}\Omega + \rho \int_\Omega (\boldsymbol{\varepsilon} \times \boldsymbol{r}) \cdot \boldsymbol{u} \mathrm{d}\Omega + \int_\Omega \nabla p \cdot \boldsymbol{u} \mathrm{d}\Omega$$

$$
\begin{aligned}
&= \frac{1}{2}\rho\frac{\mathrm{d}}{\mathrm{d}t}\int_\Omega |\boldsymbol{u}|^2\mathrm{d}\Omega + \rho\int_\Omega \left(\frac{\mathrm{d}\boldsymbol{\omega}}{\mathrm{d}t}\times\boldsymbol{r}\right)\cdot\boldsymbol{u}\mathrm{d}\Omega + \rho g\int_\Gamma (\zeta - \delta_2 x_1 + \delta_1 x_2)\frac{\partial\zeta}{\partial t}\mathrm{d}\Gamma \\
&= \frac{1}{2}\rho\frac{\mathrm{d}}{\mathrm{d}t}\int_\Omega |\boldsymbol{u}|^2\mathrm{d}\Omega + \rho\int_\Omega \left(\frac{\mathrm{d}\boldsymbol{\omega}}{\mathrm{d}t}\times\boldsymbol{r}\right)\cdot\boldsymbol{u}\mathrm{d}\Omega + \frac{1}{2}\rho g\frac{\mathrm{d}}{\mathrm{d}t}\int_\Gamma |\zeta|^2\mathrm{d}\Gamma \\
&\quad + \rho g\int_\Gamma (-\delta_2 x_1 + \delta_1 x_2)\frac{\partial\zeta}{\partial t}\mathrm{d}\Gamma \\
&= \rho\int_\Omega \boldsymbol{f}\cdot\boldsymbol{u}\mathrm{d}\Omega.
\end{aligned}
$$

Multiplying (5.4) by $\boldsymbol{\omega}$, we similarly obtain

$$
\begin{aligned}
&\frac{1}{2}\frac{\mathrm{d}}{\mathrm{d}t}(\boldsymbol{J}\boldsymbol{\omega}\cdot\boldsymbol{\omega}) + \rho\int_\Omega \left(\boldsymbol{r}\times\frac{\partial\boldsymbol{u}}{\partial t}\right)\cdot\boldsymbol{\omega}\mathrm{d}\Omega + \frac{1}{2}mga\frac{\mathrm{d}}{\mathrm{d}t}(|\delta_1|^2 + |\delta_2|^2) \\
&\quad - \rho g\int_\Gamma (\boldsymbol{e}_3\times\boldsymbol{r})\zeta\cdot\boldsymbol{\omega}\mathrm{d}\Gamma = \boldsymbol{M}\cdot\boldsymbol{\omega}.
\end{aligned}
$$

Adding up these two equations we get

$$
\begin{aligned}
&\frac{1}{2}\frac{\mathrm{d}}{\mathrm{d}t}\left\{\rho\int_\Omega |\boldsymbol{u}|^2\mathrm{d}\Omega + J\boldsymbol{\omega}\cdot\boldsymbol{\omega} + 2\rho\int_\Omega (\boldsymbol{\omega}\times\boldsymbol{r})\cdot\boldsymbol{u}\mathrm{d}\Omega\right\} \\
&\quad + \frac{1}{2}\frac{\mathrm{d}}{\mathrm{d}t}\left\{\rho g\int_\Gamma |\zeta|^2\mathrm{d}\Gamma + mga(|\delta_1|^2 + |\delta_2|^2) + 2\rho g\int_\Gamma (-\delta_2 x_1 + \delta_1 x_2)\zeta\mathrm{d}\Gamma\right\} \\
&= \rho\int_\Omega \boldsymbol{f}\cdot\boldsymbol{u}\mathrm{d}\Omega + \boldsymbol{M}\cdot\boldsymbol{\omega}. \qquad (5.6)
\end{aligned}
$$

The first term within brackets (multiplyed by two) is the kinetic energy, and the second one is the potential energy, respectively, while the right side represents the capacity of the external forces influencing the system.

3.5.3 Projecting Euler Equations

As in Section 3.3, the solution $\boldsymbol{u}(t,x)$ of the problem (5.1)–(5.5) will be looked for in the subspace $\boldsymbol{J}_{0,S}(\Omega)$, and equation (5.1) will be projected onto the subspace $\boldsymbol{J}_0(\Omega)$ and $\boldsymbol{G}_{\mathrm{h},S}(\Omega)$, respectively. Using previous notations, we have

$$
\frac{\mathrm{d}P_0\boldsymbol{u}}{\mathrm{d}t} = -P_0(\boldsymbol{\varepsilon}\times\boldsymbol{r}) + P_0\boldsymbol{f}. \qquad (5.7)
$$

We solve this equation by means of Zhukovsky potentials, in the way described in Section 3.4.

For the potential component of velocity $\nabla\Phi = (I - P_0)\boldsymbol{u}$, we obtain from (5.1) that

$$
\frac{\partial}{\partial t}\nabla\Phi + \nabla\psi + \frac{1}{\rho}\nabla p = \nabla\varphi,
$$

where $\nabla\psi = (I - P_0)(\boldsymbol{\varepsilon} \times \boldsymbol{r})$, and $\nabla\varphi = (I - P_0)\boldsymbol{f}$. Hence, we get the Cauchy-Lagrange integral (see (3.6))

$$\frac{\partial\Phi}{\partial t} + \psi + \frac{1}{\rho}p = \varphi + c(t) \quad \text{in } \Omega. \tag{5.8}$$

This equation for $x \in \Gamma$, in conjunction with condition (5.3), leads to the following,

$$\frac{\partial\Phi}{\partial t} + g\zeta + \psi + g(-\delta_2 x_1 + \delta_1 x_2) = \varphi + c(t). \tag{5.9}$$

Besides, in virtue of (5.3), $\partial\zeta/\partial t = \partial\Phi/\partial n$ on Γ.

Let us recall that function Φ satisfies the equations

$$\Delta\Phi = 0 \quad \text{in } \Omega,$$
$$\frac{\partial\Phi}{\partial n} = 0 \quad \text{on } S,$$

and, therefore, on Γ we get

$$\Phi|_\Gamma = \gamma_\Gamma(\Phi|_\Omega) = \gamma_\Gamma T\left(\frac{\partial\Phi}{\partial n}\right)_\Gamma = C\left(\frac{\partial\Phi}{\partial n}\right)_\Gamma = C\frac{\partial\zeta}{\partial t}.$$

Then, equation (5.9) can be written down as

$$C\frac{\mathrm{d}^2\zeta}{\mathrm{d}t^2} + g\zeta + \psi + g(-\delta_2 x_1 + \delta_1 x_2) = \varphi + c(t).$$

In this equation we can get rid of the arbitrary function $c(t)$ by projecting the equation onto the subspace $L_{2,\Gamma}$ of functions from $L_2(\Gamma)$ that are ortogonal to the unity. Let us denote the operator of orthogonal projection onto $L_{2,\Gamma}$ by θ. Then, taking into account that

$$\int_\Gamma \Phi \mathrm{d}\Gamma = 0, \qquad \int_\Gamma \zeta \mathrm{d}\Gamma = 0,$$

we obtain

$$C\frac{\mathrm{d}^2\zeta}{\mathrm{d}t^2} + g\zeta + \theta\psi + g\theta(-\delta_2 x_1 + \delta_1 x_2) = \theta\varphi. \tag{5.10}$$

3.5.4 Kinetic Moment Equation

In (5.4), let us replace $\partial\boldsymbol{u}/\partial t$ by the sum $(\partial/\partial t)(P_0\boldsymbol{u}) + (\partial/\partial t)(\nabla\Phi)$. Using (5.7) we obtain

$$\begin{aligned}\boldsymbol{J\varepsilon} &- \rho\int_\Omega [\boldsymbol{r} \times P_0(\boldsymbol{\varepsilon} \times \boldsymbol{r})]\mathrm{d}\Omega + \rho\int_\Omega \left(\boldsymbol{r} \times \frac{\partial}{\partial t}\nabla\Phi\right) + mga(\delta_1\boldsymbol{e}_1 + \delta_2\boldsymbol{e}_2)\\ &- \rho g\int_\Gamma (-x_2\boldsymbol{e}_1 + x_1\boldsymbol{e}_2)\zeta\mathrm{d}\Gamma\\ &= \boldsymbol{M} - \rho\int_\Omega (\boldsymbol{r} \times P_0\boldsymbol{f})\mathrm{d}\Omega.\end{aligned} \tag{5.11}$$

As in Section 3.4, the term $\rho\int_\Omega [\boldsymbol{r}\times P_0(\boldsymbol{\varepsilon}\times\boldsymbol{r})]\mathrm{d}\Omega$ can be written down as $\rho\boldsymbol{\Pi}\boldsymbol{\varepsilon}$ and then, together with the first term of (5.11), we get the expression $(\boldsymbol{J}_\mathrm{b}+\boldsymbol{J}_\mathrm{as})\boldsymbol{\varepsilon}$ and

$$\boldsymbol{J}_\mathrm{as}\boldsymbol{\varepsilon}\cdot\boldsymbol{\varepsilon} = \rho\int_\Omega |(I-P_0)(\boldsymbol{\varepsilon}\times\boldsymbol{r})|^2\mathrm{d}\Omega = \rho\int_\Omega |\nabla\psi|^2\mathrm{d}\Omega, \qquad \psi=\sum_{i=1}^3 \varepsilon_i\psi_i, \tag{5.12}$$

where ε_i are the projections of the angular acceleration $\boldsymbol{\varepsilon}$ and ψ_i are the Zhukovsky potentials.

We next transform the expression

$$\int_\Omega(\boldsymbol{r}\times\nabla\Phi)\mathrm{d}\Omega = \sum_{i=1}^3 \boldsymbol{e}_i\left\{\int_\Omega(\boldsymbol{r}\times\nabla\Phi)\cdot\boldsymbol{e}_i\mathrm{d}\Omega\right\} = \sum_{i=1}^3 \boldsymbol{e}_i\int_\Omega(\boldsymbol{e}_i\times\boldsymbol{r})\cdot\nabla\Phi\mathrm{d}\Omega.$$

Since $\operatorname{div}(\boldsymbol{e}_i\times\boldsymbol{r})=0$, the right hand side of the above equation equals

$$\sum_{i=1}^3 \boldsymbol{e}_i\int_{\partial\Omega}(\boldsymbol{e}_i\times\boldsymbol{r})_n\Phi\mathrm{d}S = \sum_{i=1}^3 \boldsymbol{e}_i\int_{\partial\Omega}\frac{\partial\psi_i}{\partial n}\Phi\mathrm{d}S. \tag{5.13}$$

Taking into that condition $\partial\Phi/\partial n=0$ on S, and the fact that the functions ψ_i and Φ are harmonic, the right hand side of the last relation becomes

$$\sum_{i=1}^3 \boldsymbol{e}_i\int_{\partial\Omega}\psi_i\frac{\partial\Phi}{\partial n}\mathrm{d}S = \sum_{i=1}^3 \boldsymbol{e}_i\int_\Gamma\psi_i\frac{\partial\Phi}{\partial n}\mathrm{d}\Gamma. \tag{5.14}$$

Therefore, in virtue of the kinematic condition (5.3) we have

$$\begin{aligned}\rho\int_\Omega\left(\boldsymbol{r}\times\frac{\partial}{\partial t}\nabla\Phi\right)\mathrm{d}\Omega &= \rho\int_\Omega\left(\sum_{i=1}^3 \mathrm{e}_i\psi_i\right)\frac{\partial}{\partial t}\left(\frac{\partial\Phi}{\partial n}\right)_\Gamma\mathrm{d}\Gamma\\ &= \rho\int_\Omega\left(\sum_{i=1}^3 \mathrm{e}_i\psi_i\right)\frac{\mathrm{d}^2\zeta}{\mathrm{d}t^2}\mathrm{d}\Gamma.\end{aligned} \tag{5.15}$$

Finally, the kinetic moment equation becomes

$$\begin{aligned}(\boldsymbol{J}_\mathrm{b}+\boldsymbol{J}_\mathrm{as})\boldsymbol{\varepsilon} &+ \rho\int_\Gamma\frac{\mathrm{d}^2\zeta}{\mathrm{d}t^2}\left(\sum_{i=1}^3\psi_i\boldsymbol{e}_i\right)\mathrm{d}\Gamma + mga(\delta_1\mathrm{e}_1+\delta_2\mathrm{e}_2) - \rho g\int_\Gamma(-x_2\mathrm{e}_1+x_1\mathrm{e}_2)\zeta\mathrm{d}\Gamma\\ &= \boldsymbol{M} - \rho\int_\Omega(\boldsymbol{r}\times P_0\boldsymbol{f})\mathrm{d}\Omega.\end{aligned} \tag{5.16}$$

3.5.5 Investigating the Complete System of Motion Equations

Let us assume that we already performed a transition to the undimensional form (in the way described in Subsection 3.1.6) upon the system of equations (5.10), (5.16), and the resulting equations have the same form as the original ones. We will consider the new system of equations (5.10), (5.16) as a single second order differential equation in the space $E := L_{2,\Gamma} \oplus \mathbb{R}^3$ of pairs $(\zeta; \boldsymbol{\delta})^{\mathrm{t}}$ that consist of the deviation functions $\zeta(t, x_1, x_2)$ and the three-dimensional vectors $\boldsymbol{\delta}(t)$ of the angular displacement of the body. After multiplying (5.10) by ρ, the system (5.10), (5.16) can be written down as

$$A\frac{\mathrm{d}^2\xi}{\mathrm{d}t^2} + B\xi = \chi(t), \tag{5.17}$$

where

$$\xi = \begin{pmatrix} \zeta \\ \boldsymbol{\delta} \end{pmatrix} \in L_{2,\Gamma} \oplus \mathbb{R}^3,$$

$$\chi(t) = \begin{pmatrix} \chi_1(t) \\ \chi_2(t) \end{pmatrix} := \begin{pmatrix} \rho\theta\varphi \\ \boldsymbol{M} - \rho \int_\Omega (\boldsymbol{r} \times P_0 \boldsymbol{f}) \mathrm{d}\Omega \end{pmatrix},$$

and the operator matrices $A = (A_{ij})_{i,j=1}^2$ and $B = (B_{ij})_{i,j=1}^2$ act in the following way,

$$A\begin{pmatrix} \zeta \\ \boldsymbol{\delta} \end{pmatrix} := \left(\rho C\zeta + \rho\theta\left(\sum_{i=1}^3 \delta_i\psi_i\right);\ \rho\int_\Gamma \theta\zeta\left(\sum_{i=1}^3 \psi_i \boldsymbol{e}_i\right)\mathrm{d}\Gamma + (\boldsymbol{J}_{\mathrm{b}} + \boldsymbol{J}_{\mathrm{as}})\boldsymbol{\delta}\right)^{\mathrm{t}},$$

$$B\begin{pmatrix} \zeta \\ \boldsymbol{\delta} \end{pmatrix} := \Bigg(\rho g\zeta - \rho g\theta(\delta_2 x_1 - \delta_1 x_2);\ -\rho g\int_\Gamma \theta(-x_2\boldsymbol{e}_1 + x_1\boldsymbol{e}_2)\zeta \mathrm{d}\Gamma$$

$$+ mga(\delta_1\boldsymbol{e}_1 + \delta_2\boldsymbol{e}_2)\Bigg)^{\mathrm{t}}. \tag{5.18}$$

In our specific problem, these operators represent the kinetic and the potential energy, respectively. Let us study the properties of operators A and B.

As it is obvious from formulas (5.18), the operators A and B are bounded on $E = L_{2,\Gamma} \oplus \mathbb{R}^3$. Moreover, A is compact because the operator C is compact and, therefore, A_{11} is compact as well, and A_{21}, A_{12}, and A_{22} are finite-dimensional.

It turns out that the operators A and B are self-adjoint. In order to prove this property, we should check that the operators A_{11} and B_{11} are self-adjoint in $L_{2,\Gamma}$, the operators A_{22} and B_{22} are self-adjoint in $\mathbb{R}^3$, and $A_{21}^* = A_{12}$, $B_{21}^* = B_{12}$.

The first assertion follows from the self-adjointness of C and the form of operator B_{11}. The second assertion can be directly checked, too. Let us show that the last conditions hold true. We have,

$$A_{21}\zeta \cdot \boldsymbol{\delta} = \rho\int_\Gamma \zeta\left(\sum_{i=1}^3 \psi_i\boldsymbol{e}_i\right)\mathrm{d}\Gamma \cdot \boldsymbol{\delta} = \rho\int_\Gamma \zeta\left(\sum_{i=1}^3 \psi_i\delta_i\right)\mathrm{d}\Gamma$$

$$= \rho\int_\Gamma \zeta\theta\left(\sum_{i=1}^3 \delta_i\psi_i\right)\mathrm{d}\Gamma = \int_\Gamma \zeta(A_{12}\boldsymbol{\delta})\mathrm{d}\Gamma.$$

Further,

$$\begin{aligned} B_{21}\zeta \cdot \boldsymbol{\delta} &= -\rho g \int_\Gamma \zeta\theta(-x_2\boldsymbol{e}_1 + x_1\boldsymbol{e}_2)\mathrm{d}\Gamma \cdot \boldsymbol{\delta} \\ &= -\rho g \int_\Gamma \zeta\theta(-\delta_1 x_2 + \delta_2 x_1)\mathrm{d}\Gamma = \int_\Gamma \zeta(B_{12}\boldsymbol{\delta})\mathrm{d}\Gamma. \end{aligned}$$

Now, let us check that operator A is positive. Indeed,

$$\begin{aligned} (A\xi,\xi)_E &= \rho\int_\Gamma (C\zeta)\zeta\mathrm{d}\Gamma + \rho\int_\Gamma \theta\left(\sum_{i=1}^3 \delta_i\psi_i\right)\zeta\mathrm{d}\Gamma + \rho\int_\Gamma \zeta\left(\sum_{i=1}^3 \psi_i\boldsymbol{e}_i\right)\mathrm{d}\Gamma\cdot\boldsymbol{\delta} \\ &\quad + (\boldsymbol{J}_\mathrm{b} + \boldsymbol{J}_\mathrm{as})\boldsymbol{\delta}\cdot\boldsymbol{\delta} \\ &= \rho\int_\Gamma (C\zeta)\zeta\mathrm{d}\Gamma + 2\rho\int_\Gamma \zeta\left(\sum_{i=1}^3 \delta_i\psi_i\right)\mathrm{d}\Gamma + (\boldsymbol{J}_\mathrm{b} + \boldsymbol{J}_\mathrm{as})\boldsymbol{\delta}\cdot\boldsymbol{\delta}. \end{aligned} \tag{5.19}$$

Let us introduce the auxiliary function $\eta(x)$, called the *dispacement potential*, that is the solution of the von Neumann problem

$$\begin{aligned} \Delta\eta &= 0 \quad \text{in } \Omega, \\ \frac{\partial\eta}{\partial n} &= 0 \quad \text{on } S, \\ \frac{\partial\eta}{\partial n} &= \zeta \quad \text{on } \Gamma, \\ \int_\Gamma \eta\mathrm{d}\Gamma &= 0, \end{aligned} \tag{5.20}$$

and the projections $\nabla\Psi_i$ onto the subspaces $\boldsymbol{G}_{\mathrm{h},S}(\Omega)$ of the fields $\nabla\psi_i$ for which

$$\begin{aligned} \Delta\Psi_i &= 0 \quad \text{in } \Omega, \\ \frac{\partial\Psi_i}{\partial n} &= 0 \quad \text{on } S, \\ \Psi_i &= \theta\psi_i \quad \text{on } \Gamma. \end{aligned} \tag{5.21}$$

Then,

$$\int_\Gamma (C\zeta)\zeta\mathrm{d}\Gamma = \int_\Omega |\nabla\eta|^2\mathrm{d}\Omega.$$

Further,

$$\int_\Gamma \zeta\left(\sum_{i=1}^3 \delta_i\psi_i\right)\mathrm{d}\Gamma = \int_\Gamma \frac{\partial\eta}{\partial n}\left(\sum_{i=1}^3 \delta_i\Psi_i\right)\mathrm{d}\Gamma = \int_\Omega \nabla\eta\cdot\nabla\left(\sum_{i=1}^3 \delta_i\psi_i\right)\mathrm{d}\Omega. \tag{5.22}$$

Since the fields $\nabla\Psi_i$ and $\nabla(\psi_i - \Psi_i)$ are orthogonal by construction, from (5.12) it follows that

$$\boldsymbol{J}_{\mathrm{as}}\boldsymbol{\delta}\cdot\boldsymbol{\delta}=\rho\int_{\Omega}\left|\nabla\left(\sum_{i=1}^{3}\delta_i\psi_i\right)\right|^2\mathrm{d}\Omega\geq\rho\int_{\Omega}\left|\nabla\left(\sum_{i=1}^{3}\delta_i\Psi_i\right)\right|^2\mathrm{d}\Omega, \tag{5.23}$$

whence, taking into account (5.22) and (5.23), we have

$$(A\xi,\xi)_E=\rho\left\{\int_{\Omega}\left|\nabla\eta+\nabla\left(\sum_{i=1}^{3}\delta_i\Psi_i\right)\right|^2\mathrm{d}\Omega+\boldsymbol{J}_{\mathrm{as}}\boldsymbol{\delta}\cdot\boldsymbol{\delta}\right.$$
$$\left.-\int_{\Omega}\left|\nabla\left(\sum_{i=1}^{3}\delta_i\Psi_i\right)\right|^2\mathrm{d}\Omega+\boldsymbol{J}_{\mathrm{b}}\boldsymbol{\delta}\cdot\boldsymbol{\delta}\right\}\geq\rho\boldsymbol{J}_{\mathrm{b}}\boldsymbol{\delta}\cdot\boldsymbol{\delta}, \tag{5.24}$$

i.e., operator A is nonnegative.If $A\xi=0$, then from (5.24) it follows that $\boldsymbol{\delta}=\mathbf{0}$ and, therefore, $\nabla\eta=0$. Hence, in virtue of (5.20), $\eta=0$. Therefore, operator A is positive.

The operator A can be inverted in an explicit form. Indeed, let

$$A\begin{pmatrix}\zeta\\ \boldsymbol{\delta}\end{pmatrix}=\begin{pmatrix}\varphi\\ \boldsymbol{\sigma}\end{pmatrix},$$

i.e.,

$$\rho C\zeta+\rho\theta\left(\sum_{i=1}^{3}\delta_i\psi_i\right)=\varphi,$$

$$\rho\int_{\Gamma}\zeta\left(\sum_{i=1}^{3}\psi_i\boldsymbol{e}_i\right)\mathrm{d}\Gamma+(\boldsymbol{J}_{\mathrm{b}}+\boldsymbol{J}_{\mathrm{as}})\boldsymbol{\delta}=\boldsymbol{\sigma}.$$

From the first equation we get

$$\zeta=(\rho C)^{-1}\varphi-C^{-1}\theta\left(\sum_{i=1}^{3}\delta_i\psi_i\right). \tag{5.25}$$

Substituting this expression into the second equation we obtain the following equation relatively to $\boldsymbol{\delta}=\sum_{i=1}^{3}\delta_i\boldsymbol{e}_i$,

$$-\rho\int_{\Gamma}\left\{C^{-1}\theta\left(\sum_{i=1}^{3}\delta_i\psi_i\right)\right\}\left(\sum_{j=1}^{3}\psi_j\boldsymbol{e}_j\right)\mathrm{d}\Gamma+(\boldsymbol{J}_{\mathrm{b}}+\boldsymbol{J}_{\mathrm{as}})\boldsymbol{\delta}$$
$$=\boldsymbol{\sigma}-\int_{\Gamma}(C^{-1}\varphi)\left(\sum_{j=1}^{3}\psi_j\boldsymbol{e}_j\right)\mathrm{d}\Gamma.$$

Identifying the coefficients of $\boldsymbol{e}_j$, we obtain the next system of three equations,

$$-\rho\sum_{i=1}^{3}\delta_i\int_{\Gamma}C^{-1}(\theta\psi_i)\psi_j\mathrm{d}\Gamma+\sum_{i=1}^{3}J_{ji}\delta_i=\sigma_j-\int_{\Gamma}(C^{-1}\varphi)\psi_j\mathrm{d}\Gamma,\qquad j=1,2,3, \tag{5.26}$$

relatively to the unknowns δ_i. The corresponding homogeneous system does not have any nontrivial solution. This fact follows from the positivity of operator A.

Finally, the domain of definition of operator A^{-1} consists of all the elements $(\zeta;\ \boldsymbol{\delta})^{\mathrm{t}}$ for which $\zeta \in \mathcal{D}(C^{-1})$.

To investigate the properties of operator B, let us note that this operator equals zero for any element $\xi_0 = (0;\ \boldsymbol{e}_3)^{\mathrm{t}}$. Physically, this fact means that the homogeneous equation has a stationary solution that describes the slow rotation of the body relatively to the axis Oy_3, which passes through the fixed point O and the mass center of the system (such small rotations are not accompanied by changes in the free surface).

Let us denote the one-dimensional subspace formed by ξ_0 by E_0, and the four-dimensional subspace formed by the linearly independent elements $(0;\ \boldsymbol{e}_1)^{\mathrm{t}}$, $(0;\ \boldsymbol{e}_2)^{\mathrm{t}}$, $(\theta x_1;\ \mathbf{0})^{\mathrm{t}}$, and $(\theta x_2;\ \mathbf{0})^{\mathrm{t}}$ by E_4. We also introduce the subspace E^1 of elements of the form $(\zeta;\ \mathbf{0})^{\mathrm{t}}$ where ζ is orthogonal to the functions θx_1 and θx_2 in $L_{2,\Gamma}$. Obviously,

$$E := L_{2,\Gamma} \oplus \mathbb{R}^3 = E_0 \oplus E_4 \oplus E^1.$$

In virtue of formula (5.18), each subspace in this decomposition is invariant relatively to operator B.

In the subspace E^1, operator B acts according the equation $B(\zeta;\ 0)^{\mathrm{t}} = \rho g(\zeta;\ 0)^{\mathrm{t}}$ and, therefore, B is positive definite. Obviously, in the four-dimensional space E_4, operator B is bounded from below. Let us find out what are the circumstances under which this operator is nonnegative in E_4. If we represent an arbitrary element $\xi = (\zeta;\ \boldsymbol{\delta})^{\mathrm{t}} \in E_4$ as

$$\xi = \xi_1 + \xi_2 := \begin{pmatrix} -(\rho g)^{-1} B_{12}\boldsymbol{\delta} \\ \boldsymbol{\delta} \end{pmatrix} + \begin{pmatrix} \zeta_1 \\ \mathbf{0} \end{pmatrix}, \qquad \zeta_1 = \zeta + (\rho g)^{-1} B_{12}\boldsymbol{\delta}, \tag{5.27}$$

then

$$B\xi_1 = \begin{pmatrix} 0 \\ -(\rho g)^{-1} B_{21}B_{12}\boldsymbol{\delta} + B_{22}\boldsymbol{\delta} \end{pmatrix}, \qquad B\xi_2 = \begin{pmatrix} \rho g \zeta_1 \\ B_{21}\zeta_1 \end{pmatrix},$$

and, therefore,

$$(B\xi_1, \xi_2) = (\xi_1, B\xi_2) = 0.$$

Further,

$$(B\xi, \xi) = (B\xi_1, \xi_1) + (B\xi_2, \xi_2) = (B\xi_1, \xi_1) + \rho g \int_\Gamma |\zeta_1|^2 \mathrm{d}\Gamma.$$

Hence, operator B is nonnegative in E_4 if and only if, for some $c \geq 0$, we have

$$(B\xi_1, \xi_1) \geq c\|\xi_1\|^2, \qquad \xi_1 \in E_4. \tag{5.29}$$

Indeed, if this inequality holds true, then

$$\begin{aligned}(B\xi,\xi) &\geq c\|\xi_1\|^2+\rho g\|\xi_2\|^2 \geq \min(c;\,\rho g)(\|\xi_1\|^2+\|\xi_2\|^2)\\ &\geq \frac{1}{2}\min(c;\,\rho g)\|\xi\|^2.\end{aligned}$$

Conversely, if for a certain $d \geq 0$ the inequality $(B\xi,\xi) \geq d\|\xi\|^2$ holds true for all $\xi \in E_4$, then, in particular, $B(\xi_1,\xi_1) \geq d\|\xi_1\|^2$.

Let us consider in more detail the quadratic form $B(\xi_1,\xi_1)$. From (5.27), (5.28), and (5.18) we get

$$\begin{aligned}(B\xi_1,\xi_1) &= -(\rho g)^{-1}B_{21}B_{12}\boldsymbol{\delta}\cdot\boldsymbol{\delta}+B_{22}\boldsymbol{\delta}\cdot\boldsymbol{\delta}\\ &= -\rho g\int_\Gamma |\theta(\delta_2x_1-\delta_1x_2)|^2\mathrm{d}\Gamma + mga(\delta_1^2+\delta_2^2).\end{aligned}$$

We introduce the notation

$$\alpha_{ij} := \int_\Gamma (\theta x_i)x_j\mathrm{d}\Gamma.$$

Then,

$$g^{-1}(B\xi_1,\xi_1) = (ma-\rho\alpha_{22})\delta_1^2+2\rho\alpha_{12}\delta_1\delta_2+(ma-\rho\alpha_{11})\delta_2^2.$$

Hence, using the well-known Silvester criterion, we obtain that operator B is nonnegative on the subspace E_4 and, therefore, on the entire space $E = L_{2,\Gamma}\oplus\mathbb{R}^3$, if and only if the following conditions hold true,

$$\begin{aligned}&\Delta_1 := ma-\rho\alpha_{11} \geq 0,\\ &\Delta_2 := (ma-\rho\alpha_{11})(ma-\rho\alpha_{22})-\rho\alpha_{12}^2 \geq 0.\end{aligned} \tag{5.30}$$

Operator B is positive definite on the subspace $\tilde{E} := (L_{2,\Gamma}\oplus\mathbb{R}^3)\ominus E_0$ if and only if $\Delta_1 > 0$ and $\Delta_2 > 0$. Finally, the kernel of operator B coincides with the subspace E_0, if and only if

$$\Delta_2 \neq 0, \tag{5.31}$$

3.5.6 Proper Oscillations

Let us consider the solutions of the homogeneous problem (5.17) that depend on time according to the law $\xi(t) = \xi\exp(\mathrm{i}\omega t)$, where ω is the frequency of free oscillations. We obtain the following spectral problem,

$$\begin{aligned}&B\xi = \lambda A\xi,\\ &\lambda = \omega^2,\\ &\xi = (\zeta;\ \boldsymbol{\delta})^{\mathrm{t}} \in L_{2,\Gamma}\oplus\mathbb{R}^3 = E.\end{aligned} \tag{5.32}$$

As we just found out in Subsection 3.5.5, operator A is compact and positive on E, and operator B is bounded from below and self-adjoint.

If condition (5.31) holds true, then the kernel $\operatorname{Ker} B$ of operator B is one-dimensional and generated by the vector $\xi_0 = (0;\ \boldsymbol{e}_3)^{\mathrm{t}}$, i.e., $\operatorname{Ker} B = E_0$. Further, we will only consider this general case.

Investigating equation (5.32), we note that it has the trivial solution

$$\lambda_0 = 0, \qquad \xi = \xi_0 = (0;\ \boldsymbol{e}_3)^{\mathrm{t}} \in E_0,$$

that is connected with a slow rotation of the system around the axis Ox_3. There are no other nonzero solutions of equation (5.32) corresponding to the value $\lambda_0 = 0$ for $\Delta_2 \neq 0$.

Let us represent now the solution ξ of problem (5.32) as $\xi = \delta_3\xi_0 + \tilde{\xi}$, where ξ_0 is the above-mentioned trivial solution, and $\tilde{\xi} \in \tilde{E} = E \ominus E_0$. Substituting this decomposition of ξ into (5.32) and taking into account the equality $B\xi_0 = 0$ we obtain

$$B\tilde{\xi} = \lambda(A\tilde{\xi} + \delta_3 A\xi_0). \tag{5.33}$$

Since $B\tilde{\xi}$ is orthogonal to ξ_0, then for $\lambda \neq 0$ we obtain $(A\tilde{\xi}, \xi_0) + \delta_3(A\xi_0, \xi_0) = 0$ and in virtue of the positivity of operator A we have $\delta_3 = -(A\tilde{\xi}, \xi_0)/(A\xi_0, \xi_0)$. This relation, written down in a more detailed form, helps us to find δ_3 if ζ, δ_1, and δ_2 are known,

$$\delta_3 = -((\boldsymbol{J}_{\mathrm{b}} + \boldsymbol{J}_{\mathrm{as}})\boldsymbol{e}_3 \cdot \boldsymbol{e}_3)^{-1}\left(\rho\int_\Gamma \theta\zeta\psi_3 \mathrm{d}\Gamma + \sum_{i=1}^{2}\delta_i(\boldsymbol{J}_{\mathrm{b}} + \boldsymbol{J}_{\mathrm{as}})\boldsymbol{e}_i \cdot \boldsymbol{e}_3\right). \tag{5.34}$$

With regard to (5.34), equation (5.33) can be written down as an equation in the space $\tilde{E} = E \ominus E_0$, namely,

$$\tilde{B}\tilde{\xi} = \lambda\tilde{A}\tilde{\xi}, \qquad \tilde{\xi} = \left(\zeta;\ \sum_{i=1}^{2}\delta_i\boldsymbol{e}_i\right)^{\mathrm{t}}. \tag{5.35}$$

Here, $\tilde{B}$ is the restriction $B|\tilde{E}$ of the operator B to $\tilde{E}$, and the operator $\tilde{A}$ acting on $\tilde{E}$ is definded by

$$\begin{aligned}\tilde{A}\tilde{\xi} := \Bigg(&\rho C\zeta + \rho\theta\sum_{i=1}^{2}\delta_i\psi_i - \rho\theta\psi_3((\boldsymbol{J}_{\mathrm{b}} + \boldsymbol{J}_{\mathrm{as}})\boldsymbol{e}_3 \cdot \boldsymbol{e}_3)^{-1}\\ &\cdot\left(\rho\int_\Gamma \theta\zeta\psi_3 \mathrm{d}\Gamma + \sum_{i=1}^{2}\delta_i(\boldsymbol{J}_{\mathrm{b}} + \boldsymbol{J}_{\mathrm{as}})\boldsymbol{e}_i \cdot \boldsymbol{e}_3\right);\rho\int_\Gamma C\zeta\left(\sum_{i=1}^{2}\psi_i\boldsymbol{e}_i\right)\mathrm{d}\Gamma\\ &+\sum_{i=1}^{2}\delta((\boldsymbol{J}_{\mathrm{b}} + \boldsymbol{J}_{\mathrm{as}})\boldsymbol{e}_i - ((\boldsymbol{J}_{\mathrm{b}} + \boldsymbol{J}_{\mathrm{as}})\boldsymbol{e}_i \cdot \boldsymbol{e}_3)\boldsymbol{e}_3)\Bigg)^{\mathrm{t}}.\end{aligned} \tag{5.36}$$

According to its construction the operator $\tilde{B}$ has a zero kernel, $\operatorname{Ker}\tilde{B} = 0$. Moreover, since E_0 is one-dimensional and the operator A is positive on E, then, according to the conclusions of Subsection 1.4.3, the operator $\tilde{A}$ ($= A_{22} - A_{21}A_{11}^{-1}A_{12}$, where A_{ij} are the matrix elements of the operator A corresponding to the orthogonal decomposition $E = E_0 \oplus \tilde{E}$) is positive on $\tilde{E}$ as well. Let us note that $\tilde{B}$ may have at most a two-dimensional subspace of vectors on which the form $(\tilde{B}\tilde{\xi}, \tilde{\xi})$ takes negative values, and such vectors have the form ξ_1 indicated in (5.27).

We say that the considered hydromechanic system is *statically stable in linear approximation* if the operator $\tilde{B}$ of potential energy is positive definite. According to the results above, this fact is true if and only if

$$\Delta_1 > 0, \qquad \Delta_2 > 0. \tag{5.37}$$

The formulas (5.30) that determine Δ_1 and Δ_2 show that condition (5.37) is satisfied for a body of sufficiently large mass with the mass center C located sufficiently far from the pole O.

If the system is statically stable in linear approximation, then the operator $\tilde{B}$ in equation (5.35) is positive definite and the operator $\tilde{A}$ is positive and compact. Therefore, according to the conclusions in Subsection 1.4.2, problem (5.35) has a discrete positive spectrum $\{\lambda_n\}_{n=1}^{\infty}$, which consists of normal eigenvalues with a unique limit point $\lambda = +\infty$. The eigenelements $\{\tilde{\xi}_n\}_{n=1}^{\infty}$ of this problem form a basis which is orthogonal with respect to the forms associated with the operators $\tilde{A}$ and $\tilde{B}$,

$$(\tilde{B}^{1/2}\tilde{\xi}_n, \tilde{B}^{1/2}\tilde{\xi}_m) = \lambda_n(\tilde{A}^{1/2}\tilde{\xi}_n, \tilde{A}^{1/2}\tilde{\xi}_m) = \lambda_n\delta_{nm}. \tag{5.38}$$

If operator $\tilde{B}$ has a negative lower bound and, therefore, the hydromechanic system is not statically stable in linear approximation, then, as it was previously shown, $\tilde{B}$ has no more than κ negative eigenvalues, where $0 < \kappa \leq 2$. According to the results in Subsection 1.4.5, in this case problem (5.35) has exactly κ negative eigenvalues $\{\lambda_k^-\}_{k=1}^{\kappa}$. The eigenelements $\{\tilde{\xi}_k^-\}_{k=1}^{\kappa}$ corresponding to these eigenvalues are such that the form $(\tilde{B}\tilde{\xi}_k^-, \tilde{\xi}_k^-)$ is negative. The other eigenvalues $\{\lambda_k^+\}_{k=1}^{\infty}$ of problem (5.35) are positive and their corresponding eigenelements $\{\xi_k^+\}_{k=\kappa+1}^{\infty}$ form an orthogonal system with respect to the form $(\tilde{A}\tilde{\xi}, \tilde{\xi})$. The whole set of eigenelements $\{\tilde{\xi}_k^-\}_{k=1}^{\kappa} \cup \{\tilde{\xi}_k^+\}_{k=\kappa+1}^{\infty}$ forms a Riesz basis in the space $E = (L_{2,\Gamma} \oplus \mathbb{R}^3) \ominus E_0$. The following relations are valid for the elements of this basis,

$$(\tilde{B}\tilde{\xi}_k^{\pm}, \tilde{\xi}_j^{\pm}) = \lambda_k^{\pm}\delta_{kj}, \qquad (\tilde{A}\tilde{\xi}_k^{\pm}, \tilde{\xi}_j^{\pm}) = \delta_{kj}. \tag{5.39}$$

Since $\lambda = \omega^2$, where ω is the frequency of proper oscillations, the above-developed reasoning leads to the following physical conclusion that is an analogue to the Lagrange nonstability theorem: *if the operator of the potential energy $\tilde{B}$ has a negative lower bound and $\Delta_2 \neq 0$, then the given hydromechanic system is unstable.*

Indeed, in this case the problem has at least one negative eigenvalue $\omega^2 = \lambda = \lambda_1^-$ and, therefore, the initial evolution problem has a solution of the form $\tilde{\xi}_1^- \exp(\sqrt{-\lambda_1^- t})$ which increases exponentially with time.

3.5.7 Solving the Evolution Problem

We address now the solvability of problem (5.1)–(5.4) with the initial conditions (5.5).

If the angular acceleration $\boldsymbol{\varepsilon} = \mathrm{d}^2\boldsymbol{\delta}/\mathrm{d}t^2$ is given, then from (5.7) we obtain (as in Section 3.4),

$$(P_0\boldsymbol{u})(t,x) = (P_0\boldsymbol{u}^0)(x) - \int_0^t P_0(\boldsymbol{\varepsilon}(\tau)\times\boldsymbol{r})\mathrm{d}\tau + \int_0^t (P_0\boldsymbol{f})(\tau,x)\mathrm{d}\tau, \tag{5.40}$$

where $\boldsymbol{u}^0(x)$ is the initial velocity field.

In order to find the angular displacement vector $\boldsymbol{\delta}(t)$ and the deviation function $\zeta(t,x_1,x_2)$ of the free surface, one should only solve the evolution problem (5.17) for the corresponding initial conditions.

Since the operator B equals zero on E_0, problem (5.17) can be simplified in this subspace. Assuming that the right hand side of equation (5.17) is given by $\chi_0(t)A\xi_0 := (\chi(t), A\xi_0)A\xi_0/\|A\xi_0\|^2$, the resulting problem has the solution

$$\begin{aligned}\zeta(t,x_1,x_2) &\equiv 0,\\ \delta_1(t) &\equiv 0,\\ \delta_2(t) &\equiv 0,\\ \delta_3(t) &= \delta_3^0 + \omega_3^0 t + \int_0^t \mathrm{d}\tau \int_0^\tau \chi_0(s)\mathrm{d}s,\end{aligned} \tag{5.41}$$

which describes a slow proper rotation of the system around the axis Ox_3 consisting of a uniform rotation with an angular velocity ω_3^0 and a forced rotation with an angular velocity $\int_0^t \chi_0(\tau)\mathrm{d}\tau$.

Taking this fact into account, we can further investigate only those solutions of problem (5.17) for which

$$\delta_3^0 = 0, \quad \omega_3^0 = 0, \quad (\chi(t), A\xi_0) \equiv 0. \tag{5.42}$$

Performing the transition from (5.17) to an equation in the space $\tilde{E} = (L_{2,\Gamma} \oplus \mathbb{R}^3) \ominus E_0$ (as it was done in Subsection 3.5.5), i.e., excluding the function $\delta_3(t)$ according to (5.34), we obtain

$$\tilde{A}\frac{\mathrm{d}^2\tilde{\xi}}{\mathrm{d}t^2} + \tilde{B}\tilde{\xi} = \tilde{\chi}(t), \tag{5.43}$$

where the operators $\tilde{A}$ and $\tilde{B}$ are already defined, and

$$\tilde{\chi}(t) = \chi(t) - \chi_0(t) A\xi_0. \tag{5.44}$$

The initial conditions for equation (5.43) are

$$\begin{aligned} \tilde{\xi}(0) &= (\zeta^0(x_1, x_2); \delta_1^0 \boldsymbol{e}_1 + \delta_2^0 \boldsymbol{e}_2)^{\mathrm{t}}, \\ \tilde{\xi}'(0) &= ((u^0 \cdot \boldsymbol{e}_3)|_\Gamma; \omega_1^0 \boldsymbol{e}_1 + \omega_2^0 \boldsymbol{e}_2)^{\mathrm{t}}. \end{aligned} \tag{5.45}$$

Let us represent the generalized solution of the Cauchy problem (5.43)–(5.45) as a series by means of the eigenelements of the corresponding spectral problem (5.35), and let us consider the most general situation when $\tilde{B}$ has κ, $0 < \kappa \leq 2$, negative eigenvalues.

Expanding the initial data (5.45) and the function $\tilde{\chi}(t)$ in (5.44) into Fourier series with respect to the system $\{\xi_k^\pm\}$ of eigenelements of problem (5.35) and using the formulas of orthogonality (5.39) we get

$$\begin{aligned} \tilde{\xi}(0) &= \sum_{k=1}^{\kappa} \alpha_k^- \tilde{\xi}_k^- + \sum_{k=\kappa+1}^{\infty} \alpha_k^+ \tilde{\xi}_k^+, \qquad \alpha_k^\pm = (\tilde{A}\tilde{\xi}(0), \tilde{\xi}_k^\pm), \\ \tilde{\xi}'(0) &= \sum_{k=1}^{\kappa} \beta_k^- \tilde{\xi}_k^- + \sum_{k=\kappa+1}^{\infty} \beta_k^+ \tilde{\xi}_k^+, \qquad \beta_k^\pm = (\tilde{A}\tilde{\xi}'(0), \tilde{\xi}_k^\pm), \\ \tilde{\chi}(t) &= \sum_{k=1}^{\kappa} \tilde{\chi}_k^-(t) \tilde{\xi}_k^- + \sum_{k=\kappa+1}^{\infty} \tilde{\chi}_k^+(t) \tilde{\xi}_k^+, \qquad \tilde{\chi}_k^\pm(t) = (\tilde{A}\tilde{\chi}(t), \tilde{\xi}_k^\pm), \end{aligned} \tag{5.46}$$

and the solution of problem (5.43)–(5.45) becomes

$$\begin{aligned} \tilde{\xi}(t) &= \tilde{\xi}^-(t) + \tilde{\xi}^+(t) = \sum_{k=1}^{\kappa} c_k^-(t) \tilde{\xi}_k^- + \sum_{k=\kappa+1}^{\infty} c_k^+(t) \tilde{\xi}_k^+, \\ c_k^+(t) &= \alpha_k^+ \cos(\omega_k^+ t) + (\beta_k^+/\omega_k^+) \sin(\omega_k^+ t) + \frac{1}{\omega_k^+} \int_0^t \sin[\omega_k^+(t-\tau)] \tilde{\chi}_k^+(\tau) d\tau, \\ c_k^-(t) &= \alpha_k^+ \cosh(\nu_k^- t) + (\beta_k^-/\nu_k^-) \sinh(\nu_k^- t) + \frac{1}{\nu_k^-} \int_0^t \sinh[\nu_k^-(t-\tau)] \tilde{\chi}_k^-(\tau) d\tau, \end{aligned}$$

where $\omega_k^+ = \sqrt{\lambda_k^+} > 0$, $\nu_k^- = \sqrt{-\lambda_k^-} > 0$.

If the system is statically stable in linear approximation and therefore $\kappa = 0$, then $\tilde{\xi}^-(t) \equiv 0$, i.e., there are no solutions of the Cauchy problem that exponentially increase with t.

As at the end of Section 1.4, let us introduce the following function according to (5.34),

$$\tilde{\delta}_3(t) = -((\boldsymbol{J}_{\rm b} + \boldsymbol{J}_{\rm as})\boldsymbol{e}_3 \cdot \boldsymbol{e}_3)^{-1} \left(\rho \int_\Gamma \theta \tilde{\zeta} \psi_3 {\rm d}\Gamma + \sum_{i=1}^{2} \tilde{\delta}_i (\boldsymbol{J}_{\rm b} + \boldsymbol{J}_{\rm as})\boldsymbol{e}_i \cdot \boldsymbol{e}_3 \right), \quad (5.48)$$

that is connected with the solution $\tilde{\xi}(t) = (\tilde{\zeta}(t); \sum_{i=1}^{2} \tilde{\delta}_i(t)\boldsymbol{e}_i)^{\rm t}$ of the problem (5.43)–(5.45).

The physical meaning of this function is the following: it equals the proper angular displacement of the system caused by deviations of the free surface of the fluid during small movements.

Now, if we introduce the function

$$\xi(t) = \tilde{\xi}(t) + (0; \tilde{\delta}_3(t)\boldsymbol{e}_3)^{\rm t},$$

then it is obvious that this function is the generalized solution of the evolution problem (5.17) for the additional conditions (5.42). Adding $\xi(t)$ to the solution $(0; \delta_3(t)\boldsymbol{e}_3)^{\rm t}$, where $\delta_3(t)$ is the function in (5.41), we obtain the solution of the initial evolution problem (5.17) for the corresponding initial conditions and the right hand side equal to $\chi(t)$. Therefore, it is the solution of the initial problem (5.1)–(5.5) for arbitrary initial conditions (5.5), too.

One can check that, if at the initial moment $t = 0$ the kinetic and potential energies are finite, the external field $\boldsymbol{f}(t,x)$ is a continuous function with values in $\boldsymbol{L}_2(\Omega)$, and the moment $\boldsymbol{M}(t)$ is a continuous function, too, then the law of full energy balance of the system is valid for the solutions of problem (5.1)–(5.4). This law follows from (5.6),

$$\begin{aligned}
&\frac{1}{2}\{\rho\|\boldsymbol{u}(t)\|^2_{\boldsymbol{L}_2(\Omega)} + \boldsymbol{J}\boldsymbol{\omega}(t)\cdot\boldsymbol{\omega}(t) + 2\rho(\boldsymbol{\omega}(t)\times\boldsymbol{r}, \boldsymbol{u}(t))_{\boldsymbol{L}_2(\Omega)}\} \\
&\qquad + \frac{1}{2}\{\rho\|\zeta(t)\|^2_{L_2(\Gamma)} + mga(\delta_1^2(t) + \delta_2^2(t)) + 2\rho g(\zeta(t), \delta_1(t)x_2 - \delta_2(t)x_1)_{L_2(\Gamma)}\} \\
&\quad = \frac{1}{2}\{\rho\|\boldsymbol{u}^0\|^2_{\boldsymbol{L}_2(\Omega)} + \boldsymbol{J}\boldsymbol{\omega}^0\cdot\boldsymbol{\omega}^0 + 2\rho(\boldsymbol{\omega}^0\times\boldsymbol{r}, \boldsymbol{u}^0)_{\boldsymbol{L}_2(\Omega)}\} \\
&\qquad + \frac{1}{2}\{\rho\|\zeta^0\|^2_{L_2(\Gamma)} + mga((\delta_1^0)^2 + (\delta_2^0)^2) + 2\rho(\zeta^0, \delta_1^0 x_2 - \delta_2^0 x_1)_{L_2(\Gamma)}\} \\
&\qquad + \int_0^t [\rho(\boldsymbol{f}(\tau), \boldsymbol{u}(\tau))_{\boldsymbol{L}_2(\Omega)} + \boldsymbol{M}(\tau)\cdot\boldsymbol{\omega}(\tau)]{\rm d}\tau. \qquad (5.49)
\end{aligned}$$

3.6 Oscillations of a System of Fluids in an Immovable Container

The problem considered in this section is a natural generalization of the problems addressed in Sections 3.2 and 3.3 in the case of a system of several nonmixing fluids. The general method discussed in Section 1.8 and the assumptions made in Section 2.1 apply to this new problem.

The exposition given in some parts of this section will be rather concise.

3.6.1 Statement of the Problem

Let us assume that Ω is an immovable closed container filled up with a system of m nonmixing heavy fluids that are located relatively to the axis Ox_3 in such a way that their densities decrease, that is, $\rho_1 > \cdots > \rho_k > \cdots > \rho_m > 0$. The constant acceleration $\boldsymbol{g}$ of the gravitation field is directed against the axis Ox_3.

We denote by Ω_k the region filled with the fluid of density ρ_k in a stable state. The corresponding part of the solid boundary S will be denoted by S_k. The lower and the upper horizontal boundaries between fluids are denoted by Γ_{k-1} and Γ_k, $k = 1, 2, \ldots, m-1$, respectively. The normals n_k to the boundaries Γ_k are directed upwards, i.e., from the kth to the $(k+1)$th fluid.

If the system is in an equilibrium state, the pressure in every fluid $P_{0k}(x)$ is defined (similarly to the way it was defined for the problems discussed in Sections 3.2 and 3.3) by the formulas

$$P_{0k}(x) = -\rho_k g x_3 + c_k, \tag{6.1}$$

where the constants c_k can be determined, in terms of one of them, from the condition that the pressures on the boundaries Γ_k, $k = 1, 2, \cdots, m-1$, are equal.

To formulate the problem on small oscillations of the system of nonmixing fluids let us write down the linearized Euler equations

$$\frac{\partial \boldsymbol{u}^k}{\partial t} = -\frac{1}{\rho_k}\nabla p_k + \boldsymbol{f}, \quad \operatorname{div} \boldsymbol{u}^k = 0 \qquad \text{in } \Omega_k, \tag{6.2}$$

where $\boldsymbol{u}^k = \boldsymbol{u}^k(t,x)$ is the velocity field of the kth fluid, $p_k = p_k(t,x)$ is the dynamic pressure of the kth fluid, and $\boldsymbol{f}(t,x)$ is a small addition to the field of external forces. The boundary conditions on the solid boundary S for the kth fluid are the nonleaking conditions

$$\boldsymbol{u}^k \cdot \boldsymbol{n} = 0 \qquad \text{on } S_k. \tag{6.3}$$

Further, the vertical normal components of velocities have to coincide on the boundaries Γ_i, $i = 1, 2, \ldots, m-1$. Hence, the kinetic conditions become

$$\boldsymbol{u}^i \cdot \boldsymbol{n}_i = \boldsymbol{u}^{i+1} \cdot \boldsymbol{n}_i = \frac{\partial \zeta_i}{\partial t} \quad \text{on } \Gamma_i, \qquad i = 1, 2, \ldots, m-1, \tag{6.4}$$

where the function $\zeta_i(t, x_1, x_2)$ is defined on Γ_i and determines the vertical deviation of the moving surface $\tilde{\Gamma}_i(t)$ from the equilibrium horizontal surface Γ_i.

The dynamic boundary conditions for the system of heavy fluids can be obtained by imposing the condition of pressure equality on the moving boundaries

$\tilde{\Gamma}_i(t)$ between two fluids. After linearization and transition to Γ_i we get

$$p_i - p_{i+1} = (\Delta\rho)_i g\zeta_i, \quad (\Delta\rho)_i := \rho_i - \rho_{i+1} > 0, \qquad i = 1, 2, \ldots, m-1. \tag{6.5}$$

In order to determine the motion of the fluid system at any moment in time, we should add up to (6.2)–(6.5) the conditions that define the velocity field and the deviation of the boundaries for each fluid at the initial moment:

$$\begin{aligned} \boldsymbol{u}^k(0, x) &= \boldsymbol{u}^{k,0}(x) \qquad k = 1, 2, \ldots, m; \\ \zeta_i(0, x_1, x_2) &= \zeta_i^0(x_1, x_2), \qquad i = 1, 2, \ldots, m-1. \end{aligned} \tag{6.6}$$

Let us note that the solenoidality conditions for each fluid with regard to (6.3) lead to the following relations,

$$\int_{\Gamma_i} \boldsymbol{u}^i \cdot \boldsymbol{n}_i \mathrm{d}\Gamma_i = 0, \qquad i = 1, 2, \ldots, m-1. \tag{6.7}$$

3.6.2 Orthogonal Projection Method

Let us consider, for every t, the collection $\hat{\boldsymbol{u}}(t) := \{\boldsymbol{u}^1, \ldots, \boldsymbol{u}^m\}$ of velocity fields of the fluid system as an element of the Hilbert space $\hat{\boldsymbol{L}}_2(\Omega)$ with the scalar product

$$(\hat{\boldsymbol{u}}, \hat{\boldsymbol{v}})_{\hat{L}_2(\Omega)} := \sum_{k=1}^{m} \rho_k \int_{\Omega_k} \boldsymbol{u}^k \cdot \boldsymbol{v}^k \mathrm{d}\Omega_k. \tag{6.8}$$

Then, based on Subsections 2.1.11 and 2.1.12, and according to the boundary conditions (6.3) and the solenoidality conditions (6.2), we conclude that the function $\hat{\boldsymbol{u}}(t)$, for every t, can be considered as an element of the subspace $\hat{\boldsymbol{J}}_{0,S}(\Omega)$. In its turn, this subspace admits an orthogonal decomposition into the subspaces of vortex and potential harmonic fields

$$\hat{\boldsymbol{J}}_{0,S}(\Omega) = \hat{\boldsymbol{J}}_0(\Omega) \oplus \hat{\boldsymbol{G}}_{\mathrm{h},S}(\Omega), \tag{6.9}$$

where

$$\hat{\boldsymbol{J}}_0(\Omega) = \bigoplus_{k=1}^{m} \boldsymbol{J}_0(\Omega_k)$$

and $\hat{\boldsymbol{G}}_{\mathrm{h},S}(\Omega)$ was defined in (2.1.41).

Similarly, considering for every t the collection

$$(\widehat{\nabla_\rho p})(t) := \{\rho_1^{-1}\nabla p_1, \cdots, \rho_m^{-1}\nabla p_m\} \tag{6.10}$$

as an element of $\hat{\boldsymbol{L}}_2(\Omega)$ we conclude, with regard to Section 2.1, that $\widehat{\nabla_\rho p}$ should belong to the subspace $\hat{\boldsymbol{G}}(\Omega) = \hat{\boldsymbol{L}}_2(\Omega) \ominus \hat{\boldsymbol{J}}_0(\Omega)$ of potential fields, with a scalar product that is generated by the norm (2.1.27). Namely, if $\hat{\boldsymbol{u}} = \widehat{\nabla\Phi} := \{\nabla\Phi_1, \ldots, \nabla\Phi_m\}$ and

$\hat{\boldsymbol{v}} = \nabla\Psi = \{\nabla\Psi_1, \dots, \nabla\Psi_m\}$ are elements from $\hat{\boldsymbol{G}}(\Omega)$, then

$$(\hat{\boldsymbol{u}}, \hat{\boldsymbol{v}})_{\hat{\boldsymbol{L}}_2(\Omega)} = \sum_{k=1}^{m} \rho_k \int_{\Omega_k} \nabla\Phi_k \cdot \nabla\Psi_k \mathrm{d}\Omega_k. \tag{6.11}$$

For elements in this subspace, condition (2.1.28) holds true:

$$\int_{\Gamma_i} (\rho_i\Phi_i - \rho_{i+1}\Phi_{i+1})\mathrm{d}\Gamma_i = 0, \qquad i = 1, 2, \cdots, m-1,$$

$$\int_{\Gamma_{m-1}} \rho_m\Phi_m \mathrm{d}\Gamma_{m-1} = 0. \tag{6.12}$$

The subspace $\hat{\boldsymbol{G}}(\Omega)$ admits the next orthogonal decomposition that follows from (2.1.34),

$$\hat{\boldsymbol{G}}(\Omega) = \hat{\boldsymbol{G}}_{0,\Gamma}(\Omega) \oplus \hat{\boldsymbol{G}}_{\mathrm{h},S}(\Omega). \tag{6.13}$$

Here, $\hat{G}_{0,\Gamma}(\Omega)$ is the set of collections $\widehat{\nabla p} = \{\nabla p_k\}_{k=1}^m$, whose elements satisfy the relations

$$\rho_i p_i - \rho_{i+1} p_{i+1} = 0, \quad \text{on } \Gamma_i \qquad i = 1, 2, \dots, m-1. \tag{6.14}$$

Let us consider now the system of equations (6.2) as a single equation,

$$\frac{\mathrm{d}\hat{\boldsymbol{u}}}{\mathrm{d}t} + \widehat{\nabla_\rho p} = \hat{\boldsymbol{f}}, \quad \hat{\boldsymbol{f}} := \{\boldsymbol{f}, \dots, \boldsymbol{f}\}, \tag{6.15}$$

in the space $\hat{\boldsymbol{L}}_2(\Omega)$. The fields $\hat{\boldsymbol{u}}$ and $\widehat{\nabla_\rho p}$ can be found by decomposing them as

$$\hat{\boldsymbol{u}} = \widehat{\nabla\Phi} + \hat{\boldsymbol{w}}, \quad \hat{\boldsymbol{w}} \in \hat{\boldsymbol{J}}_0(\Omega),\ \widehat{\nabla\Phi} \in \hat{\boldsymbol{G}}_{\mathrm{h},S}(\Omega), \tag{6.16}$$

$$\widehat{\nabla_\rho p} = \widehat{\nabla_\rho \kappa} + \widehat{\nabla_\rho \varphi}, \quad \widehat{\nabla_\rho \kappa} \in \hat{\boldsymbol{G}}_{0,\Gamma}(\Omega),\ \widehat{\nabla_\rho \varphi} \in \hat{\boldsymbol{G}}_{\mathrm{h},S}(\Omega). \tag{6.17}$$

Introducing the orthoprojections P_Γ, P_S, and P_0 onto the subspaces $\hat{\boldsymbol{G}}_{0,\Gamma}(\Omega)$, $\hat{\boldsymbol{G}}_{\mathrm{h},S}(\Omega)$, and $\hat{\boldsymbol{J}}_0(\Omega)$, respectively, substituting (6.16) and (6.17) into (6.15), and projecting both sides of (6.15) onto these subspaces we obtain

$$\widehat{\nabla_\rho \kappa} = P_\Gamma \hat{\boldsymbol{f}}, \quad \frac{\mathrm{d}\hat{\boldsymbol{w}}}{\mathrm{d}t} = P_0 \hat{\boldsymbol{f}}, \tag{6.18}$$

$$\frac{\mathrm{d}}{\mathrm{d}t}\widehat{\nabla\Phi} + \widehat{\nabla_\rho \varphi} = P_S \hat{\boldsymbol{f}} =: \widehat{\nabla F}. \tag{6.19}$$

Hence, we deduce the following conclusions, that are similar to properties pointed out for the problem in Section 3.3 on small movements of a fluid in a partially filled container:

a) The component $P_\Gamma \hat{\boldsymbol{f}}$ of the external field $\hat{\boldsymbol{f}} = \{\boldsymbol{f}, \dots, \boldsymbol{f}\}$ does not affect the velocity field $\hat{\boldsymbol{u}} = \{\boldsymbol{u}^k\}_{k=1}^m$ of fluid particles.

b) This component does not affect the value of the pressure difference $p_i - p_{i+1}$ on the boundaries Γ_i, $i = 1, 2, \dots, m-1$, as well. Indeed, since

$$p_i - p_{i+1} = (\kappa_i - \kappa_{i+1}) + (\varphi_i - \varphi_{i+1}) \quad \text{on } \Gamma_i, \qquad i = 1, 2, \dots, m-1$$

and since for $\widehat{\nabla_\rho \kappa} \in \hat{\boldsymbol{G}}_{0,\Gamma}(\Omega)$ the conditions (6.14) hold true (with the substitution $p_i \mapsto \kappa_i$), then

$$p_i - p_{i+1} = \varphi_i - \varphi_{i+1} \quad \text{on } \Gamma_i, \qquad i = 1, 2, \dots, m-1.$$

c) The component $P_0 \hat{\boldsymbol{f}}$ of the external field that is derived from the second equation in (6.18) affects only a change of the vortical component $\hat{\boldsymbol{w}} \in \hat{\boldsymbol{J}}_0(\Omega)$ of the field $\hat{\boldsymbol{u}}$. Since for the element $\hat{\boldsymbol{w}} \in \hat{\boldsymbol{J}}_0(\Omega)$ conditions

$$\boldsymbol{w}^i \cdot \boldsymbol{n}_i = \boldsymbol{w}^{i+1} \cdot \boldsymbol{n}_i = 0 \quad \text{on } \Gamma_i, \qquad i = 1, 2, \dots, m-1$$

are true, such changes of the vortex components do not cause any motion of the boundaries $\tilde{\Gamma}_i$ and, therefore, they do not cause any changes in the pressure field.

According to these conclusions, it would be sufficient to consider further only the (nontrivial) problem (6.19) in the subspace $\hat{\boldsymbol{G}}_{\mathrm{h},S}(\Omega)$.

3.6.3 Transition to Operator Equation

Let us write down problem (6.19) componentwise, with regard to boundary conditions for $\widehat{\nabla_\rho \kappa}$ and $\hat{w}$. Using the Cauchy-Lagrange integrals for each fluid we obtain the boundary value problem

$$\begin{aligned}
&\Delta\Phi_k = 0 \quad \text{in } \Omega_k, \qquad \frac{\partial \Phi_k}{\partial n} = 0 \quad \text{on } S_k,\\
&\frac{\partial \Phi_i}{\partial n_i} = \frac{\partial \Phi_{i+1}}{\partial n_i} = \frac{\partial \zeta_i}{\partial t} \quad \text{on } \Gamma_i,\\
&\frac{\partial}{\partial t}(\rho_i \Phi_i - \rho_{i+1}\Phi_{i+1}) + g(\Delta\rho)_i \zeta_i = \rho_i F_i - \rho_{i+1} F_{i+1} \quad \text{on } \Gamma_i,\\
&\int_{\Gamma_i} (\rho_i \Phi_i - \rho_{i+1}\Phi_{i+1})\, d\Gamma_i = 0 \qquad i = 1, 2, \dots, m-1, \qquad \int_{\Gamma_{m-1}} \rho_m \Phi_m \, d\Gamma_{m-1} = 0;\\
&\zeta_i(0, x_1, x_2) = \zeta_i^0(x_1, x_2) \quad i = 1, 2, \dots, m-1,\\
&\frac{\partial \zeta_i}{\partial t}(0, x_1, x_2) = v_n^{i,0}(x_1, x_2) = \nabla \Phi_i^0 \cdot \boldsymbol{n}_i,\\
&(\widehat{\nabla\Phi})|_{t=0} = P_S \hat{\boldsymbol{u}}^0, \quad \hat{\boldsymbol{u}}^0 = \{\boldsymbol{u}^{1,0}(x), \cdots, \boldsymbol{u}^{m,0}(x)\}.
\end{aligned} \tag{6.20}$$

Let us note that, according to the definition (2.1.41) of the subspace $\hat{\boldsymbol{G}}_{\mathrm{h},S}(\Omega)$, there is an isometry between this subspace and the subspace $\hat{H}^1_{\mathrm{h},S}(\Omega)$ that consists of all the collections $\hat{\Phi} = \{\Phi_k\}_{k=1}^m$ of scalar functions from the space $\hat{H}_\Gamma(\Omega)$. That is why the abstract scheme discussed in Subsection 1.8.5 and its detailed description provided in Subsection 2.1.11 can be applied to problem (6.20). According to that scheme, we should consider the Hilbert pair of spaces $(F; G)$, where $F = \hat{H}^1_\Gamma(\Omega)$ and $G = \hat{L}_{2,\Gamma} := \bigoplus_{i=1}^{m-1} L_{2,\Gamma_i}(\Gamma_i)$. For $\hat{\Phi} = \{\Phi_k\}_{k=1}^m$, we define the trace operator by

$$\gamma_\Gamma \hat{\Phi} := \{(\rho_i \Phi_i - \rho_{i+1}\Phi_{i+1})|_{\Gamma_i}\}_{i=1}^{m-1}, \tag{6.21}$$

and the operator ∂ that generates the boundary value problem (2.1.39),

$$\begin{gathered}\Delta\Phi_k = 0, \quad \text{in } \Omega_k \qquad \frac{\partial\Phi_k}{\partial n} = 0 \quad \text{on } S_k, \\ \frac{\partial\Phi_i}{\partial n_i} = \frac{\partial\Phi_{i+1}}{\partial n_i} = \psi_i \qquad \text{on } \Gamma_i, \\ \int_{\Gamma_i} (\rho_i\Phi_i - \rho_{i+1}\Phi_{i+1})\mathrm{d}\Gamma_i = 0 \quad i = 1, 2, \ldots, m-1, \qquad \int_{\Gamma_{m-1}} \rho_m\Phi_m \mathrm{d}\Gamma_{m-1} = 0.\end{gathered} \tag{6.22}$$

If the collection $\{\psi_i\}_{i=1}^{m-1}$ has the property

$$\{\psi_i\}_{i=1}^{m-1} \in (\hat{H}_\Gamma^{1/2}(\Gamma))^* = \left(\bigoplus_{i=1}^{m-1} H_{\Gamma_i}^{1/2}(\Gamma_i)\right)^* \supset \hat{L}_{2,\Gamma},$$

then problem (6.22) has a unique solution $\hat{\Phi} \in \hat{H}^1_{\mathrm{h},S}(\Omega)$.

Let us next introduce the operator $\hat{B} := g\mathrm{diag}((\Delta\rho)_i)_{i=1}^{m-1}$ acting on $\hat{L}_{2,\Gamma}$. Comparing (6.20) to (6.22), we can assume that $\psi_i = \partial\zeta_i/\partial t$. Therefore, according to the abstract scheme and to the definition of operator $\hat{B}$, we obtain that (6.20) can be consider as a Cauchy problem of the form

$$\begin{gathered}\hat{C}\frac{\mathrm{d}^2\hat{\zeta}}{\mathrm{d}t^2} + \hat{B}\hat{\zeta} = \hat{F}(t), \quad \hat{F} = \{\tilde{F}_i\}_{i=1}^{m-1}, \qquad \tilde{F}_i = \rho_i F_i - \rho_{i+1}F_{i+1}, \\ \hat{\zeta}(0) = \hat{\zeta}^0 = \{\zeta_i^0\}_{i=1}^{m-1}, \quad \frac{\mathrm{d}\hat{\zeta}(0)}{\mathrm{d}t} = \hat{v}^0 = \{v_{n_i}^{i,0}\}_{i=1}^{m-1}.\end{gathered} \tag{6.23}$$

Here, $\hat{C} = \gamma_\Gamma T$ is the compact positive operator acting on $\hat{L}_{2,\Gamma}$ that appears in the abstract scheme, and $\hat{\zeta} := \{\zeta_k\}_{k=1}^{m-1}$ is an unknown function of the deviations of the boundary, that takes values in $\hat{L}_{2,\Gamma}$.

3.6.4 Proper Osccilations

For $\hat{F}(t) \equiv 0$, the problem (6.23) on small oscillations depending on time according to the $\exp(\mathrm{i}\omega t)$-law leads to the spectral problem

$$\hat{B}\hat{\zeta} = \lambda\hat{C}\hat{\zeta}, \quad \lambda = \omega^2, \quad \hat{\zeta} \in \hat{L}_{2,\Gamma}. \tag{6.24}$$

If the conditions $\rho_1 > \cdots > \rho_k > \cdots > \rho_m$ are true, then $(\Delta\rho)_i > 0$, for each i, and operator $\hat{B}$ is bounded and uniformly positive on $\hat{L}_{2,\Gamma}$. Since $\hat{C}$ is compact and positive on $\hat{L}_{2,\Gamma}$, then with regard to the results in Subsection 1.4.2 we obtain that problem (6.24) has a discrete positive spectrum $\{\lambda_j\}_{j=1}^\infty$. The basis $\{\hat{\zeta}_j\}_{j=1}^\infty$ that corresponds to this spectrum is orthogonal with respect to the forms associated with the operators $\hat{B}$ and $\hat{C}$. For example, the following relations are valid,

$$(\hat{B}\hat{\zeta}_j, \hat{\zeta}_l)_{\hat{L}_{2,\Gamma}} = \lambda_j(\hat{C}\hat{\zeta}_j, \hat{\zeta}_l)_{\hat{L}_{2,\Gamma}} = \lambda_j\delta_{jl}. \tag{6.25}$$

The eigenvalues of the problem (6.24) can be found as consecutive minima of the variational ratio

$$\frac{(\hat{B}\hat{\zeta},\hat{\zeta})_{\hat{L}_{2,\Gamma}}}{(\hat{C}\hat{\zeta},\hat{\zeta})_{\hat{L}_{2,\Gamma}}} = g\frac{\sum_{i=1}^{m-1}(\Delta_\rho)_i\int_{\Gamma_i}|\zeta_i|^2 d\Gamma_i}{\sum_{k=1}^{m}\rho_k\int_{\Omega_k}|\nabla\Phi_k|^2 d\Omega_k}, \tag{6.26}$$

which is considered for solutions of problem (6.22), using the substitution $\psi_i \mapsto \zeta_i$. In a different form, by using the substitution $\hat{C}\hat{\zeta} = \hat{\Phi}$ in (6.24), it follows that the eigenvalues $\{\lambda_j\}_{j=1}^{\infty}$ can be found as consecutive minima of the ratio

$$\frac{(\hat{C}^{-1}\hat{\Phi},\hat{\Phi})_{\hat{L}_{2,\Gamma}}}{(\hat{B}^{-1}\hat{\Phi},\hat{\Phi})_{\hat{L}_{2,\Gamma}}} = g\frac{\sum_{k=1}^{m}\rho_k\int_{\Omega_k}|\nabla\Phi_k|^2 d\Omega_k}{\sum_{i=1}^{m-1}[(\Delta_\rho)_i]^{-1}\int_{\Gamma_i}|\rho_i\Phi_i - \rho_{i+1}\Phi_{i+1}|^2 d\Gamma_i}, \tag{6.27}$$

where $\Phi_i(x)$ satisfy the conditions (6.22), with the new kinematic conditions $\partial\Phi_i/\partial n_i = \partial\Phi_{i+1}/\partial n_i$ on Γ_i.

The variational problems (6.26), (6.22) and (6.27), (6.22), respectively, yield the following spectral asymptotics for the eigenvalues of problem (6.24) in the 3-dimensional case,

$$\lambda_n = \lambda_n(\hat{C};\hat{B}) = \left(\frac{n}{K_{\hat{C},\hat{B}}}\right)^{1/2}[1+o(1)], \qquad n\to\infty,$$
$$K_{\hat{C},\hat{B}} := \frac{1}{4\pi}\sum_{i=1}^{m-1}\frac{(\rho_i+\rho_{i+1})^2}{g^2[(\Delta\rho)_i]^2}\operatorname{mes}\Gamma_i > 0. \tag{6.28}$$

Hence, we obtain the asymptotics of the spectrum of the frequencies as

$$\omega_n^{\pm} = \pm\sqrt{\lambda_n(\hat{C},\hat{B})} = \pm\left(\frac{n}{K_{\hat{C},\hat{B}}}\right)^{1/4}[1+o(1)], \qquad n\to\infty. \tag{6.29}$$

We notice that the set of all the eigenelements $\widehat{\nabla\Phi}_j = T\hat{\zeta}_j$ found by solving problem (6.22) for $\psi_i = \zeta_{j,i}$ forms an orthogonal basis in the space $\hat{\boldsymbol{G}}_{h,S}(\Omega)$.

If the conditions of static stability of the fluid system are not valid, and there exists at least one negative difference $(\Delta\rho)_i = \rho_i - \rho_{i+1}$, then the operator $\hat{B}$ in (6.24) is not positive definite, and zero does not belong to the spectrum of $\hat{B}$. In the corresponding infinite-dimensional space $L_{2,\Gamma_i}(\Gamma_i)$, the form $(\hat{B}\hat{\zeta},\hat{\zeta})_{\hat{L}_2(\Gamma)}$ takes negative values. Therefore, according to the conclusions of Subsection 1.4.5, problem

(6.24) is equivalent to the eigenvalue problem for a J-positive compact operator acting on a J-space. The latter problem has an infinite-dimensional negative invariant subspace that generates a countable set of eigenelements of the problem (6.24) with negative eigenvalues $\lambda = \omega^2$.

Hence, the assumption $(\Delta\rho)_i < 0$ for some $i \in \{1, 2, \ldots, m-1\}$ leads to dynamic instability of the system within the linear theory. This fact is physically evident, because if some layer of fluid with a larger density is located on top of another layer with smaller density, then the layer on top will occupy the place of the lower layer. The initial (linear) stage of this process can be described by using the eigenelements corresponding to the negative eigenvalues of problem (6.24), to build a solution of the evolution problem.

3.6.5 Small Movements of Stable Systems

Let us now assume that the stability conditions $(\Delta\rho)_i > 0$, $i = 1, 2, \ldots, m-1$, hold true and consider the evolution problem (6.23). Using the initial value $\hat{\zeta}^0$ of the deviation of the boundaries, we evaluate the quadratic functional of the potential energy of the system, and get

$$\frac{1}{2}(\hat{B}\hat{\zeta}^0, \hat{\zeta}^0)_{\hat{L}_{2,\Gamma}} = \frac{g}{2}\sum_{i=1}^{m-1}(\Delta_\rho)_i \int_{\Gamma_i} |\zeta_i^0|^2 d\Gamma_i. \tag{6.30}$$

Further, using the set of initial velocities $d\hat{\zeta}^0/dt = \hat{v}^0 = \{v_{n_i}^{i,0}\}_{i=1}^{m-1}$ of the boundaries, we introduce the potential $\widehat{\nabla\Phi}^0 = \{\nabla\Phi_k^0\}_{k=1}^m$ of the velocities by means of the solution to problem (6.22) with $\psi_i = v_{n_i}^{i,0}$. In its turn, this potential is used to determine the initial kinetic energy of the potential movements of the system, namely,

$$\frac{1}{2}(\hat{C}\hat{v}^0, \hat{v}^0)_{\hat{L}_{2,\Gamma}} = \frac{1}{2}\sum_{k=1}^{m}\rho_k \int_{\Omega_k} |\nabla\Phi_k^0|^2 d\Omega_k. \tag{6.31}$$

Let us assume that $\hat{F}(t)$ is a continuous function that takes values in the space $(\hat{H}_\Gamma^{1/2}(\Gamma))^*$, and the potential and kinetic energies of the system are finite at the initial moment. Then, using the scheme developed in Subsection 1.5.8, we can prove that problem (6.23) has a unique generalized solution $\hat{\zeta}(t)$ with continuous kinetic and potential energies. In regard with the law of full energy balance, we have

$$\frac{1}{2}\left(\hat{C}\frac{d\hat{\zeta}}{dt}, \frac{d\hat{\zeta}}{dt}\right)_{\hat{L}_{2,\Gamma}} + \frac{1}{2}(\hat{B}\hat{\zeta}(t), \hat{\zeta}(t))_{\hat{L}_{2,\Gamma}}$$
$$= \frac{1}{2}(\hat{C}\hat{v}^0, \hat{v}^0)_{\hat{L}_{2,\Gamma}} + \frac{1}{2}(\hat{B}\hat{\zeta}^0, \hat{\zeta}^0)_{\hat{L}_{2,\Gamma}} + \int_0^t \left(\hat{F}(\tau), \frac{d\hat{\zeta}(\tau)}{d\tau}\right)_{\hat{L}_{2,\Gamma}} d\tau. \tag{6.32}$$

As it was mentioned previously, the first equation in (6.18) causes only a change in pressure, without any change in the energy of the system. The second equation in (6.18) can be easily integrated by using the corresponding initial condition,

$$\hat{\boldsymbol{w}}(t) = \hat{\boldsymbol{w}}^0 + \int_0^t (P_0\hat{\boldsymbol{f}})(\tau)\mathrm{d}\tau, \qquad \hat{\boldsymbol{w}}^0 := P_0\hat{\boldsymbol{u}}^0.$$

As in Section 3.2 we get for this solution that the law of kinetic energy balance for the vortex movements of the system is true, namely,

$$\frac{1}{2}(\hat{\boldsymbol{w}}(t), \hat{\boldsymbol{w}}(t))_{\hat{\boldsymbol{L}}_2(\Omega)} = \frac{1}{2}(\hat{\boldsymbol{w}}^0, \hat{\boldsymbol{w}}^0)_{\hat{\boldsymbol{L}}_2(\Omega)} + \int_0^t (\hat{\boldsymbol{f}}(\tau), \hat{\boldsymbol{w}}(t))_{\hat{\boldsymbol{L}}_2(\Omega)}\mathrm{d}\tau. \tag{6.33}$$

The following conclusion is the neat result of the previous reasonings. If, for the initial value problem (6.2)–(6.6) the potential energy (6.30) and the kinetic energy $(1/2)\sum_{k=1}^m \rho_k \|\boldsymbol{u}^{k,0}\|^2_{\boldsymbol{L}_2(\Omega_k)}$ are finite at the initial moment of time, and the field of external forces $\hat{\boldsymbol{f}}(t) = \{\boldsymbol{f}(t)\}_{k=1}^m$ is a continuous function with values in $\hat{\boldsymbol{L}}_2(\Omega)$, then that problem has a unique generalized solution with continuous potential and kinetic energies. Moreover, the law of the full energy balance takes place, that is,

$$\begin{aligned}
&\frac{1}{2}\sum_{k=1}^m \rho_k \int_{\Omega_k} |\boldsymbol{u}^k(t)|^2 \mathrm{d}\Omega_k + \frac{1}{2}g\sum_{i=1}^{m-1}(\Delta\rho)_i \int_{\Gamma_i} |\zeta_i(t)|^2 \mathrm{d}\Gamma_i \\
&\quad = \frac{1}{2}\sum_{k=1}^m \rho_k \int_{\Omega_k} |\boldsymbol{u}^{k,0}|^2 \mathrm{d}\Omega_k + \frac{1}{2}g\sum_{i=1}^{m-1}(\Delta\rho)_i \int_{\Gamma_i} |\zeta_i^0|^2 \mathrm{d}\Gamma_i \\
&\quad + \sum_{k=1}^m \rho_k \int_0^t \left(\int_{\Omega_k} \boldsymbol{f}(\tau) \cdot \frac{\mathrm{d}\boldsymbol{u}^k(\tau)}{\mathrm{d}\tau} \mathrm{d}\Omega_k \right) \mathrm{d}\tau.
\end{aligned} \tag{6.34}$$

To conclude, let us note that the problem considered here contains the problem on oscillations of a system of heavy fluids with a free upper surface as a particular case. Indeed, it is sufficient to assume in every formula that $\rho_m = 0$ and that the number of fluids equals $m - 1$.

Chapter 4

Problems on Oscillations of Capillary Fluids and Problems on Hydroelasticity in Immovable Containers

This chapter will address two problems that are different in their physical aspects, yet similar in a mathematical sense, because they both are devoted to the study of conservative hydromechanic systems. In Section 4.1, we investigate the problem of small oscillations of a capillary fluid, i.e., a fluid that has surface tension. That tension should be taken into account while considering this problem in conditions close to low-gravity. Further, in Section 4.2, we consider the problem of oscillations of a fluid wholly filling a container with one or two elastic horizontal ends. Finally, in Section 4.3, the problem on oscillations of a fluid partially filling a container with an elastic bottom is formulated and investigated.

4.1 Oscillations of a Capillary Fluid in a Rigid Container

In conditions of low-gravity, while considering various problems of oscillations of a fluid partially filling a container, one should take into account the capillary (surface) forces that act on the boundaries "fluid-gas," "fluid-rigid body," and "gas-rigid body". At long distances from the surface of Earth, i.e., in outer space, such forces could be of the same order as the gravitation and centrifugal forces, and their consideration helps to find out qualitatively new effects caused by capillarity.

4.1.1 On the Equilibrium State

Assume that an immovable rigid container partially filled with an ideal capillary fluid of density ρ is located within a potential field of external forces $\boldsymbol{F}_0 = -\nabla U_0$ with the mass density of potential $U_0 = U_0(x)$. All the subsequent considerations will be carried out in a coordinate system $0x_1x_2x_3$ rigidly connected with the container.

As it is known, even under ordinary conditions, the free surface of a fluid in capillaries (sufficiently thin pipes) is not horizontal under the influence of the gravitational force, i.e., the free surface is not orthogonal to the direction of the gravitational force, it is rather curved. The curvature of this surface is caused by the action of capillary (surface) forces. Similar phenomena take place in an ordinary container in conditions of low gravity or zero gravity.

As usual, we denote the region filled with fluid in the equilibrium state by Ω, the solid boundary of the container by S, and the equilibrium free surface of the fluid by Γ. We assume that Ω is a region with a piecewise smooth boundary $\partial\Omega$ that has nonzero external and internal dihedral angles; here, Γ is considered to be a sufficiently smooth surface, defined by the equations of equilibrium.

In equilibrium state, the fluid is stable and the existing pressure is compensated by the field of external forces. Indeed, from the Euler condition

$$\nabla P_0 = -\rho \boldsymbol{F}_0 = -\rho \nabla U_0 \qquad \text{in } \Omega, \tag{1.1}$$

it follows, that

$$P_0(x) = -\rho U_0(x) + \text{ const} \qquad \text{in } \Omega; \tag{1.2}$$

in particular, for the gravitation field, when

$$U_0(x) \equiv g x_3 + \text{ const}$$

we have

$$P_0(x) = -\rho g x_3 + \text{ const}.$$

Further, on the boundary "fluid-gas," Γ, the Laplace condition for the leap of pressure holds true,

$$P_0(x) - p_a = -\sigma(k_1 + k_2) \qquad \text{on } \Gamma. \tag{1.3}$$

Here p_a is the external constant pressure, $\sigma \geq 0$ is the coefficient of surface tension on the boundary "fluid-gas," and $k_i = k_i(x)$, $i = 1, 2$, are the main curvatures of the surface Γ assumed to be positive if the convexity of Γ is directed inside Ω.

Condition (1.3) means that the difference of pressures on the free surface Γ is not zero, it equals the capillary leap caused by the influence of surface forces.

It is also known that in stable state the dihedral angle δ between the surface Γ and S is not arbitrary; it is determined by properties of the three contacting media. Namely, if $\sigma_1 \geq 0$ is the coefficient of surface tension on the boundary "gas-rigid body," and σ_0 is the same coefficient on the boundary "fluid-rigid body," then in the stable state the Dupré-Young condition is to be valid on the contour $\partial\Gamma$,

$$\sigma \cos \delta = \sigma_1 - \sigma_0; \tag{1.4}$$

hence it follows that the angle of contact δ (also called the boundary or *wetting angle*) is constant.

Conditions (1.2)–(1.4), together with the additional condition

$$\int_\Omega d\Omega = \operatorname{mes} \Omega = V, \tag{1.5}$$

that indicates the volume V occupied by fluid, define a nonlinear boundary value problem that amounts to finding the equilibrium surface Γ and the region Ω. The investigation of this problem is not related to the issues addressed in the present book. The interested reader can find solutions, for some typical cases, in the first part of the book by V. G. Babskii et al. [1], and also in the new edition [3] of that book.

4.1.2 Statement of the Problem on Small Oscillations

Assume that the equilibrium state of a capillary fluid in a container is already defined, i.e., the surface Γ, the region Ω, and the equilibrium pressure $P_0(x)$ are known. We consider the linear problem on small movements of the fluid under the influence of an external force field $\boldsymbol{F}(t,x) = -\nabla U_0(x) + \boldsymbol{f}(t,x)$ and subject to certain initial conditions.

As we did for the problems studied in Chapter 3, the Euler equations of motion lead by linearization to the equations

$$\frac{\partial \boldsymbol{u}}{\partial t} = -\frac{1}{\rho}\nabla p + \boldsymbol{f}(t,x), \quad \operatorname{div} \boldsymbol{u} = 0 \qquad \text{in } \Omega, \tag{1.6}$$

and the boundary condition on the solid boundary is the nonleaking condition

$$u_n = 0 \quad \text{on } S. \tag{1.7}$$

In order to deduce the kinematic and dynamic conditions on Γ and to obtain the boundary condition characteristic for a capillary fluid on the contour $\partial\Gamma$, we introduce a curvilinear coordinate system $\hat{O}\xi^1\xi^2\xi^3$ in a neighborhood of the surface Γ, in such a way that surface Γ has the equation $\xi^3 = 0$, the ξ^3-coordinate lines are directed along the normals $\boldsymbol{n}$ to Γ for $\xi^3 = 0$ and have the corresponding Lame coefficients $l^3 = 1$ for each $x \in \Gamma$. Then, the moving free surface $\tilde{\Gamma}(t)$ of the fluid can be described by a function

$$\xi^3 = \zeta(t, \xi^1, \xi^2), \tag{1.8}$$

that coincides with the deviation, along the normal $\boldsymbol{n}$, of the moving surface from the equilibrium surface, with a higher order accuracy.

As in the case of a heavy fluid and a horizontal equilibrium surface Γ, from (1.8) we get the kinematic condition

$$\boldsymbol{u} \cdot \boldsymbol{n} = u_n = \frac{\partial \zeta}{\partial t} \quad \text{on } \Gamma. \tag{1.9}$$

The dynamic condition on Γ and the boundary condition on $\partial\Gamma$ can be deduced using the variational principle of Hamilton-Ostrogradsky, applied to the considered system, with an infinite number of degrees of freedom. However, the first condition can be obtained by linearizing condition (3.1.21)

$$P(t, x) - p_a = -\sigma(\tilde{k}_1 + \tilde{k}_2) \tag{1.10}$$

for the leap of pressures on the free moving surface $\tilde{\Gamma}(t)$, where $\tilde{k}_1$ and $\tilde{k}_2$ are the corresponding main curvatures of $\tilde{\Gamma}(t)$. Let us next use the well-known formula for the variation of twice the mean curvature of the surface,

$$(\tilde{k}_1 + \tilde{k}_2) - (k_1 + k_2) = (k_1^2 + k_2^2)\zeta + \Delta_\Gamma \zeta + 0(\zeta^2), \tag{1.11}$$

where Δ_Γ is the two-dimensional Leplace-Beltrami operator, calculated in the curvilinear coordinate system $\tilde{O}\xi^1\xi^2$ defined on Γ. Now, we assume that $P(t,x) = P_0(x)+p(t,x)$, where $P_0(x)$ is the equilibrium pressure (1.2) satisfying the equilibrium condition (1.3) and $p(t,x)$ is a small dynamic pressure, and we linearize it by using formula (1.11) and condition (1.10). This leads to the dynamic condition

$$\begin{aligned} p = \sigma\mathcal{L}(\zeta) &:= \sigma\left[-\Delta_\Gamma + a(x)\right]\zeta, \qquad x \in \Gamma, \\ a(x) &:= -(k_1^2 + k_2^2) + \rho\sigma^{-1}\nabla U_0 \cdot \boldsymbol{n}. \end{aligned} \tag{1.12}$$

In equation (1.12) above, $\mathcal{L}$ is an elliptic differential operator defined by the equilibrium state of the system. For a correct formulation of the problem, we should take into account the boundary condition for $\mathcal{L}$ on $\partial\Gamma$. This is actually the difference between problems on oscillations of a capillary ideal fluid and the similar problems for a heavy fluid.

By analogy to the deduction of condition (1.12), using the Hamilton-Ostrogradsky principle it can be proved that the wetting angle δ is preserved during the process of nonlinear movements of a capillary fluid in a container. The linearization of this condition for a small deviation $\zeta(t, \xi^1, \xi^2)$ gives the equation

$$\frac{\partial \zeta}{\partial \nu} + \chi\zeta = 0 \quad \text{on } \partial\Gamma, \qquad \chi := \frac{k_\Gamma \cos\delta - k_S}{\sin\delta}, \tag{1.13}$$

where k_Γ and k_S are the curvatures calculated on $\partial\Gamma$ of the cross-sections of the surfaces Γ and S by a plane orthogonal to $\partial\Gamma$, and $\partial/\partial\nu$ stands for the derivative in the direction of the normal vector to $\partial\Gamma$ in the plane tangent to Γ. A complete derivation of formula (1.13) can be found in the book by V. G. Babskii et al. [1, p. 109] or [3].

Hence, the problem on small movements of an ideal capillary fluid in an arbitrary container amounts to solving equations (1.6) with the boundary conditions (1.7), (1.9), (1.12), (1.13); moreover, the following initial conditions should be taken into account,

$$\begin{aligned} \boldsymbol{u}(0,x) &= \boldsymbol{u}^0(x) \qquad x \text{ in } \Omega, \\ \zeta(0,\xi^1,\xi^2) &= \zeta^0(\xi^1,\xi^2) \qquad (\xi^1,\xi^2) \text{ on } \Gamma. \end{aligned} \tag{1.14}$$

4.1.3 Law of Energy Balance

Let us multiply the first equation in (1.6) by $\rho\boldsymbol{u}$ and integrate the resulting scalar products over Ω by using the solenoidality condition in Ω and the nonleaking condition on S. We obtain

$$\rho\int_\Omega \frac{\partial \boldsymbol{u}}{\partial t}\cdot \boldsymbol{u}\mathrm{d}\Omega + \int_\Gamma p u_n \mathrm{d}\Gamma = \rho\int_\Omega \boldsymbol{f}\cdot\boldsymbol{u}\mathrm{d}\Omega. \tag{1.15}$$

Further, we use the dynamic condition (1.12), the kinematic condition (1.9), and Green's formula for the Laplace-Baltrami operator,

$$-\int_\Gamma \Delta_\Gamma\varphi\cdot\psi\mathrm{d}\Gamma = \int_\Gamma \nabla_\Gamma(\varphi,\psi)\mathrm{d}\Gamma - \int_{\partial\Gamma}\frac{\partial\varphi}{\partial\nu}\psi\mathrm{d}s. \tag{1.16}$$

Here, $\nabla_\Gamma(\varphi,\varphi)$ is the so-called first differential Beltrami parameter, which is a natural generalization of the square of the gradient of a function defined on an arbitrary surface Γ and $\nabla_\Gamma(\varphi,\psi)$ is just the corresponding bilinear expression. Using the boundary condition (1.13), we finally obtain

$$\begin{aligned} &\frac{1}{2}\frac{\mathrm{d}}{\mathrm{d}t}\left\{\rho\int_\Omega |\boldsymbol{u}|^2\mathrm{d}\Omega + \sigma\|\zeta\|_B^2\right\} = \rho\int_\Omega \boldsymbol{f}\cdot\boldsymbol{u}\mathrm{d}\Omega, \\ &\|\zeta\|_B^2 := \int_\Gamma \left[\nabla_\Gamma(\xi,\xi) + a|\zeta|^2\right]\mathrm{d}\Gamma + \int_{\partial\Gamma}\chi|\zeta|^2\mathrm{d}s. \end{aligned} \tag{1.17}$$

In the left hand side, within the curly brackets, one has the full energy of small movements of the system that corresponds to the velocity field $\boldsymbol{u}(t,x)$ and the deviation $\zeta(t,\xi^1,\xi^2)$ of the free surface; in the right hand side, similar to the problems studied in Chapter 3, one has the capacity of the small field of external forces influencing the system.

Let us note that for $\sigma = 0$ and $U_0 = gx_3$, i.e., for a heavy fluid and, therefore, a plane horizontal equilibrium surface Γ, the form $\sigma\|\zeta\|_B^2$ degenerates and becomes equal to $\rho g \int_\Gamma |\zeta|^2 d\Gamma$, which is twice the potential energy of small movements of a heavy fluid.

4.1.4 Transition to Operator Equation

The subsequent investigation of the problem is carried out according to the scheme developed in Section 3.3 for the case of a heavy ideal fluid. We should notice that Γ is not a plane horizontal surface in the general case. By finding the velocity field $\boldsymbol{u}(t,x)$ in the subspace $\boldsymbol{J}_{0,S}(\Omega)$ and then by projecting the Euler equation (1.6) onto $\boldsymbol{J}_0(\Omega)$, we obtain the vortical solution (3.3.5) and the Cauchy-Lagrange integral (3.3.6). Further, for the potential component $\nabla\Phi = (I - P_0)\boldsymbol{u}$ of the velocity field $\boldsymbol{u}$ we again get the boundary value problem

$$\Delta\Phi = 0 \quad \text{in } \Omega, \qquad \frac{\partial\Phi}{\partial n} = 0 \quad \text{on } S, \qquad \frac{\partial\Phi}{\partial n} = \frac{\partial\zeta}{\partial t} \quad \text{on } \Gamma. \tag{1.18}$$

Let us leave out the pressure p from the dynamic conditions (1.12) and introduce the operator $C = \gamma_\Gamma T$. Projecting the resulting relation onto the space $H = L_{2,\Gamma}$, we get the equation

$$\rho C \frac{\mathrm{d}^2\zeta}{\mathrm{d}t^2} + \sigma B\zeta = \rho f_1(t), \tag{1.19}$$

where $f_1(t)$ is the same function as in Section 3.3.2 and operator B, acting on $L_{2,\Gamma}$, is defined by the expression

$$B\zeta := P_\Gamma \mathcal{L}\zeta = \mathcal{L}\zeta - (\operatorname{mes}\Gamma)^{-1} \int_\Gamma (\mathcal{L}\zeta)\mathrm{d}\Gamma, \tag{1.20}$$

and the boundary condition (1.13).

The following initial conditions should be added up to (1.19)–(1.20),

$$\zeta(0) = \zeta^0, \quad \frac{\mathrm{d}\zeta}{\mathrm{d}t}(0) = \left[(I - P_0)\boldsymbol{u}^0\right] \cdot \boldsymbol{n} =: \zeta^1 \qquad \text{on } \Gamma. \tag{1.21}$$

Hence, the problem on small oscillations of a capillary ideal fluid in the potential subspace $\boldsymbol{G}_{h,S}(\Omega)$ is reduced to the Cauchy problem (1.19)—(1.21) in the space $H = L_{2,\Gamma}$.

4.1.5 Properties of the Potential Energy Operator

The above defined operator B is an unbounded operator acting on $L_{2,\Gamma}$. Its domain of definition, $\mathcal{D}(B)$, is the set of functions $\{\zeta\}$ from $H^2(\Gamma) \cap H$, which satisfy the boundary condition (1.13). From Green formula (1.16), it follows that for ζ_1, ζ_2 from $\mathcal{D}(B)$,

$$(B\zeta_1, \zeta_2)_{L_2(\Gamma)} = \int_\Gamma [\nabla_\Gamma(\zeta_1, \zeta_2) + a\zeta_1\zeta_2]\, d\Gamma + \int_{\partial\Gamma} \chi\zeta_1\zeta_2 ds; \tag{1.22}$$

hence, B is a symmetric operator.

For simplicity, we assume that the functions $a(x)$ and $\chi(x)$ in (1.22), defined by the equilibrium state, are continuous. Then operator B is bounded from below on $L_{2,\Gamma}$. For $\chi \geq 0$ this property is obvious, because

$$(B\zeta, \zeta)_{L_2,\Gamma} = \int_\Gamma [\nabla_\Gamma(\zeta, \zeta) + a|\zeta|^2]\, d\Gamma + \int_{\partial\Gamma} \chi|\zeta|^2 ds. \tag{1.23}$$

Whenever the condition $\chi \geq 0$ is not valid, we choose an arbitrary and sufficiently smooth function $\psi(\xi^1, \xi^2)$ on Γ such that $\partial\psi/\partial\nu \geq -\chi$ at every point of $\partial\Gamma$ (this can be done provided the contour $\partial\Gamma$ does not have any corner points with the spread 0° or 360° where $\chi < 0$). Hence, using Green formula (1.16) we obtain, for any function $\zeta \in \mathcal{D}(B)$ and any $\varepsilon > 0$,

$$\begin{aligned}
-\int_{\partial\Gamma} \chi|\zeta|^2 ds &\leq \int_{\partial\Gamma} \frac{\partial\psi}{\partial\nu}|\zeta|^2 ds = \int_\Gamma \nabla_\Gamma(\psi, \zeta^2) d\Gamma + \int_\Gamma \Delta_\Gamma\psi \cdot \zeta^2 d\Gamma \\
&\leq 2\int_\Gamma \zeta\nabla_\Gamma(\psi, \zeta) d\Gamma + c_1\|\zeta\|^2_{L_2(\Gamma)} \\
&\leq c_2 \int_\Gamma |\zeta|\sqrt{\nabla_\Gamma(\zeta, \zeta)} d\Gamma + c_1\|\zeta\|^2_{L_2(\Gamma)} \\
&\leq \varepsilon \int_\Gamma \nabla_\Gamma(\zeta, \zeta) d\Gamma + \left(c_1 + \frac{c_2^2}{4\varepsilon}\right)\|\zeta\|^2_{L_2(\Gamma)},
\end{aligned} \tag{1.24}$$

where c_1 and c_2 are positive constants depending on the function $\psi(\xi^1, \xi^2)$. The boundedness from below of the operator B follows immediately from (1.24) for $0 < \varepsilon < 1$ and (1.23). Operator B admits a selfadjoint Friedrichs extension with the same lower bound; that extension is denoted by B, too.

Since, by (1.23) and (1.17), the form $(B\zeta, \zeta)_{L_2(\Gamma)}$ coincides with the potential energy of small deviations of the system from the equilibrium state up to a multiplier, operator B will be called the *potential energy operator.*

We say that the equilibrium state of a capillary fluid in a container is *statically stable in linear approximation* provided the potential energy operator B is positive definite,

$$(B\zeta, \zeta)_{L_2(\Gamma)} \geq c^2\|\zeta\|^2_{L_2(\Gamma)}, \qquad c > 0,\ \zeta \in \mathcal{D}(B). \tag{1.25}$$

If condition (1.25) holds true, then we can introduce the energetic space $H_B \subset H = L_{2,\Gamma}$; the scalar product in this space,

$$(\zeta_1, \zeta_2)_B := (B^{1/2}\zeta_1, B^{1/2}\zeta_2)_{L_2(\Gamma)},$$

is defined by the right side of (1.23). The corresponding norm is equivalent to the Dirichlet norm

$$\|\zeta\|_\nabla^2 := \int_\Gamma \nabla_\Gamma(\zeta, \zeta) d\Gamma, \tag{1.26}$$

therefore H_B consists of all functions from $H^1(\Gamma)$ with zero mean value on Γ.

In order to prove the equivalence of the norms $\|\cdot\|_B$ and $\|\cdot\|_\nabla$, we note that according to Green formula (1.16), for $\zeta \in H_B$,

$$\begin{aligned} \|\zeta\|_B^2 &= \int_\Gamma \left[\nabla_\Gamma(\zeta,\zeta) + a|\zeta|^2\right] d\Gamma + \int_{\partial\Gamma} \chi|\zeta|^2 ds \\ &\le \|\zeta\|_\nabla^2 + c_1 \int_\Gamma |\zeta|^2 d\Gamma + c_2 \int_{\partial\Gamma} |\zeta|^2 ds \le c_3 \|\zeta\|_\nabla^2, \end{aligned} \tag{1.27}$$

where c_i, $i = 1, 2, 3$ are some constants, and the last inequality was derived from the embedding theorems $H^1(\Gamma) \hookrightarrow L_2(\Gamma)$ and $H^1(\Gamma) \hookrightarrow L_2(\partial\Gamma)$. On the other hand, using the inequalities (1.24) and (1.29) we obtain

$$\|\zeta\|_B^2 \ge (1-\varepsilon)\|\zeta\|_\nabla^2 - \frac{c_4}{c_2}\|\zeta\|_{L_2(\Gamma)}^2 \ge (1-\varepsilon)\|\zeta_\nabla^2 - \frac{c_4}{c_2}\|\zeta\|_B^2,$$

hence

$$\|\zeta\|_B^2 \ge \left(1 + \frac{c_4}{c_2}\right)^{-1} (1-\varepsilon)\|\zeta\|_\nabla^2. \tag{1.28}$$

The latter inequality, in conjunction with (1.27), proves the equivalence of the norms. From this property and the theorem on compact embedding $H^1(\Gamma) \hookrightarrow L_2(\Gamma)$, it follows that the inverse operator to B, which is bounded from below in virtue of (1.25), is a compact operator on H.

4.1.6 Proper Oscillations

We now consider proper oscillations of the system, i.e., such solutions $\zeta(t)$ of the homogeneous equation (1.19), which depend on t according to the law $\zeta(t) = e^{i\omega t}h(\xi^1, \xi^2)$, where ω is a frequency and $h(\xi^1, \xi^2)$ is a mode (shape) of proper oscillations. Equation (1.19) leads to a spectral problem of the form

$$Bh = \lambda Ch, \ \lambda = \rho\omega^2/\sigma, \tag{1.29}$$

similar to the problem investigated in Section 1.4. Indeed, operator C in (1.29) is positive and compact, and B is an unbounded selfadjoint operator bounded from below.

If the equilibrium state of the system is statically stable in linear approximation, i.e., $B \gg 0$, then, according to the conclusions of Section 1.4.2, problem (1.29) has a discrete spectrum, $\{\lambda_k\}_{k=1}^{\infty}$, that consists of positive eigenvalues, λ_k, counted with regard to multiplicities in increasing order, with the only limit point $\lambda = +\infty$. The corresponding eigenfunctions, $\{h_k(\xi^1, \xi^2)\}_{k=1}^{\infty}$, form orthogonal bases both in H_B and H_C (the latter space is the set of elements with the scalar product $(\zeta_1, \zeta_2)_C := \big(C\zeta_1, \zeta_2)_{L_2(\Gamma)}\big)$,

$$(h_i, h_j)_B = \lambda_i (h_i, h_j)_C = \lambda_i \delta_{ij}. \tag{1.30}$$

The eigenvalues λ_k of problem (1.29) can be found by considering successive minima of the variational ratio

$$\frac{(h,h)_B}{(h,h)_C} = \frac{\left\{\int_\Gamma \left[\nabla_\Gamma(h,h) + a|h|^2\right] \mathrm{d}\Gamma + \int_{\partial\Gamma} \chi |h|^2 \mathrm{d}s\right\}}{\int_\Omega |\nabla\varphi|^2 \mathrm{d}\Omega} \tag{1.31}$$

for solutions of the Neumann problem

$$\Delta\varphi = 0 \quad \text{in } \Omega, \qquad \frac{\partial\varphi}{\partial n} = 0 \quad \text{on } S,$$
$$\int_\Gamma \varphi \mathrm{d}\Gamma = 0, \qquad \frac{\partial\varphi}{\partial n} = h \qquad \text{on } \Gamma. \tag{1.32}$$

If we perform the substitution $Ch = \varphi$ in (1.29), then we get the following equivalent equation,

$$C^{-1}\varphi = \lambda B^{-1}\varphi \tag{1.33}$$

with the operator $C^{-1} \gg 0$ and the compact operator $B^{-1} > 0$. Hence, based on previous results, we obtain that the eigenvalues λ_k can be found as successive minima of the variational ratio

$$\frac{(\varphi,\varphi)_{C^{-1}}}{(\varphi,\varphi)_{B^{-1}}} = \frac{\int_\Omega |\nabla\varphi|^2 \mathrm{d}\Omega}{\int_\Gamma (B^{-1}\varphi)\varphi \mathrm{d}\Gamma}, \tag{1.34}$$

which is considered for functions $\varphi(x)$ that satisfy the first three equations in (1.32). Here, the first two conditions in (1.32) for the functional (1.34) are naturally interpreted in the sense of variational calculus, that is, the employed functions $\varphi(x)$ may not satisfy these conditions, but each such function has a finite Dirichlet integral $\int_\Omega |\nabla\varphi|^2 \mathrm{d}\Omega$ with $\int_\Gamma \varphi \mathrm{d}\Gamma = 0$.

For the variational problems (1.31), (1.32) and (1.34), (1.32), we get the spectral asymptotics

$$\lambda_k = \lambda_k(C; B) = \left(\frac{\operatorname{mes}\Gamma}{4\pi}\right)^{-3/2} k^{3/2}[1 + \mathrm{o}(1)] \ (k \to \infty), \tag{1.35}$$

which show that the dominant component of the asymptotic behavior of oscillation frequencies of the system is determined by the density coefficient of surface tension, σ, and by the area of the equilibrium surface Γ,

$$\omega_k = \omega_k^{\pm} = \pm\sqrt{\frac{\sigma\lambda_k}{\rho}} = \pm\left(\frac{\sigma}{\rho}\right)^{1/2}\left(\frac{\operatorname{mes}\Gamma}{4\pi}\right)^{-3/4} k^{3/4}[1 + \mathrm{o}(1)]. \tag{1.36}$$

Comparing this formula with the similar formula (3.3.22) for oscillation frequencies of a heavy fluid in a container, we conclude that taking into account the capillary forces leads to qualitative changes of the asymptotic behavior of oscillation frequencies.

4.1.7 Instability Conditions of the System

Let us now assume that the potential energy operator B does not satisfy the condition of statical stability (1.25) and is alternating. As long as it is bounded from below, we have

$$-\infty < \lambda_1(B) = \lambda_{\min}(B) < 0. \tag{1.37}$$

For the sake of convenience, let us suppose that B has exactly κ (with regard to multiplicity) negative eigenvalues. Then, according to conclusions in Sections 1.4.4 and 1.4.5, problem (1.29) also has exactly κ negative eigenvalues, $\lambda_1 \leq \ldots \leq \lambda_\kappa < 0$. If condition (1.37) holds true, then at least the eigenvalue $\lambda_1 = \lambda_1(C; B)$ of problem (1.29) is negative; to this negative eigenvalue there corresponds a pure imaginary frequency, $\omega_1 = \pm\sqrt{\sigma\lambda_1(C; B)/\rho}$, and at least one solution of the form $h(\xi^1, \xi^2)\exp(|\omega_1|t)$, which is increasing exponentially in time.

Hence, if condition (1.37) is valid, then the considered hydrosystem is dynamically unstable in linear approximation.

4.1.8 Solvability of the Evolution Problem

We consider again the Cauchy problem (1.19)—(1.21); for simplicity, let us assume that $B \gg 0$. (The case of static instability can be investigated in the same way as the situation described in detail in Section 3.5 for $\kappa = 2$.) Its generalized solution can be written explicitly as

$$\begin{aligned}\zeta(t) = (\rho C)^{-1/2}\Big\{&\cos(tA)\left[(\rho C)^{1/2}\zeta^0\right] + \sin(tA)A^{-1}\left[(\rho C)^{1/2}\zeta^1\right] \\ &+\rho\int_0^t \sin\left[(t-\tau)A\right]A^{-1}\left[(\rho C)^{-1/2}f_1(\tau)\right]d\tau\Big\}; \\ A := (\sigma/\rho)^{1/2}&(C^{-1/2}BC^{-1/2}) \gg 0.\end{aligned} \tag{1.38}$$

If the next conditions

$$\zeta^0 \in \mathcal{D}(B^{1/2}),\ \zeta^1 \in H_C = H_\Gamma^{-1/2} \tag{1.39}$$

are satisfied, and the function $f_1(t)$ is continuous and takes values in $H_\Gamma^{1/2}$, then for the generalized solution (1.38) the law of full energy balance, corresponding to the potential movements of the system, holds true,

$$\frac{1}{2}\left\{\rho\left\|\frac{\mathrm{d}\zeta}{\mathrm{d}t}\right\|_C^2 + \sigma\|\zeta(t)\|_B^2\right\} = \frac{1}{2}\left\{\rho\|\zeta^1\|_C^2 + \sigma\|\zeta^0\|_B^2\right\} + \rho\int_0^t \left(f_1(\tau), \frac{\mathrm{d}\zeta}{\mathrm{d}\tau}\right)_{L_2(\Gamma)} \mathrm{d}\tau. \tag{1.40}$$

Using this fact and the corresponding law of kinetic energy balance related to the vortical movements of the system, we obtain that the general law of full energy balance, which also follows from (1.17), takes place,

$$\frac{1}{2}\left\{\rho\|\boldsymbol{u}(t)\|_{\boldsymbol{L}_2(\Omega)}^2 + \sigma\|\zeta(t)\|_B^2\right\} = \frac{1}{2}\left\{\rho\|\boldsymbol{u}^0\|_{\boldsymbol{L}_2(\Omega)}^2 + \sigma\|\zeta^0\|_B^2\right\} + \rho\int_0^t (\boldsymbol{f}(\tau), \boldsymbol{u}(\tau))_{\boldsymbol{L}_2(\Omega)}\,\mathrm{d}\tau. \tag{1.41}$$

To ensure that all terms in (1.41) have finite values, it is sufficient to require in the initial problem (1.6), (1.7), (1.9), (1.12)—(1.14) that

$$\boldsymbol{u}^0 \in \boldsymbol{J}_{0,S}(\Omega), \qquad \zeta^0(\xi^1, \xi^2) \in H_B; \tag{1.42}$$

the external field of forces $\boldsymbol{f}(t,x)$ should be a continuous function of t with values in $\boldsymbol{L}_2(\Omega)$.

4.1.9 Oscillations of a System of Capillary Fluids

We next address the problem on oscillations of a system of m nonmixing fluids. This problem is similar to the problem on oscillations of a system of heavy fluids considered in Section 3.6, but has its own specific features.

First of all, the boundaries Γ_i between fluids are not plane in the equilibrium state. Similar to the case of one fluid, they can be found as a result of solving the static problem for the whole system of fluids.

As usual, we assume that the fluid of density ρ_k occupies the region Ω_k ($k = 1, 2, \ldots, m$), and label the fluids "from below-upwards," i.e., against the general direction of the external field $\boldsymbol{F}_0 = -\nabla U_0$. The direction of the normal $\boldsymbol{n}_i$ to Γ_i is chosen from the i-th to the $(i+1)$-th fluid, respectively. The dihedral angles between Γ_k and S_k, which are constant on the contact lines $\partial\Gamma_k$, are denoted by δ_k.

In the stable state, the pressure in each fluid is defined by the formulas

$$P_{0,k}(x) = -\rho_k U_0(x) + c_k \quad \text{in } \Omega_k, \tag{1.43}$$

and on the boundaries Γ_i the Laplace conditions hold true,

$$P_{0,i}(x) - P_{0,i+1}(x) = -\sigma(k_{1,i} + k_{2,i}) \quad \text{on } \Gamma_i, \tag{1.44}$$

where $k_{j,i}$ $(j = 1, 2)$ are the main curvatures of the surface Γ_i. Similar to the case of one fluid, the Dupré-Young conditions are valid on the contact lines,

$$\sigma_i \cos\delta_i = \sigma_{1,i} - \sigma_{0,i} \quad \text{on } \partial\Gamma_i. \tag{1.45}$$

Relations (1.43)–(1.45) along with the next volume-preserving conditions for each fluid,

$$\int_{\Omega_k} d\Omega_k = \operatorname{mes}\Omega_k = V_k, \tag{1.46}$$

make it possible to find the surface Γ_k and to determine the constants c_k in (1.43), i.e., to solve the static problem completely.

Considering now the problem of small oscillations of a system of capillary fluids, let us first write down (similar to Section 4.1.2) the corresponding system of equations with boundary and initial conditions. The difference from the case of a system of heavy fluids is in a modification of the dynamic conditions on Γ_i and in the appearance of boundary conditions on $\partial\Gamma_i$. Specifically, we obtain

$$\begin{aligned}
\frac{\partial \boldsymbol{u}^k}{\partial t} &= -\frac{1}{\rho_k}\nabla p_k + \boldsymbol{f}, \quad \operatorname{div}\boldsymbol{u}^k = 0 \quad \text{in } \Omega_k,\\
\boldsymbol{u}^k \cdot \boldsymbol{n} &= 0 \quad \text{on } S, \quad \int_{\Gamma_i} \boldsymbol{u}^i \cdot \boldsymbol{n}_i d\Gamma_i = 0,\\
\boldsymbol{u}^i \cdot \boldsymbol{n}_i &= \boldsymbol{u}^{i+1} \cdot \boldsymbol{n}_i = \frac{\partial \zeta_i}{\partial t} \quad \text{on } \Gamma_i,\\
p_i - p_{i+1} &= \sigma_i \mathcal{L}_i \zeta_i := \sigma_i \left[-\Delta_{\Gamma_i} + a_i(x)\right]\zeta_i \quad \text{on } \Gamma_i,\\
a_i(x) &= -(k_{1,i}^2 + k_{2,i}^2) + (\Delta\rho)_i \sigma_i^{-1} \nabla U_0 \cdot \boldsymbol{n}_i,\\
\frac{\partial \zeta_i}{\partial \nu_i} &+ \chi_i \zeta_i = 0 \quad \text{on } \partial\Gamma_i,\\
\chi_i &:= \frac{k_{\Gamma_i}\cos\delta_i - k_{S_i}}{\sin\delta_i},\\
\boldsymbol{u}^k(0, x) &= \boldsymbol{u}^{k,0}(x), \quad \zeta_i(0, \xi_i^1, \xi_i^2) = \zeta_i^0(\xi_i^1, \xi_i^2).
\end{aligned} \tag{1.47}$$

Here, the function $\zeta_i(t, \xi_i^1, \xi_i^2)$ describes the deviation along the normal $\boldsymbol{n}_i$ of the moving boundary $\tilde{\Gamma}_i$ from the equilibrium boundary Γ_i; $\tilde{O}_i \xi_i^1 \xi_i^2 \xi_i^3$ is a curvilinear coordinate system in a neighborhood of Γ_i, similar to the system $\tilde{O}\xi^1\xi^2\xi^3$ employed in Section 4.1.2.

We apply to problem (1.47) the method of orthogonal projection, in the way described in Section 3.6.2 for a system of heavy fluids. Instead of problem (3.6.20), we obtain the next generalization for the case of capillary fluids,

$$
\begin{gathered}
\Delta\Phi_k = 0 \quad \text{in } \Omega_k, \qquad \frac{\partial \Phi_k}{\partial n} = 0 \quad \text{on } S_k, \\
\frac{\partial \Phi_i}{\partial n_i} = \frac{\partial \Phi_{i+1}}{\partial n_i} = \frac{\partial \zeta_i}{\partial t} \quad \text{on } \Gamma_i, \\
\frac{\partial}{\partial t}(\rho_i \Phi_i - \rho_{i+1}\Phi_{i+1}) + \sigma_i B_i \zeta_i = \rho_i F_i - \rho_{i+1} F_{i+1} \quad \text{on } \Gamma_i, \\
\int_{\Gamma_i} (\rho_i \Phi_i - \rho_{i+1}\Phi_{i+1})\, \mathrm{d}\Gamma_i = 0 \qquad i = 1, 2, \dots, m-1, \\
\int_{\Gamma_{m-1}} \rho_m \Phi_m \mathrm{d}\Gamma_{m-1} = 0; \qquad \zeta_i\left(0, \xi_i^1, \xi_i^2\right) = \zeta_i^0\left(\xi_i^1, \xi_i^2\right), \\
\frac{\partial \zeta_i}{\partial t}\left(0, \xi_i^1 \xi_i^2\right) = v_{n_i}^{i,0}\left(\xi_i^1, \xi_i^2\right) = \nabla \Phi_i^0 \cdot \boldsymbol{n}_i, \qquad i = 1, 2, \dots, m-1, \\
(\widehat{\nabla \Phi})\,|_{t=0} = \left\{\nabla \Phi_k^0\right\}_{k=1}^m = P_S \hat{\boldsymbol{u}}^0, \qquad \hat{\boldsymbol{u}}^0 = \left\{\boldsymbol{u}^{k,0}(x)\right\}_{k=1}^m.
\end{gathered}
\tag{1.48}
$$

Here, similarly to (1.19), (1.20), the operators B_i in the dynamic boundary conditions on Γ_i are defined by the expressions

$$
B_i \zeta_i := P_i \mathcal{L}_i \zeta_i = \mathcal{L}_i \zeta_i - (\operatorname{mes} \Gamma_i)^{-1} \int_{\Gamma_i} (\mathcal{L}_i \zeta_i) \mathrm{d}\Gamma_i \tag{1.49}
$$

and the boundary conditions

$$
\frac{\partial \zeta_i}{\partial \nu_i} + \chi_i \zeta_i = 0 \quad \text{on } \partial\Gamma_i.
$$

Their properties are completely similar to properties of the operator B used in Section 4.1.4, namely, they are selfadjoint operators defined in $\mathcal{D}(B_i) = H^2(\Gamma_i) \cap L_{2,\Gamma_i}$ and bounded from below.

Further, let us introduce the operator-matrix $\hat{B} := \operatorname{diag}(\sigma_i B_i)_{i=1}^{m-1}$. It allows us (similarly to Section 3.6.3) to perform the transition from the evolution boundary value problem (1.48)–(1.49) to a Cauchy problem of the form (3.6.23), using the abstract scheme. As a result we get

$$
\begin{gathered}
\hat{C} \frac{\mathrm{d}^2 \hat{\zeta}}{\mathrm{d}t^2} + \hat{B} \hat{\zeta} = \hat{F}(t), \quad \hat{F} = \{\hat{F}_i\}_{i=1}^{m-1}, \quad \tilde{F}_i := \rho_i F_i - \rho_{i+1} F_{i+1}, \\
\hat{\zeta}(0) = \hat{\zeta}^0 = \left\{\zeta_i^0\right\}_{i=1}^{m-1}, \quad \frac{\mathrm{d}\hat{\zeta}(0)}{\mathrm{d}t} = \hat{v}^0 = \left\{v_{n_i}^{i,0}\right\}_{i=1}^{m-1},
\end{gathered}
\tag{1.50}
$$

where the potential energy operator $\hat{B}$ has a new significance. Its associated bilinear form for the system of capillary fluids equals

$$
(\hat{B}\hat{\zeta}, \hat{\zeta})_{\hat{L}_{2,\Gamma}} = \sum_{i=1}^{m-1} \left\{ \sigma_i \int_{\Gamma_i} \left[\nabla_{\Gamma_i}(\zeta_i, \zeta_i) + a_i |\zeta_i|^2\right] \mathrm{d}\Gamma_i + \sigma_i \int_{\partial\Gamma_i} \chi_i |\zeta_i|^2 \mathrm{d}s_i \right\}. \tag{1.51}
$$

We will say that the equilibrium state of a system of fluids is *statically stable in linear approximation*, if $\hat{B}$ is positive definite, i.e., its minimal eigenvalue is positive: $\lambda_{\min}(\hat{B}) > 0$. Then, as in Section 4.1.5, it can be proved that the form (1.51) generates a norm equivalent to the Dirichlet norm

$$\|\hat{\zeta}\|^2_{\hat{\nabla}} := \sum_{i=1}^{m-1} \int_{\Gamma_i} \nabla_{\Gamma_i}(\zeta_i, \zeta_i) d\Gamma_i, \int_{\Gamma_i} \zeta_i d\Gamma_i = 0.$$

Moreover, the inverse operator $\hat{B}^{-1}$ is compact and positive.

This fact means that by carrying out the reasoning developed in Sections 4.1.6–4.1.8, we can obtain all the results proved there for our new problem on oscillations of a system of capillary fluids. In particular, for the proper modes of oscillations $\hat{h} = \{h_i\}_{i=1}^{m-1}$ the orthogonality conditions of the form (1.30) hold true, and the eigenvalues λ_k can be found as successive minima of the variational ratio

$$\frac{\sum_{i=1}^{m-1} \left\{ \sigma_i \int_{\Gamma_i} \left[\nabla_{\Gamma_i}(h_i, h_i) + a_i |h_i|^2 \right] d\Gamma_i + \sigma_i \int_{\partial\Gamma_i} \chi_i |h_i|^2 ds_i \right\}}{\sum_{k=1}^{m} \rho_k \int_{\Omega_k} |\nabla \Phi_k|^2 d\Omega_k}, \tag{1.52}$$

or of the next variational ratio

$$\frac{\sum_{k=1}^{m} \rho_k \int_{\Omega_k} |\nabla \Phi_k|^2 d\Omega_k}{\sum_{i=1}^{m-1} \int_{\Gamma_i} \left[(\sigma_i B_i)^{-1} (\rho_i \Phi_i - \rho_{i+1} \Phi_{i+1}) \right] (\rho_i \Phi_i - \rho_{i+1} \Phi_{i+1}) d\Gamma_i}. \tag{1.53}$$

These ratios are considered for the solution of the boundary value problem (3.6.22), i.e., for the additional conditions

$$\Delta \Phi_k = 0 \quad \text{in } \Omega_k, \qquad \frac{\partial \Phi_k}{\partial n} = 0 \quad \text{on } S_k,$$
$$\frac{\partial \Phi_i}{\partial n_i} = \frac{\partial \Phi_{i+1}}{\partial n_i} = h_i \quad \text{on } \Gamma_i,$$
$$\int_{\Gamma_i} (\rho_i \Phi_i - \rho_{i+1} \Phi_{i+1}) d\Gamma_i = 0, \qquad i = 1, 2, \dots, m-1,$$
$$\int_{\Gamma_{m-1}} \rho_m \Phi_m d\Gamma_{m-1} = 0. \tag{1.54}$$

The variational problems (1.52), (1.54) and (1.53), (1.54) result in the following spectral asymptotics for eigenvalues,

$$\lambda_n = \lambda_n(\hat{C}; \hat{B}) = \left(\frac{n}{K_{\hat{C},\hat{B}}} \right)^{3/2} [1 + o(1)], \qquad n \to \infty,$$

$$K_{\hat{C},\hat{B}} := \frac{1}{4\pi} \sum_{i=1}^{m-1} \left(\frac{\rho_i + \rho_{i+1}}{\sigma_i} \right)^{2/3} \operatorname{mes} \Gamma_i. \tag{1.55}$$

Further, if $\lambda_{\min}(\hat{B}) < 0$, then the system of nonmixing capillary fluids is dynamically unstable (in linear approximation).

As for the solvability of the initial boundary value problem (1.47), if

$$\lambda_{\min}(\hat{B}) = \min_{1 \le i \le m-1} \{\lambda_{\min}(\sigma_i B_i)\} > 0,$$

then the problem is univalently solvable and one gets a continuous full energy (hence, the law of full energy balance is valid), provided at the initial moment $t = 0$ the potential and kinetic energies of the system are finite and the external field $\hat{\boldsymbol{f}}(t) = \{\boldsymbol{f}(t)\}_{k=1}^{m}$ is a continuous function with values in $\hat{\boldsymbol{L}}_2(\Omega)$.

4.2 Oscillations of Fluids in Containers With Elastic Ends

The problem considered in this section is similar to the problem studied in Section 4.1 and the general scheme used in its investigation remains basically unchanged. The ends, which are plane when there is no influence of an external field, are considered to be plane plates of constant thickness and are taken as elastic parts of the container. It is assumed that during the process of motion, the fluid and the bottom are not separated, i.e., the motion is noncavitational.

4.2.1 Solving the Static Problem

Let us consider a rigid container Ω that has upper and lower plane elastic ends. We assume that this container is wholly filled with an incompressible ideal fluid of density ρ. We denote the solid boundary by S, and the ends by Γ_1, Γ_2, respectively. In conditions of complete low-gravity, i.e., without any external field, the ends remain plane. If the homogeneous gravitation field $\boldsymbol{F}_0 = -\boldsymbol{g} = -g\boldsymbol{e}_3$ acts along the axis Ox_3 directed orthogonally to the ends, then the ends sag due to the weight, and we have to determine the extent of the sagging.

Let h be the height of the container and assume that p_a is the external constant pressure. We look for static saggings of the form

$$x_3 = \zeta_1^0(x_1, x_2), \quad (x_1, x_2) \in \Gamma_1, \qquad x_3 = \zeta_2^0(x_1, x_2) + h, \quad (x_1, x_2) \in \Gamma_2. \tag{2.1}$$

In the equilibrium state, the pressure in fluid is distributed according to the law

$$P_0(x_3) = -\rho g x_3 + P_0(0). \tag{2.2}$$

Therefore, the force that acts downward on unit area of the lower end—the bottom—

equals $P_0(x_3) - p_a$, and the force that acts on unit area of the upper end—the top—equals $p_a - P_0(x_3)$.

In order to determine the static saggings of the ends, we use the well-known equation of Sophie Germaine that yields the equilibrium shape of a slightly sagged elastic plate,

$$-\mathrm{d} \cdot \Delta_2^2 \zeta = q. \tag{2.3}$$

Here, $\mathrm{d} > 0$ is the rigidity coefficient of the plate, Δ_2 is the two-dimensional Laplace operator (in variables x_1, x_2), $\zeta(x_1, x_2)$ is the unknown sagging, and $q(x_1, x_2)$ is the load acting downward on unit area.

In the considered case, taking into account the expressions for q at the upper and lower ends, we obtain the equations

$$\begin{aligned} -\mathrm{d}_1 \Delta_2^2 \zeta_1^0 &= q_1 = P_0 \Big|_{x_3=\zeta_1^0} - p_a, \\ -\mathrm{d}_2 \Delta_2^2 \zeta_2^0 &= q_2 = p_a - P_0 \Big|_{x_3=h+\zeta_2^0} . \end{aligned} \tag{2.4}$$

Hence, from (2.1) and (2.2) we get the equations

$$\begin{aligned} (\mathrm{d}_1 \Delta_2^2 - \rho g)\zeta_1^0 &= p_a - P_0(0) =: c \quad \text{on } \Gamma_1, \\ (\mathrm{d}_2 \Delta_2^2 + \rho g)\zeta_2^0 &= -c - \rho g h \quad \text{on } \Gamma_2. \end{aligned} \tag{2.5}$$

Moreover, let us assume that the ends Γ_1 and Γ_2 are rigidly attached to the contours $\partial\Gamma_1$ and $\partial\Gamma_2$. It means that the following boundary conditions should be valid,

$$\zeta_i^0 = \frac{\partial \zeta_i^0}{\partial \nu_i} = 0 \quad \text{on } \partial\Gamma_i, \qquad i = 1, 2, \tag{2.6}$$

where $\boldsymbol{\nu}_i$ is the external normal vector to the region Γ_i.

By the incompressibility condition of the fluid, its volume should coincide with the volume of region Ω. Therefore, for small deviations of the ends we obtain the additional condition

$$\int_{\Gamma_1} \zeta_1^0 \mathrm{d}\Gamma_1 = \int_{\Gamma_2} \zeta_2^0 \mathrm{d}\Gamma_2. \tag{2.7}$$

Hence, the static problem on determining the saggings of the ends amounts to finding the functions ζ_1^0, ζ_2^0, and the constant c, by using the equations (2.5), the boundary conditions (2.6), and the additional condition (2.7).

In order to solve it, let us introduce the operators B_i that act on $L_2(\Gamma_i)$ and are defined by the differential expressions

$$B_i u_i := (\mathrm{d}_i \Delta_2^2 + (-1)^i \rho g) u_i =: (\mathrm{d}_i B_i^0 + (-1)^i \rho g I_i) u_i, \qquad i = 1, 2, \tag{2.8}$$

and boundary conditions of the form (2.6) for u_i. We choose the sets of functions from $H^4(\Gamma_i)$ that satisfy conditions (2.6) as the domains of definition $\mathcal{D}(B_i)$.

It can be proven by means of Green formula for the biharmonic operator Δ_2^2,

$$\int_{\Gamma_i} (\Delta_2^2 u_i)\, v_i \mathrm{d}\Gamma_i = \int_{\Gamma_i} \Delta_2 u_i \cdot \Delta_2 v_i \mathrm{d}\Gamma_i + \int_{\partial\Gamma_i} \left(\frac{\partial \Delta_2 u_i}{\partial \nu_i} v_i - \Delta_2 u_i \frac{\partial v_i}{\partial \nu_i} \right) \mathrm{d}s, \qquad (2.9)$$

that the operators B_i are symmetric on $\mathcal{D}(B_i)$. Moreover, they are selfadjoint operators bounded from below.

As it was proven in elasticity theory, the operators B_i^0 from (2.8), which are defined by the differential expressions $B_i^0 u_i = \Delta_2^2 u_i$ and the boundary conditions (2.6), are positive definite in $L_2(\Gamma_i)$,

$$(B_i^0 u_i, u_i)_{L_2(\Gamma_i)} \geq k_i (u_i, u_i)_{L_2(\Gamma_i)}, \qquad i = 1, 2,\ k_i = \lambda_{\min}(B_i^0) > 0. \qquad (2.10)$$

Therefore, the following inequalities are valid for operator B_i,

$$(B_i u_i, u_i)_{L_2(\Gamma_i)} \geq (\mathrm{d}_i k_i + (-1)^i \rho g) \|u_i\|^2_{L_2(\Gamma_i)}, \qquad u_i \in \mathcal{D}(B_i). \qquad (2.11)$$

Further, let us assume that the ends of the container are sufficiently rigid, i.e., the conditions

$$\rho g / (\mathrm{d}_i k_i) \ll 1, \qquad i = 1, 2, \qquad (2.12)$$

are satisfied. Obviously, in this case not only operator B_2 but also B_1 is positive definite, and there is for this kind of rigidity of the ends that their saggings are small, not only in the static problem but also for small movements of the fluid in the container.

Let us write down the problem (2.5)–(2.7) in the form

$$B_1 \zeta_1^0 = c, \quad B_2 \zeta_2^0 = -c - \rho g h, \quad \int_{\Gamma_1} \zeta_1^0 \mathrm{d}\Gamma_1 = \int_{\Gamma_2} \zeta_2^0 \mathrm{d}\Gamma_2. \qquad (2.13)$$

Now, we introduce the solutions w_i^0 of the following problem

$$B_i w_i^0 = 1_i, \qquad i = 1, 2,$$

as auxiliary functions, where 1_i is a function that identically equals 1 on Γ_i. In virtue of the invertibility of B_i we have $w_i^0 = B_i^{-1} 1_i$, hence by (2.13) we get

$$\zeta_1^0 = c w_1^0, \qquad \zeta_2^0 = -(c + \rho g h) w_2^0,$$

and the additional condition (2.13) helps to determine the constant c. Finally, the solution of the static problem has the form

$$\zeta_1^0 = -\rho g h \frac{\alpha_2}{\alpha_1 + \alpha_2} w_1^0, \qquad \zeta_2^0 = -\rho g h \frac{\alpha_1}{\alpha_1 + \alpha_2} w_2^0,$$
$$\alpha = (B_i^{-1} 1_i, 1_i)_{L_2 \Gamma_i)} > 0, \qquad c = -\rho g h \frac{\alpha_2}{\alpha_1 + \alpha_2}. \qquad (2.14)$$

4.2.2 Formulation of the Problem on Small Oscillations

We now consider the problem of small oscillations of a fluid in a container without the influence of any additional field of external forces. As pointed out in the previous section, the vortical movements of the fluid remain unchanged relatively to time and do not affect the moving parts of the boundary of the container, i.e., its elastic ends. That is why the velocity field $\boldsymbol{u}(t,x)$ of the fluid is assumed to be potential and to belong to the subspace $\boldsymbol{G}_{\mathrm{h},S}(\Omega)$.

If $\boldsymbol{u} = \nabla\Phi$, $\Phi = \Phi(t,x)$, then the solenoidality condition and the nonleaking condition on S lead to the relations

$$\Delta\Phi = 0 \quad \text{in } \Omega, \qquad \frac{\partial\Phi}{\partial n} = 0 \quad \text{on } S. \tag{2.15}$$

We suppose the the normal $\boldsymbol{n}_i$ to the plane Γ_i is directed vertically upward, and let us describe the vertical deviations of the ends during the fluid movement by the functions $\zeta_i(t,x_1,x_2)$. Then linearized kinematic conditions on the ends have the form

$$\frac{\partial\Phi}{\partial n_i} = \frac{\partial\zeta_i}{\partial t} \qquad \text{on } \Gamma_i,\ i = 1,2. \tag{2.16}$$

In order to obtain the dynamic conditions on Γ_i, we use the dynamic equation of Sophie Germaine that takes into account inertial forces. If the plate has linear mass M (mass of the unit area of the plate), then instead of (2.3) we have

$$-\mathrm{d}\cdot\Delta_2^2\zeta - M\frac{\partial^2\zeta}{\partial t^2} = q. \tag{2.17}$$

Pursuing a reasoning similar to that used to derive the static equations (2.4), we have

$$\begin{aligned} P|_{\Gamma_1} - p_a &= -\mathrm{d}_1\Delta_2^2\zeta_1 - M_1\frac{\partial^2\zeta_1}{\partial t^2} \qquad \text{on } \Gamma_1,\\ p_a - P|_{\Gamma_2} &= -\mathrm{d}_2\Delta_2^2\zeta_2 - M_2\frac{\partial^2\zeta_2}{\partial t^2} \qquad \text{on } \Gamma_2, \end{aligned} \tag{2.18}$$

where the meaning of the symbols is obvious.

We now represent the full pressure in fluid as

$$P(t,x) = P_0(x_3) + p(t,x), \tag{2.19}$$

where $p(t,x)$ is the dynamic pressure, substitute (2.19) into (2.18), and use formula (2.2). We have

$$\begin{aligned} p + P_0(0) - \rho g\zeta_1 - p_a &= -\mathrm{d}_1\Delta_2^2\zeta_1 - M\frac{\partial^2\zeta_1}{\partial t^2} \qquad \text{on } \Gamma_1,\\ p_a - P_0(0) - p + \rho g\zeta_2 + \rho g h &= -\mathrm{d}_2\Delta_2^2\zeta_2 - M_2\frac{\partial^2\zeta_2}{\partial t^2} \qquad \text{on } \Gamma_2. \end{aligned} \tag{2.20}$$

We look for the sagging of each end as a sum of its static and dynamic saggings,

$$\zeta_i(t,x_1,x_2) = \zeta_1^0(x_1,x_2) + w_i(t,x_1,x_2), \qquad i=1,2 \tag{2.21}$$

and use the fact that functions ζ_i^0 satisfy the equilibrium equations (2.5). A substitution of (2.21) into (2.20) leads to the conditions

$$\begin{aligned} M_1\frac{\partial^2 w_1}{\partial t^2} + B_1 w_1 + p = 0 \qquad &\text{on } \Gamma_1, \\ M_2\frac{\partial^2 w_2}{\partial t^2} + B_2 w_2 - p = 0 \qquad &\text{on } \Gamma_2, \end{aligned} \tag{2.22}$$

where B_i are the operators introduced in Section 4.2.1 according to formulas (2.8). We eliminate the dynamic pressure p from (2.22) by means of the Cauchy-Lagrange integral (with an arbitrary function $c(t)$),

$$p + \rho\frac{\partial\Phi}{\partial t} = c(t) \qquad \text{in } \Omega,$$

which takes place for the free potential flows of an ideal fluid. Taking into account (2.21) and the independence of functions ζ^0 of t, we finally obtain the following form of the boundary value problem on oscillations of a fluid in a container with elastic ends:

$$\begin{aligned} &\Delta\Phi = 0 \quad \text{in } \Omega, \qquad \frac{\partial\Phi}{\partial n} = 0 \quad \text{on } S, \\ &\frac{\partial\Phi}{\partial n_i} = \frac{\partial w_i}{\partial t} \quad \text{on } \Gamma_i, \quad i=1,2, \qquad \int_{\Gamma_1} w_1 d\Gamma_1 = \int_{\Gamma_2} w_2 d\Gamma_2, \\ &M_1\frac{\partial^2 w_1}{\partial t^2} + B_1 w_1 - \rho\frac{\partial\Phi}{\partial t} = -c(t) \qquad \text{on } \Gamma_1, \\ &M_2\frac{\partial^2 w_2}{\partial t^2} + B_2 w_2 + \rho\frac{\partial\Phi}{\partial t} = c(t) \quad \text{on } \Gamma_2, \\ &w_i(0,x_1,x_2) = w_i^0(x_1,x_2), \quad i=1,2, \qquad \int_{\Gamma_1} w_1^0 d\Gamma_1 = \int_{\Gamma_2} w_2^0 d\Gamma_2, \\ &\frac{\partial w_i}{\partial t}(0,x_1,x_2) = v_i^0(x_1,x_2), \quad i=1,2, \qquad \int_{\Gamma_1} v_1^0 d\Gamma_1 = \int_{\Gamma_2} v_2^0 d\Gamma_2. \end{aligned} \tag{2.23}$$

Here, similar to Section 4.2.1, the integral links are consequences of the volume preserving condition of an oscillating fluid, and the boundary conditions on $\partial\Gamma_i$ are taken into account by introducing the operators B_i that act on $L_2(\Gamma_i)$ and have the domains of definition $\mathcal{D}(B_i)$. The last two lines of initial conditions define the dynamic saggings of the ends and their velocities at the initial moment of time.

4.2.3 Energy-Preserving Law

Let us assume that the classical solution of problem (2.23) is found. We multiply both sides of the dynamic equations on the ends (2.23) by $\partial w_i/\partial t$, integrate the products over Γ_i, and add the resulting expressions with regard to the kinematic links on Γ_i. We obtain

$$\frac{1}{2}\frac{\mathrm{d}}{\mathrm{d}t}\sum_{i=1}^{2} M_i \left\|\frac{\partial w_i}{\partial t}\right\|^2_{L_2(\Gamma_i)} + \sum_{i=1}^{2}\left(B_i w_i, \frac{\partial w_i}{\partial t}\right)_{L_2(\Gamma_i)}$$
$$+\rho\left(\int_{\Gamma_2}\frac{\partial\Phi}{\partial t}\frac{\partial\Phi}{\partial n_2}\mathrm{d}\Gamma_2 - \int_{\Gamma_1}\frac{\partial\Phi}{\partial t}\frac{\partial\Phi}{\partial n_1}\mathrm{d}\Gamma_1\right)$$
$$= c(t)\frac{\mathrm{d}}{\mathrm{d}t}\left(\int_{\Gamma_2} w_2\mathrm{d}\Gamma_2 - \int_{\Gamma_1} w_1\mathrm{d}\Gamma_1\right). \tag{2.24}$$

For functions w_i, v_i from $\mathcal{D}(B_i)$, the bilinear functional $(B_i w_i, v_i)_{L_2(\Gamma_i)}$ can be easily found by means of the Green formula (2.10),

$$(B_i w_i, v_i)_{L_2(\Gamma_i)} = \mathrm{d}_i\int_{\Gamma_i}\Delta_2 w_i\cdot\Delta_2 v_i\mathrm{d}\Gamma_i + (-1)^i\rho g\int_{\Gamma_i} w_i v_i\mathrm{d}\Gamma_i, \qquad i = 1,2. \tag{2.25}$$

Since $B_i \gg 0$, the operators $B_i^{1/2}$ exist and the right side of (2.25) can be written down as

$$\left(B_i^{1/2}w_i, B_i^{1/2}v_i\right)_{L_2(\Gamma_i)} =: (w_i, v_i)_{B_i},$$

whence

$$\left(B_i w_i, \frac{\partial w_i}{\partial t}\right)_{L_2(\Gamma_i)} = \frac{1}{2}\frac{\mathrm{d}}{\mathrm{d}t}\left\|B_i^{1/2}w_i\right\|^2_{L_2(\Gamma_2)} = \frac{1}{2}\frac{\mathrm{d}}{\mathrm{d}t}\|w_i\|^2_{B_i}.$$

Further, the expression within brackets in the right side of (2.24) equals zero in virtue of the volume-preserving condition for oscillations. Now, we transform the last expression within brackets in the left side of (2.24) by using the Neumann boundary condition for Φ on the solid boundary S, and the harmonicity condition of function Φ in Ω.

Then,

$$\int_{\Gamma_2}\frac{\partial\Phi}{\partial t}\frac{\partial\Phi}{\partial n_2}\mathrm{d}\Gamma_2 - \int_{\Gamma_1}\frac{\partial\Phi}{\partial t}\frac{\partial\Phi}{\partial n_1}\mathrm{d}\Gamma_1 = \int_{\partial\Omega}\frac{\partial\Phi}{\partial t}\frac{\partial\Phi}{\partial n}\mathrm{d}S$$
$$= \int_\Omega \nabla\frac{\partial\Phi}{\partial t}\cdot\nabla\Phi\mathrm{d}\Omega + \int_\Omega\frac{\partial\Phi}{\partial t}\Delta\Phi\mathrm{d}\Omega$$
$$= \frac{1}{2}\frac{\mathrm{d}}{\mathrm{d}t}\int_\Omega|\nabla\Phi|^2\mathrm{d}\Omega.$$

Finally, from (2.24) we obtain

$$\frac{1}{2}\frac{\mathrm{d}}{\mathrm{d}t}\left\{\sum_{i=1}^{2}\left[M_i\left\|\frac{\partial w_i}{\partial t}\right\|^2_{L_2(\Gamma_i)} + \|w_i\|^2_{B_i}\right] + \rho\int_\Omega|\nabla\Phi|^2\mathrm{d}\Omega\right\} = 0. \tag{2.26}$$

Here, within curly brackets we have twice the full energy of the system calculated at an arbitrary moment of time. The energy-preserving law follows from (2.26).

4.2.4 Operator Equation of the Problem

In order to perform the transition from problem (2.23) to an operator equation, we need to introduce the necessary Hilbert spaces. Let us first consider the space $L_2(\Gamma) := \bigoplus_{i=1}^{2} L_2(\Gamma_i)$ of pairs of functions $w := \{w_1; w_2\}$ with the scalar product

$$(w, v)_{L_2(\Gamma)} := \sum_{i=1}^{2} (w_i, v_i)_{L_2(\Gamma_i)} \tag{2.27}$$

and its subspace $L_{2,\Gamma}$ with codimension 1 defined by imposing the following condition,

$$\int_{\Gamma_1} w_1 \mathrm{d}\Gamma_1 = \int_{\Gamma_2} w_2 \mathrm{d}\Gamma_2. \tag{2.28}$$

This condition is necessary for preserving the volume of the fluid in problem (2.23). If we introduce the element $\eta_0 := \{1_1; -1_2\}$, where 1_i is the unit function defined on Γ_i, then $L_{2,\Gamma}$ is the set of elements from $L_2(\Gamma)$ orthogonal to η_0 in the sense of the scalar product (2.27).

We next introduce the space of functions $H^1(\Omega)$ with the norm

$$\|\varphi\|^2_{H^1(\Omega)} := \int_\Omega |\nabla\varphi|^2 \mathrm{d}\Omega + \left(\int_{\Gamma_1} \varphi \mathrm{d}\Gamma_1 + \int_{\Gamma_2} \varphi \mathrm{d}\Gamma_2\right)^2, \tag{2.29}$$

and the subspace $H^1_\Gamma(\Omega)$ consisting of elements for which the second term in the right side of (2.29) equals zero and, therefore, the norm in this subspace is defined by the Dirichlet integral.

For elements φ from $H^1_\Gamma(\Omega)$, we introduce the trace operator γ_Γ according to the rule

$$\gamma_\Gamma \varphi := \{-\varphi|_{\Gamma_1}; \varphi|_{\Gamma_2}\}, \tag{2.30}$$

and consider the spaces $F = H^1_\Gamma(\Omega)$, $G = L_{2,\Gamma}$ by using the abstract scheme discussed in Section 1.8. Since in virtue of the embedding theorems, operator γ_Γ from F into G is compact, one gets that operator $C = \gamma_\Gamma T$ from the general scheme is positive and compact on $G = L_{2,\Gamma}$. Besides, operator T yields an isometry between the space $(H^{1/2}_\Gamma(\Gamma))^*$ and the subspace $H^1_{\mathrm{h},S}(\Omega)$ of harmonic functions from $H^1_\Gamma(\Omega)$ for which the normal derivative equals zero on S.

Hence we conclude that the Neumann problem

$$\Delta\varphi = 0 \quad \text{in } \Omega, \qquad \frac{\partial\phi}{\partial n} = 0 \quad \text{on } S,$$
$$\frac{\partial\phi}{\partial n_i} = \psi_i \quad \text{on } \Gamma_i, \qquad i = 1, 2 \tag{2.31}$$

has a unique solution $\varphi \in H^1_{\mathrm{h},S}(\Omega)$ for any $\psi = \{\psi_1; \psi_2\} \in (H^{1/2}_\Gamma(\Gamma))^* \supset L_{2,\Gamma}$.

Let us now consider the expression

$$\begin{aligned}\gamma_\Gamma T\psi := C\psi &= \gamma_\Gamma\varphi = \{-\varphi|_{\Gamma_1}; \varphi|_{\Gamma_2}\} \\ &=: \{C_{11}\psi_1 + C_{12}\psi_2; C_{21}\psi_1 + C_{22}\psi_2\} \\ &= \begin{pmatrix} C_{11} & C_{12} \\ C_{21} & C_{22} \end{pmatrix} \begin{pmatrix} \psi_1 \\ \psi_2 \end{pmatrix}. \end{aligned} \tag{2.32}$$

From this and from some equations of problem (2.23) we obtain

$$\begin{pmatrix} -\dfrac{\partial\Phi}{\partial t}\Big|_{\Gamma_1} \\ \dfrac{\partial\Phi}{\partial t}\Big|_{\Gamma_2} \end{pmatrix} = \frac{\mathrm{d}}{\mathrm{d}t}\begin{pmatrix} -\Phi|_{\Gamma_1} \\ \Phi|_{\Gamma_2} \end{pmatrix} = \frac{\mathrm{d}}{\mathrm{d}t}\begin{pmatrix} C_{11} & C_{12} \\ C_{21} & C_{22} \end{pmatrix}\begin{pmatrix} \dfrac{\partial\Phi}{\partial n_1} \\ \dfrac{\partial\Phi}{\partial n_2} \end{pmatrix} = \frac{\mathrm{d}^2}{\mathrm{d}t^2}Cw, \tag{2.33}$$

and therefore, (2.23) can be written down in the vector-matrix form

$$\begin{aligned} &(\tilde{M} + \rho C)\frac{\mathrm{d}^2 w}{\mathrm{d}t^2} + \tilde{B}w = -c(t)\eta_0, \qquad w(0) = w^0,\ w'(0) = v^0, \\ &\tilde{M} = \operatorname{diag}(M_1 I_1; M_2 I_2), \quad \tilde{B} = \operatorname{diag}(B_1; B_2), \\ &w^0 = \{w_1^0, w_2^0\}, \quad v^0 = \{v_1^0; v_2^0\}, \end{aligned} \tag{2.34}$$

where $\eta_0 = \{1_1; -1_2\}$. In this equation, operator C acts on $L_{2,\Gamma}$, and operators $\tilde{M}, \tilde{B}$ act from $L_{2,\Gamma}$ into $L_2(\Gamma)$. Let us apply the orthoprojector P_Γ onto the subspace $L_{2,\Gamma}$ to (2.34); for any element $w \in L_2(\Gamma)$ it acts according to the rule

$$P_\Gamma w = w - (\operatorname{mes}\Gamma_1 + \operatorname{mes}\Gamma_2)^{-1}\left(\int_{\Gamma_1} w_1 \mathrm{d}\Gamma_1 - \int_{\Gamma_2} w_2 \mathrm{d}\Gamma_2\right). \tag{2.35}$$

Then, problem (2.23) is reduced to the Cauchy problem

$$A\frac{\mathrm{d}^2 w}{\mathrm{d}t^2} + Bw = 0, \qquad w(0) = w^0,\ w'(0) = v^0, \tag{2.36}$$

$$A = M + \rho C, \quad M = P_\Gamma \tilde{M} P_\Gamma, \quad B = P_\Gamma \tilde{B} P_\Gamma, \tag{2.37}$$

in the Hilbert space $L_{2,\Gamma}$.

4.2.5 Properties of the Operators of the Problem

We next point out some properties of the operators A and B in (2.36), (2.37). As mentioned above, operator C in (2.37) is compact and positive on $L_{2,\Gamma}$. Its form for any ψ from $L_{2,\Gamma}$ (according to the representation (2.32) and based on (2.31)) equals

$$\begin{aligned}(C\psi, \psi)_{L_{2,\Gamma}} &= (\gamma_\Gamma\varphi, \psi)_{L_{2,\Gamma}} = -\int_{\Gamma_1}\varphi\psi_1 \mathrm{d}\Gamma_1 + \int_{\Gamma_2}\varphi\psi_2 \mathrm{d}\Gamma_2 \\ &= -\int_{\Gamma_1}\varphi\frac{\partial\varphi}{\partial n_1}\mathrm{d}\Gamma_1 + \int_{\Gamma_2}\varphi\frac{\partial\varphi}{\partial n_2}\mathrm{d}\Gamma_2 = \int_{\partial\Omega}\varphi\frac{\partial\varphi}{\partial n}\mathrm{d}S \\ &= \int_\Omega |\nabla\varphi|^2 \mathrm{d}\Omega. \end{aligned} \tag{2.38}$$

Hence, with regard to the kinematic relations (2.23) we have

$$\rho\left(C\frac{\partial w}{\partial t}, \frac{\partial w}{\partial t}\right)_{L_{2,\Gamma}} = \rho\int_\Omega |\nabla\Phi|^2 \mathrm{d}\Omega, \tag{2.39}$$

therefore, C can be referred to as the *kinetic energy operator of the fluid.*

Further, the form of operator M in (2.37) is obviously the following,

$$(M\psi,\psi)_{L_{2,\Gamma}} = (P_\Gamma\tilde{M}P_\Gamma\psi,\psi)_{L_{2,\Gamma}} = (\tilde{M}\psi,\psi)_{L_{2,\Gamma}} = \sum_{i=1}^{2} M_i\int_{\Gamma_i}|\psi_i|^2\mathrm{d}\Gamma_i. \tag{2.40}$$

If $\psi = \partial w/\partial t$, then

$$\left(M\frac{\partial w}{\partial t}, \frac{\partial w}{\partial t}\right)_{L_{2,\Gamma}} = \sum_{i=1}^{2} M_i\int_{\Gamma_i}\left|\frac{\partial w_i}{\partial t}\right|^2 \mathrm{d}\Gamma_i, \tag{2.41}$$

and operator M can be called the *kinetic energy operator of the elastic part* of the system. From (2.41) it follows that operator M is positive definite on $L_{2,\Gamma}$ and bounded.

Hence, operator $A = M + \rho C$ is a positive definite and bounded operator. We call it the *kinetic energy operator of the whole system.*

Now, let us determine the form of operator B. If $w \in \mathcal{D}(B)$, then as in (2.40) and based on (2.8), (2.6) we obtain

$$\begin{aligned}(Bw,w)_{L_{2,\Gamma}} &= (\tilde{B}w,w)_{L_{2,\Gamma}} = \sum_{i=1}^{2}(B_iw_i,w_i)_{L_2(\Gamma_i)}\\ &= \sum_{i=1}^{2}\int_{\Gamma_i}\left[\mathrm{d}_i\Delta_2^2 w_i + (-1)^i\rho g w_i\right] w_i \mathrm{d}\Gamma_i\\ &= \sum_{i=1}^{2}\mathrm{d}_i\int_{\Gamma_i}|\Delta_2 w_i|^2\mathrm{d}\Gamma_i + \rho g\left(\int_{\Gamma_2}|w_2|^2\mathrm{d}\Gamma_2 - \int_{\Gamma_1}|w_1|^2\mathrm{d}\Gamma_1\right).\end{aligned} \tag{2.42}$$

Here, the first term equals twice the potential energy of the elastic plates, and the second term is the potential energy of the fluid caused by the displacement of particles during the process of oscillation. Therefore, B can be naturally called the *potential energy operator of the whole system.*

Since operators B_i are unbounded and positive definite on $L_2(\Gamma_i)$, and $\tilde{B} = \mathrm{diag}(B_1; B_2)$, one gets that operator B is also positive definite and unbounded on $L_{2,\Gamma}$, with a compact positive inverse operator.

4.2.6 Proper Oscillations

For the solutions of equation (2.36) that depend on t according to the law $w(t) = \exp(i\omega t)\eta$, we get the spectral problem

$$B\eta = \lambda A\eta, \qquad \lambda = \omega^2. \tag{2.43}$$

Since operator A is positive definite and bounded, and B is positive definite and has a compact inverse operator B^{-1}, all the conclusions of Section 1.4.2 are valid for equation (2.43) because the norm on H_A is equivalent to the norm of the initial space $H = L_{2,\Gamma}$. Consequently, problem (2.43) has a discrete positive spectrum $\{\lambda_n\}_{n=1}^{\infty}$ with the limit point $\lambda = +\infty$, and the corresponding eigenelements $\{\eta_n\}_{n=1}^{\infty}$ form an orthogonal basis in the energetic spaces H_A and H_B of the operators of kinetic and potential energy. That basis can be chosen in such a way that the following relations are true,

$$\begin{aligned}
(\eta_k, \eta_j)_B &= \sum_{i=1}^{2}\left[\mathrm{d}_i \int_{\Gamma_i} \Delta_2\eta_{k,i} \cdot \Delta_2\eta_{j,i}\mathrm{d}\Gamma_i\right] + \rho g\left(\int_{\Gamma_2} \eta_{k,2}\eta_{j,2}\mathrm{d}\Gamma_2 - \int_{\Gamma_1} \eta_{k,1}\eta_{j,1}\mathrm{d}\Gamma_1\right)\\
&= \lambda_k(\eta_k, \eta_j)_A\\
&= \lambda_k\left(\sum_{i=1}^{2} M_i \int_{\Gamma_i} \eta_{k,i}\eta_{j,i}\mathrm{d}\Gamma_i + \rho\int_{\Omega} \nabla\varphi_k \cdot \nabla\varphi_j \mathrm{d}\Omega\right)\\
&= \lambda_k\delta_{kj},
\end{aligned} \tag{2.44}$$

where φ_k are solutions of the Neumann problem (2.31) for $\psi_i = \eta_{k,i}$, $i = 1, 2$.

The eigenvalues λ_k of problem (2.43) can be found as successive minima of the variational ratio

$$\|\eta\|_B^2/\|\eta\|_A^2 = \frac{\left\{\sum_{i=1}^{2} \mathrm{d}_i \int_{\Gamma_i} |\Delta_2\eta_i|^2\mathrm{d}\Gamma_i + \rho g\left(\int_{\Gamma_2} |\eta_2|^2\mathrm{d}\Gamma_2 - \int_{\Gamma_1} |\eta_1|^2\mathrm{d}\Gamma_1\right)\right\}}{\left\{\sum_{i=1}^{2} M_i \int_{\Gamma_i} |\eta_i|^2\mathrm{d}\Gamma_i + \rho\int_{\Omega} |\nabla\varphi|^2\mathrm{d}\Omega\right\}} \tag{2.45}$$

for solutions of the problem

$$\Delta\varphi = 0 \quad \text{in } \Omega, \qquad \frac{\partial\varphi}{\partial n} = 0 \quad \text{on } S, \qquad \frac{\partial\varphi}{\partial n_i} = \eta_i \quad \text{on } \Gamma_i, \qquad 1 \le i \le 2, \tag{2.46}$$

from the space $H_\Gamma^1(\Omega)$.

As $n \to \infty$, the asymptotic behavior of the eigenvalues λ_n of the variational problem (2.45), (2.46) is given by

$$\lambda_n = \left(\frac{n}{C_{A,B}}\right)^2 [1 + \mathrm{o}(1)], \qquad C_{A,B} := \frac{1}{4\pi}\sum_{i=1}^{2}(M_i/\mathrm{d}_i)^{1/2} \operatorname{mes}\Gamma_i, \tag{2.47}$$

and does not contain any information in regard with the fluid presence in the container. This conclusion can be explained by observing that in the variational ratio (2.45), written down as

$$\|\eta\|_B^2/\|\eta\|_A^2 = \|\eta\|_B^2/\left(\|M^{1/2}\eta\|_{L_{2,\Gamma}}^2 + \rho(C\eta, \eta)_{L_{2,\Gamma}}\right),$$

the second term in the denominator is quite dominated (in virtue of the compactness of operator C) by the first term and, therefore, it does not affect the relevant part of the asymptotics. Hence, the modes of proper oscillations η_n of problem (2.43) for large n are connected with oscillations of the lower and upper ends; the influence of the fluid presence on these modes is insignificant. Physically, this conclusion is obvious. To large numbers n, there correspond oscillations of the fluid only in a neighborhood of the surfaces Γ_i; these oscillation wither fast (i.e., exponentially fast) while getting inside the fluid from Γ_i. Therefore, for a large n, the corresponding upper and lower attached layers of fluid which are engaged in oscillations of the ends are very small, that is, they do not affect the asymptotic behavior of the eigenvalues λ_n.

4.2.7 Evolution Problem

Considering now the Cauchy problem (2.36), let us represent its solution by a series in terms of the eigenfunctions of the spectral problem (2.43), normalized by the conditions (2.44), namely,

$$w(t) = \sum_{k=1}^{\infty} c_k(t)\eta_k.$$

A substitution into (2.36), and the expansions of the initial data

$$w^0 = \sum_{k=1}^{\infty} \alpha_k \eta_k = \sum_{k=1}^{\infty} (w^0, \eta_k)_A \eta_k,$$
$$v^0 = \sum_{k=1}^{\infty} \beta_k \eta_k = \sum_{k=1}^{\infty} (v^0, \eta_k)_A \eta_k, \tag{2.48}$$

lead to the following generalized solution,

$$w(t) = \sum_{k=1}^{\infty} \eta_k \left[\alpha_k \cos(\omega_k t) + \beta_k \omega_k^{-1} \sin(\omega_k t)\right], \qquad \omega_k = \sqrt{\lambda_k}. \tag{2.49}$$

If the initial kinetic and potential energies are finite, i.e., the following conditions are true,

$$\|v^0\|_A^2 = \sum_{k=1}^{\infty} |\beta_k|^2 < \infty, \quad \|w^0\|_B^2 = \sum_{k=1}^{\infty} |\alpha_k|^2 \lambda_k < \infty, \tag{2.50}$$

then the kinetic and potential energies for the solution (2.49) are continuous at any moment of time, and the full energy preserving law that follows from (2.26) takes the form

$$\frac{1}{2}\left\{\left\|\frac{dw(t)}{dt}\right\|_A^2 + \|w(t)\|_B^2\right\} = \frac{1}{2}\left\{\|v^0\|_A^2 + \|w^0\|_B^2\right\}. \tag{2.51}$$

Let us note that the existence of a generalized solution with continuous full energy follows from the conclusions of Section 1.5.8.

4.2.8 Oscillations of a Fluid in a Container With One Elastic End

We now assume that only one end, Γ_1, is elastic (for definiteness, let it be the lower end, i.e., the bottom), and the other part, $\partial\Omega \setminus \Gamma_1$, of the boundary of the container is rigid. Though this problem is not a special case of the one already considered, some small transformations help to obtain the Cauchy problem (2.36), where operators A and B have the same properties.

While addressing the static problem, one should take into account only the first equation in (2.5), and instead of (2.7) we impose the new volume-preserving condition

$$\int_{\Gamma_1} \zeta_1^0 d\Gamma_1 = 0. \tag{2.52}$$

Then, from the first equation in (2.13) and (2.52) we obtain

$$B_1\zeta_1^0 = c, \ \zeta_1^0 = cB_1^{-1}1_1, \quad \int_{\Gamma_1} \zeta_1^0 d\Gamma_1 = c\left(B_1^{-1}1_1, 1_1\right)_{L_2(\Gamma_1)} =: c\alpha_1 = 0,$$

hence, based on the inequality $\alpha_1 > 0$, we have $c = 0$ and, therefore, $\zeta_1^0 = 0$. Consequently, in the considered case the static sagging of the bottom equals zero.

For the problem of small oscillations, we can repeat the scheme discussed in Section 4.2.2 and derive equations (2.23), where now we assume that $i = 1$. Then, by repeating the line of argument developed in Section 4.2.4 that produced equation (2.34), we can formally consider Γ_2 and $\partial\Gamma_2$ as the empty sets, and the corresponding boundary conditions should not be taken into account. The result of this reasoning is equation (2.34), where $w(t)$ is no longer a column-function with two components, $w_1(t)$ and $w_2(t)$, but a function with values in $L_{2,\Gamma} := L_2(\Gamma_1) \ominus \{1_1\}$; in this new equation $\eta_0 \equiv 1_1$.

Applying the orthoprojector P_Γ to the new equation (2.34), we get a problem of the form (2.36), where $M = M_1 I_1$, $B = P_\Gamma B_1 P_\Gamma$, and the operator A still equals $M + \rho C$. There is, of course, only one term in the corresponding forms (2.41) and (2.42) of operators M and B, but otherwise their general properties remain unchanged.

This reasoning shows that the general conclusions of Sections 4.2.5–4.2.7 are true for the problem on oscillations of a fluid in a container with one elastic bottom. In particular, the eigenvalues $\lambda_k = \omega_k^2$ of this problem can be found as successive minima of the variational ratio

$$\|\eta\|_B^2 / \left(M\|\eta\|_{L_{2,\Gamma}}^2 + \rho\|\varphi\|_{H_\Gamma^1(\Omega)}^2\right), \tag{2.53}$$

with regard to the linkage conditions

$$\Delta\varphi = 0 \quad \text{in } \Omega, \qquad \frac{\partial\varphi}{\partial n} = 0 \quad \text{on } S, \qquad \frac{\partial\varphi}{\partial n_1} = \eta \quad \text{on } \Gamma_1. \tag{2.54}$$

To obtain estimates for λ_k, instead of using the minimum principle (2.53), (2.54), we employ the minimax principle obtained in Section 1.4.1. That principle applies to the inverse ratio

$$\frac{\|\eta\|_A^2}{\|\eta\|_B^2} = \frac{M_1 \int_{\Gamma_1} |\eta|^2 d\Gamma_1 + \rho \int_\Omega |\nabla\varphi|^2 d\Omega}{\int_{\Gamma_1} [d_1|\Delta_2\eta|^2 - \rho g|\eta|^2] d\Gamma_1} \tag{2.55}$$

and produces the inverse quantities λ_k^{-1}. We first notice the obvious inequalities

$$\begin{aligned}
&\max_{\eta\in U_{k-1}} \frac{M_1 \int_{\Gamma_1} |\eta|^2 d\Gamma_1}{\int_{\Gamma_1} [d_1|\Delta_2\eta|^2 - \rho g|\eta|^2] d\Gamma_1} \\
&\quad \le \max_{\eta\in U_{k-1}} \frac{\|\eta\|_A^2}{\|\eta\|_B^2} \\
&\quad \le \max_{\eta\in U_{k-1}} \frac{M_1 \int_{\Gamma_1} |\eta|^2 d\Gamma_1}{\int_{\Gamma_1} [d_1|\Delta_2\eta|^2 - \rho g|\eta|^2] d\Gamma_1} + \max_{\eta\in H_B} \frac{\rho \int_\Omega |\nabla\varphi|^2 d\Omega}{\int_{\Gamma_1} [d_1|\Delta_2\eta|^2 - \rho g|\eta|^2] d\Gamma_1},
\end{aligned} \tag{2.56}$$

where U_{k-1} is an arbitrary subspace in H_B of codimension $k-1$.

Now, by taking the minimum in (2.56) for all the subspaces of codimension $k-1$, we obtain the inequalities

$$\frac{1}{\check{\lambda}_k} \le \frac{1}{\lambda_k} \le \frac{1}{\mu_1} + \frac{1}{\check{\lambda}_k}, \qquad k = 1, 2, \ldots, \tag{2.57}$$

where $\check{\lambda}_k$ are the eigenvalues of the problem on successive minima of the ratio

$$\int_{\Gamma_1} [d_1|\Delta_2\eta|^2 - \rho g|\eta|^2] d\Gamma_1 \Big/ \left(M_1 \int_{\Gamma_1} |\eta|^2 d\Gamma_1\right), \tag{2.58}$$

and μ_k are the eigenvalues of the problem on successive minima of the ratio

$$\int_{\Gamma_1} [d_1|\Delta_2\eta|^2 - \rho g|\eta|^2]\, d\Gamma_1 \Big/ \left(\rho \int_\Omega |\nabla\varphi|^2 d\Omega\right) \tag{2.59}$$

for the linkage conditions (2.54).

Obviously, $\check{\lambda}_k = \tilde{\lambda}_k - \rho g/M_1$, where $\tilde{\lambda}_k$ are the eigenvalues of the problem on oscillations of an elastic bottom. In turn, these eigenvalues can be expressed as $\tilde{\lambda}_k = (d_1/M_1)\lambda_k^0$, where λ_k^0 are the eigenvalues of the biharmonic operator Δ_2^2 for the boundary conditions of rigid sizing of the bottom. The asymptotic behavior of λ_k^0 is the following,

$$\lambda_k^0 = \left(\frac{4\pi}{\operatorname{mes}\Gamma_1}\right)^2 k^2[1+\mathrm{o}(1)], \qquad k\to\infty. \tag{2.60}$$

To the variational problem (2.59), (2.54) one associates the spectral problem

$$B\eta = \lambda C\eta, \qquad \eta\in L_{2,\Gamma}, \tag{2.61}$$

that can be formally obtained from the initial spectral problem of the form (2.43) for $A=\rho C$ ($M_1=0$, i.e., the case of an elastic massless bottom Γ_1). Since operator $C>0$ is compact and $B\gg 0$, the general conclusions of Section 1.4 can be applied to (2.61); in particular, it follows that this problem has a positive discrete spectrum $\{\mu_k\}_{k=1}^{\infty}$, with the limit point $\mu=+\infty$. The asymptotic behavior of the eigenvalues μ_k is the following

$$\mu_k = \left[(4\pi)^{5/2}\,\mathrm{d}_1/\rho\right] k^{5/2}[1+o(1)], \qquad k\to\infty. \tag{2.62}$$

From the inequalities (2.57) we obtain for any natural k the double inequalities

$$\theta_k\check{\lambda}_k \le \lambda_k \le \check{\lambda}_k, \qquad \theta_k \ge 1/(1+\check{\lambda}_k/\mu_1). \tag{2.63}$$

These inequalities help to estimate the oscillation frequencies of the system by oscillation frequencies of the bottom and the first eigenvalue of the auxiliary problem (2.61).

Inequalities (2.63) show that in the presence of fluid in a container, the oscillation frequencies of the system tend to decrease compared to oscillation frequencies of the bottom.

Let us remark in conclusion that $\theta_k\to 1$ as $k\to\infty$ in (2.63), and the asymptotic formula

$$\lambda_k = \check{\lambda}_k[1+\mathrm{o}(1)], \qquad k\to\infty \tag{2.64}$$

takes place for problem (2.53) [see also (2.47)].

4.2.9 Oscillations of a System of Fluids in a Container With Elastic Plane Partitions

The problem on joint oscillations of a system of m fluids located in a container and separated from each other by horizontal elastic plates—called, in what follows, ends—is a natural generalization of the problem studied in Sections 4.2.1–4.2.7. We label the fluids starting from the bottom and moving against the vector $\boldsymbol{g}$. The region occupied by the fluid of density ρ_k is denoted by Ω_k, its rigid boundary is denoted by S_k, and the lower and upper ends for that fluid are Γ_k and Γ_{k+1}, respectively, $k=1,2,\ldots,m$. The normals $\boldsymbol{n}_k$ to Γ_k are directed upwards.

Let us first solve the static problem. We denote by h_k, $k = 1, 2, \ldots, m$, the height of the plate Γ_{k+1} in the nonsagged state, and look for the pressure P_{0k} in the fluid of density ρ_k with sagged bottom $\tilde{\Gamma}_k$ in the form

$$P_{0k} = -\rho_k g x_3 + c_k + p_a,$$

where p_a is the external constant pressure and c_k are some unknown constants. The equation of the sagged plate $\tilde{\Gamma}_k$ can be found in the form $x_3 = h_{k-1} + \zeta_k^0(x_1, x_2)$. Using the equilibrium conditions of Sophie Germaine for each end $\tilde{\Gamma}_k$, we obtain the equations

$$\begin{aligned}
B_k\zeta_k^0 &:= \mathrm{d}_k\Delta_2^2\zeta_k^0 + g(\Delta\rho)_k\zeta_k^0 = g\beta_k - (c_k - c_{k-1}),\\
(\Delta\rho)_k &:= \rho_{k-1} - \rho_k, \qquad k = 2, 3, \ldots, m;\\
B_1\zeta_1^0 &:= \mathrm{d}_1\Delta_2^2\zeta_1^0 - \rho_1 g\zeta_1^0 = -c_1,\\
B_{m+1}\zeta_{m+1}^0 &:= \mathrm{d}_{m+1}\Delta_2^2\zeta_{m+1}^0 + \rho_m g\zeta_{m+1}^0 = c_m - g\beta_{m+1}\\
\beta_k &= -h(\Delta\rho)_k, \qquad k = 2, 3\ldots, m, \ \beta_{m+1} = \rho_m h_m.
\end{aligned} \tag{2.66}$$

Here, the operators B_k, which are determined by their differential expressions on Γ_k and by the conditions of rigid attachment to $\partial\Gamma_k$, are defined in the subspaces $\mathcal{D}(B_k)$ dense in $L_2(\Gamma_k)$. They are selfadjoint operators bounded from below. Further, we assume that conditions of sufficient rigidity for each bottom (which are similar to condition (2.12)) hold true,

$$(|\rho_{i-1} - \rho_i|g)/(k_i d_i) << 1, \qquad i = 1, 2, \ldots, m+1. \tag{2.67}$$

From equations (2.66) we have

$$\begin{aligned}
\zeta_k^0 &= [g\beta_k - (c_k - c_{k-1})]\, B_k^{-1}1_k =: [g\beta_k - (c_k - c_{k-1})]\, w_k^0, \qquad k = 2, 3, \ldots, m;\\
\zeta_1^0 &= -c_1 B_1^{-1}1_1 =: -c_1 w_1^0,\\
\zeta_{m+1}^0 &= (c_m - g\beta_{m+1})B_{m+1}^{-1}1_{m+1} =: (c_m - g\beta_{m+1})w_{m+1}^0.
\end{aligned} \tag{2.68}$$

Substituting these solutions into the volume-preserving conditions for every region Ω_k,

$$\int_{\Gamma_k} \zeta_k^0 \mathrm{d}\Gamma_k = \int_{\Gamma_{k+1}} \zeta_{k+1}^0 \mathrm{d}\Gamma_{k+1}, \qquad k = 1, 2, \ldots, m, \tag{2.69}$$

we obtain a system of linear algebraic equations of the form

$$\begin{aligned}
(\alpha_1 + \alpha_2)c_1 - \alpha_2 c_2 &= -g\beta_2\alpha_2,\\
-\alpha_k c_{k-1} + (\alpha_k + \alpha_{k+1})c_k - \alpha_{k+1}c_{k+1} &= (\alpha_k\beta_k - \alpha_{k+1}\beta_{k+1})g, \qquad k = 2, 3\ldots, m,\\
-\alpha_m c_{m-1} + (\alpha_m + \alpha_{m+1})c_m &= (\alpha_m\beta_m + \alpha_{m+1}\beta_{m+1})g,
\end{aligned} \tag{2.70}$$

where $\alpha_i = \left(B_i^{-1}1_i, 1_i\right)_{L_2(\Gamma_i)} > 0$. We claim that the coefficient determinant

$$\delta_m := \begin{pmatrix} \alpha_1+\alpha_2 & -\alpha_2 & & \\ -\alpha_2 & \ddots & \ddots & 0 \\ & \ddots & \ddots & \ddots \\ 0 & \ddots & \ddots & -\alpha_m \\ & & -\alpha_m & \alpha_m+\alpha_{m+1} \end{pmatrix} \tag{2.71}$$

of the system (2.70) is positive for each m and any positive constants α_k, because it equals the sum $\tilde{\delta}_m$ that consists of $m+1$ terms with each term equal to a product of m different numbers α_k.

Indeed, our claim is true for $m=1$ and $m=2$. Further, obviously

$$\begin{aligned}\delta_{m+1} &= (\alpha_{m+1}+\alpha_{m+2})\delta_m - \alpha_{m+1}^2\delta_{m-1}\\ &= \alpha_{m+2}\delta_m + \alpha_{m+1}\left(\delta_m - \alpha_{m+1}\delta_{m-1}\right).\end{aligned}$$

By the induction assumption, $\delta_m = \tilde{\delta}_m$, $\delta_{m-1} = \tilde{\delta}_{m-1}$, hence

$$\begin{aligned}\alpha_{m+2}\delta_m + \alpha_{m+1}(\delta_m - \alpha_{m+1}\delta_{m-1}) &= \alpha_{m+2}\tilde{\delta}_m + \alpha_{m+1}\left(\tilde{\delta}_m - \alpha_{m+1}\tilde{\delta}_{m-1}\right)\\ &= \left(\tilde{\delta}_{m+1} - \prod_{i=1}^{m+1}\alpha_i\right) + \alpha_{m+1}\prod_{i=1}^{m}\alpha_i = \tilde{\delta}_{m+1}.\end{aligned}$$

Since $\delta_m > 0$ for each m, the linear system (2.70) is univalently solvable, and therefore, problem (2.66) has a unique solution of the form (2.68) with the coefficients c_k determined by (2.70). For $g=0$, that is, for zero gravity, the right side of (2.70) equals zero, hence $c_k = 0$ for each k, and from (2.68) we get the trivial solution $\zeta_k^0 = 0$, $k = 1, 2, \ldots, m+1$.

Before formulating the problem on small oscillations, let us note that the unknown velocity field $\hat{\boldsymbol{u}} = \{\boldsymbol{u}^1; \ldots; \boldsymbol{u}^m\}$ should be a potential harmonic field that takes values in the subspace $\hat{\boldsymbol{G}}_{h,S}(\Omega) \subset \hat{\boldsymbol{L}}_2(\Omega)$ as in Section 3.6. The complete formulation of the boundary value problem on small movements of the considered hydroelastic system is the following:

$$\begin{gathered}\Delta\Phi_k = 0 \quad \text{in } \Omega_k, \qquad \frac{\partial\Phi_k}{\partial n} = 0 \quad \text{on } S_k,\\ \frac{\partial\Phi_1}{\partial n_1} = \frac{\partial w_1}{\partial t} \quad \text{on}\Gamma_1, \qquad \frac{\partial\Phi_{i-1}}{\partial n_i} = \frac{\partial\Phi_i}{\partial n_i} = \frac{\partial w_i}{\partial t} \quad \text{on } \Gamma_i, \qquad i = 2, 3, \ldots, m-1;\\ \frac{\partial\Phi_m}{\partial n_{m+1}} = \frac{\partial w_{m+1}}{\partial t} \quad \text{on } \Gamma_{m+1}, \qquad w_i = \frac{\partial w_i}{\partial \nu_i} = 0 \quad \text{on } \partial\Gamma_i, \qquad i = 1, 2, \ldots, m+1;\end{gathered}$$

$$M_i \frac{\partial^2 w_i}{\partial t^2} + \frac{\partial}{\partial t}(\rho_{i-1}\Phi_{i-1} - \rho_i \Phi_i) + B_i w_i = c_{i-1}(t) - c_i(t);$$

$$\int_{\Gamma_j} w_j \mathrm{d}\Gamma_j = \int_{\Gamma_{j+1}} w_{j+1} \mathrm{d}\Gamma_{j+1}, \qquad j = 1, 2, \ldots, m,$$

$$w_j(0, x_1, x_2) = w_j^0(x_1, x_2), \qquad \left(\int_{\Gamma_j} w_j^0 \mathrm{d}\Gamma_j = \int_{\Gamma_{j+1}} w_{j+1}^0 \mathrm{d}\Gamma_j \right),$$

$$v_j(0, x_1, x_2) = v_j^0(x_1, x_2), \qquad \left(\int_{\Gamma_j} v_j^0 \mathrm{d}\Gamma_j = \int_{\Gamma_{j+1}} v_{j+1}^0 \mathrm{d}\Gamma_{j+1} \right),$$

$$j = 1, 2, \ldots, m, \tag{2.72}$$

where we formally assume that $\rho_0 = \rho_{m+1} = 0$, $c_0(t) \equiv c_{m+1}(t) \equiv 0$, and the operators B_i are introduced according to (2.66).

The full energy preserving law for the problem (2.72) has the form

$$\sum_{i=1}^{m+1} \left\{ M_i \int_{\Gamma_i} \left| \frac{\partial w_i}{\partial t} \right|^2 \mathrm{d}\Gamma_i + \int_{\Gamma_i} \left[\mathrm{d}_i |\Delta_2 w_i|^2 + g(\rho_{i-1} - \rho_i)|w_i|^2 \right] \mathrm{d}\Gamma_i \right\}$$
$$+ \sum_{k=1}^{m} \rho_k \int_{\Omega_k} |\nabla \Phi_k|^2 \mathrm{d}\Omega_k = \mathrm{const}. \tag{2.73}$$

We may now repeat the general reasoning carried out in Section 4.2.4. First, we introduce the space $\hat{H}^1(\Omega)$ consisting of elements $\hat{\Phi} = \{\Phi_k\}_{k=1}^m$, with the squared norm defined by

$$\left\| \hat{\Phi} \right\|_{\hat{H}^1(\Omega)}^2 := \sum_{k=1}^{m} \rho_k \int_{\Omega_k} |\nabla \Phi_k|^2 \mathrm{d}\Omega$$
$$+ \sum_{j=1}^{m} (\rho_j / \mathrm{mes}\, \Gamma_j) \left(\int_{\Gamma_j} \Phi_j \mathrm{d}\Gamma_j + \int_{\Gamma_{j+1}} \Phi_j \mathrm{d}\Gamma_{j+1} \right)^2; \tag{2.74}$$

its subspace of codimension m formed by elements for which the second sum in (2.74) becomes zero is denoted by $\hat{H}^1_\Gamma(\Omega)$. We next define the trace operator γ_Γ by the equations

$$(\gamma_\Gamma \hat{\Phi})_i := (\rho_{i-1}\Phi_{i-1} - \rho_i \Phi_i)\,|_{\Gamma_i}, \qquad i = 1, 2, \ldots, m+1, \tag{2.75}$$

and take the spaces $F = \hat{H}^1_\Gamma(\Omega)$, $\hat{G} = \hat{L}_{2,\Gamma}$ according to the abstract scheme described in Section 1.8, where $\hat{L}_{2,\Gamma}$ is the set of all the elements $\{w_k\}_{k=1}^{m+1}$ from $L_2(\Gamma) = \bigoplus_{k=1}^{m+1} L_2(\Gamma_k)$, for which the volume preserving conditions for Ω_k hold true,

$$\int_{\Gamma_j} w_j \mathrm{d}\Gamma_j = \int_{\Gamma_{j+1}} w_{j+1} \mathrm{d}\Gamma_{j+1}, \qquad j = 1, 2, \ldots, m. \tag{2.76}$$

Finally, we end up once more with a problem of the form (2.34),

$$
\begin{aligned}
&(\tilde{M}+\hat{C})\frac{\mathrm{d}^2\hat{w}}{\mathrm{d}t^2}+\tilde{B}\hat{w}=\sum_{j=1}^{m}c_j(t)\hat{\eta}_j^0, \qquad \hat{w}(0)=\hat{w}^0,\ \hat{w}'(0)=\hat{v}^0,\\
&\qquad\qquad \tilde{M}=\operatorname{diag}(M_iI_i)_{i=1}^{m+1}, \quad \tilde{B}=\operatorname{diag}(B_i)_{i=1}^{m+1},\\
&\qquad\qquad \hat{w}^0=\{w_i^0\}_{i=1}^{m+1}, \quad \hat{v}^0=\{v_i^0\}_{i=1}^{m+1},\\
&\qquad\qquad \hat{\eta}_j^0=\{0,\dots,0,\underbrace{-1}_{j},\underbrace{+1}_{j+1},0,\dots,0\}, \qquad j=1,2,\dots,m.
\end{aligned}
\tag{2.77}
$$

Here, the matrix operator $\hat{C}$ is positive, compact and has a three-diagonal structure. It is "glued together" from matrix blocks of the form (2.32), multiplied by ρ_k; for example, if $m=3$, then matrix $\hat{C}$ has the form

$$
\hat{C}=\begin{pmatrix}
\rho_1C_{11}^{(1)} & \rho_1C_{12} & 0 & 0\\
\rho_1C_{21} & \rho_1C_{22}^{(1)}+\rho_2C_{22}^{(2)} & \rho_2C_{23} & 0\\
0 & \rho_2C_{32} & \rho_2C_{33}^{(2)}+\rho_3C_{33}^{(3)} & \rho_3C_{34}\\
0 & 0 & \rho_3C_{43} & \rho_3C_{44}^{(3)}
\end{pmatrix}. \tag{2.78}
$$

It can be proven based on the definition of the operator γ_Γ and the relation $\hat{C}=\gamma_\Gamma T$ that

$$
\left(\hat{C}\frac{\mathrm{d}\hat{w}}{\mathrm{d}t},\frac{\mathrm{d}\hat{w}}{\mathrm{d}t}\right)_{\hat{L}_{2,\Gamma}}=\sum_{k=1}^{m}\rho_k\int_{\Omega_k}|\nabla\Phi_k|^2\mathrm{d}\Omega_k,
$$

that is, $\hat{C}$ is the kinetic energy operator for potential movements of the fluids.

The formulas in (2.76) show that the solution $\hat{w}(t)$ should be orthogonal in $\hat{L}_2(\Gamma)$ to the linearly independent elements $\hat{\eta}_j^0$, $j=1,2,\dots,m$, that is, to the subspace formed by these elements. Denoting by P_Γ the orthoprojector onto the orthogonal complement in $\hat{L}_2(\Gamma)$ to $\hat{L}_{2,\Gamma}$, and applying this operator to equation (2.77), we obtain

$$
\hat{A}\frac{\mathrm{d}^2\hat{w}}{\mathrm{d}t^2}+\hat{B}\hat{w}=0, \qquad \hat{w}(0)=\hat{w}^0,\ \hat{w}'(0)=\hat{v}^0,\ \hat{A}=P_\Gamma\tilde{M}P_\Gamma+\hat{C},\ \hat{B}=P_\Gamma\tilde{B}P_\Gamma, \tag{2.79}
$$

where $\hat{A}$ is the bounded positive definite operator of kinetic energy and $\hat{B}$ is a positive definite operator with a compact positive inverse operator $\hat{B}^{-1}$.

Therefore, the conclusions derived in Section 4.2.6 are valid for the considered spectral problem, as are the conclusions in Section 4.2.7 with regard to the solvability of the Cauchy problem (2.79). In particular, the eigenvalues λ_k of the problem

$$
\hat{B}\hat{\eta}=\lambda\hat{A}\hat{\eta}, \qquad \lambda=\omega^2, \tag{2.80}
$$

can be found as successive minima of the variational ratio

$$\frac{\sum_{i=1}^{m+1}\left[\mathrm{d}_i\int_{\Gamma_i}|\Delta_2\eta_i|^2\mathrm{d}\Gamma_i+g(\rho_{i-1}-\rho_i)\int_{\Gamma_i}|\eta_i|^2\mathrm{d}\Gamma_i\right]}{\sum_{i=1}^{m+1}M_i\int_{\Gamma_i}|\eta_i|^2\mathrm{d}\Gamma_i+\sum_{k=1}^{m}\rho_k\int_{\Omega_k}|\nabla\Phi_k|^2\mathrm{d}\Omega_k}, \tag{2.81}$$

which is considered for solutions $\hat{\Phi}=\{\Phi_k\}_{k=1}^{m}\in\hat{H}^1_\Gamma(\Omega)$ of the problem

$$\begin{aligned}
&\Delta\Phi_k=0 \text{ in } \Omega_k, \quad \frac{\partial\Phi_k}{\partial n}=0 \text{ on } S_k, \qquad k=1,2,\dots,m;\\
&\frac{\partial\Phi_1}{\partial n_1}=\eta_1 \text{ on } \Gamma_1, \quad \frac{\partial\Phi_{i-1}}{\partial n_i}=\frac{\partial\Phi_i}{\partial n_i}=\eta_i \text{ on } \Gamma_i, \qquad i=2,3,\dots,m;\\
&\frac{\partial\Phi_m}{\partial n_{m+1}}=\eta_{m+1} \text{ on } \Gamma_{m+1}, \qquad \hat{\eta}=\{\eta_k\}_{k=1}^{m+1}\in\hat{L}_{2,\Gamma}.
\end{aligned} \tag{2.82}$$

For $n\to\infty$, the asymptotic behavior of the numbers λ_n is the following,

$$\lambda_n=\left(\frac{n}{C_{\hat{A},\hat{B}}}\right)^2[1+\mathrm{o}(1)], \qquad C_{\hat{A},\hat{B}}:=\frac{1}{4\pi}\sum_{i=1}^{m+1}(M_i/\mathrm{d}_i)^{1/2}\,\mathrm{mes}\,\Gamma_i. \tag{2.83}$$

Similar to Sections 4.2.1–4.2.7, here we get that the dominant component of the asymptotic behavior of squared oscillation frequencies does not depend on the presence or absence of fluids (with densities ρ_k) between partitions.

In conclusion, we would like to notice that the methods used so far can be applied as well to the problem on oscillations of a fluid in a container with elastic plate-partitions that have variable slowly changing characteristics $\mathrm{d}_i=\mathrm{d}_i(x_1,x_2)$, $M_i=M_i(x_1,x_2)$, $i=1,2,\dots,m+1$. Also, the important differential expression Δ_2^2, defining the operators B_i on Γ_i, should be added up here to the smaller summands that contain derivatives up to the third order. Respectively, the solutions of both the static and the dynamic problems change insignificantly, the general conclusions remain unchanged, and the asymptotic formulas (2.47), (2.83) for the eigenvalues λ_n have the same form, but with constants

$$C_{\hat{A},\hat{B}}=\frac{1}{4\pi}\sum_{i=1}^{m+1}\int_{\Gamma_i}\left[\frac{M_i(x_1,x_2)}{\mathrm{d}_i(x_1,x_2)}\right]^{1/2}\mathrm{d}\Gamma_i>0, \tag{2.84}$$

where $m+1$ is the number of elastic plates.

We now outline some modifications needed in the considered problem when one of the plates, for example, Γ_q, is clutched, that is, rigid. In this situation, from the volume-preserving conditions for each fluid, instead of (2.69) we get the following conditions,

$$\int_{\Gamma_k} \zeta_k^0 d\Gamma_k = 0, \qquad k \neq q,\ k = 1, 2, \ldots, m+1.$$

Further, the qth equation in the system of equations (2.66) should be omitted; its solution is trivial in this case, $\zeta_k^0 \equiv 0$ $k \neq q$, $k = 1, 2, \ldots, m+1$, and the constants c_k are uniquely defined and give the distribution of the equilibrium pressure in Ω_k by the formulas

$$P_{0k} = -\rho_k g x_3 + p_a + c_k,$$

according to the laws of hydrostatics for the nonstaged ends Γ_k.

The problem (2.77) on small movements of the system needs to be changed in the following way: we should assume $w_q \equiv 0$, omit the string number q everywhere, and consider $\hat{\eta}_j^0$ as a column that has a unit function in the j-th spot and all of its other components equal zero. The scheme that yields the transition from (2.77) to (2.79) remains unchanged except that in this case the operator P_Γ is given by $P_\Gamma := \operatorname{diag}(P_{\Gamma_j})_{j=1}^m$, $j \neq q$, where P_{Γ_j} is the orthoprojector onto $L_{2,\Gamma_j} := L_2(\Gamma_j) \ominus \{1_j\}$. In the problem (2.81)—(2.82) we should assume that $w_q \equiv 0$, and in the asymptotic formulas (2.83)—(2.84) the term for which $i = q$ is not taken into account.

4.3 Oscillations of a Fluid in a Partially Filled Container With an Elastic Bottom

In this section we address problems on oscillations of a heavy or capillary fluid partially filling a container with an elastic bottom.

4.3.1 Determining the Equilibrium State

Let us assume that a heavy ideal fluid partially fills an open container with a horizontal elastic bottom Γ_1, which is a rigidly fixed homogeneous plate of constant thickness. If the acceleration, g, of the gravity field that acts "downwards," is sufficiently small, then the bottom Γ_1 almost does not sag, and the free surface Γ_2 of the fluid is horizontal, that is, it is parallel to the unsagged bottom Γ_1. As the gravitaty field becomes more intense, the bottom Γ_1 starts sagging, and the free surface $\tilde{\Gamma}_2$ of the fluid moves downward, being still horizontal. Hence, similar to Sections 4.1–4.2, the problem of determining the equilibrium state of the system comes forward.

We seek the static sagging of Γ_1 in the form $x_3 = \zeta_1^0(x_1, x_2)$, $(x_1, x_2) \in \Gamma_1$, where the origin O of the coordinate system $Ox_1x_2x_3$ is chosen on the horizontal surface Γ_1 and ζ_1^0 is considered to have a small magnitude of the first order. The vertical deviation of the surface $\tilde{\Gamma}_2$ from its initial position under conditions of small gravity, obviously equals $\zeta_2^0 = c_2^0$, where c_2^0 is a constant that needs to be determined.

Using the nonperturbed volume-preserving condition for the region Ω enclosed between the solid boundary S, the horizontal nonperturbed surface (bottom) Γ_1, and the free surface Γ_2, we obtain

$$\int_{\Gamma_1} \zeta_1^0 \mathrm{d}\Gamma_1 = c_2^0 \operatorname{mes} \Gamma_2. \tag{3.1}$$

Let us denote the height of fluid in the container (for small gravitaty) by h, and the constant external pressure by p_a. Then, in the equilibrium state, the pressure in fluid is distributed according to the law

$$P_0(x_3) = -\rho g(x_3 - h - c_2^0) + p_a. \tag{3.2}$$

Using the equilibrium condition of the bottom Γ_1 and the equation of Sophie Germain, we obtain

$$-\mathrm{d}_1 \Delta_2^2 \zeta_1^0 = P_0 \Big|_{x_3=\zeta_1^0} - p_a = -\rho g(\zeta_1^0 - h - c_2^0) \qquad \text{on } \Gamma_1. \tag{3.3}$$

Hence, with regard to the conditions of rigid attachment, we obtain the equation

$$B_1 \zeta_1^0 = -\rho g(h + c_2^0), \tag{3.4}$$

where the operator B_1 has been defined already by the formulas (2.8), (2.6) and described in detail in Section 4.2. From (3.4) we have

$$\zeta_1^0 = -\rho g(h + c_2^0) B_1^{-1} 1_1 =: -\rho g(h + c_2^0) w_1^0.$$

Now, condition (3.1) helps us to determine the constant c_2^0, and we can finally write down the solution to the static problem as

$$\begin{aligned} c_2^0 &= -\rho g h\, \alpha_1 / (\rho g\, \alpha_1 + \operatorname{mes} \Gamma_2), \\ \zeta_1^0 &= -\rho g h \operatorname{mes} \Gamma_2 / (\rho g\, \alpha_1 + \operatorname{mes} \Gamma_2) w_1^0, \end{aligned} \tag{3.5}$$

where $\alpha_1 := (B_1^{-1} 1_1, 1_1)_{L_2(\Gamma_1)}$. Here and further (as in Section 4.2), we assume that condition (2.12) of sufficient rigidity of the elastic bottom Γ_1 holds true; then $B_1 \gg 0$ and, therefore, $\alpha_1 > 0$.

Let us note that solutions (3.5) can be formally obtained from formulas (2.14) that give us the solution of the problem on static saggings for two ends Γ_1 and Γ_2, if we assume that $c_2^0 = \zeta_2^0$ and $B_2 \zeta_2^0 := \rho g \zeta_2^0$. In other words, the free surface Γ_2 of the fluid can be considered as a nonelastic ($\mathrm{d}_2 = 0$) and massless (in dynamics) plate. Then, from (2.41) we have

$$w_2^0 = (\rho g)^{-1}, \quad \alpha_2 = (B_2^{-1} 1_2, 1_2)_{L_2(\Gamma_2)} = \operatorname{mes} \Gamma_2 / (\rho g).$$

4.3.2 Formulation of the Problem on Small Oscillations

We next write down without any proof the system of equations and the boundary and initial conditions of the problem on small movements of the system. They can be obtained similarly to the problem in Section 3.3 on oscillations of a heavy fluid in a rigid container and the problem in Section 4.1.8 on oscillations of a fluid in an entirely filled container with an elastic bottom. Specifically, the equations and the conditions have the form

$$
\begin{aligned}
&\Delta\Phi = 0 \text{ in } \Omega, \quad \frac{\partial\Phi}{\partial n} = 0 \quad \text{on } S \\
&\frac{\partial\Phi}{\partial n_i} = \frac{\partial w_i}{\partial t} \text{ on } \Gamma_i, \quad i = 1,2, \quad \int_{\Gamma_1} w_1 \mathrm{d}\Gamma_1 = \int_{\Gamma_2} w_2 \mathrm{d}\Gamma_2, \\
&M_1 \frac{\partial^2 w_1}{\partial t^2} - B_1 w_1 - \rho\frac{\partial\Phi}{\partial t} = -c(t) \quad \text{on } \Gamma_1, \\
&\rho g w_2 + \rho\frac{\partial\Phi}{\partial t} = c(t) \quad \text{on } \Gamma_2, \\
&w_i(0,x_1,x_2) = w_i^0(x_1,x_2) \quad \left(\int_{\Gamma_1} w_1^0 \mathrm{d}\Gamma_1 = \int_{\Gamma_2} w_2^0 \mathrm{d}\Gamma_2\right), \\
&\frac{\partial w_i}{\partial t}(0,x_1,x_2) = v_i^0(x_1,x_2) \quad \left(\int_{\Gamma_1} v_1^0 \mathrm{d}\Gamma_1 = \int_{\Gamma_2} v_2^0 \mathrm{d}\Gamma_2\right), \qquad i = 1,2. \qquad (3.6)
\end{aligned}
$$

Here, $w_1(t,x_1,x_2)$ is the dynamic sagging of the bottom Γ_1, and $w_2(t,x_1,x_2)$ is an additional term added to c_2^0 that describes the vertical displacement of the free surface $\tilde{\Gamma}_2$ of the fluid during the process of oscillations.

Concluding in a similar way as in Section 4.1.3, we obtain the energy-preserving law in the form (2.6), where it is formally assumed that $M_2 = 0$, and $\mathrm{d}_2 = 0$, i.e., Γ_2 is considered as a massless plate with a zero rigidity coefficient.

4.3.3 On Solvability of the Evolution and Spectral Problems

We can now apply the reasoning developed in Section 4.2.4 to problem (3.6). As a result, we obtain the following problem

$$\left(\tilde{M} + \rho C\right)\frac{\mathrm{d}^2 w}{\mathrm{d}t^2} + \tilde{B}w = -c(t)\eta_0, \qquad w(0) = w^0,\ w'(0) = v^0, \qquad (3.7)$$

where now

$$\tilde{M} = \mathrm{diag}(M_1 I_1; 0), \quad \tilde{B} := \mathrm{diag}(B_1; \rho g I_2). \qquad (3.8)$$

The element η_0 is still given by $\eta_0 = \{1_1; -1_2\}^{\mathrm{t}}$, $c(t)$ is an arbitrary function of time, and the unknown function $w(t) = \{w_1(t); w_2(t)\}^{\mathrm{t}}$ satisfies the volume-preserving condition

$$\int_{\Gamma_1} w_1 \mathrm{d}\Gamma_1 = \int_{\Gamma_2} w_2 \mathrm{d}\Gamma_2. \qquad (3.9)$$

The operators $\tilde{M}$ and $\tilde{B}$ in (3.8) differ from the two similar operators in (2.34) by some of their properties. Namely, the new operator $\tilde{M}$ is bounded and nonnegative, operator $\tilde{B}$ is positive and unbounded in virtue of the unboundedness of B_1, and the inverse operator $\tilde{B}^{-1} := \operatorname{diag}\left(B_1^{-1}; (\rho g)^{-1} I_2\right)$ is not compact. That is why we cannot directly apply the general conclusions derived in Section 1.4 to the spectral problem corresponding to (3.7) and to the Cauchy problem (3.7).

In order to overcome this difficulty, we rewrite problem (3.7) in the vector-matrix form

$$\begin{gathered} M_1 \frac{\mathrm{d}^2 w_1}{\mathrm{d}t^2} + \rho \left(C_{11} \frac{\mathrm{d}^2 w_1}{\mathrm{d}t^2} + C_{12} \frac{\mathrm{d}^2 w_2}{\mathrm{d}t^2} \right) + B_1 w_1 = -c(t), \\ \rho \left(C_{21} \frac{\mathrm{d}^2 w_1}{\mathrm{d}t^2} + C_{22} \frac{\mathrm{d}^2 w_2}{\mathrm{d}t^2} \right) + \rho g w_2 = c(t), \end{gathered} \tag{3.10}$$

and perform the substitution

$$B_1^{1/2} w_1 = u_1. \tag{3.11}$$

Applying the operator $B_1^{-1/2}$ to the first equation in (3.10), we obtain the problem

$$\begin{gathered} (\check{M} + \rho \check{C}) \frac{\mathrm{d}^2 u}{\mathrm{d}t^2} + \check{B} u = -c(t) u_o, \qquad u(0) = u^0, \; u'(0) = u^1 \\ \check{M} = \operatorname{diag}\left(M_1 B_1^{-1}; 0\right), \quad \check{B} = \operatorname{diag}(I_1; \rho g I_2), \\ \check{C} = \begin{pmatrix} B_1^{-1/2} C_{11} B_1^{-1/2} & B_1^{-1/2} C_{12} \\ C_{21} B_1^{-1/2} & C_{22} \end{pmatrix}, \quad u = \begin{pmatrix} u_1 \\ w_2 \end{pmatrix}, \\ u^0 = \left\{ B_1^{-1/2} w_1^0; w_2^0 \right\}^{\mathrm{t}}, \quad u^1 = \left\{ B_1^{-1/2} v_1^0; v_2^0 \right\}^{\mathrm{t}}, \quad u_0 = \left\{ B_1^{-1/2} 1_1; -1_2 \right\}^{\mathrm{t}}. \end{gathered} \tag{3.12}$$

After the substitution (3.11), condition (3.9) takes the form

$$\int_{\Gamma_1} B_1^{-1/2} u_1 \mathrm{d}\Gamma_1 = \int_{\Gamma_2} w_2 \mathrm{d}\Gamma_2, \tag{3.13}$$

which shows that the unknown solution $u(t)$ should be orthogonal to the subspace formed by the element u_0 in the space $L_2(\Gamma) = \bigoplus_{i=1}^2 L_2(\Gamma_i)$. Let us next denote $L_{2,\Gamma} := L_2(\Gamma) \ominus \{u_0\}$, introduce the orthogonal projector P_Γ onto $L_{2,\Gamma}$ and apply this operator to both sides of equation (3.12). Finally, we obtain the Cauchy problem

$$\begin{gathered} A \frac{\mathrm{d}^2 u}{\mathrm{d}t^2} + B u = 0, \quad u(0) = u^0, \quad u'(0) = u^1, \\ A = P_\Gamma \left(\check{M} + \rho \check{C}\right) P_\Gamma, \quad B = P_\Gamma \check{B} P_\Gamma. \end{gathered} \tag{3.14}$$

Since operator $\check{M}$ in (3.12) is compact and nonnegative due to the compactness and positiveness of operator B_1^{-1}, and operator $\check{C}$ is compact and positive, we conclude that $\check{M} + \rho \check{C}$ is a compact positive operator. Hence we get that operator A is also compact and positive. Obviously, operator $\check{B}$ is positive definite, therefore, operator

$B = P_\Gamma \check{B} P_\Gamma$ also possesses this property in $L_{2,\Gamma}$, so it has a bounded positive inverse operator B^{-1}.

Based on the just proved properties of the kinetic energy operator A and the potential energy operator B, the general conclusions of Section 1.4.2 on the structure of the spectrum and on properties of orthogonality of the eigenelements in the energetic spaces H_B, H_A are now applicable to the spectral problem $Bu = \lambda Au$, $\lambda = \omega^2$, that follows from (3.14). Similar to Section 1.4.2, we infer the solvability of the Cauchy problem (3.14); its generalized solution $u(t)$ has continuous kinetic and potential energies provided that at the initial moment of time both the potential and the kinetic energies are finite, and the energy-preserving law holds true for this generalized solution. Therefore, the boundary value problem (3.6) is univalently solvable and has a generalized solution with continuous full energy under the conditions $\{v_1^0; v_2^0\}^t \in H_A$, $\{w_1^0; w_2^0\}^t \in H_B$.

In conclusion, we note that the eigenvalues λ_n of the spectral problem $Bu = \lambda Au$ can be found as successive minima of the variational ratio

$$\frac{\|u\|_B^2}{\|u\|_A^2} = \frac{d_1 \int_{\Gamma_1} |\Delta_2 w_1|^2 d\Gamma_1 + \rho g \left(\int_{\Gamma_2} |w_2|^2 d\Gamma_2 - \int_{\Gamma_1} |w_1|^2 d\Gamma_1 \right)}{M_1 \int_{\Gamma_1} |w_1|^2 d\Gamma_1 + \rho \int_\Omega |\nabla \Phi|^2 d\Omega}, \tag{3.15}$$

considered for solutions of the problem

$$\Delta \Phi = 0 \text{ in } \Omega, \quad \frac{\partial \Phi}{\partial n} = 0 \text{ on } S, \quad \frac{\partial \Phi}{\partial n_i} = w_i \text{ on } \Gamma_i, \qquad i = 1, 2 \tag{3.16}$$

from the space $H_\Gamma^1(\Omega)$ (for each of them, $\int_{\Gamma_1} \Phi d\Gamma_1 + \int_{\Gamma_2} \Phi d\Gamma_2 = 0$); $w = \{w_1; w_2\}^t \in L_2(\Gamma)$ and satisfies the necessary condition (3.9) of solvability of the Neumann problem (3.16).

For $n \to \infty$, the asymptotic behavior of the eigenvalues λ_n of the problem (3.15), (3.16) is the following,

$$\lambda_n = \left[\frac{g}{(\text{mes}\,\Gamma_2 / 4\pi)^{1/2}} \right] n^{1/2} [1 + o(1)]. \tag{3.17}$$

The same asymptotic behavior took place in the problem from Section 3.3 on oscillations of a heavy fluid in a rigid container. It means that while finding the dominant component of the asymptotic behavior of λ_n one should not take into account the property of elasticity of the bottom, because the asymptotics of λ_n is defined by the surface oscillations at Γ_2. This is the main difference between the problems on oscillations of a heavy fluid in wholly or partially filled containers.

4.3.4 Oscillations of a Capillary Fluid in a Partially Filled Elastic Container

Whenever a fluid is located in a container with a plane elastic bottom under conditions of complete low-gravity, the bottom does not sag due to the absence of the gravitational field influence. We assume that if the gravitational field is small, the gravitation and capillary forces that act on a hydroelastic system are of the same order, and the condition of sufficient rigidity of the bottom (2.12) holds true relatively to the small gravitation $\boldsymbol{g}$.

While solving the static problem for $\boldsymbol{g} \neq \boldsymbol{0}$, we suppose that the bottom is rigid and the curved free surface Γ_2 of the fluid (similar to Section 4.1) occurs under the joint influence of capillary and gravitational forces. Now, let us diminish the rigidity coefficient of the bottom from $+\infty$ to a given value d_1. Then, the bottom sags and the free surface Γ_2 deviates. Looking for the static sagging in the form $x_3 = \zeta_1^0(x_1, x_2)$ (and choosing the coordinate system $Ox_1x_2x_3$ on the nonsagged bottom Γ_1 with the axis Ox_3 directed against the vector $\boldsymbol{g}$), and the deviation of the surface Γ_2 in the form $\xi^3 = \zeta_2^0(\xi^1, \xi^2)$ in a curvilinear coordinate system $\tilde{O}\xi^1\xi^2\xi^3$ chosen in a neighborhood of Γ_2 similar to Section 4.1, we obtain the following relation from the volume-preserving condition,

$$\int_{\Gamma_1} \zeta_1^0 \mathrm{d}\Gamma_1 = \int_{\Gamma_2} \zeta_2^0 \mathrm{d}\Gamma_2. \tag{3.18}$$

In the nonperturbed state, the pressure $P_0(x_3)$ in fluid equals $P_0(x_3) = -\rho g x_3 + c_0$, and on the free surface Γ_2 the Laplace condition is satisfied,

$$P_0|_{\Gamma_2} - p_a = -\sigma_2(k_1 + k_2), \tag{3.19}$$

where p_a is the external constant pressure and σ_2 is the coefficient of surface tension on Γ_2, which is the boundary "fluid–gas." On $\partial\Gamma_2$, the Dupré-Young condition holds true,

$$\sigma_2 \cos\delta_2 = \sigma_{12} - \sigma_{02}, \tag{3.20}$$

where δ_2 is the dihedral angle between Γ_2 and S (wetting angle), and σ_{i2} are the coefficients of surface tension for the pair of two media at points on $\partial\Gamma_2$.

To make the law of modification of pressure $P_0(x_3)$ more specific, let us choose a point on Γ_2, for example, the center $\tilde{O}$ of the curvilinear coordinate system $\tilde{O}\xi^1\xi^2\xi^3$; the distance from this point to the bottom plate is denoted by h, and the curvatures at this point by k_1^0 and k_2^0. Then, in virtue of (3.19), the next relation is true at that point,

$$-\rho g h + c_0 - p_a = -\sigma_2(k_1^0 + k_2^0),$$

hence

$$P_0(x_3) = -\rho g(x_3 - h) + p_a - \sigma_2\left(k_1^0 + k_2^0\right). \tag{3.21}$$

For the sagged lower bottom the equilibrium pressure $\tilde{P}_0(x_3)$ differs by a certain constant c from the function (3.21), that is, $\tilde{P}_0(x_3) = P_0(x_3) + c$, and the mean curvature of the deviated free surface $\tilde{\Gamma}_2$ is close to the curvature of the surface Γ_2. On the new surface $\tilde{\Gamma}_2$, the Laplace condition holds true,

$$P_0|_{\tilde{\Gamma}_2} - p_a = -\sigma_2\left(\tilde{k}_1 + \tilde{k}_2\right). \tag{3.22}$$

Taking into account the correlation between $\tilde{P}_0$, P_0, and formula (1.11) for the variation of the mean curvature of a surface, we obtain from (3.22)

$$\begin{aligned} &-\rho g\left(x_3|_{\Gamma_2} + \cos(\widehat{n_2, e_3})\zeta_2^0 - h - \sigma_2(k_1^0 + k_2^0)\right) + c \\ &\quad = -\sigma_2(k_1 + k_2) - \sigma_2\Delta_{\Gamma_2}\zeta_2^0 - \sigma_2\left(k_1^2 + k_2^2\right)\zeta_2^0 \qquad \text{on } \Gamma_2. \end{aligned} \tag{3.23}$$

Here, the quadratic terms of ζ_2^0 are omitted, and the zero order terms equal zero in virtue of (3.19). Therefore, from (3.23) we get

$$B_2\zeta_2^0 := -\sigma_2\Delta_{\Gamma_2}\zeta_2^0 + \rho g\cos(\widehat{n_2, e_3})\zeta_2^0 - \sigma_2\left(k_1^2 + k_2^2\right)\zeta_2^0 = c. \tag{3.24}$$

The variation of relation (3.20) connected with the transition from Γ_2 to $\tilde{\Gamma}_2$ gives the boundary condition (as mentioned in Section 4.1)

$$\frac{\partial\zeta_2^0}{\partial\nu_2} + \chi_2\zeta_2^0 = 0 \qquad \text{on } \partial\Gamma_2, \tag{3.25}$$

which along with (3.24) defines the operator B_2, investigated in Section 4.1; B_2 is an unbounded, bounded from below, selfadjoint operator, acting in $L_2(\Gamma_2)$ (in Section 4.1, instead of B_2 we employed the operator $P_{\Gamma_2}B_2P_{\Gamma_2}$, where P_{Γ_2} is the projector onto L_{2,Γ_2}).

We may now repeat the reasoning in Section 4.2.1 related with the derivation of the static condition on the bottom Γ_1. Instead of the first equation (2.5), we have

$$-\mathrm{d}_1\Delta_2^2\zeta_1^0 = \tilde{P}_0|_{x_3=\zeta_1^0} - p_a = -\rho g\zeta_1^0 + \rho g h - \sigma_2\left(k_1^0 + k_2^0\right) + c,$$

which leads to the equation

$$B_1\zeta_1^0 := \mathrm{d}_1\Delta_2^2\zeta_1^0 - \rho g\zeta_1^0 = \left[-\rho g h + \sigma_2(k_1^0 + k_2^0)\right] - c =: \beta - c. \tag{3.26}$$

Here, B_1 is a positive definite, unbounded, selfadjoint operator (already investigated in Section 4.2), which is defined by the differential expression (3.26) and by the conditions of rigid attachment

$$\zeta_1^0 = \frac{\partial \zeta_1^0}{\partial \nu_1} = 0 \qquad \text{on } \partial\Gamma_1; \tag{3.27}$$

B_1 acts in $L_2(\Gamma_1)$.

From (3.24), (3.26) we obtain the equations

$$\zeta_2^0 = cB_2^{-1}1_2 =: cw_2^0, \quad \zeta_1^0 = (\beta - c)B_1^{-1}1_1 =: (\beta - c)w_1^0. \tag{3.28}$$

The constant c can be found from the latter and from (3.18) and, therefore, we get the solution of the static problem for a capillary fluid located in a container with an elastic bottom:

$$\begin{aligned}
&\zeta_1^0 = \beta\frac{\alpha_2}{\alpha_1+\alpha_2}w_1^0, \quad \zeta_2^0 = \beta\frac{\alpha_1}{\alpha_1+\alpha_2}w_2^0,\\
&\beta = -\rho g h + \sigma_2(k_1+k_2), \quad \alpha_i = (B_i^{-1}1_i, 1_i)_{L_2(\Gamma_i)} > 0, \quad c = \beta\frac{\alpha_1}{\alpha_1+\alpha_2}.
\end{aligned} \tag{3.29}$$

We should note that when $\sigma_2 = 0$, that is, for a heavy fluid, solution (3.29) leads to formulas (3.5) if we take into account that in this case $B_2^{-1} = (\rho g)^{-1}I_2$, $\alpha_2 = (\rho g)^{-1}\operatorname{mes}\Gamma_2$, and $w_2^0 = (\rho g)^{-1}$. If the surface Γ_2 is plane (and, therefore, horizontal), then $k_1 + k_2 = 0$ and formulas (3.29) formally coincide with the solution (2.14) of the static problem for two elastic ends; however, here operator B_2 has a different sense.

Considering now the problem on small oscillations of a capillary fluid in an elastic container, let us state it in complete form (it differs from the problem (3.6) for a heavy fluid only by the presence of capillary terms on Γ_2 and the boundary condition on $\partial\Gamma_2$) by relying on the derivation of the equations from Sections 4.1, 4.2. As a final answer, we have

$$\begin{aligned}
&\Delta\Phi = 0 \text{ in } \Omega, \quad \frac{\partial\Phi}{\partial n} = 0 \text{ on } S,\\
&\frac{\partial\Phi}{\partial n_i} = w_i \text{ on } \Gamma_i, \qquad i = 1,2, \quad \int_{\Gamma_1} w_1 \mathrm{d}\Gamma_1 = \int_{\Gamma_2} w_2 \mathrm{d}\Gamma_2,\\
&M_1\frac{\partial^2 w_1}{\partial t^2} + B_1 w_1 - \rho\frac{\partial\Phi}{\partial t} = -c(t) \quad \text{on } \Gamma_1,\\
&B_2 w_2 + \rho\frac{\partial\Phi}{\partial t} = c(t) \quad \text{on } \Gamma_2,\\
&w_i(0,x_1,x_2) = w_i^0(x_1,x_2) \quad \left(\int_{\Gamma_1} w_1^0 \mathrm{d}\Gamma_1 = \int_{\Gamma_2} w_2^0 \mathrm{d}\Gamma_2\right),\\
&\frac{\partial w_i}{\partial t}(0,x_1,x_2) = v_i^0(x_1,x_2) \quad \left(\int_{\Gamma_1} v_1^0 \mathrm{d}\Gamma_1 = \int_{\Gamma_2} v_2^0 \mathrm{d}\Gamma_2\right), \qquad i = 1,2.
\end{aligned} \tag{3.30}$$

Here (as in (3.6)), $w_1(t,x_1,x_2)$ is the dynamic sagging of the bottom Γ_1 and $w_2(t,\xi^1,\xi^2)$ is the dynamic displacement of the free surface, that is, a complement to the function ζ_2^0 from (3.29). Further, the operators B_i, $i = 1, 2$ are the ones that resulted in the process of solving the static problem and they are related to this process by the formulas (3.24)–(3.27). In particular, each of these operators is positive definite and has a compact inverse operator in the corresponding space $L_2(\Gamma_i)$. We note that the assumption of positive definiteness of B_2 is related to the static stability of the capillary fluid in linear approximation.

We rely once again on the conclusions of Section 4.2.4 with regard to problem (3.30). Now, instead of (2.34) and (3.7) we get a problem of the form (2.34),

$$\left(\tilde{M} + \rho C\right) \frac{\mathrm{d}^2 w}{\mathrm{d}t^2} + \tilde{B} w = -c(t)\eta_0, \qquad w(0) = w^0, \ w'(0) = v^0, \tag{3.31}$$

where

$$\tilde{M} = \operatorname{diag}(M_1 I_1; 0), \quad \tilde{B} = \operatorname{diag}\,(B_1; B_2), \tag{3.32}$$

and η_0 and the other quantities have the previous meaning.

Here, the operator matrix C is positive and compact and $\tilde{M}$ is a nonnegative bounded operator, therefore, the operator $\tilde{M} + \rho C$ is positive and bounded, but not compact. The operator $\tilde{B}$ in (3.32) is positive definite and has a compact inverse operator $\tilde{B}^{-1}$, because the diagonal elements B_i possess these properties.

Let us next introduce the subspace $L_{2,\Gamma} := [\bigoplus_{i=1}^2 L_2(\Gamma_i)] \ominus \{\eta_0\}$ and the orthogonal projector P_Γ onto $L_{2,\Gamma}$. Applying this orthoprojector to both sides of equation (3.31), we finally obtain the following Cauchy problem in the space $L_{2,\Gamma}$,

$$\begin{gathered} A \frac{\mathrm{d}^2 w}{\mathrm{d}t^2} + Bw = 0, \qquad w(0) = w^0, \ w'(0) = v^0, \\ A = M + \rho C, \quad M = P_\Gamma \tilde{M} P_\Gamma, \quad B = P_\Gamma \tilde{B} P_\Gamma. \end{gathered} \tag{3.33}$$

Here, the kinetic energy operator A is positive and bounded, and operator B is unbounded and positive definite with a compact inverse operator. From these properties and also based on the general conclusions of Section 1.4, we conclude that the spectral problem

$$Bw = \lambda A w, \qquad \lambda = w^2, \tag{3.34}$$

which corresponds to the evolution problem (3.33), has a discrete positive spectrum $\{\lambda_n\}_{n=1}^\infty$, $\lim_{n\to\infty} \lambda_n = +\infty$, and the eigenelements form an orthogonal basis in H_A and H_B.

Further, if $w^0 \in H_B$ and $v^0 \in H_A$, then problem (3.30) has a generalized solution with continuous kinetic and potential energies, and the energy-preserving law is valid for that solution.

The eigenvalues λ_n of the spectral problem (3.34) can be found as successive minima of the variational ratio

$$\frac{\|w\|_B^2}{\|w\|_A^2} = \Big\{ \int_{\Gamma_1} \left[\mathrm{d}_1 |\Delta w_1|^2 - \rho g |w_1|^2 \right] d\Gamma_1 + \int_{\Gamma_2} \left[\sigma_2 \nabla_{\Gamma_2}(w_2, w_2) + \rho g \cos(\widehat{n_2, e_3}) |w_2|^2 \right] d\Gamma_2 + \sigma_2 \int_{\partial\Gamma_2} \chi_2 |w_2|^2 ds \Big\} \Big/ \Big\{ M_1 \int_{\Gamma_1} |w_1|^2 d\Gamma_1 + \rho \int_\Omega |\nabla\Phi|^2 d\Omega \Big\}, \qquad (3.35)$$

which is considered, as in Section 4.3.3, for solutions of the problem (3.16). Let us note that for $\sigma_2 \to 0$, relation (3.35) transforms into relation (3.17), because, along with that, Γ_2 transforms into a horizontal plane, and therefore $\cos(\widehat{n_2, e_3}) \to 1$.

The asymptotic behavior of the eigenvalues of the variational problem (3.35) is expressed by the formula

$$\lambda_n = \left(\frac{\operatorname{mes} \Gamma_2}{4\pi} \right)^{-3/2} n^{3/2} \left[1 + o(1) \right], \qquad n \to \infty, \qquad (3.36)$$

that is, it has the form (1.35), the same as for a capillary fluid in a rigid container. Here again, the oscillations of the fluid near the free surface Γ_2 affect the dominant component of the asymptotic behavior of λ_n, and the high-frequency oscillations of the bottom Γ_1 produce an effect of a lower order.

4.3.5 Systems of Heavy Fluids in a Container With Elastic Plates

We now address problems of hydroelasticity that are connected with oscillations of a system of m nonmixing fluids. We begin by assuming that these fluids are heavy, have the densities $\rho_1 > \rho_2 > \cdots > \rho_m > 0$, and entirely fill a container Ω which has the lower and upper horizontal elastic ends Γ_1 and Γ_{m+1}, respectively. Let us denote the solid boundary of the container by S, the region occupied by the k-th fluid between the horizontal plates, that is, in the nonperturbed state, by Ω_k, and the corresponding boundaries by Γ_k, Γ_{k+1}, respectively.

The solution to the resulting static problem can be formally obtained as a solution of the problem in Section 4.2.9 on finding the saggings of the plates $\Gamma_1, \ldots, \Gamma_{m+1}$, if we assume that the rigidity coefficients of the plates $\Gamma_2, \ldots, \Gamma_m$, which are identified with boundaries, equal zero. Further, we should use equations (2.66), where the operators B_k, for $k = 2, 3, \ldots, m$, are defined by the formulas

$$B_k \zeta_k^0 := g(\Delta\rho)_k \zeta_k^0, \qquad (\Delta\rho)_k = \rho_{k-1} - \rho_k > 0, \qquad (3.37)$$

and by the equations (2.68), where the numbers c_k can be found from the system of linear algebraic equations (2.70) with coefficients

$$\alpha_k = \left(B_k^{-1} 1_k, 1_k\right)_{L_2(\Gamma_k)} = [g(\Delta\rho)_k]^{-1} \operatorname{mes}\Gamma_k, \qquad k = 2, 3, \ldots, m, \tag{3.38}$$

and β_k have the previous meaning.

The complete formulation of the problem on small oscillations of a system of fluids in a neighborhood of the known static position can be formally obtained from the corresponding problem (2.72) on oscillations of fluids located between a system of elastic plates. By assuming in (2.72) that $\mathrm{d}_i = M_i = 0$ for $i = 2, 3, \ldots, m$, and repeating the same reasoning as in Section 4.2.9, we obtain the modified form of problem (2.77),

$$(\tilde{M} + \hat{C})\frac{\mathrm{d}^2\hat{w}}{\mathrm{d}t^2} + \tilde{B}\hat{w} = \sum_{j=1}^{m} c_j(t)\hat{\eta}_j^0, \qquad \hat{w}(0) = \hat{w}^0,\ \hat{w}'(0) = \hat{v}^0, \tag{3.39}$$

where now

$$\tilde{M} = \operatorname{diag}(M_1 I_1, 0, \ldots, 0, M_{m+1} I_{m+1}), \quad \tilde{B} = \operatorname{diag}(B_i)_{i=1}^{m+1}. \tag{3.40}$$

The operator-matrix $\hat{C}$ is compact and positive; its form for $m = 3$ was shown in (2.78). The elements $\hat{\eta}_j^0$ have the same sense as in (2.77). The operators B_1 and B_{m+1} are as before, and the operators B_k for $k = 2, 3, \ldots, m$ are defined by formulas (3.37).

Since the operator $\tilde{M} + \hat{C}$ in (3.39) is positive and bounded but not compact, and $\tilde{B}$ is a positive definite operator for which $\tilde{B}^{-1}$ is not a compact operator, the present situation is similar to the one considered in Section 4.3.3 for the problem (3.7), (3.8). Let us perform in problem (3.39) a substitution of the first and last variable according to the formulas $B_0^{1/2}\hat{w} = \hat{u}$, $B_0 :=\operatorname{diag}(B_1; I_2; \ldots; I_m; B_{m+1})$, that is,

$$B_1^{1/2} w_1 = u_1, \quad B_{m+1}^{1/2} w_{m+1} = u_{m+1}, \tag{3.41}$$

and apply the operators $B_1^{-1/2}$ and $B_{m+1}^{-1/2}$ to the first and last lines of the system of equations (3.39) written in a vector-matrix form. Finally, we obtain the problem

$$\begin{aligned}
&(\breve{M} + \breve{C})\frac{\mathrm{d}^2\hat{u}}{\mathrm{d}t^2} + \breve{B}\hat{u} = \sum_{j=1}^{m} c_j(t)\hat{\xi}_j^0, \qquad \hat{u}(0) = \hat{u}^0,\ \hat{u}'(0) = \hat{u}^1,\\
&\breve{M} = \operatorname{diag}\left(M_1 B_1^{-1}; 0; \ldots; 0; M_{m+1} B_{m+1}^{-1}\right),\\
&\breve{B} = \operatorname{diag}(I_1; B_2; \ldots; B_m; I_{m+1}),\\
&\hat{u} = \{u_1; w_2; \ldots; w_m; u_{m+1}\}^{\mathrm{t}}, \quad \breve{C} = B_0^{-1/2}\hat{C}B_0^{-1/2},\\
&B_0 = \operatorname{diag}(B_1; I_2; \ldots; I_m; B_{m+1}),\\
&\hat{u}^0 = B_0^{-1/2}\hat{w}_0, \hat{u}_1 = B_0^{-1/2} v^0, \quad \hat{\xi}_j^0 = B_0^{-1/2}\eta_j^0, \qquad j = 1, 2, \ldots, m,
\end{aligned} \tag{3.42}$$

where operator $\check{C}$ is positive and compact, because $B_0^{-1/2}$ is compact, and $\hat{C}$ is a compact positive operator. Further, operator $\check{M}$ is a nonnegative compact operator, because operators B_1^{-1}, B_{m+1}^{-1} are positive and compact. Operator $\check{B}$ is positive definite and bounded, because in virtue of (3.37) operators B_k are also positive definite and bounded.

The volume-preserving conditions

$$\int_{\Gamma_i} w_i \mathrm{d}\Gamma_i = \int_{\Gamma_{i+1}} w_{i+1} \mathrm{d}\Gamma_{i+1}, \qquad i = 1, 2, \ldots, m \tag{3.43}$$

after the substitution (3.41) become orthogonality conditions of the elements $\hat{\xi}_j^0$ from (3.42) to solutions $\hat{u}$ of the problem (3.42). Let us denote by $L_{2,\Gamma}$ the subspace of the space $L_2(\Gamma) := \bigoplus_{i=1}^{m+1} L_2(\Gamma_i)$ which is orthogonal to the subspace formed by the elements $\hat{\xi}_j^0$, $j = 1, 2, \ldots, m$, and introduce the orthoprojector P_Γ onto $L_{2,\Gamma}$. Applying P_Γ to both sides of (3.42), we obtain

$$\begin{gathered}\hat{A}\frac{\mathrm{d}^2\hat{u}}{\mathrm{d}t^2} + \hat{B}\hat{u} = 0, \qquad \hat{u}(0) = \hat{u}^0, \ \hat{u}'(0) = \hat{u}^1, \\ \hat{A} = P_\Gamma\left(\check{M} + \check{C}\right)P_\Gamma, \quad \hat{B} = P_\Gamma \check{B} P_\Gamma.\end{gathered} \tag{3.44}$$

Here, the kinetic energy operator $\hat{A}$ is compact and positive, and the potential energy operator $\hat{B}$ is positive definite and bounded in $L_{2,\Gamma}$.

Using the just proved properties of operators $\hat{A}$ and $\hat{B}$, we obtain in a similar way as in Section 4.3.4 the solvability of the evolution problem (3.44) and, therefore, of the initial boundary value problem on small oscillations, as well as the discreteness of the spectrum of the eigenvalue problem $\hat{B}\hat{u} = \lambda\hat{A}\hat{u}$ that corresponds to the Cauchy problem (3.44).

In the present case, the eigenvalues λ_n of the spectral problem are successive minima of the variational ratio

$$\begin{aligned}\frac{\|\hat{u}\|_{\hat{B}}^2}{\|\hat{u}\|_{\hat{A}}^2} = \Bigg\{&\int_{\Gamma_1}\left[\mathrm{d}_1|\Delta_2 w_1|^2 - \rho_1 g|w_1|^2\right]\mathrm{d}\Gamma_1 + \sum_{k=2}^{m}(\Delta\rho)_k\int_{\Gamma_k}|w_k|^2\mathrm{d}\Gamma_k \\ &+ \int_{\Gamma_{m+1}}\left[\mathrm{d}_{m+1}|\Delta_2 w_{m+1}|^2 + \rho_m g|w_{m+1}|^2\right]\mathrm{d}\Gamma_{m+1}\Bigg\} \\ &\div\left\{M_1\int_{\Gamma_1}|w_1|^2\mathrm{d}\Gamma_1 + \sum_{k=1}^{m}\rho_k\int_{\Omega_k}|\nabla\Phi_k|^2\mathrm{d}\Omega_k + M_{m+1}\int_{\Gamma_{m+1}}|w_{m+1}|^2\mathrm{d}\Gamma_{m+1}\right\},\end{aligned} \tag{3.45}$$

which is considered for the solutions $\hat{\Phi} \in \hat{H}^1_\Gamma(\Omega)$ of the Neumann problem

$$\begin{aligned}
&\Delta\Phi_k = 0 \text{ in } \Omega_k, \quad \frac{\partial \Phi_k}{\partial n} = 0 \text{ on } S_k, \qquad k = 1, 2, \ldots, m;\\
&\frac{\partial \Phi_1}{\partial n_1} = w_1 \text{ on } \Gamma_1, \quad \frac{\partial \Phi_{i-1}}{\partial n_i} = \frac{\partial \Phi_i}{\partial n_i} = w_i \text{ on } \Gamma_i, \qquad i = 2, 3, \ldots, m\\
&\frac{\partial \Phi_m}{\partial n_{m+1}} = w_{m+1} \text{ on } \Gamma_{m+1},
\end{aligned} \tag{3.46}$$

where the element $\hat{w} = \{w_1; \ldots; w_{m+1}\}$ from $L_2(\Gamma)$ should satisfy the necessary conditions (3.43).

For $n \to \infty$, the asymptotic behavior of the eigenvalues λ_n is expressed by the next formula, which is similar to (3.6.28) and generalizes formula (3.17) for the case of several fluids,

$$\lambda_n = g\left(n/K_{\hat{A},\hat{B}}\right)^{1/2}[1 + o(1)], \qquad K_{\hat{A},\hat{B}} := \frac{1}{4\pi}\sum_{i=2}^{m}\frac{(\rho_{k-1}+\rho_i)^2}{(\rho_{i-1}-\rho_i)^2}\operatorname{mes}\Gamma_i > 0. \tag{3.47}$$

Here, similar to the case of one fluid, the influence of oscillations of the plates does not affect the dominant component of asymptotics.

4.3.6 Systems of Capillary Fluids in a Container With Elastic Ends

Once more we consider the problem from Section 4.3.5, but now we assume that a system of m capillary fluids is located between the two elastic ends Γ_1 and Γ_{m+1}.

First, let us suppose that the ends Γ_1 and Γ_{m+1} are rigid. As usual, the fluids in container occupy the regions Ω_k, $k = 1, 2, \ldots, m$, and the boundaries between fluids are Γ_i, $i = 2, 3, \ldots, m$. As we did in Section 4.3.4, we diminish the rigidity coefficients of the bottoms Γ_1 and Γ_{m+1} from $+\infty$ to certain finite positive values. Then, the bottoms sag and the boundaries deviate; their deviations from the normals $\boldsymbol{n}_k$ to Γ_k are described by the functions ζ_k^0 in the corresponding variables defined as follows: for $k = 1$ and $k = m + 1$, ζ_k^0 are functions of the Cartesian variables x_1, x_2, and for $k = 2, 3, \ldots, m$, they are functions of the variables ξ_k^1, ξ_k^2 in some curvilinear coordinate system $\tilde{O}_k\xi_k^1\xi_k^2\xi_k^3$, which are chosen in neighborhoods of Γ_k according to the above mentioned method.

Further, by repeating the reasoning developed in Section 4.3.5 for one fluid, and in Section 4.2.9 for a system of fluids with partitions, we obtain a system of equations of the form (2.66) that can be used in determining the functions ζ_k^0, where the operators B_k for $k = 2, 3, \ldots, m$ have a form similar to (3.24)–(3.25), namely,

$$B_k\zeta_k^0 := -\sigma_k\Delta_{\Gamma_k}\zeta_k^0 + a_k(x)\zeta_k^0, \quad \text{on } \Gamma_k,$$

$$a_k(x) = g\left(\rho_{k-1} - \rho_k\right)\cos(\widehat{n_k, e_3}) - \sigma_k\left(k_{1,k}^2 + k_{2,k}^2\right),$$
$$\frac{\partial \zeta_k^0}{\partial \nu_k} + \chi_k \zeta_k^0 = 0 \quad \text{on } \partial\Gamma_k. \tag{3.48}$$

The constants $\beta_k g$ have a new connotation that is expressed (as in (3.28), (3.29)) by $\rho_k g h_k$ and the quantities $\sigma_k(k_{1,k}^0 + k_{2,k}^0)$ calculated at the points $\tilde{O}_k$ on Γ_k.

Assuming that the equilibrium state of the system is stable in linear approximation, that is, $B_k \gg 0$ for $k = 2, 3, \ldots, m$, we can now go from the new system (2.66) to its solution (2.68), with the constants c_k found from the system (2.70), where $\alpha_k = (B_k^{-1} 1_k, 1_k)_{L_2(\Gamma_k)} > 0$ and hence the determinant of the system is positive.

Considering small oscillations of the system of capillary fluids and repeating the conclusions of Section 4.2.9, we obtain the problem (2.76), (2.77), where operators $B_k, k = 2, 3, \ldots, m$, have a new meaning. However, these operators (as in (2.77)) are positive definite, and the inverse operators B_k^{-1} are compact. Therefore, the so obtained problem can be reduced to the form (2.79), where $\hat{A}$ is a positive bounded operator and $\hat{B}$ is a positive definite operator, which has a compact inverse operator $\hat{B}^{-1}$. Hence we get all the general conclusions on the discreteness of the spectrum and the solvability of the evolution problem that have already been outlined earlier in regard with the problems in Sections 4.1–4.3. Let us note that the eigenvalues λ_n of the spectral problem can be found as successive minima of the ratio (3.45), where instead of $\sum_{k=2}^m (\Delta\rho)_k \int_{\Gamma_k} |w_k|^2 d\Gamma_k$ we have to use

$$\sum_{k=1}^m \left\{ \int_{\Gamma_k} \left[\sigma_k \nabla_{\Gamma_k}(w_k, w_k) + a_k(x)|w_k|^2\right] d\Gamma_k + \sigma_k \int_{\partial\Gamma_k} \chi_k |w_k|^2 ds_k \right\}, \tag{3.49}$$

and conditions (3.46) are taken into account.

The asymptotic behavior of λ_n for $n \to \infty$ is expressed by formula (1.55), where the summation is performed from $i = 2$ to $i = m$.

Here again the presence of elastic (not rigid) ends does not affect the dominant component of asymptotics.

Now, let us briefly point out certain modifications in the problems from Sections 4.3.5 and 4.3.6, which should occur when one of the ends, for example the top one, is rigid, and the other one is elastic. First of all, in the equilibrium state the bottom is also horizontal, and the boundaries are the same as in a system of heavy or capillary fluids in a rigid container. Further, the transition schemes from equations (3.39) to (3.42) and (3.44) are to be repeated for heavy fluids—and also, for the corresponding equations for capillary fluids—with regard to the fact that a solution should satisfy volume-preserving conditions of the form

$$\int_{\Gamma_k} w_k d\Gamma_k = 0 \qquad k = 1, 2, \ldots, m. \tag{3.50}$$

4.3.7 Compound Systems of Plate-Partitions and Nonmixing Fluids

Assume that a container Ω is divided by certain horizontal plates $\Gamma_1, \ldots, \Gamma_{m+1}$ into some regions, and located between any two consecutive plates Γ_k, Γ_{k+1} there is a system of nonmixing fluids with densities $\rho_{k,1}, \ldots, \rho_{k,q_k}$. The region within the nonsagged ends Γ_k, Γ_{k+1} occupied by the fluid with density $\rho_{k,i}$ is denoted by $\Omega_{k,i}$, its solid boundary by $S_{k,i}$, and the boundaries between fluids are counted upward and denoted by $\Gamma_{k,1}, \ldots, \Gamma_{k,q_k-1}$.

Without solving the corresponding problems on static and small potential movements of such a compound hydroelastic system, let us note some of the specific features of these problems. The arguments developed in the previous sections of this chapter show that in addressing the problem on the equilibrium state of the system for sagged plates $\tilde{\Gamma}_k$, $k = 1, 2, \ldots, m+1$, we associate to each Γ_k the "elastic" operator B_k defined in $L_2(\Gamma_k)$ by the differential expression $B_k\zeta := \mathrm{d}_k\Delta_2^2\zeta + (\Delta\rho)_k\zeta$, where $(\Delta\rho)_k$ is the corresponding leap of density, and by conditions of rigid attachment to $\partial\Gamma_k$. On the boundary Γ_k between capillary fluids, we make use of the "capillary" operator B_k defined by formulas (3.48). Finally, on the boundary Γ_k between heavy fluids we employ the "gravitation" operator B_k; it is the multiplication operator defined by (3.37).

As long as each of these operators is positive definite, or is supposed to be—as in the statically stable state—positive definite, then in the process of solving the static problem for a compound system we get a system of algebraic equations of the form (2.70), with positive coefficients $\alpha_i = (B_i^{-1}1_i, 1_i)_{L_2(\Gamma_i)}$ and a positive determinant. Therefore, the static problem always has a unique solution, which in concrete situations can be written down using, for example, Cramer's formulas.

In regard with the problem of small oscillations, we note that for heavy fluids the situation considered in this section is a special case of the problem in Section 4.2, because the boundary $\Gamma_{k,i}$ can be formally thought of as a plate with the characteristics $\mathrm{d}_{k,i} = 0$, $M_{k,i} = 0$. Therefore, the evolution problem, on the one hand can be reduced to an operator equation of the form (2.77), but on the other hand the operators $\tilde{M}$ and $\tilde{B}$ may degenerate due to the equations $M_{k,i} = 0$ on the boundaries $\Gamma_{k,i}$ and transitions from the "elastic" to "gravitation" operators $B_{k,i}$ on $\Gamma_{k,i}$. The last circumstance was pointed out in Sections 4.3.3 and 4.3.5. The results related to the evolution problems are summarized in equations (3.14) and (3.44) with a compact positive operator of kinetic energy of the system and a positive definite operator of potential energy, respectively.

Accordingly, for a compound system with heavy fluids and elastic partitions the general conclusions of Sections 4.3.3 and 4.3.5 on the spectrum and solvability of the evolution problem are valid. The eigenvalues λ_n of the spectral problem can be found as successive minima of the ratio

$$\frac{\sum_{k=1}^{m+1}\int_{\Gamma_k}\left[\mathrm{d}_k|\Delta_2 w_k|^2+(\Delta\rho)_k|w_k|^2\right]\mathrm{d}\Gamma_k+\sum_{k=1}^{m}\sum_{i=1}^{q_k}(\Delta\rho)_{k,i}\int_{\Gamma_{k,i}}|w_{k,i}|^2\mathrm{d}\Gamma_{k,i}}{\sum_{k=1}^{m+1}M_k\int_{\Gamma_k}|w_k|^2\mathrm{d}\Gamma_k+\sum_{k=1}^{m}\sum_{i=1}^{q_k}\rho_{k,i}\int_{\Omega_{k,i}}|\nabla\Phi_{k,i}|^2\mathrm{d}\Omega_{k,i}} \tag{3.51}$$

where $(\Delta\rho)_k$ and $(\Delta\rho)_{k,i}$ are the leaps of densities on Γ_k and $\Gamma_{k,i}$, respectively, and $w_k, w_{k,i}$ are the dynamic saggings of the bottoms Γ_k and the dynamic deviations of the boundaries $\Gamma_{k,i}$. In the variational problem (3.51), we should consider functions $\Phi_{k,i}(x)$ that are harmonic in $\Omega_{k,i}$ and satisfy a trivial Neumann condition on $S_{k,i}$, as well as certain kinematic conditions on Γ_k and $\Gamma_{k,i}$ that connect, as in (3.46), the normal derivatives with respect to $\boldsymbol{n}_k$ and $\boldsymbol{n}_{k,i}$ of potentials from adjacent subregions and the deviations w_k and $w_{k,i}$ of the boundaries between these subregions. For problem (3.51), the asymptotic formula (3.47) remains unchanged except we now assume that

$$K_{\hat{A},\hat{B}}=\frac{1}{4\pi}\sum_{k=1}^{m}\sum_{i=1}^{q_k}\frac{(\rho_{k,i-1}+\rho_{k,i})^2}{(\rho_{k,i-1}-\rho_{k,i})^2}\operatorname{mes}\Gamma_{k,i}>0,$$

that is, the summation is performed for all the boundaries $\Gamma_{k,i}$ without regard to the elastic plates Γ_k.

Whenever a system of capillary fluids is located between the plates, then assuming that the static problem is already solved for the case of nonsagged plates, and, in the case of elastic plates, assuming that the system deviates only slightly from that state, we again can use the reasoning developed in Sections 4.3.4 and 4.3.6. Here, to $\Gamma_{k,i}$ we associate the "capillary" operators $B_{k,i}$, which are assumed to be positive definite in $L_2(\Gamma_{k,i})$ and to have compact inverse operators. Once more, the algebraic system of equations of the form (2.70) has a unique solution that makes it possible to write down the solution of the static problem.

The evolution problem for a system of capillary fluids is similar to problem (2.79) (with a new meaning for $B_{k,i}$), or to the problem in Section 4.3.6. Hence, similar conclusions on its solvability can be derived, and for the spectral problem the discreteness of the spectrum follows. The eigenvalues λ_n of the spectral problem can be found by considering the variational ratio (3.51), where the second sum in the numerator should be replaced by the quadratic functional

$$\sum_{k=1}^{m}\sum_{i=1}^{q_k}\left\{\int_{\Gamma_{k,i}}\left[\sigma_{k,i}\nabla_{\Gamma_{k,i}}(w_{k,i},w_{k,i})+a_{k,i}(x)|w_{k,i}|^2\right]\mathrm{d}\Gamma_{k,i}\right.$$

$$\left.+\sigma_{k,i}\int_{\partial\Gamma_{k,i}}\chi_{k,i}|w_{k,i}|^2\mathrm{d}S_{k,i}\right\}, \tag{3.53}$$

subject to all the links mentioned with regard to (3.51). In this case, the asymptotic formula as $n \to \infty$ for the eigenvalues λ_n has the form (1.55), where

$$K_{\hat{C},\hat{B}} = \frac{1}{4\pi}\sum_{k=1}^{m}\sum_{i=1}^{q_k}\left(\frac{\rho_{k,i}+\rho_{k,i-1}}{\sigma_{k,i}}\right)^{2/3}\operatorname{mes}\Gamma_{k,i}. \tag{3.54}$$

If one of the ends is rigid, then some modifications in the general operator scheme should be performed; these modifications are essentially the same as those already described at the end of Section 4.2 and in Section 4.3.6.

Chapter 5

Other Operator Approaches to Hydrodynamics Problems of Ideal Fluids

As already mentioned in the introduction to Chapter 2, this chapter addresses contain problems that in addition to the general scheme developed in Section 1.8 can be approached by using several well-known methods of solving spectral problems of mathematical physics. All these problems have a spectral parameter involved either in their boundary condition or in both their equation and boundary condition. We are going to use a variety of operators associated with some auxiliary boundary value problems that can be naturally introduced in connection with such spectral problems.

5.1 Plane Problems on Proper Oscillations of a Heavy Fluid in a Channel. An Application of the Stream Function

The so-called *stream function* is used in analyzing two-dimensional (plane) problems on oscillations of an incompressible fluid to describe the solenoidal field of velocity. In this section, the stream-function approach is applied to the problem of proper oscillations of an ideal fluid in a plane channel with an arbitrary cross section.

5.1.1 Spectral Problem for the Stream Function

We reexamine the problem discussed in Section 3.3 when the region Ω occupied by an ideal incompressible fluid is plane, that is, $\Omega \subset \mathbb{R}^2$. To begin with, we assume that Ω is a cross section of a channel with the generatrix parallel to the axis Ox_3. Furthermore, since we are interested only in plane (two-dimensional) movements of the fluid in that channel, the field of velocities is defined by

$$\boldsymbol{u}(t,x) = u_1(t,x)\boldsymbol{e}_1 + u_2(t,x)\boldsymbol{e}_2, \qquad x = (x_1, x_2) \in \Omega.$$

As pointed out in Section 3.3.4, the problem of proper oscillations of a fluid in a partially filled container Ω can be reduced to the spectral problem

$$\Delta\Phi = 0 \quad \text{in } \Omega, \qquad \frac{\partial\Phi}{\partial n} = 0 \quad \text{on } S,$$
$$\frac{\partial\Phi}{\partial n} = \lambda\Phi \quad \text{on } \Gamma, \qquad \lambda = \omega^2/g, \qquad \int_\Gamma \Phi \mathrm{d}\Gamma = 0, \tag{1.1}$$

where $\Phi = \Phi(x)$ is the velocity potential, ω is a frequency of proper oscillations, Γ is the free surface of the fluid—which in this case is an interval on the axis Ox_1—and S is the solid boundary of the container—a curve that bounds the plane region Ω together with Γ. The points $x = (x_1, x_2)$ of Ω satisfy $x_2 < 0$.

As it was shown in Section 3.3.4, problem (1.1) has a discrete spectrum, $0 < \lambda_1 \leq \cdots \leq \lambda_k \leq \cdots$, $\lambda_k \to \infty$ as $k \to \infty$, and the system of corresponding eigenfunctions satisfy the orthogonality conditions (3.3.20).

We now introduce the stream function $\Psi = \Psi(x)$, which is related to the velocity potential $\Phi = \Phi(x)$ by the Cauchy-Riemann conditions,

$$\frac{\partial\Phi}{\partial x_1} = \frac{\partial\Psi}{\partial x_2}, \quad \frac{\partial\Phi}{\partial x_2} = -\frac{\partial\Psi}{\partial x_1}, \tag{1.2}$$

and provides the solenoidality of the velocity field. Hence, it is obvious that $\Psi(x)$ is a harmonic function conjugate to $\Phi(x)$ that takes a constant value on S. Since the components of the velocity field are given by the formulas

$$u_1 = \frac{\partial\Psi}{\partial x_2}, \quad u_2 = -\frac{\partial\Psi}{\partial x_1}, \tag{1.3}$$

that is, the derivatives of the stream function, we can assume that $\Psi(x)$ equals zero on S.

We next formulate the spectral problem for the stream function $\Psi(x)$. To do this, we first transform the boundary condition on Γ in (1.1). We begin by assuming that Γ is the interval $[a_-, a_+]$. Then,

$$\Phi|_\Gamma = c + \int_{a_-}^{x_1} \frac{\partial\Phi}{\partial\xi}(\xi,0)\mathrm{d}\xi = c + \int_{a_-}^{x_1} \frac{\partial\Psi}{\partial x_2}(\xi,0)\mathrm{d}\xi, \tag{1.4}$$

where the constant c can be obtained from the normalization condition $\int_\Gamma \Phi \mathrm{d}\Gamma = 0$. Since $\partial\Phi/\partial n = \partial\Phi/\partial x_2 = -\partial\Psi/\partial x_1$ on Γ, the boundary condition on Γ has the form

$$\frac{\partial\Psi}{\partial x_1} = \lambda\left(c + \int_{a_-}^{x_1} \frac{\partial\Psi}{\partial x_2}(\xi,0)\mathrm{d}\xi\right).$$

If $(\partial\Psi/\partial x_2)_\Gamma$ is continuous, then the function $(\partial\Psi/\partial x_1)_\Gamma$ has a derivative with respect to x_1 and $-\partial^2\Psi/\partial x_1^2 = \lambda\partial\Psi/\partial x_2$ on Γ. We also note that since the function $\Psi(x)$ equals zero on S, we have $\Psi(a_\pm, 0) = 0$.

Hence, the classical formulation of the spectral problem for the stream function $\Psi(x)$ is

$$\begin{aligned} &\Delta\Psi = 0 \quad \text{in } \Omega, \qquad \Psi = 0 \quad \text{on } S, \\ &-\frac{\partial^2\Psi}{\partial x_1^2} = \lambda\frac{\partial\Psi}{\partial x_2} \quad \text{on } \Gamma, \qquad \Psi(a_\pm, 0) = 0 \quad \text{on } \partial\Gamma. \end{aligned} \tag{1.5}$$

Problem (1.5) has an operator interpretation. We introduce the operator A defined by $Au := -\mathrm{d}^2u/\mathrm{d}x_1^2$ with the domain $\mathcal{D}(A) := H_0^2(\Gamma)$ dense in $L_2(\Gamma)$. From the equations

$$\begin{aligned} (Au, v) &:= -\int_\Gamma u''(x_1)v(x_1)\mathrm{d}x_1 = \int_\Gamma u'(x_1)v'(x_1)\mathrm{d}\Gamma, \\ (Au, u) &= \int_\Gamma |u'(x_1)|^2\mathrm{d}x_1 \geq \frac{2}{(a_+ + a_-)^2}\int_\Gamma |u(x_1)|^2\mathrm{d}x_1, \end{aligned}$$

that hold true for any $u(x_1)$ and $v(x_1)$ in $\mathcal{D}(A)$, we find that operator A is positive definite in $\mathcal{D}(A)$, its energetic space $H_A = \mathcal{D}(A^{1/2})$ coincides with $H_0^1(\Gamma)$, and its energetic norm is equivalent to the standard norm in $H^1(\Gamma)$.

Let us now consider the auxiliary problem

$$\Delta\Psi = 0 \quad \text{in } \Omega, \qquad \Psi = 0 \quad \text{on } S, \qquad \Psi = \psi \quad \text{on } \Gamma, \tag{1.6}$$

for a given function $\psi = \psi(x_1) \in H^{1/2}(\Gamma)$. Similarly to Example 3° in Section 1.8.6 (see problem (1.8.37)), it can be proved that for any $\psi(x_1) \in H^{1/2}(\Gamma)$ problem (1.6) has a unique solution $\Psi(x)$ in the space of harmonic functions from $H^1(\Omega)$ that are equal to zero on S. Employing this unique solution of problem (1.6), we introduce the operator B defined as

$$B\psi := \left(\frac{\partial\Psi}{\partial x_2}\right)_\Gamma. \tag{1.7}$$

If $\psi(x_1) \in H^{1/2}(\Gamma)$, then $(\partial\Psi/\partial x_2)_\Gamma \in H^{-1/2}(\Gamma)$ and

$$\int_\Omega |\nabla\Psi|^2\mathrm{d}\Omega = \int_\Gamma (B\psi)\psi\mathrm{d}\Gamma \geq c^2\int_\Gamma |\psi|^2\mathrm{d}\Gamma,$$

where the last inequality is a corollary of the boundedness (compactness) of the trace operator from $H^1(\Omega)$ to $L_2(\Gamma)$. Hence, we get that operator B is positive definite on $L_2(\Gamma)$ with a domain $\mathcal{D}(B)$ dense in $L_2(\Gamma)$, and $H_B = H^{1/2}(\Gamma)$.

Using the so defined operators A and B, problem (1.5) is equivalent to the problem

$$A\psi = \lambda B\psi, \qquad \psi := (\Psi)_\Gamma \in \mathcal{D}(A) \subset \mathcal{D}(B) \subset L_2(\Gamma), \tag{1.8}$$

in the Hilbert space $L_2(\Gamma)$, where A and B are unbounded positive definite operators and $H_A = H_0^1(\Gamma)$ is compactly embedded into $H_B = H^{1/2}(\Gamma)$.

According to these properties, the eigenvalues $\{\lambda_k\}_{k=1}^{\infty}$ for problem (1.8) form a discrete spectrum. These eigenvalues can be calculated as successive minima of the variational ratio

$$\|\psi\|_A^2/\|\psi\|_B^2 = \frac{\displaystyle\int_\Gamma \left|\frac{\partial \Psi}{\partial x_1}\right|^2 \mathrm{d}\Gamma}{\displaystyle\int_\Omega |\nabla\Psi|^2 \mathrm{d}\Omega}, \tag{1.9}$$

considered in the space of functions $\Psi(x)$ that satisfy the first two equations in (1.6). The corresponding eigenfunctions $\Psi_k(x)$ satisfy the orthogonality conditions

$$\int_\Gamma \frac{\partial \Psi_k}{\partial x_1} \cdot \frac{\partial \Psi_m}{\partial x_1} \mathrm{d}x_1 = \lambda_k \int_\Omega \nabla\Psi_k \cdot \nabla\Psi_m \mathrm{d}\Omega = \lambda_k \delta_{km}. \tag{1.10}$$

5.1.2 Properties of Nodal Lines of Stream Eigenfunctions

Applying the stream-function approach to our problem makes it possible to derive some interesting properties of its solutions, in particular, properties of the nodal lines of stream functions.

To be specific, the *nodal line* of the stream function $\Psi_k(x)$ corresponding to the eigenvalue λ_k is defined as the set of points $x \in \Omega$ for which $\Psi_k(x) = 0$. From the maximum principle for harmonic functions it follows that an eigenfunction $\Psi_k(x)$ cannot have any isolated zeros in Ω.

A region $U \subset \Omega$ is called a *nodal region* of the function $\Psi_k(x)$ provided $\Psi_k(x) \neq 0$ in U and ∂U consists of nodal lines of the function $\Psi_k(x)$ and, maybe, of some intervals of the free surface Γ and arcs of the curve S, the solid boundary of the container.

In what follows we discuss some of the main properties of the nodal lines of the stream eigenfunctions $\Psi_k(x)$.

1°. *A nodal line of the function $\Psi_k(x)$ cannot be closed and cannot have both its ends on S.*

Indeed, in the opposite case, the theorem on uniqueness of the Dirichlet problem shows that for the harmonic function $\Psi_k(x)$ we must have $\Psi_k(x) \equiv 0$ in the region located within the nodal line, or between S and the nodal line. Therefore, $\Psi_k(x) \equiv 0$ in Ω.

2°. *Any nodal line of the stream function $\Psi_k(x)$ has one of its ends in Γ and the other end on S, and two nodal lines do not intersect each other.*

Based on Property 1°, in order to prove this new property, we need to check that a nodal line of the function $\Psi_k(x)$ cannot connect two points of the free surface Γ.

Assuming that the opposite is true, we define

$$\Phi_1(x) := \begin{cases} \Psi_k(x), & x \in U', \\ 0, & x \in \Omega \setminus U', \end{cases}$$

where U' is the nodal region enclosed between Γ and a nodal line that has both its ends in Γ. Together with function $\Phi_1(x)$, we introduce the functions $\Phi_j(x) = \Phi_j(x_1, x_2) := \Phi_1(x_1 + t_j, x_2)$, $j = 2, \ldots, k$, where $t_i \neq t_j$ for $i \neq j$, and the numbers t_j are chosen in such a way that by shifting the origin to t_j along the axis Ox_1 the region U' is transformed into a region located within Ω. We set every function $\Phi_j(x)$, $j = 2, \ldots, k$, equal to zero outside its domain of definition, so each such function is now defined on the whole Ω. Since the functions $\Phi_j(x)$, $j = 1, 2 \ldots, k$, are linearly independent, there exists a nontrivial combination $\Phi := \sum_{j=1}^{k} a_j \Phi_j$ for which

$$\int_\Gamma \Phi \frac{\partial \Psi_i}{\partial x_2} dx_1 = \int_\Omega \nabla\Phi \cdot \nabla\Psi_i d\Omega = 0, \qquad i = 1, 2, \cdots, k-1. \tag{1.11}$$

As a result of our reasoning, function Φ belongs to the energetic space $H_A \cap H_B$ and satisfies the boundary condition $\partial^2\Phi/\partial x_1^2 + \lambda_k \partial\Phi/\partial x_2 = 0$ on Γ, therefore, according to the variational principle for the functional (1.9), function Φ provides a minimum to that functional. Hence, we conclude that $\Phi(x)$ is an eigenfunction corresponding to the eigenvalue λ_k, and therefore, $\Phi(x)$ is harmonic in Ω. Since $\Phi(x) \equiv 0$ in a subregion located within Ω, then $\Phi \equiv 0$ in Ω. We just reached a contradiction, hence Property 2° is proved.

3°. *The stream eigenfunction $\Psi_k(x)$ corresponding to the eigenvalue λ_k can change its sign on Γ no more than $k-1$ times. Its nodal lines split Ω into no more than k subregions.*

Indeed, according to Property 2°, the nodal lines of function $\Psi_k(x)$ split the region Ω into a certain number n of subregions Ω_i. Let us assume that $n > k$, and introduce the functions

$$\psi_i(x) := \begin{cases} \Psi_k(x) & \text{in } \Omega_i \\ 0 & \text{in } \Omega/\Omega_i \end{cases}, \qquad i = 1, 2, \ldots, n,$$

which belong to the set $H_A \cap H_B$. These functions are linearly independent and there exists a nontrivial linear combination $\Psi := \sum_{j=1}^{k} b_j \psi_j(x)$ that satisfies orthogonality conditions of the form (1.11),

$$\int_\Omega \nabla\Psi \cdot \nabla\Psi_i d\Omega = 0, \qquad i = 1, 2, \ldots, k-1.$$

Since $\partial^2\Psi/\partial x_1^2 + \lambda_k \partial\Psi/\partial x_2 = 0$ on Γ, function $\Psi(x)$ provides a minimum to the functional (1.9), hence $\Psi(x)$ is harmonic in Ω. Because $n > k$ we conclude that $\Psi_k(x) \equiv 0$ in a subregion located within Ω. We reached a contradiction that proves Property 3°.

As a remark in regard to Property 3°, we note that its proof is based on the idea used in *Courant's theorem* on nodal lines of an oscillating membrane, when the inequality $n_k < k$ is true for any sufficiently large k, where n_k is the number of nodal regions of the kth eigenfunction of the membrane. However, for proper oscillations of a fluid in a channel, the number n_k of nodal regions of eigenfunction $\Psi_k(x)$ may be equal to k for each $k = 1, 2, \ldots$. Indeed, if $\Gamma = (0, l)$, $(-d, 0)$ and $(-d, l)$ are the lateral boundaries of a rectangular channel, then

$$\Psi_k(x) = \sin\frac{\pi k x_1}{l}\sinh\frac{\pi k(x_2+d)}{l}, \quad \lambda_k = \frac{\pi k}{l}\tanh\frac{\pi k d}{l}, \qquad k = 1, 2, \ldots. \tag{1.12}$$

4°. *The trace on Γ of the eigenfunction $\Psi_k(x)$ corresponding to the eigenvalue λ_k has no more than k points of extremum. In particular, $\Psi_1(x)$—if we assume it to be positive on Γ—has exactly one maximum on Γ that coincides with* $\max\{\Psi_1(x) : x = (x_1, x_2) \in \bar{\Omega}\}$.

Indeed, at every point of extremum $(x_1^0, 0) \in \Gamma$, the derivative $\partial\Psi_k/\partial x_1$ changes its sign. Based on the Cauchy-Riemann conditions (1.2) for the conjugate harmonic function $\Phi_k(x)$ and using the condition on Γ from (1.1), it follows that function $\Phi_k(x_1, 0)$ also changes its sign at the point $(x_1^0, 0)$. However, for the functions $\Phi_k(x)$ it is known, and we can easily check for the example discussed above in (1.12) where

$$\Phi_k(x) = \cos\frac{\pi k x_1}{l}\cosh\frac{\pi k(x_2+d)}{l}, \tag{1.13}$$

that in the case when $k > 1$ the changes of sign on Γ occur no more than k times, and for $k = 1$ the sign changes exactly once. Therefore, $\Psi_k(x_1, 0)$ has no more than k points of extremum on Γ, and $\Psi_1(x_1, 0)$ has exactly one maximum if $\Psi_1(x_1, 0) > 0$. We should note that $\Psi_1(x_1, 0)$ has at least one maximum, because $\Psi_1(a\pm, 0) = 0$. In concluding the proof of Property 4°, we observe that the equality $\max\{\Psi_1(x_1, 0) : x_1 \in (a_-, a_+)\} = \max\{\Psi_1(x) : x = (x_1, x_2) \in \bar{\Omega}\}$ holds true as a consequence of the maximum principle for the harmonic function $\Psi_1(x)$ and the boundary condition (1.5).

5°. *The eigenvalue λ_1 is simple.*

To prove this property, let us assume that the eigenvalue λ_1 corresponds to two linearly independent eigenfunctions $\Psi_1'(x)$ and $\Psi_1''(x)$. According to Property 3°, these functions can be assumed positive. We denote the maxima of the functions $\Psi_1'(x)$ and $\Psi_1''(x)$ by M' and M'', respectively; these maxima can be reached at certain points $(x_1', 0)$ and $(x_1'', 0)$ (property 4°). If $x_1' \neq x_2'$, then the function

$$\varphi(x_1) := M''\Psi_1'(x_1, 0) - M'\Psi_1''(x_1, 0)$$

changes its sign on Γ in contradiction to Property 3°. If $x_1' = x_2'$, then function $\varphi(x_1)$ either equals zero, which is a contradiction to the linear independence of $\Psi_1'(x_1, 0)$ and $\Psi_1''(x_1, 0)$, or has three zeroes on $\bar{\Gamma}$, and therefore, has at least two extrema on Γ, which is a contradiction to Property 4°.

6°. *If function $\Psi_1(x_1, 0)$—that is, the trace of function $\Psi_1(x)$ on Γ—does not have any inflection points on Γ, then $\Psi_1(x)$ is the only stream eigenfunction that preserves its sign in Ω. In this case, $\Psi_2(x)$ has exactly one nodal line.*

Indeed, according to our assumption, the derivative $\partial^2\Psi_1/\partial x_1^2$, and therefore, $\partial\Psi_1/\partial x_2$, preserve the sign on Γ. Since the stream eigenfunctions $\Psi_k(x)(k \geq 2)$ should satisfy the orthogonality conditions

$$\int_\Omega \nabla\Psi_k \cdot \nabla\Psi_1 \mathrm{d}\Omega = \int_\Gamma \Psi_k \left(\frac{\partial\Psi_1}{\partial x_2}\right) \mathrm{d}x_1 = 0, \qquad k = 2, 3, \ldots,$$

the functions $\Psi_k(x, 0)$ should change their signs on Γ, because the derivative $\partial\Psi_1/\partial x_2$ has a constant sign on Γ. According to Property 3°, function $\Psi_2(x)$ can have no more than one nodal line, hence it has exactly one such line.

7°. *If every stream eigenfunction that corresponds to the eigenvalue λ_2 possesses nodal lines, then the eigenvalue λ_2 is simple.*

In the opposite case, we may assume that the eigenvalue λ_2 has two linearly independent eigenfunctions $\Psi_2'(x)$ and $\Psi_2''(x)$ that possess nodal lines. According to Property 3°, each of them changes its sign on Γ only once, and these functions can be chosen in such a way that

$$\int_\Gamma \Psi_2'(x_1, 0)\Psi_2''(x_1, 0)\mathrm{d}x_1 = 0.$$

Therefore, the points x_1' and x_1'' where $\Psi_2'(x_1, 0)$ and $\Psi_2''(x_1, 0)$ change their sign cannot coincide. To be specific, let us assume that $\Psi_2'(x_1, 0) > 0$ for $x_1 > x_1'$ and $\Psi_2'(x_1, 0) < 0$ for $x_1 < x_1'$, and at the same time $\Psi_2''(x_1, 0) < 0$ for $x_1 > x_1''$ and $\Psi_2''(x_1, 0) > 0$ for $x_1 < x_1''$, with $x_1' < x_1''$.

We define the function $w(t, x_1) := (1 - t)\Psi_2'(x_1, 0) + t\Psi_2''(x_1, 0)$, $t \in [0, 1]$. It is obvious that the inequality $w(t, x_1) > 0$ holds true for $x_1 \in (x_1', x_1'')$ and any $t \in (0, 1]$. Let us introduce the sets $T' := \{t \in [0, 1] : w(t, x_1) < 0 \text{ for some } x_1 < x_1'\}$, $T'' := \{t \in [0, 1] : w(t, x_1) < 0 \text{ for some } x_1 > x_1''\}$. It is obvious that $T' = [0, t')$ and $T'' = (t'', 1]$ for some t' and t'' from $[0, 1]$. If $t'' < t'$, then for $t \in (t'', t')$ the function $w(t, x_1)$ changes its sign on Γ no more than two times, which is a contradiction to Property 3°. If $t' < t''$, then for $t \in [t', t'']$ and all $x_1 \in (a_-, a_+)$ the inequality $w(t, x_1) \geq 0$ holds true, which is a contradiction to our assumption, because $w(t, x_1)$ is the trace on Γ of the eigenfunction $W(t, x) := (1 - t)\Psi_2'(x) + t\Psi_2''(x)$.

5.1.3 Estimates of Eigenvalues

The preceding properties of stream eigenfunctions can be also used to obtain estimates for the first and second eigenvalues of problem (1.5). To do this, we first introduce some notations. We set

$$\mathrm{d}(x_1) := \min\{|x_2| : x = (x_1, x_2) \in S, x_1 \in (a_-, a_+)\},$$
$$\mathrm{d} := \sup\{\mathrm{d}(x_1) : x_1 \in (a_-, a_+)\}.$$

Let $\Psi(x) \in H^1(\Omega)$ be a function equal to zero on S. Then $\Psi(x)$ is absolutely continuous in x_2 for almost all $x_1 \in (a_-, a_+)$, and therefore, for these x_1 we have

$$|\Psi(x_1, 0)| = \left| \int_{-\mathrm{d}(x_1)}^{0} \frac{\partial \Psi}{\partial x_2}(x_1, x_2) \mathrm{d}x_2 \right| \le \left(\mathrm{d} \int_{-\mathrm{d}(x_1)}^{0} \left| \frac{\partial \Psi}{\partial x_2}(x_1, x_2) \right|^2 \mathrm{d}x_2 \right)^{1/2}.$$

Integrating over Ω and using the minimum principle for the eigenvalue λ_1 and the variational ratio (1.9), we have

$$\lambda_1 \le \mathrm{d} \min \left\{ \int_\Gamma \left| \frac{\partial \Psi}{\partial x_1} \right|^2 \mathrm{d}x_1 \Big/ \int_\Gamma |\Psi|^2 \mathrm{d}\Gamma : \Psi(x) \in H_B \cap H_A \right\}. \tag{1.14}$$

Since $H_A = H_0^1(\Gamma)$ and because any function from $H_0^1(\Gamma)$ can be extended within the region Ω—with the Lipschitz boundary $\partial\Omega$—in such a way that the extension belongs to H_B, from (1.14) it follows that

$$\lambda_1 \le \mathrm{d} \min \left\{ \int_\Gamma \left| \frac{\mathrm{d}\psi}{\mathrm{d}x_1} \right|^2 \mathrm{d}\Gamma \Big/ \int_\Gamma |\psi|^2 \mathrm{d}\Gamma : \psi = \Psi(x_1, 0) \in H_0^1(\Gamma) \right\}.$$

The minimum in this inequality equals the first eigenvalue of the operator A of problem (1.8), that is, the number $(\pi/l)^2$ where $l := a_+ - a_-$.

Hence, the following estimate for the first eigenvalue of problem (1.5) is proved,

$$\lambda_1 \le \mathrm{d}(\pi/l)^2. \tag{1.15}$$

We notice that inequality (1.15) is sharp in the following sense. If the cross section of the channel is a rectangle with width l and depth h, then the first eigenvalue (1.12) is equivalent to the right side of (1.15) as $d/l \to 0$. We note also that whenever the region Ω is located within a rectangle with the same free surface Γ and depth d, then estimate (1.15) follows from the following result that can be easily proved in the three-dimensional case (see Section 3.3). Namely, if the free surface Γ of fluid is the same for two regions Ω_1 and Ω_2 and one of the regions is located within the other one, then the larger region has a larger eigenvalue. As for the estimate (1.15), it has a more general feature because quantity d characterizes the maximal depth of the channel only for points of the free surface Γ, and not the maximal depth of the whole channel.

We mention without proof one more estimate that connects the first eigenvalue λ_1 of problem (1.5) and the first stream eigenfunction $\Psi_1(x)$,

$$\frac{\pi}{4}\lambda_1 \int_\Omega |\Psi_1|^2 \,\mathrm{d}\Omega \leq \max\left\{\left|\frac{\partial\Psi_1}{\partial x_1}(x_1,0)\right| : x_1 \in [a_-, a_+]\right\} \cdot \int_\Omega \Psi_1(x)\mathrm{d}\Omega \qquad (1.16)$$

We next point out another property of the solutions of problem (1.5), which can be considered as a corrolary to inequality (1.16). Let $\Psi_2(x)$ be a stream eigenfunction corresponding to the eigenvalue λ_2. We claim that inequality (1.16) holds true with λ_2 replacing λ_1 and $|\Psi_2(x)|$ replacing $\Psi_1(x)$.

Indeed, if function $\Psi_2(x)$ preserves its sign and is positive in Ω, then by Property 4°, it has a unique maximum located on Γ. In this case the proof of our claim is quite similar to the proof of inequality (1.16). If function $\Psi_2(x)$ has exactly one—according to property 3°—nodal line, then the proof of the corresponding inequality can be carried out in each of the nodal regions. Summing up all the resulting inequalities we get a proof of our claim for the whole region Ω.

5.2 Shallow Water Theory in Problems on Oscillations of Heavy Ideal Fluids in Bounded Regions

In the present section, the problem of proper oscillations of an ideal heavy fluid is reconsidered (see Sections 3.3 and 5.1), but now we assume that the depth of fluid is considerably smaller than the diameter of its free surface. This makes it possible to introduce a small parameter and to obtain the first and second components of the asymptotic solution of the problem. The advantage of this approach is that the considered three-dimensional problem can be reduced to a two-dimensional problem on the free surface of a fluid.

As specific applications we mention the theory of standing waves in shallow lakes and the theory of proper oscillations of fuel in tanks when the fuel occupies a small volume. Similar problems with various applications will also be considered for systems of nonmixing fluids in the case when the total depth is considerably smaller than the characteristic horizontal size of the region.

5.2.1 Formulation of the Problem with a Small Parameter

We reexamine the problem of proper oscillations of a heavy homogeneous ideal fluid filling the region $\Omega \subset \mathbb{R}^3$ bounded by a solid surface S and a free surface Γ. We assume that surface Γ has the equation $x_3 = 0$ in a coordinate system $Ox_1x_2x_3$, and for each point $x = (x_1, x_2, x_3)$ of the region Ω we have $x_3 < 0$. In terms of the amplitude functions $\Phi = \Phi(x)$ of the velocity potential and in dimensionless variables, the corresponding spectral problem is problem (1.1), namely

$$\Delta_3\Phi := \Delta\Phi + \frac{\partial^2\Phi}{\partial x_3^2} = 0 \quad \text{in } \Omega, \qquad \frac{\partial\Phi}{\partial n} = 0 \quad \text{on } S,$$
$$\frac{\partial\Phi}{\partial x_3} = \tilde{\lambda}\Phi \quad \text{on } \Gamma, \qquad \int_\Gamma \Phi d\Gamma = 0, \qquad \tilde{\lambda} = \omega^2 d/g. \tag{2.1}$$

Here, $\Delta = \partial^2/\partial x_1^2 + \partial^2/\partial x_2^2$ is the two-dimensional Laplace operator in the variables x_1 and x_2, $\partial/\partial n$ is the derivative along the external normal to S, ω is a frequency of proper oscillations, $g > 0$ is the acceleration of gravitational force, and d is the diameter of the free surface Γ chosen as the characteristic size of the problem.

We assume that the solid boundary S consists of a vertical side wall S_0, with the normal $\boldsymbol{n}_0 = n_{0,1}\boldsymbol{e}_1 + n_{0,2}\boldsymbol{e}_2$, and of a bottom S_1 defined by an equation of the form $x_3 = -\tilde{h}(x_1, x_2)$, $(x_1, x_2) =: \hat{x} \in \Gamma$. Further, we will assume that $\tilde{h}(\hat{x})$ is sufficiently small. More precisely, the following condition holds true,

$$0 < \varepsilon := \left(\tilde{h}_{\max}/d\right)^2 \ll 1 \qquad \tilde{h}_{\max} := \max_{\hat{x}\in\Gamma} \tilde{h}(\hat{x}). \tag{2.2}$$

Hence, problem (2.1) is being considered in a region $\Omega = \Omega(\varepsilon)$, which is a narrow belt of a variable depth of order $\sqrt{\varepsilon}$ and a width of order 1 as $\varepsilon \to 0$. To obtain the asymptotic solution of this problem as $\varepsilon \to 0$, we pursue an approach usually applied in the method of narrow belts or in the theory of shallow water. Namely, we perform a "strain" of region $\Omega(\varepsilon)$ in the vertical direction in such a way that in the new variables the characteristic vertical size is also of order 1 as $\varepsilon \to 0$. With the same purpose, we perform two substitutions in (2.1) according to the formulas

$$z = \sqrt{\varepsilon}\, x_3, \qquad \tilde{h}(\hat{x}) = \sqrt{\varepsilon}\, h(\hat{x}). \tag{2.3}$$

Instead of (2.1), we obtain the next problem in a new region Ω_z defined by $\Omega_z := \{(\hat{x}, z) : \hat{x} \in \Gamma,\ -h(\hat{x}) < z < 0\}$, namely,

$$\varepsilon\Delta\Phi + \frac{\partial^2\Phi}{\partial z^2} = 0 \quad \text{in } \Omega_z, \qquad \frac{\partial\Phi}{\partial n_0} = 0 \quad \text{on } S_{0,z},$$
$$\frac{\partial\Phi}{\partial z} = \lambda\Phi \quad \text{on } \Gamma,$$
$$\frac{\partial\Phi}{\partial z} = -\varepsilon\nabla\Phi\cdot\nabla h \quad \text{on } S_{1,z}, \qquad \text{for } z = -h(\hat{x}),$$
$$\int_\Gamma \Phi d\Gamma = 0, \qquad \lambda := \sqrt{\varepsilon}\,\tilde{\lambda} = \sqrt{\varepsilon}\,\omega^2 d/g. \tag{2.4}$$

The just formulated spectral problem (2.4) contains the spectral parameter λ on Γ, and a small positive parameter ε in the equation over the region Ω_z and in the boundary condition on the bottom $S_{1,z}$.

5.2.2 Asymptotic Solution in First Approximation

We seek a solution of problem (2.4) in the form of an asymptotic power series in the parameter ε:

$$\Phi(\hat{x}, z; \varepsilon) = \Phi_0(\hat{x}, z) + \varepsilon\Phi_1(\hat{x}, z) + \varepsilon^2\Phi_2(\hat{x}, z) + \cdots$$
$$\lambda(\varepsilon) = \lambda_0 + \varepsilon\lambda_1 + \varepsilon^2\lambda_2 + \cdots . \tag{2.5}$$

For the approximation of order zero we get the problem

$$\frac{\partial^2\Phi_0}{\partial z^2} = 0 \quad \text{in } \Omega_z, \qquad \frac{\partial\Phi_0}{\partial z} = \lambda_0\Phi_0 \quad \text{on } \Gamma, \qquad \frac{\partial\Phi_0}{\partial \boldsymbol{n}_0} = 0 \quad \text{on } \partial\Gamma,$$
$$\frac{\partial\Phi_0}{\partial z} = 0 \quad \text{for } z = -h(\hat{x}), \qquad \int_\Gamma \Phi_0 d\Gamma = 0, \tag{2.6}$$

where $\boldsymbol{n}_0$ is the normal to $\partial\Gamma$, which coincides with the normal to S_0.

From the first equation it follows that $\Phi(\hat{x}, z)$ is a linear function of the variable z, that is, $\Phi_0(\hat{x}, z) = v(\hat{x}) + zw(\hat{x})$. The other conditions in (2.6) give $w(\hat{x}) \equiv 0$, $\lambda_0 v(\hat{x}) = 0$. It can be proved that the assumption $v(\hat{x}) \equiv 0$ leads to the trivial asymptotic solution $\Phi(\hat{x}, z; \varepsilon) \equiv 0$ (with an arbitrary $\lambda = \lambda(\varepsilon)$); further, assuming $v(\hat{x}) \not\equiv 0$ we obtain the following approximation of order zero

$$\Phi_0(\hat{x}, z) = v(\hat{x}), \quad \lambda_0 = 0, \quad \frac{\partial v}{\partial n_0} = 0 \quad \text{on } \partial\Gamma, \quad \int_\Gamma v(\hat{x})\, d\Gamma = 0, \tag{2.7}$$

where $v(\hat{x})$ is to be determined later on.

For the first approximation we have

$$\frac{\partial^2\Phi_1}{\partial z^2} = -\Delta\Phi_0 \quad \text{in } \Omega_z, \qquad \frac{\partial\Phi_1}{\partial z} = \lambda_1\Phi_0 \quad \text{on } \Gamma,$$
$$\frac{\partial\Phi_1}{\partial z} = -\nabla\Phi_0 \cdot \nabla h \quad \text{for } z = -h(\hat{x}),$$
$$\frac{\partial\Phi_1}{\partial n_0} = 0 \quad \text{on } \partial\Gamma, \qquad \int_\Gamma \Phi_1 d\Gamma = 0. \tag{2.8}$$

With regard to (2.7), the first two equations in (2.8) give

$$\Phi_1(\hat{x}, z) = -1/2\Delta v(\hat{x})\, z^2 + \lambda_1 v(\hat{x})\, z + w_1(\hat{x}), \int_\Gamma w_1(\hat{x})\, d\Gamma = 0, \tag{2.9}$$

where $w_1(\hat{x})$ is a new unknown function. A substitution of (2.9) into the third condition in (2.8), in conjunction with the condition on $\partial\Gamma$ from (2.6), leads to the next spectral problem for the function $v(\hat{x})$,

$$\mathcal{L}v := -\mathrm{div}(h\nabla v) = \lambda_1 v \quad \text{on } \Gamma,$$
$$\frac{\partial v}{\partial n_0} = 0 \quad \text{on } \partial\Gamma, \qquad \int_\Gamma v d\Gamma = 0, \tag{2.10}$$

where $h = h(\hat{x})$ is a given function that describes the profile of the bottom.

Further, let us assume that function $h(\hat{x})$ is continuously differentiable on $\bar{\Gamma}$ and is bounded from below by the positive constant, $h(\hat{x}) \geq h_0 > 0 \left(\hat{x} \in \bar{\Gamma}\right)$, and the curve $\partial\Gamma$ is a Lipshitz curve.

For an operator interpretation of problem (2.10), we introduce the Hilbert space $H = L_2(\Gamma) \ominus \{1\}$ and the set $\mathcal{D}(L)$ defined as

$$\mathcal{D}(L) := \left\{ v(\hat{x}) \in H \cap H^2(\Gamma) : \frac{\partial v}{\partial n_0} = 0 \quad \text{on } \partial\Gamma \right\},$$

which is dense in H. We also introduce an operator L acting on $\mathcal{D}(L)$ according to the law

$$Lv := \mathcal{L}v, \qquad v \in \mathcal{D}(L).$$

Then, problem (2.10) can be written down in the form

$$Lv = \lambda_1 v, \qquad v \in \mathcal{D}(L) \subset H. \tag{2.11}$$

It is easy to prove that operator L does not take away any element $v \in \mathcal{D}(L)$ from $H = L_2(\Gamma) \ominus \{1\}$ because

$$(Lu, 1) := \int_\Gamma (Lu) \cdot 1 \mathrm{d}\Gamma = -\int_\Gamma \operatorname{div}(h\nabla u) \mathrm{d}\Gamma = -\int_{\partial\Gamma} h \frac{\partial u}{\partial n_0} \mathrm{d}s = 0.$$

Similarly, it can be proved that L is symmetric and positive definite on $\mathcal{D}(L)$, and its energetic norm is equivalent to the standard norm of the space $H^1(\Gamma)$.

Indeed, if $u(\hat{x})$ and $v(\hat{x})$ belong to $\mathcal{D}(L)$, then

$$\begin{aligned}
(Lu, v) &= -\int_\Gamma \operatorname{div}(h\nabla u) v \mathrm{d}\Gamma \\
&= -\int_\Gamma \operatorname{div}\left((h\nabla u)v\right) \mathrm{d}\Gamma + \int_\Gamma h\nabla u \cdot \nabla v \mathrm{d}\Gamma \\
&= \int_\Gamma h(\hat{x}) \nabla u \cdot \nabla v \mathrm{d}\Gamma - \int_{\partial\Gamma} h \frac{\partial u}{\partial n_0} v \mathrm{d}s \\
&= \int_\Gamma h(\hat{x}) \nabla u \cdot \nabla v \mathrm{d}\Gamma,
\end{aligned} \tag{2.12}$$

and the symmetry property for the operator L clearly follows from the last form.

To prove the other already formulated properties, we introduce the norm in $H^1(\Gamma)$ using one of the existing equivalent forms, namely,

$$\|u\|^2_{H^1(\Gamma)} := \int_\Gamma |\nabla u|^2 \mathrm{d}\Gamma + \left(\int_\Gamma u \mathrm{d}\Gamma\right)^2.$$

Then the norm squared of each element $u \in H^1(\Gamma) \cap H =: H^1_\Gamma$ coincides with the Dirichlet integral $\int_\Gamma |\nabla u|^2 \mathrm{d}\Gamma$. Assuming in (2.12) that $v = u \in \mathcal{D}(L)$, we have

$$h_0 \int_\Gamma |\nabla u|^2 \mathrm{d}\Gamma \leq (Lu, u) = \int_\Gamma h(\hat{x}) |\nabla u|^2 \mathrm{d}\Gamma \leq h_{\max} \int_\Gamma |\nabla u|^2 \mathrm{d}\Gamma. \tag{2.13}$$

From (2.13) we obviously get the equivalence of the norms in the energetic space H_L and in H^1_Γ, and therefore, we also prove the positive definiteness of operator L. Further, L will be understood as the Friedrichs' extension of the operator investigated above.

Based on already proved facts and the theorem on compact embedding of $H^1(\Gamma)$ into $L_2(\Gamma)$, we get that H_L is compactly enbedded into $H = L_2(\Gamma) \ominus \{1\}$. Therefore, according to Section 1.4.1, the inverse operator L^{-1} is compact and problem (2.11) has the discrete spectrum $\{\lambda_{1k}\}_{k=1}^\infty$, with

$$0 < \lambda_{11} \leq \lambda_{12} \leq \ldots \leq \lambda_{1k} \leq \ldots, \lambda_{1k} \to \infty, \qquad k \to \infty,$$

and a system of eigenelements $\{v_k\}_{k=1}^\infty$ that form an orthogonal basis in H,

$$(v_k, v_j) = \int_\Gamma v_k(\hat{x})\, v_j(\hat{x})\, \mathrm{d}\Gamma = \delta_{kj},$$
$$(v_k, v_j)_L := \int_\Gamma h(\hat{x})\, \nabla v_k \cdot \nabla v_j \cdot \mathrm{d}\Gamma = \lambda_{1k}\gamma_{kj}. \tag{2.14}$$

Hence, the first order approximation in ε shows that the coefficients $\lambda_1 = \lambda_{1k}$, $k = 1, 2, \ldots$, in the decomposition (2.5) form a discrete positive spectrum, and the eigenfunctions $\Phi_0 = \Phi_{0k} = v_k(\hat{x})$ from the zero order approximation form a basis in H with the properties (2.14). Also, let us note that for a given profile of the bottom, $h = h(\hat{x})$, solutions to problem (2.10) (for horizontal Γ) can be found approximately using the Ritz method and employing the variational ratio

$$\|u\|_L^2 / \|u\|^2 = \int_\Gamma h(\hat{x}) |\nabla u|^2 \mathrm{d}\Gamma \Big/ \int_\Gamma |u|^2 \mathrm{d}\Gamma, \int_\Gamma u(\hat{x})\, \mathrm{d}\Gamma = 0. \tag{2.15}$$

In this problem, the boundary condition $\partial u / \partial n_0 = 0$ on $\partial\Gamma$ for solutions holds true automatically, because it is a natural condition.

5.2.3 Formulas for Calculating Second Order Approximations

We can write down the problem for the second order approximation in ε as

$$\frac{\partial^2 \Phi_2}{\partial z^2} = -\Delta\Phi_1 \quad \text{in } \Omega_z, \qquad \frac{\partial \Phi_2}{\partial z} = -\nabla\Phi_1 \cdot \nabla h, \qquad z = -h(\hat{x}),$$
$$\frac{\partial \Phi_2}{\partial z} = \lambda_2 \Phi_0 + \lambda_1 \Phi_1 \quad \text{on } \Gamma, \qquad \frac{\partial \Phi_2}{\partial n_0} = 0 \quad \text{on } \partial\Gamma, \qquad \int_\Gamma \Phi_2 \mathrm{d}\Gamma = 0. \tag{2.16}$$

From the first equation and from (2.7), (2.9) we obtain

$$\Phi_2(\hat{x}, z) = \left(\Delta^2 v\right) \frac{z^4}{24} - \lambda_1 (\Delta v) \frac{z^3}{6} - (\Delta w_1) \frac{z^2}{2} + \varphi(\hat{x})\, z + \psi(\hat{x}) \tag{2.17}$$

with two unknown functions $\varphi(\hat{x})$ and $\psi(\hat{x})$. The condition on Γ in virtue of (2.16) leads to

$$\varphi(\hat{x}) = \lambda_1 w_1(\hat{x}) + \lambda_2 v(\hat{x}), \tag{2.18}$$

and the condition on the bottom $S_{1,z}$ in conjunction with (2.18) leads to the relation

$$\begin{aligned} &- \left(\Delta^2 v\right) \frac{h^3}{6} - \lambda_1 (\Delta v) \frac{h^2}{2} + h \Delta w_1 + \lambda_2 v + \lambda_1 w_1 \\ &= \nabla h \cdot \nabla \left((\Delta v) \frac{h^2}{2} + \lambda_1 v h - w_1 \right). \end{aligned}$$

With regard to the last two conditions in (2.8), from the previous equation we get the following problem,

$$\begin{aligned} &\mathcal{L} w_1 - \lambda_1 w_1 = f(\hat{x}) := \lambda_2 v - \lambda_1 \ell_1 v - \ell_0 v \quad \text{on } \Gamma, \\ &\quad \ell_1 v := (\Delta v) \frac{h^2}{2} + \nabla h \cdot \nabla (h v), \\ &\quad \ell_0 v := \left(\Delta^2 v\right) \frac{h^3}{6} + \nabla h \cdot \nabla \left((\Delta v) \frac{h^2}{2} \right), \\ &\quad \frac{\partial w_1}{\partial n_0} = 0 \quad \text{on } \partial\Gamma, \qquad \int_\Gamma w_1 \, d\Gamma = 0, \qquad \lambda_1 = \lambda_{1k}, \; k = 1, 2, \ldots . \end{aligned} \tag{2.19}$$

As it easily follows from our reasoning in Section 5.2.2, problem (2.19) can be considered as an operator equation of the form

$$L w_1 - \lambda_1 w_1 = f, \tag{2.20}$$

where $\lambda_1 = \lambda_{1k}$, $k = 1, 2, \ldots$, that is, λ_1 is an eigenvalue in the spectrum of operator L. Since the homogeneous problem has nontrivial solutions and $L = L^*$, equation (2.20) has solutions if and only if the right side, $f(\hat{x})$, is orthogonal in $H = L_2(\Gamma) \ominus \{1\}$ to all the eigenfunctions $v_k(\hat{x})$ of problem (2.11) that correspond to the eigenvalue $\lambda_1 = \lambda_{1k}$.

For simplicity, let us assume further that region Γ and function $h(\hat{x})$ possess properties necessary for all the eigenvalues λ_{1k} of problem (2.11) to be simple, that is, of multiplicity 1. Then, using the equations $(v_k, v_k) = 1$ that follow from (2.14), and the above mentioned orthogonality condition $(f, v_k) = 0$, we obtain the next calculation formula for determinig the second approximation to the eigenvalue $\lambda_k(\varepsilon)$,

$$\lambda_2 = \lambda_{2k} = \lambda_{1k} (\ell_1 v_k, v_k) + (\ell_0 v_k, v_k), \qquad k = 1, 2, \ldots . \tag{2.21}$$

Formulas (2.21) with the differential expressions for ℓ_0 and ℓ_1 from (2.19), do not have a completely symmetrical form relatively to the function $v_k(\hat{x})$. Using Green's formula for the Laplace operator Δ and properties of the solutions of problem (2.10), after a simple though clumsy reasoning we have

$$\begin{aligned}\lambda_2 = \lambda_{2k} = &-\frac{5}{6}\lambda_{1k}^2 \int_\Gamma h|v_k|^2 \mathrm{d}\Gamma + \lambda_{1k} \\ &\times \left\{ \int_\Gamma \left(\frac{1}{12}|\nabla h|^2 - \frac{11}{12} h\Delta h \right) |v_k|^2 \mathrm{d}\Gamma \right. \\ &\left. \quad + \frac{1}{2}\int_\Gamma h^2 |\nabla v_k|^2 \mathrm{d}\Gamma + \frac{7}{4}\int_{\partial\Gamma} h\frac{\partial h}{\partial n_0}|v_k|^2 \mathrm{d}s \right\} \\ &+ \int_\Gamma \left[\frac{1}{12}\Delta^2 \left(h^3\right) + \frac{1}{4}\nabla \cdot (h\Delta h \Delta h) \right] |v_k|^2 \mathrm{d}\Gamma \\ &- \int_\Gamma \left(\frac{11}{6} h|\nabla h|^2 + \frac{1}{2}h^2\Delta h \right) |\nabla v_k|^2 \mathrm{d}\Gamma \\ &- \int_{\partial\Gamma} \left[\frac{1}{12}\frac{\partial}{\partial n_0} \left(\Delta \left(h^3\right)\right) + \frac{1}{4}(\Delta h) h \frac{\partial h}{\partial n_0} \right] |v_k|^2 \mathrm{d}s. \end{aligned} \tag{2.22}$$

Obviously, we assume that $h(\hat{x})$ is a sufficiently smooth function defined on Γ.

5.2.4 Plane Problems

The method of narrow belts can be also applied to the plane (two-dimensional) problem, that is, when $\Omega \subset \mathbb{R}^2$, $\Gamma \subset \mathbb{R}$. In problem (2.1) we can formally assume that $x = (x_1, x_3) \in \Omega$, Δ_3 should be replaced by the operator $\partial^2/\partial x_1^2 + \partial^2/\partial x_3^2$, operator Δ is replaced by $\partial^2/\partial x_1^2$, and the line width of the free surface Γ should be d. Assuming again that condition (2.2) holds true and performing the substitutions (2.3), we obtain problem (2.4), where now

$$h = h(x_1), \quad \nabla h \cdot \nabla \Phi = h'(x_1)\frac{\partial \Phi}{\partial x_1}, \quad \Omega_z = \{(x_1, z) : x_1 \in \Gamma,\ -h(x_1) < z < 0\}.$$

For the first order approximation in ε we have the problem

$$\begin{aligned} &\mathcal{L}v := -\left[h(x_1)v'(x_1)\right]' = \lambda_1 v(x_1), \qquad -\frac{d}{2} < x_1 < \frac{d}{2}, \\ &v'\left(\pm\frac{d}{2}\right) = 0, \qquad \int_{-d/2}^{d/2} v(x_1)\mathrm{d}x_1 = 0, \end{aligned} \tag{2.23}$$

which has the discrete spectrum $\{\lambda_{1k}\}_{k=1}^\infty$ and a complete orthogonal system of eigenfunctions $\{v_k(x_1)\}_{k=1}^\infty$ with

$$\int_{-d/2}^{d/2} v_k(x_1)v_j(x_1)\mathrm{d}x_1 = \delta_{kj}, \qquad \int_{-d/2}^{d/2} h(x_1)v_k'(x_1)v_j'(x_1)\mathrm{d}x_1 = \lambda_{1k}\delta_{kj}. \tag{2.24}$$

Formulas for the second order approximation can be obtained from (2.22) after the above mentioned formal simplifications, and the result has the form

$$
\begin{aligned}
\lambda_2 = \lambda_{2k} = & -\frac{5}{6}\lambda_{1k}^2 \int_{-d/2}^{d/2} h|v_k|^2 \mathrm{d}x_1 + \lambda_{1k} \\
& \times \left\{ \int_{-d/2}^{d/2} \left[\frac{1}{12}(h')^2 - \frac{11}{12}hh'' \right] |v_k|^2 \mathrm{d}x_1 \right. \\
& \left. + \frac{1}{2}\int_{-d/2}^{d/2} h^2 |v_k'|^2 \mathrm{d}x_1 + \left(\frac{7}{4}hh'|v_k|^2 \right) \Big|_{-d/2}^{d/2} \right\} \\
& + \int_{-d/2}^{d/2} \left[\frac{1}{12}(h^3)^{(4)} + \frac{1}{4}(hh'h'')' \right] |v_k|^2 \mathrm{d}x_1 \\
& - \int_{d/2}^{d/2} \left[\frac{11}{6}h(h')^2 + \frac{1}{2}h^2h'' \right] |v_k'|^2 \mathrm{d}x_1 \\
& - \left(\left[\frac{1}{12}(h^3)''' = \frac{1}{4}hh'h'' \right] |v_k|^2 \right) \Big|_{-d/2}^{d/2} .
\end{aligned}
\tag{2.25}
$$

5.2.5 Examples

In this section we discuss two examples of plane and three-dimensional axially symmetrical problems, which however address the case when the minimal depth of fluid equals zero.

PROBLEM 1. *Plane oscillations of a fluid in a channel with a parabolic profile of the bottom.*

We choose the halfwidth $d/2$ of the channel as a characteristic size and assume that $h(x_1) := 1 - x_1^2$. Then, problem (2.23) takes the form

$$
\begin{aligned}
& Lv := -\left[(1-x_1)^2 v'(x_1)\right]' = \lambda_1 v(x_1), \qquad -1 < x_1 < 1, \\
& |v_1(\pm 1)| < \infty, \quad \int_{-1}^{1} v(x_1)\mathrm{d}x_1 = 0,
\end{aligned}
\tag{2.26}
$$

where instead of the boundary conditions at $x_1 = \pm 1$ we require boundedness conditions of solutions.

Equation (2.26) is the well-known Legendre equation, and its solutions have the form

$$
v_k(x_1) = P_k(x_1), \quad \lambda_{1k} = k(k+1), \qquad k = 1, 2, \dots, \tag{2.27}
$$

where $P_k(x_1)$ are the Legendre polynomials.

PROBLEM 2. *Three-dimensional oscillations of a fluid in a region of the form of a paraboloid of revolution.*

We choose the radius of the free surface of a fluid as a characteristic size d and assume that $h = h(r) := (1 - r^2)/2$, where r is the radial coordinate in the cylindrical coordinate system (r, θ, x_3).

After determining the variable θ in problem (2.10), we obtain a countable set of one-dimensional problems,

$$\begin{aligned} L_m v_m :&= \frac{1}{2r}\left[r(r^2-1)v_m'\right]' + \frac{m^2}{2r^2}(1-r^2)v_m \\ &= \lambda_{1m} v_m, \qquad 0 < r < 1, \ m = 0, 1, 2, \ldots, \end{aligned}$$

$$\int_0^1 v_0(r) r \mathrm{d}r = 0, \quad |v_m(0)| < \infty, \quad |v_m(1)| < \infty. \tag{2.28}$$

Here again there is degeneration for $r = 0$ and $r = 1$, and therefore, we impose conditions of boundedness of solutions at these points.

It can be proved that for every $m = 1, 2, \ldots$, operator L_m is positive definite in the Hilbert space H_m with the scalar product $\int_0^1 r u(r) v(r) \mathrm{d}r =: (u, v)$, and for $m = 0$ it has the same property in the subspace orthogonal to 1. Each of these operators has the discrete spectrum

$$\lambda_{1mp} = m + 2(p+m)(p-1), \qquad m = 0, 1, 2 \ldots, \ p = 1, 2, \ldots, \tag{2.29}$$

and a system of eigenelements of the form

$$\begin{aligned} v_{mp}(r) &= \sum_{k=1}^{p} a_{mp}^{(k)} r^{m+2k-2}, a_{mp}^{(1)} = 1, \\ a_{mp}^{k+1} &= a_{mp}^{(k)} \frac{(k-p)(k+p+m-1)}{k(m+k)}. \end{aligned} \tag{2.30}$$

For any given m, these polynomials provide an orthogonal basis in H_m.

The previous examples show that in some cases the theory of shallow water helps in obtaining asymptotic solutions of the problem on proper oscillations of a heavy fluid in a container of a form that makes it impossible to use the standard method of separation of variables.

5.2.6 Systems of Nonmixing Fluids

We now assume that the container, which is flattened out in the vertical direction, is filled with several heavy ideal fluids with different densities. The corresponding problem on proper oscillations of such a system in an ordinary container was already considered in Section 3.6, and in dimensionless variables it has the next form that follows from (3.6.20),

$$
\begin{aligned}
&\Delta_3\Phi_k = 0 \quad \text{in } \Omega_k, \qquad k = 1,2,\ldots,m, \\
&\frac{\partial\Phi_k}{\partial n} = 0 \quad \text{on } S_k, \\
&\frac{\partial\Phi_j}{\partial x_3} = \frac{\partial\Phi_{j+1}}{\partial x_3} \quad \text{on } \Gamma_j, \qquad j = 1,2,\ldots,m-1, \\
&\frac{\partial\Phi_k}{\partial n_0} = 0 \quad \text{on } S_0, \qquad k = 1,2,\ldots,m, \\
&(\Delta\rho)_j\frac{\partial\Phi_j}{\partial x_3} = \tilde{\lambda}\left(\rho_j\Phi_j - \rho_{j+1}\Phi_{j+1}\right) \quad \text{on } \Gamma_j, \qquad j = 1,2,\ldots,m-1, \\
&(\Delta\rho)_j := \rho_j - \rho_{j+1} > 0, \\
&\frac{\partial\Phi_1}{\partial n} = 0 \quad \text{on } S_1, \qquad \frac{\partial\Phi_m}{\partial n} = 0 \quad \text{on } S_2, \\
&\int_{\Gamma_j}(\rho_j\Phi_j - \rho_{j+1}\Phi_{j+1})\,\mathrm{d}\Gamma_j = 0, \qquad j = 1,2,\ldots,m-1, \\
&\int_{\Gamma_{m-1}}\rho_m\Phi_m \mathrm{d}\Gamma_{m-1} = 0.
\end{aligned} \tag{2.31}
$$

Here, S_0 is a cylindrical surface that forms the lateral boundary of region Ω, S_1 is the bottom and S_2 is the upper boundary of the container, ρ_j is the density of the fluid filling the region Ω_j bounded from above by the horizontal surface Γ_j, from below by Γ_{j-1}, and enclosed by the wall S_0, $j = 2,3,\ldots,m-1$, and $\tilde{\lambda} = \omega^2 d/g$ where d is the diameter of the boundary Γ_1. In addition, we assume that the equation of the bottom S_1 has the form $x_3 = -\tilde{h}_0(x_1,x_2)$, $(x_1,x_2) = \hat{x} \in \Gamma_1$, the equation of the boundary Γ_1 is $x_3 = \tilde{h}_1 \equiv 0$, the equation of Γ_j is $x_3 = \tilde{h}_j$, $j = 2,3,\ldots,m-1$, and for the upper surface S_2 we take $x_3 = \tilde{h}_m(\hat{x})$, where

$$
-\tilde{h}_0(\hat{x}) < 0 < \tilde{h}_2 < \ldots < \tilde{h}_{m-1} < \tilde{h}_m(\hat{x}), \qquad \hat{x} \in \Gamma_1. \tag{2.32}
$$

Similar to the problem in Sections 5.2.1–5.2.5 on oscillations of one fluid, we assume that the vertical length of the container is considerably smaller than its horizontal length, that is,

$$
0 < \varepsilon := \left(\tilde{h}_{\max}/d\right)^2 \ll 1, \qquad \tilde{h}_{\max} := \max_{\hat{x}\in\Gamma_1}\left\{\tilde{h}_0(\hat{x}) + \tilde{h}_m(\hat{x})\right\}. \tag{2.33}
$$

Performing the substitutions

$$
x_3 = \sqrt{\varepsilon}\, z, \quad \lambda = \sqrt{\varepsilon}\,\tilde{\lambda}, \quad \tilde{h}_i = \sqrt{\varepsilon}\, h_i, \qquad i = 0,1,\ldots,m,
$$

in (2.31), we obtain the spectral problem

$$
\begin{aligned}
&\frac{\partial^2 \Phi_k}{\partial z^2} + \varepsilon \Delta \Phi_k = 0 \quad \text{in } \Omega_{k,z}, \qquad k = 1,2,\ldots,m;\\
&\frac{\partial \Phi_j}{\partial z} = \frac{\partial \Phi_{j+1}}{\partial z} \quad \text{on } \Gamma_{j,z}, \qquad j = 1,2,\ldots,m-1,\\
&\frac{\partial \Phi_k}{\partial n_0} = 0 \quad \text{on } S_{0,z}, \qquad k = 1,2,\ldots,m;\\
&(\Delta\rho)_j \frac{\partial \Phi_j}{\partial z} = \lambda\left(\rho_j \Phi_j - \rho_{j+1}\Phi_{j+1}\right) \quad \text{on } \Gamma_{j,z}, \qquad j = 1,2,\ldots,m-1;\\
&\frac{\partial \Phi_1}{\partial z} = -\varepsilon \nabla h_1 \cdot \nabla \Phi_1 \quad \text{for } z = -h_0(\hat{x}), \qquad \hat{x} \in \Gamma_1,\\
&\frac{\partial \Phi_m}{\partial z} = \varepsilon \nabla h_m \cdot \nabla \Phi_m \quad \text{for } z = h_m(\hat{x}), \qquad \hat{x} \in \Gamma_1,\\
&\int_{\Gamma_{j,z}} \left(\rho_j \Phi_j - \rho_{j+1}\Phi_{j+1}\right) d\Gamma_{j,z} = 0, \qquad j = 1,2,\ldots,m-1,\\
&\int_{\Gamma_{m-1,z}} \rho_m \Phi_m d\Gamma_{m-1,z} = 0.
\end{aligned}
\tag{2.34}
$$

Similar to Section 5.2.2, we are looking for an asymptotic solution to problem (2.34) of the form

$$
\begin{aligned}
\Phi_k(\hat{x}, z; \varepsilon) &= \Phi_{0,k}(\hat{x}, z) + \varepsilon \Phi_{1k}(\hat{x}, z) + O(\varepsilon^2), \qquad k = 1,2,\ldots,m,\\
\lambda(\varepsilon) &= \lambda_0 + \varepsilon\lambda_1 + O(\varepsilon^2), \qquad \varepsilon \to 0.
\end{aligned}
\tag{2.35}
$$

The zero order approximation leads to the solutions

$$
\begin{aligned}
&\Phi_{0,k}(\hat{x}, z) = v_k(\hat{x}), \qquad k = 1,2,\ldots,m,\\
&\frac{\partial v_k}{\partial n_0} = 0 \quad \text{on } \partial\Gamma_1, \quad \int_{\Gamma_1} v_k(\hat{x})\, d\Gamma_1 = 0, \qquad k = 1,2,\ldots,m,
\end{aligned}
\tag{2.36}
$$

with the unknown functions $v_k(\hat{x})$ defined on Γ_1, and to the relations

$$
\lambda_0\left(\rho_j v_j(\hat{x}) - \rho_{j+1} v_{j+1}(\hat{x})\right) = 0, \qquad j = 1,2,\ldots,m-1,
\tag{2.37}
$$

The assumption $\lambda_0 \neq 0$ in (2.37) and the equation $\lambda(\varepsilon) = \sqrt{\varepsilon}\tilde{\lambda}(\varepsilon) = \sqrt{\varepsilon}\omega^2(\varepsilon)\cdot d/g$ lead to the conclusion—unnatural from a physical point of view—that the frequencies of oscillations $\omega(\varepsilon)$ in the considered system of fluids go to infinity when the thickness of the layer decreases in the vertical direction.

Assuming $\lambda_0 = 0$, we can now obtain the equations of the first order approximation,

$$
\begin{aligned}
&\frac{\partial^2\Phi_{1,k}}{\partial z^2}+\Delta v_k=0,\quad \frac{\partial\Phi_{1,k}}{\partial n_0}=0 \quad \text{on } \partial\Gamma_1, \qquad k=1,2,\dots,m;\\
&\qquad \frac{\partial\Phi_{1,j}}{\partial z}=\frac{\partial\Phi_{1,j+1}}{\partial z},\quad z=h_j,\qquad j=1,2,\dots,m-1,\\
&(\Delta\rho)_j\frac{\partial\Phi_{1,j}}{\partial z}=\lambda_1\left(\rho_j v_j-\rho_{j+1}v_{j+1}\right),\qquad z=h_j,\ j=1,2,\dots,m-1,\\
&\qquad \frac{\partial\Phi_{1,1}}{\partial z}=-\nabla h_0\cdot\nabla v_1,\qquad z=-h_0\left(\hat{x}\right),\\
&\qquad \frac{\partial\Phi_{1,m}}{\partial z}=\nabla h_m\cdot\nabla v_m,\qquad z=h_m\left(\hat{x}\right).
\end{aligned}
\tag{2.38}
$$

Hence, we have

$$
\Phi_{1,k}(\hat{x},z)=-\frac{z^2}{2}\Delta v_k+w_k\left(\hat{x}\right)z+\psi_k\left(\hat{x}\right),\qquad k=1,2,\dots,m.
$$

Substituting these solutions into the boundary conditions we obtain the following relations,

$$
\begin{aligned}
u_j&=-h_j\Delta v_j+w_j=-h_j\Delta v_{j+1},\qquad j=1,2,\dots,m\\
(\Delta\rho)_j u_j&=\lambda_1\left(\rho_j v_j-\rho_{j+1}v_{j+1}\right),\qquad j=1,2,\dots,m,\\
\mathcal{L}_1 v_1 &:=-\operatorname{div}\left((h_1-h_0)\nabla v_1\right)=w_1,\\
\mathcal{L}_m v_m &:=-\operatorname{div}\left((h_m-h_{m-1})\nabla v_m\right)=-w_m+h_{m-1}\Delta v_m,\\
\frac{\partial v_k}{\partial n_0}&=0 \quad \text{on } \partial\Gamma_1,\quad \int_{\Gamma_1} v_k\left(\hat{x}\right)\mathrm{d}\Gamma_1=0,\qquad k=1,2,\dots,m.
\end{aligned}
\tag{2.39}
$$

Problem (2.39) can be reduced to an eigenvalue problem for the vector-column $\{u_j\left(\hat{x}\right)\}_{j=1}^{m-1}$ consisting of the unknown functions $u_j\left(\hat{x}\right)$. To this end, let us introduce the Hilbert space $L_{2,\Gamma_1} := L_2(\Gamma_1)\ominus\{1\}$, and the operators L_k, defined on the sets $\mathcal{D}(L_k)=\{v\left(\hat{x}\right)\in L_{2,\Gamma_1}\cap H^2(\Gamma_1) : \partial v/\partial n_0=0 \text{ on } \partial\Gamma_1\}$ by the equations

$$
L_k v := -\operatorname{div}\left((\Delta h)_k\nabla v\right),\quad (\Delta h)_k := h_k-h_{k-1},\qquad k=1,2,\dots,m. \tag{2.40}
$$

If $k=2,3,\dots,m-1$, the differences $(\Delta h)_k$ are certain constants, and for $k=1$ or $k=m$ they are the given functions that determine the shape of the bottom and of the upper cover of the container.

Similar to the reasoning in Section 5.2.2 for a single operator L, it can be proven that for each $k=1,2,\dots,m$, operator L_k from (2.40) is a positive definite unbounded selfadjoint operator, which acts in the (real) Hilbert space $H := L_{2,\Gamma_1}$, and has a compact positive inverse operator L_k^{-1}.

From those equations in (2.39) that do not contain the parameter λ_1, after eliminating w_j $(j = 1, 2, \dots, m)$ we can obtain by means of the introduced operators the following system of equations,

$$\begin{aligned} u_1 &= L_1 v_1, \\ u_{m-1} &= -L_m v_m, \\ u_j - u_{j-1} &= L_j v_j, \qquad j = 2, 3, \dots, m-1. \end{aligned} \tag{2.41}$$

Therefore, we get

$$\begin{aligned} v_1 &= L_1^{-1} u_1, \\ v_j &= L_j^{-1}(u_j - u_{j-1}), \qquad j = 2, 3, \dots, m-1, \\ v_m &= -L_m^{-1} u_{m-1}, \end{aligned}$$

and the remaining equations of the system (2.39) lead to the vector-matrix equation

$$\mathcal{B}u = \lambda_1 \mathcal{A}u, \tag{2.42}$$

where

$$\begin{aligned} u &= (u_1; \dots; u_{m-1})^{\mathrm{t}} \in \tilde{H} := (L_{2,\Gamma_1})^{m-1}, \\ \mathcal{B} &:= \operatorname{diag}\left((\Delta\rho)_k I\right)_{k=1}^{m-1}, \end{aligned} \tag{2.43}$$

and the operator matrix $\mathcal{A} := (A_{jk})_{j,k=1}^{m-1}$ has a three-diagonal structure with the nonzero elements

$$\begin{aligned} A_{kk} &= \rho_k L_k^{-1} + \rho_{k+1} L_{k+1}^{-1}, \qquad k = 1, 2, \dots, m-1, \\ A_{k,k+1} &= A_{k+1,k} = -\rho_{k+1} L_{k+1}^{-1}, \qquad k = 1, 2, \dots, m-2. \end{aligned} \tag{2.44}$$

Since $(\Delta\rho)_k = \rho_k - \rho_{k+1} > 0$, $k = 1, 2, \dots, m-1$, then operator $\mathcal{B}$ in (2.43) is a bounded and positive definite operator acting on $\tilde{H}$. Further, it is obvious that $\mathcal{A}$ is a compact operator, because all the operators L_k^{-1}, $k = 1, 2, \dots, m$, are compact, and therefore, the matrix coefficients A_{jk} are compact, too. Let us show that $\mathcal{A} > 0$ in $\tilde{H}$. It can be directly proved that the quadratic form associated with $\mathcal{A}$ is

$$\begin{aligned} (\mathcal{A}u, u)_{\tilde{H}} &= \rho_1 \left(L_1^{-1} u_1, u_1\right)_{L_{2,\Gamma_1}} + \sum_{j=2}^{m-1} \rho_j (L_j^{-1}(u_{j-1} - u_j), u_{j-1} - u_j)_{L_{2,\Gamma_1}} \\ &\quad + \rho_m \left(L_m^{-1} u_{m-1}, u_{m-1}\right)_{L_{2,\Gamma_1}}. \end{aligned} \tag{2.45}$$

Since $L_k^{-1} > 0$ for every k, operator $\mathcal{A}$ is nonnegative. If the form $(\mathcal{A}u, u)_{\tilde{H}}$ equals zero, then $u_1 = 0$, $u_{j-1} - u_j = 0$ for each $j = 2, 3, \dots, m-1$, and $u_{m-1} = 0$; therefore, $u_j = 0$ for all $j = 1, 2, \dots, m-1$, which shows that $\mathcal{A} > 0$.

Consequently, equation (2.42) is analogous to the eigenvalue problem discussed in Section 1.4.2, whence it follows that (2.42) has a discrete positive spectrum, $\lambda_1 = \lambda_{1n}(\mathcal{B};\mathcal{A})$, $n = 1,2,\ldots$, where $0 < \lambda_{11} \leq \lambda_{12} \leq \ldots \leq \lambda_{1n} \leq \ldots$, and $\lambda_{1n} \to \infty$ as $n \to \infty$, and the system of eigenelements $\{u^n\}_{n=1}^{\infty} = \{u_1^n;\ldots;u_{m-1}^n\}_{n=1}^{\infty} \subset \tilde{H}$, which form orthogonal bases for the forms $(\mathcal{A}u,u)_{\tilde{H}}$ and $(\mathcal{B}u,u)_{\tilde{H}}$, namely,

$$(\mathcal{B}u^n,u^q)_{\tilde{H}} = \delta_{nq}\lambda_{1n}, \quad (\mathcal{A}u^n,u^q)_{\tilde{H}} = \delta_{nq}. \tag{2.46}$$

The eigenvalues λ_{1n} can be found as successive minima of the variational ratio $(\mathcal{B}u,u)_{\tilde{H}}/(\mathcal{A}u,u)_{\tilde{H}}$, where

$$(\mathcal{B}u,u)_{\tilde{H}} = \sum_{j=1}^{m-1}(\Delta\rho)_j \int_{\Gamma_1} |u_j(\hat{x})|^2 d\Gamma_1, \tag{2.47}$$

and the quadratic form $(\mathcal{A}u,u)_{\tilde{H}}$ is defined as in (2.45).

The solution found in the first approximation can be used to formulate and solve the problem on determining a solution in the second approximation with respect to ε. This can be done similarly to Section 5.2.3 where we addressed the case of a single fluid. We leave the task of solving this problem to the reader.

The previous investigations of the spectral problem of proper oscillations of a single fluid and a system of nonmixing heavy fluids based on the theory of shallow water demonstrate that the corresponding dimensionless eigenvalues, $(\tilde{\lambda})_k = \omega_k^2 d/g$, which depend on a small parameter $\varepsilon > 0$, have the following asymptotic behavior,

$$\tilde{\lambda}_k = \lambda_{1k}\varepsilon^{1/2} + \lambda_{2k}\varepsilon^{3/2} + O(\varepsilon^{5/2}) \quad \text{as } \varepsilon \to 0, \qquad k = 1,2,\ldots, \tag{2.48}$$

where $\tilde{\lambda}_{1k}$ are eigenvalues of the problem in first approximation (i.e., problems (2.11), (2.23), (2.42)) and λ_{2k} are eigenvalues in second approximation (i.e., formulas (2.22), (2.25)). In particular, from (2.48) it follows that for any $k \geq 1$ the frequencies of proper oscillations of the considered hydrodynamic systems approach zero as $\varepsilon \to 0$, that is, when the vertical size of the container decreases.

5.3 Oscillations of a System "Fluid-Gas" in a Bounded Region

The "fluid-gas" systems are quite interesting, because in the mathematical formulation of the problem on proper oscillations, the spectral parameter is involved both in one of the equations and in a boundary condition, though in general the problem is self-adjoint.

5.3.1 Formulation of the Initial Boundary Value Problem

Let $\Omega \subset \mathbb{R}^3$ be a region filled with an ideal incompressible fluid with density $\rho_1 > 0$, and an ideal gas with density $\rho_2 < \rho_1$ located above the fluid. In the equilibrium state, the fluid occupies a region $\Omega_1 \subset \Omega$ bounded by a part S_1 of the solid boundary $S = \partial\Omega$ of region Ω and by the horizontal boundary Γ between fluid and gas. Accordingly, the gas occupies the region $\Omega_2 = \Omega \setminus \bar{\Omega}_1$ bounded by the surface Γ and by the remaining part $S_2 = S \setminus \bar{S}_1$ of the boundary S.

We choose O on Γ as the origin of a coordinate system $Ox_1x_2x_3$ with the axis Ox_3 directed upward, that is, against the acceleration vector of the homogeneous gravitational field $\boldsymbol{g} = -g\boldsymbol{e}_3$. In the equilibrium state, the pressure in fluid equals

$$P_{0,1}(x_3) = -\rho_1 g x_3 + P_{0,1}(0). \tag{3.1}$$

The pressure $P_{0,2}$ in gas can be determined using the condition of its static equilibrium, $\rho_{0,2}^{-1}\nabla P_{0,2} + g\boldsymbol{e}_3 = \boldsymbol{0}$. Further, we assume that the gas is a barotropic compressible fluid, that is, $P_2 = P_2(\rho_2)$, either during the time of its motion or in the equilibrium state, and we obtain $\nabla P_{0,2} = (\mathrm{d}P_2/\mathrm{d}\rho_2)_0 \nabla\rho_{0,2} =: c^2\nabla\rho_{0,2}$, where $c > 0$ is the sound velocity in gas. Then relation $\rho_{0,2}^{-1}c^2\nabla\rho_{0,2} + g\boldsymbol{e}_3 = 0$ leads to the next law of density distribution in the immovable state,

$$\rho_{0,2} = \rho_{0,2}(x_3) = \rho_{0,2}(0)\exp\left[-gx_3/c^2\right]. \tag{3.2}$$

As for the pressure, with regard to (3.1) we have

$$P_{0,2} = P_{0,2}(x_3) = P_{0,1}(0) - g\int_0^{x_3} \rho_{0,2}(\xi)\mathrm{d}\xi. \tag{3.3}$$

If the maximal height of the gas column in region Ω is relatively small ($gx_3/c^2 \ll 1$), then from (3.2) it follows that $\rho_{0,2}(x_3) \approx \rho_{0,2}(0) =: \rho_2$, and according to (3.2), (3.3), we get

$$\left(\frac{\mathrm{d}P_{0,2}(x_3)}{\mathrm{d}x_3}\right)_0 = -g\rho_2. \tag{3.4}$$

Let us now analyze small free movements of the given system "fluid-gas" close to the equilibrium state. For fluid we have the linearized Euler equations

$$\frac{\partial \boldsymbol{u}_1}{\partial t} = -\rho_1^{-1}\nabla p_1, \quad \operatorname{div}\boldsymbol{u}_1 = 0 \qquad \text{in } \Omega_1, \tag{3.5}$$

where $\boldsymbol{u}_1 = \boldsymbol{u}_1(t,x)$ is the velocity field and $p_1 = p_1(t,x)$ is the deviation of the field of pressures from the equilibrium field (3.1). The equation of motion for gas has a similar form, namely,

$$\frac{\partial \boldsymbol{u}_2}{\partial t} = -\rho_2^{-1}\nabla p_2 \qquad \text{in } \Omega_2, \tag{3.6}$$

and the solenoidality condition of the velocity field is replaced now by the relation

$$c^{-2}\frac{\partial p}{\partial t} + \rho_2 \operatorname{div}\boldsymbol{u}_2 = 0 \qquad \text{in } \Omega_2, \tag{3.7}$$

that becomes the solenoidality condition when $c^2 = \infty$.

To deduce (3.7), it is sufficient to notice that $c^{-2}\partial p/\partial t = \partial\rho_2'/\partial t$, where $\rho_2'(t,x)$ is the deviation of density from the equilibrium value ρ_2, and then (3.7) can be transformed into the linearized continuity equation for gas.

On the solid boundary $S = \partial\Omega$ we impose the nonleaking conditions for fluid and gas,

$$\boldsymbol{u}_1 \cdot \boldsymbol{n} = 0 \quad \text{on } S_1, \qquad \boldsymbol{u}_2 \cdot \boldsymbol{n} = 0 \quad \text{on } S_2, \tag{3.8}$$

where $\boldsymbol{n}$ is an external normal to $\partial\Omega$.

Let us next assume that during the process of oscillations the boundary between fluid and gas can be described by the function $x_3 = \zeta(t,\hat{x}), \hat{x} := (x_1, x_2) \in \Gamma$. Then, the kinematic and dynamic conditions on Γ have the form

$$\boldsymbol{u}_1 \cdot \boldsymbol{n}_1 = \boldsymbol{u}_2 \cdot \boldsymbol{n}_1 = \partial\zeta/\partial t, \qquad \boldsymbol{n}_1 := \boldsymbol{e}_3, \tag{3.9}$$

$$p_1 - p_2 = g(\Delta\rho)\zeta, \qquad \Delta\rho := \rho_1 - \rho_2 > 0. \tag{3.10}$$

Condition (3.10) is a linearized equation expressing the equality of pressures in the two media on their boundary $x_3 = \zeta(t,\hat{x})$ with regard to the formulas (3.1) and (3.4).

For $t = 0$, that is, at the initial moment of time, we should state the initial conditions

$$\boldsymbol{u}_1(0,x) = \boldsymbol{u}_1^0(x) \quad \text{in } \Omega_1, \qquad \boldsymbol{u}_2(0,x) = \boldsymbol{u}_2^0(x) \quad \text{in } \Omega_2, \qquad \zeta(0,\hat{x}) = \zeta^0(\hat{x}) \quad \text{on } \Gamma. \tag{3.11}$$

Hence, the problem on small oscillations of a system “fluid-gas” amounts to finding the functions $\boldsymbol{u}_i(t,x)$, $p_i(t,x)$, $i = 1,2$, and $\zeta(t,\hat{x})$ from the equations (3.5)–(3.7), boundary conditions (3.8)–(3.10), and the intial conditions (3.11).

Also, let us note that the volume preserving condition for region Ω_1 leads to the relation

$$\int_\Gamma \zeta(t,\hat{x})d\Gamma = 0. \tag{3.12}$$

5.3.2 Formulation of the Spectral Problem

We now consider the proper oscillations of the system, that is, solutions that depend on t according to the $\exp(i\omega t)$—law, where ω is a frequency of oscillations; the corresponding amplitude functions will be denoted by the same symbols. It is easy to observe that the case $\omega = 0$ leads to trivial solutions of the form

$$\begin{gathered} p_1(x) \equiv p_2(x) = \text{const}, \qquad \zeta(\hat{x}) \equiv 0, \\ \operatorname{div} \boldsymbol{u}_i(x) = 0 \quad \text{in } \Omega_i, \qquad \boldsymbol{u}_i \cdot \boldsymbol{n} = 0 \quad \text{on } \partial\Omega_i, \qquad i = 1,2. \end{gathered} \tag{3.13}$$

These conditions show that the boundary Γ does not oscillate, and the velocity fields $\boldsymbol{u}_i(x)$ in the regions Ω_i are arbitrary functions from the subspaces $\boldsymbol{J}_0(\Omega_i)$. Hence, for

$\omega = 0$ the problem has trivial vortex solutions and the subspace of such solutions is infinite-dimensional.

To formulate the spectral problem for $\omega \neq 0$, it is convenient to introduce the fields of displacements $\boldsymbol{w}_i(t,x)$ of fluid particles in each region Ω_i. The fields of displacements are connected with the velocity fields by the obvious relations $\partial \boldsymbol{w}_i/\partial t = \boldsymbol{u}_i(t,x)$. As it follows from (3.5), (3.6), the fields $\boldsymbol{w}_i$ are potential. Assuming $\boldsymbol{w}_i(t,x) = \exp(i\omega t)\nabla\Phi_i(x)$ and eliminating all unknown functions except the potentials $\Phi_i(x)$ in equations and boundary conditions, we obtain the following spectral problem,

$$\begin{aligned}
&\Delta\Phi_1 = 0 \quad \text{in } \Omega_1, && \frac{\partial\Phi_1}{\partial n} = 0 \quad \text{on } S_1,\\
&-\Delta\Phi_2 = \lambda c^{-2}\Phi_2 \quad \text{in } \Omega_2, && \frac{\partial\Phi_2}{\partial n} = 0 \quad \text{on } S_2,\\
&\frac{\partial\Phi_1}{\partial n_1} = \frac{\partial\Phi_2}{\partial n_1} = \zeta \quad \text{on } \Gamma, && \int_\Gamma \zeta \,\mathrm{d}\Gamma = 0,\\
&g(\Delta\rho)\zeta = \lambda P_\Gamma(\rho_1\Phi_1 - \rho_2\Phi_2) \quad \text{on } \Gamma, && \lambda := \omega^2,
\end{aligned} \tag{3.14}$$

where $P_\Gamma v := v - (\operatorname{mes}\Gamma)^{-1}\int_\Gamma v\,(\hat{x})\,\mathrm{d}\Gamma$ is the orthoprojector onto $L_{2,\Gamma} := L_2(\Gamma)\ominus\{1\}$. The occurence of P_Γ is due to property (3.12). Let us note that function $\Phi_1(x)$ in (3.14) is determined up to an arbitrary constant and, therefore, the following relations should be kept in mind along with (3.14),

$$\int_\Gamma \Phi_1(x)\mathrm{d}\Gamma = 0, \qquad \int_{\Omega_2} \Phi_2(x)\mathrm{d}\Omega_2 = 0. \tag{3.15}$$

The first of these relations is a normalization condition for function $\Phi_1(x)$, and the second one follows from

$$\begin{aligned}
-\int_{\Omega_2} \Delta\Phi_2 \mathrm{d}\Omega_2 &= \lambda c^{-2}\int_{\Omega_2}\Phi_2\mathrm{d}\Omega_2 = \int_{\Omega_2}\nabla\Phi_2\cdot\nabla 1\mathrm{d}\Omega_2 - \int_{\partial\Omega_2}\frac{\partial\Phi_2}{\partial n}\mathrm{d}S\\
&= \int_\Gamma \frac{\partial\Phi_2}{\partial n_1}\mathrm{d}\Gamma = \int_\Gamma \zeta\mathrm{d}\Gamma = 0.
\end{aligned}$$

Let us note also that the spectral parameter λ in (3.14) is present both in the equation for the region Ω_2 and in the boundary condition on Γ. If we assume that region Ω_2 is filled with an incompressible fluid of density $\rho_2 < \rho_1$ instead of gas, then in (3.19) one should set $c^2 = \infty$. The corresponding problem on oscillations of a system of two (or several) heavy fluids completely filling a container Ω was already considered in Section 3.6.

The eigenvalues λ associated with solutions of the spectral problem (3.14) are positive and, therefore, as it will follow from our subsequent reasoning, it is sufficient to determine the spectrum in a real Hilbert space. Indeed, multiplying both sides of the Laplace equation in Ω_1 by $\rho_1\Phi_1$ and integrating over Ω_1, and then multiplying the similar equation in Ω_2 by $\rho_2\Phi_2$ and integrating over Ω_2, by using Green's formula for the Laplace operator in each region Ω_1 and Ω_2 and the boundary conditions in (3.14), we get that the eigenvalues λ are among values of the functional

$$F_1\{\Phi_1,\Phi_2\} := \frac{\rho_1 \int_{\Omega_1} |\nabla\Phi_1|^2 d\Omega_1 + \rho_2 \int_{\Omega_2} |\nabla\Phi_2|^2 d\Omega_2}{(g(\Delta\rho))^{-1} \int_\Gamma |P_\Gamma(\rho_1\Phi_1 - \rho_2\Phi_2)|^2 d\Gamma + \rho_2 c^{-2} \int_{\Gamma_2} |\Phi_2|^2 d\Omega_2} \tag{3.16}$$

and, therefore, every λ is positive. Hence we conclude that the proper frequencies $\omega = \sqrt{\lambda}$ of oscillations of the system are real. This fact is physically obvious, because the considered system "fluid-gas" is conservative.

5.3.3 Transition to a System of Operator Equations

We first consider three auxiliary boundary value problems closely related to the spectral problem (3.14), (3.15).

Problem 1. *Given a function $\zeta \in L_{2,\Gamma} := L_2(\Gamma) \ominus \{1\}$, it is required to find the function $\Phi_1(x) \in H^1_\Gamma(\Omega_1)$ as a solution of the boundary value problem*

$$\Delta\Phi_1 = 0 \quad in\ \Omega_1, \qquad \frac{\partial\Phi_1}{\partial n} = 0 \quad on\ S_1,$$
$$\frac{\partial\Phi_1}{\partial n_1} = \zeta \quad on\ \Gamma, \qquad \int_\Gamma \zeta d\Gamma = 0, \qquad \int_\Gamma \Phi_1 d\Gamma = 0. \tag{3.17}$$

As it follows from the reasoning in Section 1.8 (see Section 1.8.6 and problem (1.8.40)), problem (3.17) has a unique generalized solution $\Phi_1(x) = T_1\zeta$ in $H^1_{h,S_1}(\Omega_1) \subset H^1_\Gamma(\Omega_1)$ and $\Phi_1(x)|_\Gamma = \gamma_1 T_1\zeta = C_1\zeta$, where C_1 is a compact positive operator acting on $H = L_2(\Gamma) \ominus \{1\}$ and γ_1 is the trace operator on Γ. Here,

$$\int_{\Omega_1} \nabla\Phi_1 \cdot \nabla\Phi d\Omega_1 = \int_\Gamma \zeta\gamma_1\Phi d\Gamma, \qquad \text{for all } \Phi \in H^1_\Gamma(\Omega_1). \tag{3.18}$$

Problem 2. *Given a function $\zeta(\hat{x}) \in L_{2,\Gamma}$, it is required to find the function $\Phi_{21}(x) \in H^1_{\Omega_2}$ where*

$$H^1_{\Omega_2} := \left\{\Phi(x) \in L_2(\Omega_2) : \int_{\Omega_2} \Phi d\Omega_2 = 0\right\}, \qquad (\Phi,\Psi)_{H^1_{\Omega_2}} := \int_{\Omega_2} \nabla\Phi \cdot \nabla\Psi d\Omega_2$$

as a solution of the boundary value problem

$$\Delta\Phi_{21} = 0 \quad in\ \Omega_2, \qquad \frac{\partial\Phi_{21}}{\partial n} = 0 \quad on\ S_2,$$
$$\frac{\partial\Phi_{21}}{\partial n_1} = -\zeta \quad on\ \Gamma, \qquad \int_\Gamma \zeta d\Gamma = 0, \qquad \int_{\Omega_2} \Phi_{21} d\Omega_2 = 0. \tag{3.19}$$

Similarly to Problem 1, there exists a unique generalized solution $\Phi_{21}(x) = T_2\zeta$ in $H^1_{h,S_2}(\Omega_2) \subset H^1_{\Omega_2}$; here,

$$\int_{\Omega_2} \nabla\Phi_{21} \cdot \nabla\Phi d\Omega_2 = -\int_{\Gamma} \zeta(\gamma_2\Phi) d\Gamma, \qquad \text{for all } \Phi \in H^1_{\Omega_2}, \tag{3.20}$$

where $\gamma_2\Phi := \Phi|_\Gamma$ for each $\Phi \in H^1_{\Omega_2}$.

PROBLEM 3. *Given a function* $f(x) \in L_{2,\Omega_2}$, *it is required to find the function* $\Phi_{22}(x) \in H^1_{\Omega_2}$ *as a solution of the problem*

$$\begin{aligned} &-\Delta\Phi_{22} = f \quad \textit{in } \Omega_2, \qquad \frac{\partial\Phi_{22}}{\partial n} = 0 \quad \textit{on } S_2, \\ &\frac{\partial\Phi_{22}}{\partial n} = 0 \quad \textit{on } \Gamma, \qquad \int_{\Omega_2} f d\Omega_2 = 0, \qquad \int_{\Omega_2} \Phi_{22} d\Omega_2 = 0. \end{aligned} \tag{3.21}$$

As it follows from the Example 2° in Section 1.8.2, problem (3.21) is uniquely solvable and $\Phi_{22} = A^{-1}f \in H^1_{\Omega_2}$, where $A \gg 0$ and $0 < A^{-1} \in \mathcal{S}_\infty$; moreover, $\mathcal{D}(A^{1/2}) = H^1_{\Omega_2}$ and for any $\Phi(x) \in H^1_{\Omega_2}$ we have

$$\int_{\Omega_2} \nabla\Phi_{22} \cdot \nabla\Phi d\Omega_2 = \left(A^{1/2}\Phi_{22}, A^{1/2}\Phi\right)_{L_{2,\Omega_2}} = \int_{\Omega_2} f\Phi d\Omega_2. \tag{3.22}$$

Using the previous auxiliary problems, let us represent the solution of problem (3.14), (3.15) in the form $\Phi_1(x) \in H^1_\Gamma(\Omega_1)$, $\Phi_2(x) = \Phi_{21}(x) + \Phi_{22}(x)$, where $\Phi_1(x)$ is the generalized solution of Problem 1 for $\zeta := \lambda\,(g(\Delta\rho))^{-1}\, P_\Gamma(\rho_1\Phi_1 - \rho_2\Phi_2)$, $\Phi_{21}(x)$ is the generalized solution of Problem 2 with the same $\zeta \in L_{2,\Gamma}$, and $\Phi_{22}(x)$ is the generalized solution of Problem 3 for $f(x) = \lambda c^{-2}\Phi_2(x) \in L_{2,\Omega_2}$. We note that the properties (3.15) are true automatically due to properties of the spaces $L_{2,\Gamma}$, $H^1_\Gamma(\Omega_1)$, L_{2,Ω_2}, and $H^1_{\Omega_2}$. Consequently, the equations and boundary conditions of problem (3.14), (3.15) lead to a system of operator equations,

$$\begin{aligned} &\eta = \lambda c^{-2} A^{-1}\,(\eta + T_2\zeta)\,, \qquad \eta := \Phi_{22}|_{\Omega_2} \in L_{2,\Omega_2}, \\ &\zeta = \lambda\,(g(\Delta\rho))^{-1}\,(-\rho_2 P_\Gamma\gamma_2\eta + (\rho_1 C_1 - \rho_2 P_\Gamma\gamma_2 T_2)\zeta)\,, \end{aligned} \tag{3.23}$$

in the Hilbert space $H := L_{2,\Omega_2} \oplus L_{2,\Gamma}$.

If $(\eta;\zeta)^{\mathrm{t}} \in H$ is a solution to the system (3.23), then from the first equation it follows that $\eta \in \mathcal{D}(A)$. With regard to this fact, to transform (3.23) into a system that has a symmetrical form, we multiply the first equation by ρ_2 and make the substitutions

$$\rho_2^{1/2} cA\eta = \psi, \qquad (g(\Delta\rho))^{1/2}\,\zeta = \psi. \tag{3.24}$$

With these new variables we obtain the following system of equations :

$$\psi = \lambda\left(c^{-2}A^{-1}\psi + \rho_2^{1/2}\left(g(\Delta\rho)\right)^{1/2}c^{-1}A^{1/2}\left(A^{1/2}T_2\right)\varphi\right),$$
$$\varphi = \lambda\Big(-\rho_2^{1/12}\left(g(\Delta\rho)\right)^{-1/2}c^{-1}P_\Gamma\left(\gamma_2 A^{-1/2}\right)A^{-1/2}\psi$$
$$+\left(g(\Delta\rho)\right)^{-1}(\rho_1 C_1 - \rho_2 P_\Gamma\gamma_2 T_2)\varphi\Big).$$

In its turn, this system is equivalent to the equation

$$y = \lambda\mathcal{A}y, \qquad y = (\psi;\varphi)^{\mathrm{t}} \in H = L_{2,\Omega_2} \oplus L_{2,\Gamma}, \tag{3.25}$$

with the operator matrix

$$\mathcal{A} = \begin{pmatrix} c^{-2}A^{-1} & \rho_2^{1/2}(g\Delta\rho)^{-1/2}A^{-1/2}\left(A^{1/2}T_2\right) \\ -\rho_2^{1/2}\left(g(\Delta\rho)\right)^{-1/2}c^{-1}P_\Gamma\left(\gamma_2A^{-1/2}\right)A^{-1/2} & \left(g(\Delta\rho)\right)^{-1}(\rho_1C_1 - \rho_2P_\Gamma\gamma_2T_2) \end{pmatrix}. \tag{3.26}$$

5.3.4 Theorem on Spectrum

We next show that operator $\mathcal{A}$ acting in H is compact and positive. First of all we note that the properties $0 < C_1 \in \mathcal{S}_\infty$ (Problem 1) and $0 < A^{-1} \in \mathcal{S}_\infty$ (Problem 3) hold true. Further, it is easy to prove that for $C_2 := -P_\Gamma\gamma_2T_2$ the properties $0 < C_2 \in \mathcal{S}_\infty$ are also true. Indeed, let ζ and ζ_1 be from $L_{2,\Gamma}$. Setting $\Phi_{21} = T_2\zeta$ and $\Phi = T_2\zeta_1$ in (3.20) we obtain ($\zeta = P_\Gamma\zeta$):

$$\begin{aligned}(\zeta, C_2\zeta_1)_{L_{2,\Gamma}} &= -\int_\Gamma \zeta\,(P_\Gamma\gamma_2T_2\zeta_1)\,d\Gamma \\ &= -\int_\Gamma \zeta\,(\gamma_2T_2\zeta_1)\,d\Gamma \\ &= \int_{\Omega_2} \nabla\Phi_{21}\cdot\nabla\Phi d\Omega_2.\end{aligned} \tag{3.27}$$

The latter form implies the symmetry of operator C_2. Moreover, setting $\zeta_1 = \zeta$ we have

$$(\zeta, C_2\zeta)_{L_{2,\Gamma}} = \int_{\Omega_2} |\nabla\Phi_{21}|^2 d\Omega_2 \geq 0. \tag{3.28}$$

However, if $(\zeta, C_2\zeta)_{L_{2,\Gamma}} = 0$, from (3.28) it follows that $\Phi_{21} = T_2\zeta = 0$ and therefore, $\zeta = 0$.

Now, let us prove that the operators $A^{1/2}T_2 : L_{2,\Gamma} \to L_{2,\Omega_2}$ and $-P_\Gamma\left(\gamma_2A^{-1/2}\right) : L_{2,\Omega_2} \to L_{2,\Gamma}$ in (3.26) are mutually adjoint and compact, hence $\mathcal{A} = \mathcal{A}^* \in \mathcal{S}_\infty$. If $\Phi = A^{-1/2}\eta \in H^1_{\Omega_2}$, $\eta \in L_{2,\Omega_2}$, from (3.20) in conjunction with equation (3.22) we obtain

$$\begin{aligned}\left(A^{1/2}T_2\zeta,\eta\right)_{L_{2,\Omega_2}} &= -\left(\zeta,\gamma_2 A^{-1/2}\eta\right)_{L_{2,\Gamma}} \\ &= -\left(\zeta, P_\Gamma\gamma_2 A^{-1/2}\eta\right)_{L_{2,\Gamma}} \\ &= \left(\zeta, -P_\Gamma\left(\gamma_2 A^{-1/2}\right)\eta\right)_{L_{2,\Gamma}}, \qquad \zeta\in L_{2,\Gamma},\ \eta\in L_{2,\Omega_2}, \end{aligned} \tag{3.29}$$

that is, $\left(A^{1/2}T_2\right)^* = -P_\Gamma\left(\gamma_2 A^{-1/2}\right)$. Since operator $A^{-1/2}$ acts from L_{2,Ω_2} to $H^1_{\Omega_2}$, operator γ_2 acts from $H^1_{\Omega_2}$ to $H^{1/2}_\Gamma$, and $H^{1/2}_\Gamma$ is compactly embedded into $L_{2,\Gamma}$, we get that $\gamma_2 A^{-1/2}$ and $-P_\Gamma\left(\gamma_2 A^{-1/2}\right)$ are compact operators. Therefore, $A^{1/2}T_2 = \left(-P_\Gamma(\gamma_2 A^{-1/2})\right)^*$ is a compact operator, too.

Hence, operator $\mathcal{A}$ is compact and self-adjoint. We next prove that this operator is positive. For any $y=(\psi,\varphi)^{\mathrm{t}}\in H$ we have

$$\begin{aligned}(\mathcal{A}y,y)_H &= c^{-2}(A^{-1}\psi,\psi)_{L_{2,\Omega_2}} + \rho_2^{1/2}(g(\Delta\rho))^{-1/2}c^{-1}(T_2\varphi,\psi)_{L_{2,\Omega_2}} \\ &- \rho_2^{1/2}(g(\Delta\rho))^{-1/2}c^{-1}\left(P_\Gamma(\gamma_2 A^{-1})\psi,\varphi\right)_{L_{2,\Gamma}} \\ &+ (g(\Delta\rho))^{-1}\left(\rho_1(C_1\varphi,\varphi)_{L_{2,\Gamma}} + \rho_2(C_2\varphi,\varphi)_{L_{2,\Gamma}}\right).\end{aligned}$$

Performing the inverse substitutions (3.24) we have

$$\begin{aligned}(\mathcal{A}y,y)_H = \rho_2&\left[(A\eta,\eta)_{L_{2,\Omega_2}} + (T_2\zeta,A\eta)_{L_{2,\Omega_2}} - (P_\Gamma\gamma_2\eta,\zeta)_{L_{2,\Gamma}}\right. \\ &\left.+(C_2\zeta,\zeta)_{L_{2,\Gamma}}\right] + \rho_1(C_1\zeta,\zeta)_{L_{2,\Gamma}}.\end{aligned} \tag{3.30}$$

Further, we use the properties

$$\begin{aligned}(A\eta,\eta)_{L_{2,\Omega_2}} &= (A^{1/2}\eta,A^{1/2}\eta)_{L_{2,\Omega_2}} = \int_{\Omega_2}|\nabla\Phi_{22}|^2 d\Omega_2, \\ (T_2\zeta,A\eta)_{L_{2,\Omega_2}} &= (A^{1/2}T_2\zeta,A^{1/2}\eta)_{L_{2,\Omega_2}} = \int_{\Omega_2}\nabla\Phi_{21}\cdot\nabla\Phi_{22} d\Omega_2, \\ -(P_\Gamma\gamma_2\eta,\zeta)_{L_{2,\Gamma}} &= \int_{\Omega_2}\nabla\Phi_{22}\cdot\nabla\Phi_{21} d\Omega_2, \\ (C_2\zeta,\zeta)_{L_{2,\Gamma}} &= \int_{\Omega_2}|\nabla\Phi_{21}|^2 d\Omega_2, \\ (C_1\zeta,\zeta)_{L_{2,\Gamma}} &= \int_{\Omega_1}|\nabla\Phi_1|^2 d\Omega_1,\end{aligned}$$

which follow from (3.22), (3.20) and (3.28), respectively. Finally, from (3.30) we obtain

$$(\mathcal{A}y,y)_H = \rho_1\int_{\Omega_1}|\nabla\Phi_1|^2 d\Omega_1 + \rho_2\int_{\Omega_2}|\nabla\Phi_2|^2 d\Omega_2,\ \Phi_2 = \Phi_{21}+\Phi_{22}; \tag{3.31}$$

the property $\mathcal{A} > 0$ follows from the latter equation.

To conclude, we observe that equation (3.25) amounts to the problem of determinig the eigenvalues λ^{-1} and eigenelements y of the compact positive operator $\mathcal{A}$. According to the Hilbert theorem, this problem has a discrete spectrum $\{\lambda_k\}_{k=1}^{\infty}$, with $0 < \lambda_1 \leq \lambda_2 \leq \ldots \leq \lambda_k \leq \ldots$, $\lambda_k \to +\infty$ as $k \to \infty$, and a system of eigenelements $\{y_k\}_{k=1}^{\infty}$ which form an orthogonal basis in $H = L_{2,\Omega_2} \oplus L_{2,\Gamma}$,

$$(y_k, y_m)_H = \delta_{km}, \qquad (\mathcal{A}y_k, y_m)_h = \lambda_k^{-1}\delta_{km}. \tag{3.32}$$

5.3.5 Variational Principles for Eigenvalues

The eigenvalues of problem (3.25) are successive minima of the variational ratio $(y,y)_H/(\mathcal{A}y,y)_H$. For $y = (\psi;\varphi)^t \in H = L_{2,\Omega_2} \oplus L_{2,\Gamma}$, after performing the inverse substitutions introduced above we have

$$\begin{aligned}(y,y)_H &= \|\psi\|^2_{L_{2,\Omega_2}} + \|\varphi\|^2_{L_{2,\Gamma}} = \rho_2 c^2 \|A\eta\|^2_{L_{2,\Omega_2}} + g(\Delta\rho)\|\zeta\|^2_{L_{2,\Gamma}} \\ &= \rho_2 c^2 \int_{\Omega_2} |\Delta\Phi_{22}|^2 d\Omega_2 + g(\Delta\rho)\int_\Gamma \left|\frac{\partial\Phi_1}{\partial n_1}\right|^2 d\Gamma \\ &= \rho_2 c^2 \int_{\Omega_2} |\Delta\Phi_2|^2 d\Omega_2 + g(\Delta\rho)\int_\Gamma \left|\frac{\partial\Phi_1}{\partial n_1}\right|^2 d\Gamma;\end{aligned} \tag{3.33}$$

in the last step we used the properties $\Phi_2 = \Phi_{21} + \Phi_{22}$ and $\Delta\Phi_{21} = 0$ in Ω_2.

From (3.33) and (3.31) we can easily deduce that the eigenvalues λ of the problem (3.14), (3.15) are successive minima of the variational ratio

$$F_2\{\Phi_1;\Phi_2\} := \frac{g(\Delta\rho)\int_\Gamma \left|\frac{\partial\Phi_1}{\partial n_1}\right|^2 d\Gamma + \rho_2 c^2 \int_{\Omega_2} |\Delta\Phi_2|^2 d\Omega_2}{\rho_1 \int_{\Omega_1} |\nabla\Phi_1|^2 d\Omega_1 + \rho_2 \int_{\Omega_2} |\nabla\Phi_2|^2 d\Omega_2}, \tag{3.34}$$

which is considered for the set of functions

$$\Phi_1(x) \in H^1_{h,S_1}(\Omega_1) \subset H^1_\Gamma(\Omega_1), \qquad \Phi_2(x) \in H^1_{\Omega_2}.$$

The eigenvalues λ can be also calculated as successive minima of the variational ratio (3.16) for the functional $F_1\{\Phi_1;\Phi_2\}$. Indeed, by performing the substitution $\mathcal{A}y = z$ in (3.25) we obtain the eigenvalue problem

$$\mathcal{A}^{-1}z = \lambda z \tag{3.35}$$

for the operator $\mathcal{A}^{-1} \gg 0$ with the compact inverse operator $\mathcal{A}$. Therefore, its eigenvalues can be calculated as successive minima of the variational ratio

$$\frac{(\mathcal{A}^{-1}z,z)_H}{(z,z)_H} = \frac{(y,\mathcal{A}y)_H}{(\mathcal{A}y,\mathcal{A}y)_H}. \tag{3.36}$$

Direct calculations show that

$$(\mathcal{A}y, \mathcal{A}y)_H = \rho_2 c^{-2} \int_{\Omega_2} |\Phi_2|^2 \mathrm{d}\Omega_2 + (g(\Delta\rho))^{-1} \int_\Gamma |P_\Gamma(\rho_1\Phi_1 - \rho_2\Phi_2)|^2 \mathrm{d}\Gamma. \qquad (3.37)$$

The variational principle (3.16) follows from (3.31), (3.36), (3.37). Using it we can prove the following asymptotic formula,

$$\lambda_k = \frac{2g(\Delta\rho)(\operatorname{mes}\Gamma)^{1/2}}{[(\rho_1+\rho_2)\pi^{1/2}]} \cdot k^{1/2}[1+o(1)], \qquad k \to \infty. \qquad (3.38)$$

This formula demonstrates that the main contribution to the asymptotic behavior of eigenvalues is provided by boundary waves on the boundary Γ and not by acustic waves propagating in Ω_2. In other words, here the asymptotic behavior of eigenvalues is the same as for a system of two heavy incompressible fluids which completely fill the container Ω (see Section 3.6).

In conclusion, we want to point out that the operator approach applied above to the problem on normal oscillation of a system "fluid-gas" can be also applied to the initial boundary value problem (3.5)–(3.12). Instead of (3.25) we have the Cauchy problem

$$\frac{\mathrm{d}^2}{\mathrm{d}t^2}\mathcal{A}y + y = 0, \qquad y(0) = y^0, \qquad y'(0) = y^1. \qquad (3.39)$$

As it follows from the results of Section 1.5.8, this problem has a generalized solution with finite full energy provided that at the initial moment of time $t = 0$ in problem (3.5)–(3.12) the kinetic and potential energies of the system are finite. In this case, the law of full energy conservation holds true for the solutions of problem (3.39) and (3.5)–(3.12), namely,

$$\left\| \mathcal{A}^{1/2}\frac{\mathrm{d}y}{\mathrm{d}t} \right\|_H^2 + \|y(t)\|_H^2 = \text{const} = \left\| \mathcal{A}^{1/2} y^1 \right\|_H^2 + \|y^0\|_H^2. \qquad (3.40)$$

Chapter 6

Oscillations of an Ideal Rotating Fluid

This chapter will address problems on oscillations of an ideal fluid in an arbitrary wholly or partially filled rotating container, which are interesting from both a mathematical and a physical point of view.

We will investigate evolution and spectral problems with particular emphasis on finding the frequencies and modes of surface waves.

6.1 Motion of Fluid in a Rotating Container

In this chapter, as in Chapter 3, we begin the study of problems on oscillations of rotating fluids with the most simple example of a problem of this type.

6.1.1 Statement of the Problem and the Main Equations

Let us assume that a rigid container Ω is wholly filled with an ideal incompressible fluid and rotates uniformly around the axis Ox_3 with the angular velocity $\boldsymbol{\omega}_0 = \omega_0 \boldsymbol{e}_3$. We should note that if the rotation takes place for a long time, then even for very small viscous forces acting in a real fluid, the motion of fluid is close to a uniform rotation. In other words, in the state of relative equilibrium, the system "body + fluid" rotates uniformly with the angular velocity $\boldsymbol{\omega}_0$. Assuming that the gravitation field with acceleration $\boldsymbol{g} = -g\boldsymbol{e}_3$ acts along the axis Ox_3, the equilibrium pressure $P_0 = P_0(x)$, $x = (x_1, x_2, x_3) \in \Omega$, is defined by the formula

$$P_0(x) = -\rho g x_3 + \frac{1}{2}\rho\omega_0^2 \left(x_1^2 + x_2^2\right) + p_a,$$

where p_a is the pressure at the origin of the coordinate system.

Without regard to viscous forces, we will consider small movements of an ideal fluid close to a uniform rotation. In the coordinate system $Ox_1x_2x_3$, which is rigidly connected with the container and rotates with the constant angular velocity $\boldsymbol{\omega}_0 = \omega_0 \boldsymbol{e}_3$, the linearized Euler equations for determining the relative velocity $\boldsymbol{u}(t,x)$ and dynamic pressure $p(t,x)$ have the form

$$\frac{\partial \boldsymbol{u}}{\partial t} - 2\boldsymbol{u} \times \boldsymbol{\omega}_0 + \frac{1}{\rho}\nabla p = \boldsymbol{f}, \qquad \operatorname{div} \boldsymbol{u} = 0 \text{ in } \Omega, \tag{1.1}$$

where $\boldsymbol{f} = \boldsymbol{f}(t,x)$ is a field of small external forces and the second term $2\boldsymbol{u} \times \boldsymbol{\omega}_0$ takes into account the influence of the Coriolis force on fluid particles.

On the solid boundary S of container Ω we have the nonleaking condition

$$u_n = 0 \qquad \text{on } S. \tag{1.2}$$

As the initial moment of time the velocity field is defined as

$$\boldsymbol{u}(0,x) = \boldsymbol{u}^0(x). \tag{1.3}$$

6.1.2 Existence of Solutions

From (1.1) and (1.2) it follows that the field $\boldsymbol{u}$ can be assumed to be an element of the subspace $\boldsymbol{J}_0(\Omega)$. Applying the projection operator P_0 to both sides of the first equation in (1.1), we obtain

$$\frac{\mathrm{d}\boldsymbol{u}}{\mathrm{d}t} - K\boldsymbol{u} = P_0\boldsymbol{f}, \qquad K\boldsymbol{u} := 2P_0(\boldsymbol{u} \times \boldsymbol{\omega}_0). \tag{1.4}$$

(Here, as usual, the transition from the symbol $\partial/\partial t$ to the symbol $\mathrm{d}/\mathrm{d}t$ points out the fact that (1.4) is considered as an abstract differential equation in the Hilbert space $\boldsymbol{J}_0(\Omega)$).

The operator $K : \boldsymbol{J}_0(\Omega) \to \boldsymbol{J}_0(\Omega)$ used in (1.4) is called the *gyroscopic*, or *Coriolis operator*, and has some important properties. First of all, for any $\boldsymbol{u}, \boldsymbol{v} \in \boldsymbol{J}_0(\Omega)$ we have

$$\begin{aligned}\int_\Omega K\boldsymbol{u} \cdot \boldsymbol{v}\mathrm{d}\Omega &= 2\int_\Omega P_0(\boldsymbol{u} \times \boldsymbol{\omega}_0) \cdot \boldsymbol{v}\mathrm{d}\Omega = 2\int_\Omega (\boldsymbol{u} \times \boldsymbol{\omega}_0) \cdot \boldsymbol{v}\mathrm{d}\Omega \\ &= -2\int_\Omega \boldsymbol{u} \cdot (\boldsymbol{v} \times \boldsymbol{\omega}_0)\mathrm{d}\Omega = -2\int_\Omega \boldsymbol{u} \cdot P_0(\boldsymbol{v} \times \boldsymbol{\omega}_0)\mathrm{d}\Omega \\ &= -\int_\Omega \boldsymbol{u} \cdot K\boldsymbol{v}\mathrm{d}\Omega.\end{aligned}$$

Therefore, K is skew-symmetric, that is, $K^* = -K$. Obviously, K is a bounded operator because

$$\begin{aligned}\|K\boldsymbol{u}\|^2_{\boldsymbol{J}_0(\Omega)} =&\|2P_0\left(\boldsymbol{u}\times\boldsymbol{\omega}_0\right)\|^2_{\boldsymbol{J}_0(\Omega)}\\ =&4\omega_0^2\int_\Omega |P_0\left(\boldsymbol{u}\times\boldsymbol{\omega}_0\right)|^2\mathrm{d}\Omega\\ \le&4\omega_0^2\int_\Omega |\boldsymbol{u}|^2\mathrm{d}\Omega,\end{aligned}$$

hence,

$$\|K\| \le 2\omega_0. \tag{1.5}$$

(Further, we will assume that $\omega_0 > 0$; a transition to the case $\omega_0 < 0$ is not difficult.)

As mentioned in Section 1.5, a differential equation of the form (1.4) with a bounded operator K, an initial condition (1.3) with $\boldsymbol{u}^0 \in \boldsymbol{J}_0(\Omega)$, and a continuous function $(P_0\boldsymbol{f})(t)$, has a continuous solution that can be written down in the form

$$\boldsymbol{u}(t) = \mathrm{e}^{tK}\boldsymbol{u}^0 + \int_0^t \mathrm{e}^{(t-\tau)K}(P_0\boldsymbol{f})(\tau)\mathrm{d}\tau. \tag{1.6}$$

Now we apply the operator $I - P_0$ to both sides of (1.1) and get

$$-2(I-P_0)\left(\boldsymbol{u}\times\boldsymbol{\omega}_0\right) + \frac{1}{\rho}\nabla p = (I-P_0)\boldsymbol{f}. \tag{1.7}$$

Hence, the field of pressures ∇p can be obtained, by determing the field $(I-P_0)(\boldsymbol{u}\times\boldsymbol{\omega}_0)$ for every t. Let us note that ∇p is a continuous function of t with values in $\boldsymbol{G}(\Omega)$ whenever $\boldsymbol{f}(t)$ is continuous in $\boldsymbol{L}_2(\Omega)$ and $\boldsymbol{u}^0 \in \boldsymbol{J}_0(\Omega)$.

Summing up the previous comments on the evolution problem (1.1)–(1.3), we conclude that for any continuous field $\boldsymbol{f}(t) \in \boldsymbol{L}_2(\Omega)$ and $\boldsymbol{u}^0 \in \boldsymbol{J}_0(\Omega)$ that problem has a solution for which $\boldsymbol{u}(t)$ is continuous in $\boldsymbol{J}_0(\Omega)$ and $(\nabla p)(t)$ is continuous in $\boldsymbol{G}(\Omega)$. Let us also note, that for problem (1.1)–(1.3) in the presence of gyroscopic forces (and also without them, as in Section 3.1), the law of kinetic energy balance takes place,

$$\frac{1}{2}\rho\int_\Omega |\boldsymbol{u}|^2\mathrm{d}\Omega = \frac{1}{2}\rho\int_\Omega |\boldsymbol{u}^0|^2\mathrm{d}\Omega + \rho\int_0^t\left(\int_\Omega \boldsymbol{f}(\tau)\cdot\boldsymbol{u}(\tau)\mathrm{d}\Omega\right)\mathrm{d}\tau. \tag{1.8}$$

Indeed, as far as

$$\int_\Omega K\boldsymbol{u}\cdot\boldsymbol{u}\mathrm{d}\Omega = 2\int_\Omega\left(\boldsymbol{u}\times\boldsymbol{\omega}_0\right)\cdot\boldsymbol{u}\mathrm{d}\Omega = 0, \tag{1.9}$$

then after a scalar multiplication of (1.1) by $\boldsymbol{u}$ (in $\boldsymbol{L}_2(\Omega)$) we obtain equation (3.2.6).

6.1.3 Normal Oscillations

Let us assume in (1.1) that $\boldsymbol{f}(t,x) \equiv \mathbf{0}$ and determine the proper oscillations of an ideal rotating fluid. First, it is useful to complexificate all the function Hilbert spaces by assuming their elements are complex-valued functions.

Since there is no energy dissipation in an ideal fluid, it is natural to look for normal solutions of the form

$$\boldsymbol{u}(t,x) = \mathrm{e}^{i\omega t}\boldsymbol{u}(x), \;\; p(t,x) = \mathrm{e}^{i\omega t}p(x), \tag{1.10}$$

where ω is a frequency of oscillations. Substitutions of (1.10) into (1.1), (1.2) lead to the problem

$$\begin{gathered} i\omega\boldsymbol{u} - 2\boldsymbol{u}\times\boldsymbol{\omega}_0 + \frac{1}{\rho}\nabla p = \mathbf{0}, \\ \operatorname{div}\boldsymbol{u} = 0 \quad \text{in } \Omega, \qquad u_n = 0 \quad \text{on } S, \end{gathered} \tag{1.11}$$

and equation (1.4) gives

$$2P_0\left(\boldsymbol{u}\times\boldsymbol{\omega}_0\right) =: K\boldsymbol{u} = i\omega\boldsymbol{u}, \qquad \boldsymbol{u}\in \boldsymbol{J}_0(\Omega). \tag{1.12}$$

Since operator K remains skew-symmetric after complexification, its spectrum is purely imaginary, and in virtue of (1.5), it is located in the interval $[-2i\omega_0, 2i\omega_0]$. Hence, the frequencies of proper oscillations can only be real and $|\omega| \le 2\omega_0$.

It turns out that the spectrum $\sigma(K)$ of operator K fills the whole interval,

$$\sigma(K) = [-2i\omega_0, 2i\omega_0]. \tag{1.13}$$

To prove this quite important fact, it is sufficient to show that for each ω, $0 < |\omega| < 2\omega_0$, there exists a sequence of fields $\{\boldsymbol{u}_l\}_{l=1}^{\infty}$ (Weyl sequence) such that

$$\|K\boldsymbol{u}_l - i\omega\boldsymbol{u}_l\|_{\boldsymbol{J}_0(\Omega)}/\|\boldsymbol{u}_l\|_{\boldsymbol{J}_0(\Omega)} \to 0, \qquad l\to\infty. \tag{1.14}$$

To this end, we consider a finite function $q(x)$ defined in Ω and a complex parameter a. Let us define the field

$$\boldsymbol{u} = \left(a\frac{\partial\Delta q}{\partial x_3}; \frac{\partial\Delta q}{\partial x_3}; -a\frac{\partial\Delta q}{\partial x_1} - \frac{\partial\Delta q}{\partial x_2}\right)^{\mathrm{t}}. \tag{1.15}$$

Obviously, $\operatorname{div}\boldsymbol{u} = 0$ and

$$\boldsymbol{u}\times\boldsymbol{e}_3 = \left(\frac{\partial\Delta q}{\partial x_3}; -a\frac{\partial\Delta q}{\partial x_3}; 0\right)^{\mathrm{t}}.$$

In order to define the potential component of the field $\boldsymbol{u} \times \boldsymbol{e}_3$, it is necessary to solve the boundary value problem

$$\Delta\varphi = \operatorname{div}(\boldsymbol{u} \times \boldsymbol{e}_3) \quad \text{in } \Omega, \qquad \frac{\partial\varphi}{\partial n} = (\boldsymbol{u} \times \boldsymbol{e}_3)_n \quad \text{on } S. \tag{1.16}$$

Then $P_0(\boldsymbol{u} \times \boldsymbol{e}_3) = \boldsymbol{u} \times \boldsymbol{e}_3 - \nabla\varphi$. Problem (1.16) can be solved in an explicit form. Indeed,

$$\operatorname{div}(\boldsymbol{u} \times \boldsymbol{e}_3) = \frac{\partial^2 \Delta q}{\partial x_1 \partial x_3} - a\frac{\partial^2 \Delta q}{\partial x_2 \partial x_3},$$

hence the function

$$\varphi = \frac{\partial^2 q}{\partial x_1 \partial x_3} - a\frac{\partial^2 q}{\partial x_2 \partial x_3}$$

satisfies the equation from (1.16) in Ω. The boundary condition on S holds true in virtue of the finiteness of function q and the field $\boldsymbol{u} \times \boldsymbol{e}_3$. With regard to the fact that $\boldsymbol{\omega}_0 = \omega_0 \boldsymbol{e}_3$ we obtain

$$\begin{aligned}\frac{1}{\omega_0} K\boldsymbol{u} &= 2P_0(\boldsymbol{u} \times \boldsymbol{e}_3) \\ &= 2\left(\frac{\partial \Delta q}{\partial x_3} - \frac{\partial^3 q}{\partial x_1^2 \partial x_3} + a\frac{\partial^3 q}{\partial x_1 \partial x_2 \partial x_3}; -a\frac{\partial \Delta q}{\partial x_3} - \frac{\partial^3 q}{\partial x_1 \partial x_2 \partial x_3} + a\frac{\partial^3 q}{\partial x_2^2 \partial x_3};\right. \\ &\quad \left. -\frac{\partial^3 q}{\partial x_1 \partial x_3^2} + a\frac{\partial^3 q}{\partial x_2 \partial x_3^2}\right)^{\mathrm{T}}.\end{aligned} \tag{1.17}$$

Now, we choose $q = q(x)$ in a special way, namely,

$$q = q_{mn}(x) = \mathrm{e}^{i(nx_1 + mx_3)}\psi(x) \qquad n, m = 1, 2, \ldots, \tag{1.18}$$

where $\psi(x)$ does not depend on n and m and, moreover, $\psi(x)$ is a finite function, which equals 1 on a certain sphere $|\boldsymbol{r} - \boldsymbol{r}_0| \leq \rho$ located in Ω. We assume that n and m increase unboundedly and, in addition, impose a certain relationship between their values that will be indicated later on.

According to (1.15), we get the family of fields

$$\boldsymbol{u}_{mn} = \left(a\frac{\partial \Delta q_{nm}}{\partial x_3}; \frac{\partial \Delta q_{nm}}{\partial x_3}; -a\frac{\partial \Delta q_{nm}}{\partial x_1} - \frac{\partial \Delta q_{nm}}{\partial x_3}\right)^{\mathrm{T}}. \tag{1.19}$$

Obviously, differentiation of the function q_{nm} with respect to x_1 and x_3 introduces the increasing multipliers n and m, respectively, whereas differentiation with respect x_2 does not produce such multipliers. Taking this into account and separing terms in (1.19), we obtain

$$\begin{aligned}\boldsymbol{u}_{nm} &= -i\left(am(n^2 + m^2); m(m^2 + n^2); -an(n^2 + m^2)\right)^{\mathrm{T}} \cdot \mathrm{e}^{i(nx_1 + mx_3)}\psi(x) \\ &\quad + \mathrm{O}(n^2 + m^2).\end{aligned} \tag{1.20}$$

Here $\mathrm{O}(n^2+m^2)$ is a quantity for which the ratio $\mathrm{O}(n^2+m^2)/(n^2+m^2)$ is uniformly bounded as $x \in \Omega$ for all n and m.

Performing a similar calculation for $(1/\omega_0)K\boldsymbol{u}_{nm}$, from (1.17) we obtain

$$\frac{1}{\omega_0}K\boldsymbol{u}_{nm} = -2\mathrm{i}\left(m^3; -am(n^2+m^2); -nm^2\right)^{\mathrm{T}} \cdot \mathrm{e}^{\mathrm{i}(nx_1+mx_3)}\psi(x) + \mathrm{O}(n^2+m^2).$$

Now we take the sum $(1/\omega_0)K\boldsymbol{u}_{nm} + 2a\boldsymbol{u}_{nm}$ and choose the number a such that all the terms of the third order in this combination cancel out. The number $a = a_{nm} = \mathrm{i}m/(n^2+m^2)^{1/2}$ for instance has this property. Then

$$\frac{1}{\omega_0}K\boldsymbol{u}_{nm} + \frac{2\mathrm{i}m}{(n^2+m^2)^{1/2}}\boldsymbol{u}_{nm} = \mathrm{O}(n^2+m^2). \tag{1.21}$$

Let us estimate the norms of the fields $\boldsymbol{u}_{nm}$ from (1.20) with the chosen $a = a_{nm}$. Calculations show that

$$|\boldsymbol{u}_{nm}|^2 = 2m^2(n^2+m^2)^2 + \mathrm{O}\left((n^2+m^2)^2\right). \tag{1.22}$$

Hence we get the estimate

$$|\boldsymbol{u}_{nm}|^2 \le c_1(n^2+m^2)^3. \tag{1.23}$$

Further, the function $\psi(x)$ is constant and equal to 1 on the sphere $|\boldsymbol{r} - \boldsymbol{r}_0| \le \rho$. Thus, for points of this sphere, $\mathrm{O}((n^2+m^2)^2) = 0$ in (1.22) and, therefore, $|\boldsymbol{u}_{nm}|^2 = 2m^2(n^2+m^2)^2$. From the latter equation and (1.23) we obtain the double-sided estimate

$$c_0 m^2(n^2+m^2)^2 \le \|\boldsymbol{u}_{mn}\|^2 \le c_2(n^2+m^2)^3, \qquad c_0 = 8\pi\rho^3/3. \tag{1.24}$$

Now, let $\mu \in (0,2)$ be fixed. For every given $\varepsilon > 0$, we choose a rational number $\bar{n}/\bar{m}$ in such a way that the inequalities $\mu < 2\bar{m}/(\bar{n}^2+\bar{m}^2)^{1/2} < \mu+\varepsilon$ hold true. Then, for all $n = l\bar{n}$ and $m = l\bar{m}$, where $l = 1, 2, \ldots$, we have

$$\mu < \frac{2m}{\sqrt{n^2+m^2}} < \mu + \varepsilon. \tag{1.25}$$

Using (1.21), (1.24), and (1.25), get the next norm estimate

$$\begin{aligned} &\left\|\frac{1}{\omega_0}K\boldsymbol{u}_{nm} + \mathrm{i}\mu\boldsymbol{u}_{nm}\right\|^2 \\ &\le 2\left\{\left\|\frac{1}{\omega_0}K\boldsymbol{u}_{nm} + \frac{2\mathrm{i}m}{(n^2+m^2)^{1/2}}\boldsymbol{u}_{nm}\right\|^2 + \left|\mu - \frac{2m}{(n^2+m^2)^{1/2}}\right|^2 \|\boldsymbol{u}_{nm}\|^2\right\} \\ &\le c_3(n^2+m^2)^2 + 2c_2\varepsilon^2(n^2+m^2)^3. \end{aligned}$$

Then,

$$\frac{\left\|\frac{1}{\omega_0}K\boldsymbol{u}_{nm} + \mathrm{i}\mu\boldsymbol{u}_{nm}\right\|^2}{\|\boldsymbol{u}_{nm}\|^2} \le \frac{c_3(n^2+m^2)^2 + 2c_2\varepsilon^2(n^2+m^2)^3}{c_0 m^2(n^2+m^2)^2}$$
$$= \frac{c_3(\bar{n}^2+\bar{m}^2)^2 l^4 + 2c_2\varepsilon^2(\bar{n}^2+\bar{m}^2)^3 l^6}{c_0\bar{m}^2(\bar{n}^2+\bar{m}^2)^2 l^6}$$

For a sufficiently large l, the right side does not exceed the number $3c_2(\bar{n}^2 + \bar{m}^2)\varepsilon^2/(c_0\bar{m}^2) < 12c_2c_0^{-1}\mu^{-2}\varepsilon^2$, and in virtue of the arbitrary choice of ε, property (1.14) is proved for $\omega = -\mu\omega_0 (0 < \mu < 2)$, i.e. the point $-\mu\omega_0$ belongs to the spectrum of operator K. Similarly, by substituting $-m$ for m we can prove that the point $\mu\omega_0$, $0 < \mu < 2$ belongs to the spectrum, too. As far as the spectrum is a closed set, assertion (1.13) is completely proved.

6.2 Motion of a Gyrostate Similar to Uniform Rotation About a Fixed Axis

In this section we investigate a type of problem first addressed in the celebrated work of S. L. Sobolev.

6.2.1 Statement of the Problem

Let us consider a body that has a cavity wholly filled with an ideal fluid and moves around a fixed point O. It is assumed that the nonperturbed motion is a uniform rotation of the system "body + fluid" about a fixed axis, which passes through the point O and the mass center C of the system, with the angular velocity $\boldsymbol{\omega}_0$. We should note that if the body rotates for a long time, then due to the presence of viscous forces in a real fluid, such an initial state inevitably takes place. As in Section 6.1, we are going to consider small perturbations of this motion, but now the possible changes of position of the axis of rotation are taken into account, and the fluid is assumed to be ideal.

As far as the considered system is a gyrostate, we can use the linearized equations of motion (3.1.28), (3.1.29), which in this case have the form

$$\boldsymbol{J}\frac{\mathrm{d}\boldsymbol{\omega}}{\mathrm{d}t} + \boldsymbol{\omega}\times\boldsymbol{J}\boldsymbol{\omega}_0 + \boldsymbol{\omega}_0\times\boldsymbol{J}\boldsymbol{\omega} + \frac{\mathrm{d}}{\mathrm{d}t}\left(\rho\int_\Omega(\boldsymbol{r}\times\boldsymbol{u})\mathrm{d}\Omega\right) + \boldsymbol{\omega}_0\times\left(\rho\int_\Omega(\boldsymbol{r}\times\boldsymbol{u})\mathrm{d}\Omega\right)$$
$$= \boldsymbol{M}(t), \tag{2.1}$$

$$\frac{\partial \boldsymbol{u}}{\partial t} + 2\boldsymbol{\omega}_0 \times \boldsymbol{u} + \frac{\mathrm{d}\boldsymbol{\omega}}{\mathrm{d}t} \times \boldsymbol{r} = -\frac{1}{\rho}\nabla p + \boldsymbol{f}, \qquad \operatorname{div} \boldsymbol{u} = 0 \text{ in } \Omega, \tag{2.2}$$

$$u_n = 0 \text{ on } S, \qquad \boldsymbol{u}(0,x) = \boldsymbol{u}^0(x), \ \boldsymbol{\omega}(0) = \boldsymbol{\omega}^0. \tag{2.3}$$

Here, Ω is the region, occupied by an ideal fluid with density ρ, $\boldsymbol{J}$ is the tensor of inertia of the system "body + fluid" relatively to the point O, $\boldsymbol{\omega}_0 = \omega_0 \boldsymbol{e}_3$ is the initial angular velocity that corresponds to the rotation of the gyrostate in the coordinate system $Ox_1x_2x_3$, which is rigidly connected with the body, $\boldsymbol{\omega}(t)$ is the deviation of the angular velocity from the vector $\boldsymbol{\omega}_0$, $\boldsymbol{u}(t,x)$ is the field of relative velocity of the fluid at time t and the point $x = (x_1, x_2, x_3)$, $\boldsymbol{M}(t)$ is a given principal moment of the external forces influencing the body, and $\boldsymbol{f}(t,x)$ is the field of small mass forces.

6.2.2 Transition to the Evolution Equation in a Hilbert Space

As it follows from (2.2) and (2.3), the field $\boldsymbol{u}(t,x)$ can be thought of as a function of variable t with values in $\boldsymbol{J}_0(\Omega)$. Applying the orthoprojector P_0 onto the subspace $\boldsymbol{J}_0(\Omega) \subset \boldsymbol{L}_2(\Omega)$ to equation (2.2), we obtain

$$\frac{\mathrm{d}\boldsymbol{u}}{\mathrm{d}t} + P_0\left(\frac{\mathrm{d}\boldsymbol{\omega}}{\mathrm{d}t} \times \boldsymbol{r}\right) = -2\mathrm{i}\omega_0 T\boldsymbol{u} + P_0\boldsymbol{f}, \tag{2.4}$$

$$T\boldsymbol{u} := \mathrm{i}P_0(\boldsymbol{u} \times \boldsymbol{e}_3), \quad T = T^*, \quad \sigma(T) = [-1, 1] \tag{2.5}$$

We will consider equations (2.1) and (2.4) as a system of ordinary differential equations relative to the column vector $v(t) := (\boldsymbol{u}(t,x); \boldsymbol{\omega}(t))^{\mathrm{t}}$ which is supposed to be a function of t with values in the Hilbert space $H := \boldsymbol{J}_0(\Omega) \oplus \mathbb{R}^3$. (In addition, as in Section 3.5.5, it is assumed that a transition to undimensional variables have been already performed somehow in the system (2.1), (2.4) and, therefore, $v(t) = (\boldsymbol{u}(t,x); \boldsymbol{\omega}(t))^{\mathrm{t}}$ and all the other physical quantities in this section are undimensional.) In short, this system can be written down as a single equation

$$\tilde{I}\frac{\mathrm{d}v}{\mathrm{d}t} = -\omega_0 Bv + \varphi(t), \quad v(0) = v^0, \tag{2.6}$$

where

$$\tilde{I}v := \begin{pmatrix} \boldsymbol{u} + P_0(\boldsymbol{\omega} \times \boldsymbol{r}) \\ \rho\displaystyle\int_\Omega (\boldsymbol{r} \times \boldsymbol{u})\mathrm{d}\Omega + \boldsymbol{J}\boldsymbol{\omega} \end{pmatrix}, \qquad \varphi(t) = \begin{pmatrix} P_0\boldsymbol{f}(t) \\ \boldsymbol{M}(t) \end{pmatrix}, \quad v^0 = \begin{pmatrix} P_0\boldsymbol{u}^0 \\ \boldsymbol{\omega}^0 \end{pmatrix},$$

$$Bv := \begin{pmatrix} 2iT\boldsymbol{u} \\ \boldsymbol{e}_3 \times \left(\rho\displaystyle\int_\Omega (\boldsymbol{r} \times \boldsymbol{u})\mathrm{d}\Omega\right) + \boldsymbol{\omega} \times \boldsymbol{J}\boldsymbol{e}_3 + \boldsymbol{e}_3 \times \boldsymbol{J}\boldsymbol{\omega} \end{pmatrix}. \tag{2.7}$$

Hence, the initial boundary value problem (2.1)–(2.3) is reduced to the Cauchy problem (2.6).

6.2.3 Properties of the Operators in the Problem

First of all, we note that operator B in (2.6) and (2.7) is bounded on $H = \boldsymbol{J}_0(\Omega) \oplus \mathbb{R}^3$, because all its matrix elements are bounded operators.

We next describe the properties of operator $\tilde{I}$. To this end, we define the squared norm of every element v in H by the formula

$$\|v\|_H^2 := \rho \int_\Omega |\boldsymbol{u}|^2 \mathrm{d}\Omega + |\boldsymbol{\omega}|^2 \tag{2.8}$$

and calculate the quadratic form corresponding to operator $\tilde{I}$. We have

$$\begin{aligned} \left(\tilde{I}v, v\right)_H &= \rho \int_\Omega |\boldsymbol{u}|^2 \mathrm{d}\Omega + \rho \int_\Omega P_0(\boldsymbol{\omega} \times \boldsymbol{r}) \cdot \boldsymbol{u} \mathrm{d}\Omega + \rho \int_\Omega (\boldsymbol{r} \times \boldsymbol{u}) \mathrm{d}\Omega \cdot \boldsymbol{\omega} + \boldsymbol{J}\boldsymbol{\omega} \cdot \boldsymbol{\omega} \\ &= (\boldsymbol{J}_b + \boldsymbol{J}_f)\boldsymbol{\omega} \cdot \boldsymbol{\omega} + \rho \int_\Omega |\boldsymbol{u}|^2 \mathrm{d}\Omega + 2\rho \int_\Omega (\boldsymbol{\omega} \times \boldsymbol{r}) \cdot \boldsymbol{u} \mathrm{d}\Omega \\ &= \boldsymbol{J}_b \boldsymbol{\omega} \cdot \boldsymbol{\omega} + \rho \int_\Omega |\boldsymbol{\omega} \times \boldsymbol{r} + \boldsymbol{u}|^2 \mathrm{d}\Omega. \end{aligned} \tag{2.9}$$

Here we used the equation $\boldsymbol{J} = \boldsymbol{J}_b + \boldsymbol{J}_f$ for the tensor of inertia of the whole system and formula (3.1.25) for $\boldsymbol{J}_f$, from which it follows that

$$\boldsymbol{J}_f \boldsymbol{\omega} \cdot \boldsymbol{\omega} = \rho \int_\Omega |\boldsymbol{\omega} \times \boldsymbol{r}|^2 \mathrm{d}\Omega. \tag{2.10}$$

As long as the tensor of inertia $\boldsymbol{J}_b$ of the rigid body without fluid is positive definite on $\mathbb{R}^3$, from (2.9) it follows that operator $\tilde{I}$ is nonnegative. If $(\tilde{I}v, v) = 0$, then $\boldsymbol{\omega} = \boldsymbol{0}$ and, therefore, from (2.9) we obtain that $\boldsymbol{u} \equiv \boldsymbol{0}$. Hence, operator $\tilde{I}$ is positive on H. According to definition (2.7), this operator equals the sum of the positive definite operator $I_0 := \operatorname{diag}(I; J)$ and a finite dimensional operator. Therefore, $\tilde{I}$ is a positive definite bounded operator acting on H. Hence, we get that the inverse operator $\tilde{I}^{-1}$ is also bounded and positive definite on H.

6.2.4 Existence of Solution to the Boundary Value Problem

We apply operator $\tilde{I}^{-1}$ to both sides of equation (2.6) and obtain

$$\frac{\mathrm{d}v}{\mathrm{d}t} = Av + \tilde{I}^{-1}\varphi(t), \qquad v(0) = v^0, \tag{2.11}$$

where $A := -\omega_0 \tilde{I}^{-1} B$ is a bounded operator on H. From Section 1.5, it follows that for every continuous function $\varphi(t)$ from H and any $v^0 \in H$, problem (2.11) has a continuously differentiable solution that can be expressed by the formula

$$v(t) = \mathrm{e}^{tA} v^0 + \int_0^t \mathrm{e}^{(t-\tau)A} \tilde{I}^{-1} \varphi(t) \mathrm{d}\tau, \tag{2.12}$$

where e^{tA} is the group corresponding to operator A.

Hence we obtain the following conclusion.

If the following three conditions hold true:

a) $\boldsymbol{u}^0 \in \boldsymbol{J}_0(\Omega)$, $\omega^0 \in \mathbb{R}^3$;

b) the moment of external forces $\boldsymbol{M}(t)$ is a continuous function on $\mathbb{R}^3$;

c) the field of external forces $\boldsymbol{f}(t,x)$ is a continuous function of t with values in $\boldsymbol{L}_2(\Omega)$,

then the initial boundary value problem (2.1)–(2.3) has a unique solution $\{\boldsymbol{u}(t,x);\boldsymbol{\omega}(t)\}$ that is continuously differentiable in t and takes values from $H = \boldsymbol{J}_0(\Omega)\oplus\mathbb{R}^3$.

Now let us consider a more complicated situation, when the moment of external forces $\boldsymbol{M}(t)$ and the field $\boldsymbol{f}(t,x)$ are unknown, but they depend linearly on the angular displacement $\boldsymbol{\delta}(t)$ of the gyrostate. Then, instead of the function $\varphi(t)$ in (2.11), we should use the expression $F\boldsymbol{\delta}$, where F is a bounded operator acting from $\mathbb{R}^3$ to H. Adding the obvious equation $\mathrm{d}\boldsymbol{\delta}/\mathrm{d}t = \boldsymbol{\omega}(t)$ and the initial condition $\boldsymbol{\delta}(0) = \boldsymbol{\delta}^0$, we obtain the problem

$$\frac{\mathrm{d}}{\mathrm{d}t}\begin{pmatrix} v \\ \boldsymbol{\delta} \end{pmatrix} = \begin{pmatrix} A & \tilde{I}^{-1} & F \\ P_3 & 0 & \end{pmatrix}\begin{pmatrix} v \\ \boldsymbol{\delta} \end{pmatrix}, \qquad \begin{pmatrix} v \\ \boldsymbol{\delta} \end{pmatrix}(0) = \begin{pmatrix} v^0 \\ \boldsymbol{\delta}^0 \end{pmatrix}, \tag{2.13}$$

where P_3 is the orthoprojector onto $\mathbb{R}^3$ on $H = \boldsymbol{J}_0(\Omega)\oplus\mathbb{R}^3$.

From the above mentioned facts it follows that the Cauchy problem (2.13) has a continuously differentiable solution $\{\boldsymbol{u}(t,x);\boldsymbol{\omega}(t);\boldsymbol{\delta}(t)\}$ for any intial data $\boldsymbol{u}^0 \in \boldsymbol{J}_0(\Omega)$, $\boldsymbol{\omega}^0 \in \mathbb{R}^3$, $\delta^0 \in \mathbb{R}^3$.

6.3 Rotation of a Fluid in a Partially Filled Container

In this section and further on in the sequel, the method of orthogonal projection will be systematically used as a main tool that will allows us to omit unnecessary details and to formulate the problem on motion of an ideal rotating fluid in the form of a second order differential equation in a Hilbert space defined as an orthogonal sum of the vortical subspace and the corresponding potential harmonic subspace.

6.3.1 On the Equilibrium State

We suppose that an ideal incompressible fluid partially fills an arbitrary container and, in the nonperturbed state, uniformly rotates with the angular velocity $\boldsymbol{\omega}_0 = \omega_0\boldsymbol{e}_3$, where $\boldsymbol{e}_3$ is the axial vector of the rotation axis Ox_3 (as above, we assume that $\omega_0 > 0$). The coordinate system $Ox_1x_2x_3$ is rigidly connected to the container, and the external stationary field of forces (if gravitational) acts along the axis Ox_3, that is, $\boldsymbol{F}_0 = -\nabla U_0$, $U_0 = gx_3 + \text{const}$.

In the state of relative equilibrium, the pressure $P_0(x)$ in the fluid is distributed according to the following law

$$P_0(x) = -\rho g x_3 + \frac{1}{2}\rho\omega_0^2\left(x_1^2 + x_2^2\right) + c \qquad \text{in } \Omega. \tag{3.1}$$

If the gravitation forces are sufficiently big, that is, the fluid is assumed to be heavy, then the condition $(P_0)_\Gamma - p_a = 0$ holds true on the equilibrium surface Γ, where p_a is an external constant pressure. Hence, we obtain that Γ is a paraboloid of rotation, whose equation has the form

$$x_3 = \frac{1}{2g}\omega_0^2\left(x_1^2 + x_2^2\right) - \frac{c - pa}{\rho g}. \tag{3.2}$$

The constant c in (3.2) can be found by equating the volume of the region Ω filled with fluid with some given value V,

$$\int_\Omega d\Omega = \text{mes}\Omega = V. \tag{3.3}$$

In a weak gravitation field and for a slow rotation of the system, the fluid is assumed to be capillar, that is, the surface forces are taken into account. Then, the shape of the free surface Γ is determined from the Laplace equation

$$(P_0)_\Gamma - p_a = -\sigma(k_1 + k_2) \qquad \text{on } \Gamma; \tag{3.4}$$

the latter condition with regard to (3.1) leads to the following equation

$$-\sigma(k_1 + k_2) = -\rho g x_3 + \frac{1}{2}\rho\omega_0^2(x_1^2 + x_2^2) + c - p_a, \tag{3.5}$$

which is a nonlinear equation in the partial derivatives of the function $x_3 = f(x_1, x_2)$ describing the equation of the free surface Γ. Similarly to the static problem from Section 3.1 on finding the shape of the surface Γ of a nonrotating fluid, the Dupre-Young condition (3.1.4) should be valid on its boundary $\partial\Gamma$.

In principale, conditions (3.3), (3.5), and (3.1.4) make it possible to define the configuration of the region Ω occupied by fluid and, in particular, to find the equation of the free surface Γ based on the given volume V of fluid, characteristics of the three media on the line of contact $\partial\Gamma$, and the intensity of the gravitation field g. Solving this complicated nonlinear problem does not make the objective of this book. Whenever the container is axially symmetric and its axis of symmetry coincides with the axis of rotation, the problem on finding the equilibrium state of the system can be essentially simplified; this issue is considered in detail in the first part of the book by V. G. Babskii, et al. [3].

6.3.2 Statement of the Problem on Small Oscillations

Let us assume that the static problem is already solved and consider the movements of the fluid in a container that are close to the solid rotation. The pressure $P(t,x)$ in fluid can be found in the form $P(t,x) = P_0(x) + p(t,x)$, where $p(t,x)$ is the dynamic pressure. Instead of the small velocity field $\boldsymbol{u}(t,x)$, which describes the motion of the fluid in the coordinate system $Ox_1x_2x_3$ rigidly connected with the uniformly rotating container, we introduce the field of displacements $\boldsymbol{w}(t,x)$. Obviously, this field is connected with the velocity field by the equation

$$\frac{\partial \boldsymbol{w}}{\partial t}(t,x) = \boldsymbol{u}(t,x). \tag{3.6}$$

With regard to (3.6), the linearized Euler equations for an ideal fluid in the coordinate system $Ox_1x_2x_3$ have the form

$$\frac{\partial^2 \boldsymbol{w}}{\partial t^2} - 2\omega_0 \frac{\partial \boldsymbol{w}}{\partial t} \times \boldsymbol{e}_3 + \frac{1}{\rho}\nabla p = \boldsymbol{f}, \quad \operatorname{div} \boldsymbol{w} = 0 \qquad \text{in } \Omega, \tag{3.7}$$

where Ω is the region occupied by fluid in the equilibrium state, and $\boldsymbol{f} = \boldsymbol{f}(t,x)$ is a small field of external forces, superposed on $\boldsymbol{F}_0 = -\nabla U_0$.

The boundary condition on the solid boundary S for an ideal fluid has the obvious form

$$w_n := \boldsymbol{w} \cdot \boldsymbol{n} = 0 \qquad \text{on } S. \tag{3.8}$$

This means that the displacement of fluid particles along the normal $\boldsymbol{n}$ to S equals zero.

The dynamic boundary condition on the moving free surface $\Gamma(t)$ is the Laplace condition

$$P|_{\Gamma(t)} - p_a = -\sigma\left(\tilde{k}_1 + \tilde{k}_2\right) \qquad \text{on } \Gamma(t).$$

Its linearization, which can be derived as in Section 4.1.2, in conjunction with the representation of P as a sum $P_0 + p$, formula (3.1), and formula (3.1.11) for the variation of the mean curvature of the surface Γ, leads to the condition

$$\begin{aligned} &p = \sigma\mathcal{L}\zeta := -\sigma\Delta_\Gamma\zeta + a_\Gamma\zeta, \qquad \zeta := w_n \quad \text{on } \Gamma, \\ &a_\Gamma(\xi^1,\xi^2) := -\sigma\left(k_1^2 + k_2^2\right) + \rho g \cos\left(\widehat{\boldsymbol{n},\boldsymbol{e}_3}\right) - \rho\omega_0^2 \cos\left(\widehat{\boldsymbol{n},\boldsymbol{e}_r}\right), \end{aligned} \tag{3.9}$$

where $r = (x_1^2 + x_2^2)^{1/2}$ is the radius in the cylindrical coordinate system, and $\boldsymbol{e}_r$ is the axial vector of the axis.

As mentioned in Section 4.1, the linearization of the Dupre-Young condition gives the boundary condition

$$\frac{\partial \zeta}{\partial \nu} + \chi \zeta = 0 \quad \text{on } \partial\Gamma. \tag{3.10}$$

Together with the initial conditions

$$\boldsymbol{w}(0,x) = \boldsymbol{w}^0(x), \quad \frac{\partial \boldsymbol{w}}{\partial t}(0,x) = \boldsymbol{u}^0(x), \tag{3.11}$$

the equations and boundary conditions (3.7)–(3.10) define the boundary value problem on small movements of the fluid in a container that are close to a uniform rotation.

By the solenoidality condition (3.7) and the nonleaking condition (3.8), we get the volume-preserving condition on Γ, that is,

$$\int_\Gamma w_n d\Gamma = \int_\Gamma \zeta d\Gamma = 0. \tag{3.12}$$

Hence, it follows that during the process of oscillation, the deviation $w_n = \zeta$ of the free surface of the fluid, which is considered for every fixed t as an element of the space $L_2(\Gamma)$, actually belongs to the subspace $L_{2,\Gamma} = L_2(\Gamma) \ominus \{1\}$. As long as the pressure $p(t,x)$ is also defined up to an arbitrary function of time, the dynamic condition (3.9) can be written down in the form

$$p = \sigma B_1 \zeta, \qquad \zeta = w_n \quad \text{on } \Gamma, \tag{3.13}$$

where $B_1 = P_\Gamma \mathcal{L} P_\Gamma$, P_Γ is the orthoprojector onto $L_{2,\Gamma}$, and $p|_\Gamma \in L_{2,\Gamma}$. The operator B_1 is defined by the differential expression $\mathcal{L}$ and the boundary condition (3.10).

Further, let us consider only such equilibrium states that are stable in linear approximation, that is, states such that the operator B_1, defined on the dense set $\mathcal{D}(B_1)$ in $L_{2,\Gamma}$, is self-adjoint and positive definite, that is, $B_1 \gg 0$.

It can be proved that for the classical solution of the problem (3.7)–(3.11), the law of full energy balance (4.1.10) holds true, where H_B has a new meaning and $\boldsymbol{u} = \partial \boldsymbol{w}/\partial t$. Indeed, by the equation $\boldsymbol{u} \times \boldsymbol{e}_3 \cdot \boldsymbol{u} = 0$, the work of Coriolis forces in the system equals zero and, therefore, the deduction of the balance relation for the considered problem is completely similar to the deduction of formula (4.1.17) for a nonrotating fluid.

6.3.3 Method of Orthogonal Projection

We begin by performing a transition from the classical form (3.7)–(3.11) of the considered problem to its operator form and assume that $\boldsymbol{w}(t,x)$ and $(\nabla p)(t,x)$ are continuous functions of t with values in the Hilbert space $\boldsymbol{L}_2(\Omega)$. Then, by the solenoidality and the nonleaking conditions for $\boldsymbol{w}$ and by using the potentiality of the field ∇p, we obtain $\boldsymbol{w} \in \boldsymbol{J}_0(\Omega) \oplus \boldsymbol{G}_{h,S}(\Omega)$, $\nabla p \in \boldsymbol{G}_{h,S}(\Omega) \oplus \boldsymbol{G}_{0,\Gamma}(\Omega)$, according to the orthogonal decomposition (2.1.24). Therefore, these fields can be expressed as

$$\begin{aligned} \boldsymbol{w} &= \boldsymbol{v} + \nabla\Phi, \qquad && \boldsymbol{v} \in \boldsymbol{J}_0(\Omega),\ \nabla\Phi \in \boldsymbol{G}_{h,S}(\Omega), \\ \nabla p &= \nabla\varphi + \nabla\kappa, \qquad && \nabla\varphi \in \boldsymbol{G}_{h,S}(\Omega), \quad \nabla\kappa \in \boldsymbol{G}_{0,\Gamma}(\Omega). \end{aligned} \tag{3.14}$$

We now introduce the orthoprojectors P_0, P_S, and P_Γ onto $\boldsymbol{J}_0(\Omega)$, $\boldsymbol{G}_{h,S}(\Omega)$, and $G_{0,\Gamma}(\Omega)$, respectively. Substituting (3.14) into the first equation (3.7) and applying the orthoprojectors just defined to the both sides of the resulting equation, we have

$$\frac{d^2\boldsymbol{v}}{dt^2} - 2\omega_0 \frac{d}{dt}\left[P_0(\boldsymbol{v} \times \boldsymbol{e}_3) + P_0(\nabla\Phi \times \boldsymbol{e}_3)\right] = \boldsymbol{f}_0 := P_0\boldsymbol{f}, \tag{3.15}$$

$$\frac{d^2}{dt^2}\nabla\Phi - 2\omega_0 \frac{d}{dt}\left[P_S(\boldsymbol{v} \times \boldsymbol{e}_3) + P_S(\nabla\Phi \times \boldsymbol{e}_3)\right] + \rho^{-1}\nabla\varphi = \nabla F := P_S\boldsymbol{f}, \tag{3.16}$$

$$-2\omega_0 \frac{d}{dt}\left[P_\Gamma(\boldsymbol{v} \times \boldsymbol{e}_3) + P_\Gamma(\nabla\Phi \times \boldsymbol{e}_3)\right] + \rho^{-1}\nabla\kappa = \nabla Q := P_\Gamma\boldsymbol{f}. \tag{3.17}$$

The last equation shows that for known fields $\boldsymbol{v}(t), (\nabla\Phi)(t)$ that can be found based on (3.15) and (3.16), the field $(\nabla\kappa)(t)$ can be directly calculated. Consequently, we may consider only equations (3.15) and (3.16). (Here one can see the advantages of the method of orthogonal projection as mentioned in the introduction to this section.)

For further transformations, let us introduce the notations

$$P_S(\boldsymbol{v} \times \boldsymbol{e}_3) =: \nabla\psi \in \boldsymbol{G}_{h,S}(\Omega), \quad P_S(\nabla\Phi \times \boldsymbol{e}_3) =: \nabla\Psi \in \boldsymbol{G}_{h,S}(\Omega). \tag{3.18}$$

As long as all the fields in (3.16) are assumed to be continuous functions of t with values in $\boldsymbol{G}_{h,S}(\Omega)$, then their potentials are continuous functions of t with values in the space $H^1_{h,S}(\Omega)$, already introduced in Section 1.8.7. Therefore, from (3.16) the next equation for these potentials follows,

$$\frac{d^2\Phi}{dt^2} - 2\omega_0 \frac{d}{dt}(\psi + \Psi) + \rho^{-1}\varphi + F \qquad \text{in } \Omega, \tag{3.19}$$

which is a Cauchy-Lagrange integral for a rotating fluid.

Now, let us note that for $\nabla\kappa \in \boldsymbol{G}_{0,\Gamma}(\Omega)$ the condition $\kappa = 0$ on Γ is satisfied, and for $\boldsymbol{v} \in \boldsymbol{J}_0(\Omega)$ the condition $v_n = \boldsymbol{v} \cdot \boldsymbol{n} = 0$ on S holds true. Therefore, the kinematic and dynamic conditions (3.13) with regard to the representations (3.14) give

$$\varphi = \sigma B_1 \frac{\partial \Phi}{\partial n} \qquad \text{on } \Gamma. \tag{3.20}$$

Considering now equation (3.19) on Γ and taking into account (3.20), we obtain the transformed dynamic condition

$$\rho \frac{\mathrm{d}^2 \Phi}{\mathrm{d}t^2} - 2\omega_0 \rho \frac{\mathrm{d}}{\mathrm{d}t}(\psi + \Psi) + \sigma B_1 \frac{\partial \Phi}{\partial n} = \rho F \qquad \text{on } \Gamma. \tag{3.21}$$

Equations (3.15) and (3.21) contain the field $\boldsymbol{v}$ from the vortical subspace $\boldsymbol{J}_0(\Omega)$ and the potential $\Phi|_\Gamma \in H_\Gamma^{1/2}$ as unknown functions, with $\nabla\Phi \in \boldsymbol{G}_{h,S}(\Omega)$. In addition, the potentials ψ and Ψ can be found in terms of the mentioned unknown functions according to the formulas (3.18). Hence, (3.15) and (3.21) can be considered as a system of evolution equations for $\boldsymbol{v}(t)$ and $(\Phi)_\Gamma(t)$ that can be written down in a different form.

As it follows from the results in Section 1.8, which have been already used in Chapters 3 and 4, the trace $\gamma_\Gamma\Phi = \Phi|_\Gamma$ of the function $\Phi(x)$ from $H^1_{h,S}(\Omega)$ is obtained by applying the operator $C = \gamma_\Gamma T$ to the normal derivative of the function Φ, that is, $\Phi|_\Gamma = C(\partial\Phi/\partial n)_\Gamma$. Here, for $\Phi \in H^1_{h,S}(\Omega)$ we have $\Phi|_\Gamma \in H_\Gamma^{1/2} = G_+$ and $(\partial\Phi/\partial n)_\Gamma \in H_\Gamma^{-1/2} = (H_\Gamma^{1/2})^* = G_+^*$. (Let us note that for $\Phi \in H^1_{h,S}(\Omega)$ the equation $\int_\Gamma \Phi \mathrm{d}\Gamma = 0$ is true and, therefore, in the general scheme of Section 1.8 we have $G = L_{2,\Gamma} = L_2(\Gamma) \ominus \{1\}$). Further, we observe that operator C isometrically maps the space $H_\Gamma^{-1/2}$ on $H_\Gamma^{1/2}$, and $C^{1/2}$ isometrically maps $H_\Gamma^{-1/2}$ on $L_{2,\Gamma}$. As far as the embedding $H^{1/2} = G_+ \hookrightarrow G = L_{2,\Gamma}$ is compact, then operator C acting on $L_{2,\Gamma}$ is compact.

According to (3.14), $(\nabla\Phi)(t)$ is a function with values in $\boldsymbol{G}_{h,S}(\Omega)$, so Φ is a function with values in $H^1_{h,S}(\Omega)$ and, therefore, $(\partial\Phi/\partial n)_\Gamma \in H_\Gamma^{-1/2}$. Then

$$\eta := C^{1/2}\left(\frac{\partial\Phi}{\partial n}\right)_\Gamma = C^{-1/2}(\Phi)_\Gamma \tag{3.22}$$

is a function with values in $L_{2,\Gamma}$. It can be assumed that $\nabla\Phi = V\eta$, where V is an isometric operator that acts from $L_{2,\Gamma}$ to $\boldsymbol{G}_{h,S}(\Omega)$. Indeed,

$$\begin{aligned}(\eta,\eta)_{L_{2,\Gamma}} &= \left(C^{-1/2}(\Phi)_\Gamma,\ C^{1/2}\left(\frac{\partial\Phi}{\partial n}\right)_\Gamma\right)_{L_{2,\Gamma}} = \int_\Gamma \Phi \frac{\partial\Phi}{\partial n}\mathrm{d}\Gamma \\ &= \int_\Omega |\nabla\Phi|^2 \mathrm{d}\Omega = (\nabla\Phi,\nabla\Phi)_{\boldsymbol{L}_2(\Omega)} = (V\eta, V\eta)_{\boldsymbol{L}_2(\Omega)}.\end{aligned} \tag{3.23}$$

According to (3.22), we introduce the new unknown function $\eta(t)$ with values in $L_{2,\Gamma}$ in the equations (3.21) and (3.15). Equation (3.21) will become

$$\rho C^{1/2}\frac{\mathrm{d}^2\eta}{\mathrm{d}t^2} - 2\omega_0\rho\frac{\mathrm{d}}{\mathrm{d}t}(\psi+\Psi) + \sigma B_1 C^{-1/2}\eta = \rho F \quad \text{on } \Gamma. \tag{3.24}$$

Let us assume that $\boldsymbol{w}(t)$ is a two times continuously differentiable function of t with values in $\boldsymbol{L}_2(\Omega)$, and $\boldsymbol{f}(t)$ is a continuous function with values in $\boldsymbol{L}_2(\Omega)$. Then, from (3.14), (3.16), and (3.18) it follows that for each t the function $\mathrm{d}^2\eta/\mathrm{d}t^2$ belongs to $L_{2,\Gamma}$, and $\mathrm{d}(\psi)_\Gamma/\mathrm{d}t$, $\mathrm{d}(\Psi)_\Gamma/\mathrm{d}t$, F belong to the space $H_\Gamma^{1/2} = \mathcal{D}(C^{-1/2})$. Hence, we get that if condition (3.24) holds true, then $B_1 C^{-1/2}\eta \in \mathcal{D}(C^{-1/2})$ and, therefore,

$$\rho\frac{\mathrm{d}^2\eta}{\mathrm{d}t^2} - 2\omega_0\rho\frac{\mathrm{d}}{\mathrm{d}t}\left(C^{-1/2}(\psi)_\Gamma + C^{-1/2}(\Psi)_\Gamma\right) + \sigma B_0\eta = \rho C^{-1/2}F, \tag{3.25}$$

with $B_0 := C^{-1/2}B_1C^{-1/2}$.

We next introduce the notations

$$\begin{aligned} P_0(\boldsymbol{v}\times\boldsymbol{e}_3) &=: \mathrm{i}A_{11}v, \quad & P_0(V\eta\times\boldsymbol{e}_3) &=: \mathrm{i}A_{12}\eta,\\ C^{-1/2}(\psi)_\Gamma &=: \mathrm{i}A_{21}\boldsymbol{v}, \quad & C^{-1/2}(\Psi)_\Gamma &=: \mathrm{i}A_{22}\eta, \end{aligned} \tag{3.26}$$

and write down equation (3.15) multiplied by ρ and equation (3.25) as a single vector-matrix equation. (From this point onward, all Hilbert spaces are supposed to be complex spaces and, moreover, in (3.15) and (3.16) a transition to undimensional variables is performed.) Then, the unknown equation has the form

$$\begin{aligned} &\rho\frac{\mathrm{d}^2}{\mathrm{d}t^2}\begin{pmatrix}\boldsymbol{v}\\ \eta\end{pmatrix} - 2i\omega_0\rho\begin{pmatrix}A_{11} & A_{12}\\ A_{21} & A_{22}\end{pmatrix}\frac{\mathrm{d}}{\mathrm{d}t}\begin{pmatrix}\boldsymbol{v}\\ \eta\end{pmatrix} + \sigma\begin{pmatrix}0 & 0\\ 0 & B_0\end{pmatrix}\begin{pmatrix}\boldsymbol{v}\\ \eta\end{pmatrix}\\ &\quad = \begin{pmatrix}\rho\boldsymbol{f}_0\\ \rho C^{-1/2}(F)_\Gamma\end{pmatrix}. \end{aligned} \tag{3.27}$$

The following statement summarizes the reasoning carried out so far: *The initial boundary value problem* (3.7)–(3.11) *is equivalent to the Cauchy problem*

$$\begin{aligned} &\rho\frac{\mathrm{d}^2y}{\mathrm{d}t^2} - 2i\omega_0\rho A\frac{\mathrm{d}y}{\mathrm{d}t} + \sigma By = \rho f(t), \quad y(0)=y^0, \quad y'(0)=y^1,\\ &A = \begin{pmatrix}A_{11} & A_{12}\\ A_{21} & A_{22}\end{pmatrix}, \quad B = \mathrm{diag}\,(0;B_0), \quad f(t) = \left\{\boldsymbol{f}_0; C^{-1/2}(F)_\Gamma\right\}^{\mathrm{t}}, \end{aligned} \tag{3.28}$$

in the Hilbert space $H = \boldsymbol{J}_0(\Omega)\oplus L_{2,\Gamma}$.

6.3.4 Properties of the Operators of the Problem

We first note that according to the definition (3.26), for any $\boldsymbol{v} \in \boldsymbol{J}_0(\Omega)$ we have

$$2i\omega_0 A_{11}\boldsymbol{v} = 2\omega_0 P_0(\boldsymbol{v} \times \mathrm{e}_3) = K\boldsymbol{v}, \tag{3.29}$$

where K is the skew-symmetric gyroscopic operator introduced and investigated in Section 6.1. Since $K^* = -K$ and $\sigma(K) = [-2\mathrm{i}\omega_0, 2\mathrm{i}\omega_0]$, we get $A_{11}^* = A_{11}$ and

$$\sigma(A_{11}) = [-1, 1]. \tag{3.30}$$

Let us show that the whole operator matrix A is a self-adjoint bounded operator and its norm equals one. Let $y_1 = \{\boldsymbol{v}_1; \eta_1\}^{\mathrm{t}}$ and $y_2 = \{\boldsymbol{v}_2; \eta_2\}^{\mathrm{t}}$ be two arbitrary elements of $H = \boldsymbol{J}_0(\Omega) \oplus L_{2,\Gamma}$ and observe that

$$\begin{aligned}
(Ay_1, y_2)_H &= (A_{11}\boldsymbol{v}_1, \boldsymbol{v}_2)_{\boldsymbol{L}_2(\Omega)} + (A_{12}\eta_1, \boldsymbol{v}_2)_{\boldsymbol{L}_2(\Omega)} + (A_{21}\boldsymbol{v}_1, \eta_2)_{L_{2,\Gamma}} + (A_{22}\eta_2, \eta_2)_{L_{2,\Gamma}} \\
&= -\mathrm{i}\Big[(P_0(\boldsymbol{v}_1 \times \boldsymbol{e}_3), \boldsymbol{v}_2)_{\boldsymbol{L}_2(\Omega)} + (P_0(V\eta \times \boldsymbol{e}_3), \boldsymbol{v}_2)_{\boldsymbol{L}_2(\Omega)} \\
&\qquad + \left(C^{-1/2}(\psi_1)_\Gamma, \eta_2\right)_{L_{2,\Gamma}} + \left(C^{-1/2}(\Psi_1)_\Gamma, \eta_2\right)_{L_{2,\Gamma}} \Big] \\
&= -\mathrm{i}\left[(P_0((\boldsymbol{v}_1 + \nabla\Phi_1) \times \boldsymbol{e}_3), \boldsymbol{v}_2)_{\boldsymbol{L}_2(\Omega)} + \left((\psi_1 + \Psi_1)_\Gamma, \left(\frac{\partial \Phi_2}{\partial n}\right)_\Gamma \right)_{L_{2,\Gamma}} \right] \\
&= -\mathrm{i}\left((\boldsymbol{v}_1 + \nabla\Phi_1) \times \boldsymbol{e}_3, \boldsymbol{v}_2\right)_{\boldsymbol{L}_2(\Omega)} - \mathrm{i}\left(P_S((\boldsymbol{v}_1 + \nabla\Phi_1) \times \boldsymbol{e}_3), \nabla\Phi_2\right)_{\boldsymbol{L}_2(\Omega)} \\
&= -\mathrm{i}\left((\boldsymbol{v}_1 + \nabla\Phi_1) \times \boldsymbol{e}_3, \boldsymbol{v}_2 + \nabla\Phi_2\right)_{\boldsymbol{L}_2(\Omega)} \\
&= -\mathrm{i}\int_\Omega (\boldsymbol{v}_1 + \nabla\Phi_1) \times \boldsymbol{e}_3 \cdot \overline{(\boldsymbol{v}_2 + \nabla\Phi_2)}\,\mathrm{d}\Omega \\
&= \ldots = (y_1, Ay_2)_H
\end{aligned} \tag{3.31}$$

and, therefore, $A^* = A$. Let us assume now that $y_1 = y_2 = y$ in (3.31) and use the elementary inequality $|(\boldsymbol{a} \times \boldsymbol{e}_3) \cdot \boldsymbol{a}| \le |\boldsymbol{a}|^2$; we obtain

$$\begin{aligned}
|(Ay, y)_H| &= \left| \int_\Omega (\boldsymbol{v} + \nabla\Phi) \times \boldsymbol{e}_3 \cdot \overline{(\boldsymbol{v} + \nabla\Phi)}\,\mathrm{d}\Omega \right| \le \int_\Omega |\boldsymbol{v} + \nabla\Phi|^2 \mathrm{d}\Omega \\
&= \int_\Omega |\boldsymbol{v}|^2 \mathrm{d}\Omega + \int_\Omega |\nabla\Phi|^2 \mathrm{d}\Omega = \|\boldsymbol{v}\|^2_{\boldsymbol{L}_2(\Omega)} + \|\eta\|^2_{L_{2,\Gamma}} =: \|y\|^2_H.
\end{aligned} \tag{3.32}$$

Here, we used the orthogonality of $\boldsymbol{v}$ and $\nabla\Phi$ as elements of the subspaces $\boldsymbol{J}_0(\Omega)$ and $\boldsymbol{G}_{h,S}(\Omega)$ and the property of isometry, (3.23), of operator V.

Form (3.22) we get that $\|A\| \leq 1$. In virtue of (3.30), we know that $\|A_{11}\| = 1$, so for the whole operator matrix A the following property takes place

$$\|A\| = 1. \tag{3.33}$$

Using (3.31) it can be proved that for any real elements y_1 and y_2 of H next property is true

$$(Ay, y)_H = 0, \qquad y = y_1 + iy_2,\ y_2 = \beta y_1,\ \beta \in \mathbb{R}, \tag{3.34}$$

which is just a corollary to the elementary equation $(\boldsymbol{a} \times \boldsymbol{e}_3) \cdot \boldsymbol{a} = 0$, if the vector $\boldsymbol{a}$ is real.

Now, let us describe several properties of the operator B from (3.28). We note first that by the assumption of static stability of the system in linear approximation, that is, the property $B_1 \gg 0$, the operator $B_0 = C^{-1/2}B_1C^{-1/2}$ has an inverse operator, $B_0^{-1} = C^{1/2}B_1^{-1}C^{1/2}$. Obviously, this operator is compact, because operator C is compact, and B_1^{-1} is at least bounded. Namely, if the fluid is heavy, then

$$(B_1\zeta, \zeta)_{L_{2,\Gamma}} = (\mathcal{L}\zeta, \zeta)_{L_{2,\Gamma}} = (a_\Gamma^0\zeta, \zeta)_{L_{2,\Gamma}} \geq a_0\|\zeta\|^2_{L_{2,\Gamma}}, \tag{3.35}$$

where $a_\Gamma^0(\xi^1, \xi^2) \geq a_0 > 0$ is a function on Γ, defined by formula (3.9) for $\sigma = 0$. Therefore, we conclude that B_1^{-1} is a positive and bounded, but not a compact operator.

According to (3.9), the definition of operator B_1, and the Green formula (4.1.16), for a capillary fluid we obtain

$$\sigma(B_1, \zeta, \zeta)_{L_{2,\Gamma}} = \int_\Gamma \left[\sigma\nabla_\Gamma(\zeta, \zeta) + a_\Gamma|\zeta|^2\right] d\Gamma + \sigma \oint_{\partial\Gamma} \chi|\zeta|^2 ds \geq a_\sigma\|\zeta\|^2_{L_{2,\Gamma}}, a_\sigma > 0, \tag{3.36}$$

for each $\zeta \in \mathcal{D}(B_1)$. As at the end of Section 4.1.5, it can be proved that under the assumption (3.36), the norms in the energetic space H_{B_1} (with the squared norm equal to the quadratic functional from the left side of (3.36)) and in the space H_Γ^1 consisting of elements from $L_{2,\Gamma} = L_2(\Gamma) \ominus \{1\}$ (with the squared norm equal to the Dirichlet integral over Γ) are equivalent. Hence, for $\sigma > 0$, operator B_1^{-1} is compact.

Finally, both for heavy and capillary rotating fluids, the operator $B_0^{-1} = C^{1/2}B_1^{-1}C^{1/2}$ is compact and obviously positive. Therefore, its inverse operator, $B_0 = C^{-1/2}B_1C^{-1/2}$, is positive definite on $L_{2,\Gamma}$ and defined on the set $\mathcal{D}(B_0) = \mathcal{R}(B_0^{-1})$ dense in $L_{2,\Gamma}$.

In order to obtain asymptotics of the eigenvalues $\{\lambda_n(B_0)\}_{n=1}^{\infty}$ of operator B_0 for $n \to \infty$, let us consider the spectral problem $B_0\eta = \lambda\eta$, which after the substitution $C^{-1/2}\eta = \zeta$, equivalent to the substitution (3.22), is converted into the problem $B_1\zeta = \lambda C\zeta$. The eigenvalues of this problem are consecutive minima of the variational ratio

$$\|\zeta\|^2_{B_1}/\|\zeta\|^2_C = \|B_1^{1/2}\zeta\|^2_{L_{2,\Gamma}} / \int_\Omega |\nabla\Phi|^2 d\Omega \tag{3.37}$$

for solutions $\Phi(x)$ from the space $H^1_{h,S}(\Omega)$ of the Neumann problem

$$\Delta\Phi = 0 \quad \text{in } \Omega, \qquad \frac{\partial\Phi}{\partial n} = 0 \quad \text{on } S, \qquad \frac{\partial\Phi}{\partial n} = \zeta \quad \text{on } \Gamma. \tag{3.38}$$

In the numerator in equation (3.37) we use the quadratic functional (3.35) for a heavy fluid ($\sigma = 0$) and the quadratic functional (3.36) for a capillary fluid ($\sigma > 0$), respectively.

The spectral asymptotics for λ_n in the problems (3.35), (3.37) (3.38) and (3.36)–(3.38) are different, and the corresponding formulas for $n \to \infty$ are the following:

$$\lambda_n(B_0) = K_0^{-1/2} n^{1/2}\left[1 + o(1)\right], K_0 = \frac{1}{4\pi}\int_\Gamma |a^0_\Gamma|^{-2} d\Gamma > 0 \qquad (\sigma = 0), \tag{3.39}$$

$$\lambda_n(B_0) = K_\sigma^{-3/2} n^{3/2}\left[1 + o(1)\right], K_\sigma = \frac{1}{4\pi}\text{mes}\,\Gamma > 0 \qquad (\sigma > 0). \tag{3.40}$$

In particular, from these formulas it follows that the compact positive operator B_0^{-1} belongs to the class $\mathfrak{S}_p$, with $p > 2$ for a heavy fluid and with $p > 2/3$ in the case of a capillary fluid.

Taking into account only this difference between the problems on oscillations of a heavy or a capillary fluid in an arbitrary container, let us look at these problems simultaneously by means of the evolution equation (3.28) and the operator pencil in the corresponding spectral problem. We note that from the properties of operator B_0 defined above and the definition of operator B in (3.28), it follows that B is a nonnegative and unbounded operator (due to the unboundedness of B_0) and its kernel $\ker B$ consists of elements of the form $y = \{\boldsymbol{v}; 0\}^t$, with $\boldsymbol{v} \in \boldsymbol{J}_0(\Omega)$ arbitrary.

6.3.5 Systems of Nonmixing Fluids

The main purpose of our further considerations is to show that for a system of nonmixing rotating fluids, in an arbitrary rigid container Ω, the boundary value problem on small oscillations can be converted to a Cauchy problem of the form (3.28) in the corresponding Hilbert space. The line of reasoning developed below is completely similar to the scheme used in the first place for a system of fluids in Section 3.6 and Chapter 4, and in the second place for a single rotating fluid in Sections 6.3.1–6.3.4.

We assume that in the nonperturbed state, the system of m fluids with densities $\rho_1, \rho_2, \dots, \rho_m$ and the container are rotating uniformly as a rigid body with angular velocity ω_0 (> 0). In the coordinate system $Ox_1x_2x_3$ rigidly connected with the container, the pressure $P_{0k}(x)$ in the k-th fluid equals the following (by (3.1))

$$P_{0k}(x) = -\rho_k g x_3 + \frac{1}{2}\rho_k \omega_0^2(x_1^2 + x_2^2) + c_k \qquad \text{in } \Omega_k, \quad k = 1, 2, \dots, m, \tag{3.41}$$

where c_k are some constants. After equating the pressures (3.41) of two adjacent fluids on the boundary Γ_i for a system of heavy fluids, we obtain the equation of Γ_i in the form

$$x_{3,i} = \frac{1}{2g}\omega_0^2(x_1^2 + x_2^2) + \frac{c_i - c_{i+1}}{g(\Delta\rho)_i}, \quad (\Delta\rho)_i = \rho_i - \rho_{i+1} > 0, \qquad i = 1, 2, \dots, m-1. \tag{3.42}$$

The constants c_i can be found, up to an arbitrary constant, from the next $m-1$ conditions of volume-preserving for each region Ω_k occupied by the k-th fluid in the nonperturbed state,

$$\int_{\Omega_k} d\Omega_k = \operatorname{mes}\Omega_k = V_k \qquad k = 1, 2, \dots, m-1. \tag{3.43}$$

(If the $m-1$ conditions (3.43) hold true, then the volume V_m of the fluid with density ρ_m is implicitly defined because $V = \operatorname{mes}\Omega$ is given.)

For a system of capillary fluids, the Laplace conditions on Γ_i should be satisfied; with regard to (3.41), this fact leads to equations of the form

$$\begin{aligned} &-\sigma_i(k_{1,i} + k_{2,i}) \\ &\qquad = -g(\Delta\rho)_i x_3 + \frac{1}{2}(\Delta\rho)_i\omega_0^2(x_1^2 + x_2^2) + (c_i - c_{i+1}), \qquad i = 1, 2, \dots, m-1. \end{aligned} \tag{3.44}$$

These nonlinear equations in the partial derivatives of the functions $x_{3,i} = f_i(x_1, x_2)$ that define the equations of the surfaces Γ_i, together with the conditions (3.43) and the Dupre-Young conditions

$$\sigma_i \cos\delta_i = \sigma_{1,i} - \sigma_{0,i} \quad \text{on } \partial\Gamma_i, \qquad i = 1, 2, \dots, m-1, \tag{3.45}$$

define the nonlinear boundary value problem of determining the state of uniform rotation of a system of capillary fluids in the container Ω.

Let us now assume that problem (3.43)–(3.45) is already solved, and address the problem on small movements of a system of capillary or heavy fluids relative to a uniform rotation. By expressing the pressure in each region in the form $P_k(t,x) = P_{ok}(x) + p_k(t,x)$, where $p_k(t,x)$ is the dynamic pressure in the k-th fluid and introducing for each fluid the field of displacements $\boldsymbol{w}^k(t,x)$ connected with the velocity field $\boldsymbol{u}^k(t,x)$ by the equation $\partial \boldsymbol{w}^k/\partial t = \boldsymbol{u}^k$, we obtain the following boundary value problem (similar to the problems discussed in Section 6.3.2 and Section 4.1)

$$\begin{aligned}
&\frac{\partial^2 \boldsymbol{w}^k}{\partial t^2} - 2\omega_0 \frac{\partial \boldsymbol{w}^k}{\partial t} \times \boldsymbol{e}_3 + \rho_k^{-1}\nabla p_k = \boldsymbol{f}, \quad \operatorname{div} \boldsymbol{w}^k = 0 \quad \text{in } \Omega_k,\\
&\boldsymbol{w}^k \cdot \boldsymbol{n} = 0 \quad \text{on } S_k; \quad \int_{\Gamma_i} \boldsymbol{w}^i \cdot \boldsymbol{n}_i d\Gamma_i = 0 \qquad i = 1,2,\dots,m-1;\\
&\boldsymbol{w}^i \cdot \boldsymbol{n}_i = \boldsymbol{w}^{i+1} \cdot \boldsymbol{n}_i =: \zeta_i, \quad p_i - p_{i+1} = \sigma_i \mathcal{L}_i \zeta_i,\\
&\mathcal{L}_i \zeta_i := -\Delta_{\Gamma_i}\zeta_i + a_{\Gamma_i}\zeta_i,\\
&a_{\Gamma_i} := (k_{1,i}^2 + k_{2,i}^2)\\
&\qquad + \sigma_i^{-1}(\Delta\rho)_i \left[g \cos(\widehat{\boldsymbol{n}_i, \boldsymbol{e}_3}) - \omega_0^2 r \cos(\widehat{\boldsymbol{n}_i, \boldsymbol{e}_r})\right] \quad \text{on } \Gamma_i, \qquad i = 1,2,\dots,m-1;\\
&\frac{\partial \zeta_i}{\partial \nu_i} + \chi_i \zeta_i = 0 \quad \text{on } \partial\Gamma_i, \qquad i = 1,2,\dots,m-1;\\
&\boldsymbol{w}^k(0,x) = \boldsymbol{w}^{k,0}(x), \quad \frac{\partial \boldsymbol{w}^k}{\partial t}(0,x) = \boldsymbol{u}^{k,0}(x) \quad \text{in } \Omega_k, \qquad k = 1,2,\dots,m.
\end{aligned} \tag{3.46}$$

Here, similar to problem (4.1.46), the functions $\zeta_i(t,\xi_i^1,\xi_i^2)$ defined on Γ_i describe deviations along the normal $\boldsymbol{n}_i$ to Γ_i of the free moving surface $\tilde{\Gamma}_i$ from the equilibrium surface Γ_i.

As in Section 6.3.2, we introduce the operators B_i according to the formulas $B_i = P_{\Gamma_i}\mathcal{L}_i P_{\Gamma_i}$, where P_{Γ_i} is an orthoprojector onto $L_{2,\Gamma_i} = L_2(\Gamma_i) \ominus \{1_i\}$. As far as each pressure p_i is defined up to an arbitrary function of time, the dynamic conditions on Γ_i from (3.46) and the boundary conditions on $\partial\Gamma_i$ can be shortly written down in the form

$$p_i - p_{i+1} = \sigma_i B_i \zeta_i. \tag{3.47}$$

We note that the operators B_i above possess the general properties considered in detail for the operator B_1 in Section 6.3.4. We assume that B_i are positive definite in L_{2,Γ_i} and their quadratic forms are expressed by analogues of the formulas (3.35) and (3.36) for heavy and capillary fluids, respectively. Moreover, for heavy fluids all the operators B_i^{-1} are bounded, and for capillary fluids all operators B_i^{-1} are compact in L_{2,Γ_i}.

6.3.6 Transition to an Operator Equation and Properties of the Operators of the Problem

We combine all the equations of problem (3.46) for the regions Ω_k into a single equation that connects the collection of fields $\hat{\boldsymbol{w}} = \{\boldsymbol{w}^k\}_{k=1}^m$ and $\widehat{\nabla_\rho p} := \{\rho_k^{-1}\nabla p_k\}_{k=1}^m$. We will use the space $\hat{\boldsymbol{L}}_2(\Omega)$ with the squared norm

$$\|\hat{\boldsymbol{w}}\|^2_{\hat{\boldsymbol{L}}_2(\Omega)} := \sum_{k=1}^m \rho_k \|\boldsymbol{w}^k\|^2_{\boldsymbol{L}_2(\Omega_k)} \tag{3.48}$$

as the Hilbert space in which problem (3.46) will be further considered. Let us recall that, according to the conclusions of Section 2.1, the space $\hat{\boldsymbol{L}}_2(\Omega)$ has the orthogonal decomposition

$$\hat{\boldsymbol{L}}_2(\Omega) = \hat{\boldsymbol{G}}_{0,\Gamma}(\Omega) \oplus \hat{\boldsymbol{G}}_{h,S}(\Omega) \oplus \hat{\boldsymbol{J}}_0(\Omega), \tag{3.49}$$

where

$$\hat{\boldsymbol{J}}_0(\Omega) = \oplus_{k=1}^m \boldsymbol{J}_0(\Omega_k),$$

and the two other subspaces are

$$\hat{\boldsymbol{G}}_{h,S}(\Omega) = \left\{\hat{v} = \{\nabla q_k\}_{k=1}^m : \{q_k\}_{k=1}^m \in \hat{H}^1_{h,S}(\Omega)\right\},$$
$$\hat{\boldsymbol{G}}_{0,\Gamma}(\Omega) = \left\{\hat{v} = \{\nabla p_k\}_{k=1}^m : \{p_k\}_{k=1}^m \in \hat{H}^1_{0,\Gamma}(\Omega)\right\}. \tag{3.50}$$

The spaces $\hat{H}^1_{h,S}(\Omega)$ and $\hat{H}^1_{0,\Gamma}(\Omega)$ used above are the same as in Section 2.1.11. For every t, we assume that the solutions $\hat{\boldsymbol{w}} = \{\boldsymbol{w}^k\}_{k=1}^\infty$, $\widehat{\nabla_\rho p} = \{\rho_k^{-1}\nabla p_k\}_{k=1}^m$ of problem (3.46) are elements of the subspace $\hat{\boldsymbol{L}}_2(\Omega)$. Then, based on the equations and boundary conditions (3.46), and using the orthogonal decomposition (3.49) we obtain $\hat{\boldsymbol{w}} \in \hat{\boldsymbol{G}}_{h,S}(\Omega) \oplus \hat{\boldsymbol{J}}_0(\Omega)$ and $\widehat{\nabla_\rho p} \in \hat{\boldsymbol{G}}_{0,\Gamma}(\Omega) \oplus \hat{\boldsymbol{G}}_{h,S}(\Omega)$. Therefore, these solutions can be found in the form

$$\hat{\boldsymbol{w}} = \widehat{\nabla\Phi} + \hat{\boldsymbol{v}}, \quad \widehat{\nabla\Phi} = \{\nabla\Phi_k\}_{k=1}^m \in \hat{\boldsymbol{G}}_{h,S}(\Omega), \qquad \hat{v} \in \hat{\boldsymbol{J}}_0(\Omega),$$
$$\widehat{\nabla_\rho p} = \widehat{\nabla_\rho \kappa} + \widehat{\nabla_\rho \varphi}, \quad \widehat{\nabla_\rho \kappa} \in \hat{\boldsymbol{G}}_{0,\Gamma}(\Omega), \quad \widehat{\nabla_\rho \varphi} \in \hat{\boldsymbol{G}}_{h,S}(\Omega). \tag{3.51}$$

Substituting (3.51) into (3.46) and then applying the orthoprojectors $\hat{P}_\Gamma$, $\hat{P}_S$, $\hat{P}_0$ onto the orthogonal subspaces (3.49), (3.50) to the resulting equations, we end up with some equations that are identical to the relations (3.15)–(3.17), with the obvious change in notation. In these new equations, the vector-column of the form $\{\boldsymbol{v}^k \times \boldsymbol{e}_3\}_{k=1}^m$ is denoted by the symbol $\hat{\boldsymbol{v}} \times \boldsymbol{e}_3$.

As in (3.18), we next introduce the notations

$$P_S(\hat{\boldsymbol{v}} \times \boldsymbol{e}_3) =: \widehat{\nabla\psi} = \{\nabla\psi_k\}_{k=1}^m \in \hat{\boldsymbol{G}}_{h,S}(\Omega),$$
$$P_S(\widehat{\nabla\Phi} \times \boldsymbol{e}_3) =: \widehat{\nabla\Psi} = \{\nabla\Psi_k\}_{k=1}^m \in \hat{\boldsymbol{G}}_{h,S}(\Omega), \tag{3.52}$$

and perform transformations similar to those used in deriving formulas (3.19)–(3.21). Here, by taking into account the relations

$$p_i - p_{i=1} = \varphi_i - \varphi_{i+1}, \quad \boldsymbol{v}^i \cdot \boldsymbol{n}_i = 0 \quad \text{on } \Gamma_i, \qquad i = 1, 2, \dots, m-1,$$

we finally obtain the system of equations

$$\frac{\mathrm{d}^2\boldsymbol{v}^k}{\mathrm{d}t^2} - 2\omega_0 \frac{\mathrm{d}}{\mathrm{d}t}\left[P_{0k}(\boldsymbol{v}^k \times \boldsymbol{e}_3) + P_{0k}(\nabla\Phi_k \times \boldsymbol{e}_3)\right] = f_{0k} \quad \text{in } \Omega_k, \qquad k = 1, 2, \dots m; \tag{3.53}$$

$$\frac{\mathrm{d}^2}{\mathrm{d}t^2}(\rho_i\Phi_i - \rho_{i+1}\Phi_{i+1}) - 2\omega_0\frac{\mathrm{d}}{\mathrm{d}t}\left[(\rho_i\psi_i - \rho_{i+1}\psi_{i+1}) + (\rho_i\Psi_i - \rho_{i+1}\Psi_{i+1})\right]$$
$$+\sigma_i B_i \frac{\partial\Phi_i}{\partial n_i} = \rho_i F_i - \rho_{i+1}F_{i+1} \quad \text{on } \Gamma_i, \qquad i = 1, 2, \dots, m. \tag{3.54}$$

The operators P_{0k} in (3.53) are the orthoprojectors onto $J_0(\Omega_k)$ and they appear because $\hat{P}_0 = \operatorname{diag}(P_{0k})_{k=1}^m$, $f_{0k} = P_{0k}f$.

The first group of terms in (3.54) can be written down in a form that has already been used in Sections 3.6 and 4.1, namely, here again, we take the canonical operator $\hat{C} = \gamma_\Gamma\hat{T}$ that corresponds to the Hilbert pair of spaces $(F; G)$, where $F = \hat{H}^1_\Gamma(\Omega)$ and $\hat{G} = \hat{L}_{2,\Gamma}$. Therefore, this group of terms is given by

$$\left\{\frac{\mathrm{d}^2}{\mathrm{d}t^2}(\rho_i\Phi_i - \rho_{i+1}\Phi_{i+1})_{\Gamma_i}\right\}_{i=1}^{m-1} = \hat{C}\frac{\mathrm{d}^2\hat{\zeta}}{\mathrm{d}t^2}, \qquad \hat{\zeta} = \{\zeta_i\}_{i=1}^{m-1}. \tag{3.55}$$

Introducing next the operator $\hat{B} := \operatorname{diag}(\sigma_i B_i)_{i=1}^{m-1}$, similar to (4.1.9), we notice that the last group of terms in (3.54) can be written down in the form

$$\left\{\sigma_i B_i\left(\frac{\partial\Phi_i}{\partial n_i}\right)_{\Gamma_i}\right\}_{i=1}^{m-1} = \hat{B}\hat{\zeta}. \tag{3.56}$$

As before, for the right side of (3.54), we use the notation

$$\hat{F} := \{(\rho_i F_i - \rho_{i+1}F_{i+1})_{\Gamma_i}\}_{i=1}^{m-1} \tag{3.57}$$

We need to rememeber that in virtue of the facts pointed out in Section 2.1.11, for the elements $\{\rho_k^{-1}\nabla\varphi_k\}_{k=1}^m$ from the space $\hat{\boldsymbol{G}}_{h,S}(\Omega)$ the following conditions hold true on Γ_i,

$$\int_{\Gamma_i}(\rho_i\varphi_i - \rho_{i+1}\varphi_{i+1})\mathrm{d}\Gamma_i = 0 \qquad i = 1, 2, \ldots, m-1, \tag{3.58}$$

hence these differences are elements of the spaces L_{2,Γ_i}. It follows that the expressions (3.55)–(3.57) and the groups of terms

$$\left\{(\rho_i\varphi_i - \rho_{i+1}\varphi_{i+1})_{\Gamma_i}\right\}_{i=1}^{m-1}, \quad \left\{(\rho_i\Psi_i - \rho_{i+1}\Psi_{i+1})_{\Gamma_i}\right\}_{i=1}^{m-1} \tag{3.59}$$

are elements of the space $\hat{L}_{2,\Gamma} := \oplus_{i=1}^{m-1} L_{2,\Gamma_i}$.

To shape the terms from (3.53) and (3.54) that contain time derivatives of the first order into a more convenient form, we introduce several new operators and blocks of operators.

Obviously, expression $P_{0k}(\boldsymbol{v}^k \times \boldsymbol{e}_3)$ in (3.53) is the result of applying a linear bounded operator acting on $\boldsymbol{J}_0(\Omega_k)$ to the element $\boldsymbol{v}^k$. Thus, we can set

$$-\mathrm{i}P_{0k}(\boldsymbol{v}^k \times \boldsymbol{e}_3) =: A_{kk}\boldsymbol{v}^k \qquad k = 1, 2, \ldots, m. \tag{3.60}$$

Since $\hat{\boldsymbol{J}}_0(\Omega) = \oplus_{k=1}^m \boldsymbol{J}_0(\Omega_k)$ and the corresponding orthoprojector $\hat{P}_0$ onto $\hat{\boldsymbol{J}}_0(\Omega)$ has the form $\hat{P}_0 = \operatorname{diag}(P_{0k})_{k=1}^m$, it is natural to introduce the notation $\hat{A}_{11} := \operatorname{diag}(A_{kk})_{k=1}^m$. Then, in (3.53), the group of terms $\{-2\omega_0 \mathrm{d}/\mathrm{d}t P_{0k}(\boldsymbol{v}^k \times \boldsymbol{e}_3)\}_{k=1}^m$ can be written down in the form $-2\mathrm{i}\omega_0\hat{A}_{11}\mathrm{d}\hat{\boldsymbol{v}}/\mathrm{d}t$, $\hat{\boldsymbol{v}} = \{\boldsymbol{v}^k\}_{k=1}^m$. The operator matrix $\hat{A}_{11}$ plays the same role as the gyroscopic operator A_{11} introduced at the beginning of Section 6.3.4 for the problem on oscillations of a single fluid. Obviously, the condition $\sigma(A_{kk}) = [-1, 1]$ holds true for any $k = 1, 2, \ldots, m$ (as in (3.30)) and, therefore,

$$\sigma(\hat{A}_{11}) = [-1, 1]. \tag{3.61}$$

Further, by considering the second group of terms from (3.53) with time derivatives of the first order, let us note that $\{\nabla\Phi_k\}_{k=1}^m = \widehat{\nabla\Phi} \in \hat{\boldsymbol{G}}_{h,S}(\Omega)$ and the functions Φ_k are solutions of the Neumann problem

$$\Delta\Phi_k = 0 \quad \text{in } \Omega_k, \qquad \frac{\partial\Phi_k}{\partial n} = 0 \quad \text{on } S_k, \qquad \frac{\partial\Phi_j}{\partial n_j} = \frac{\partial\Phi_{j+1}}{\partial n_j} = \zeta_j \quad \text{on } \Gamma_j,$$

$$\int_{\Gamma_j}(\rho_j\Phi_j - \rho_{j+1}\Phi_{j+1})\mathrm{d}\Gamma_j = 0 \qquad \int_{\Gamma_{m-1}} \rho_m\Phi_m \mathrm{d}\Gamma_{m-1} = 0. \tag{3.62}$$

Hence, it follows that $\nabla\Phi_k$ depends linearly on ζ_{k-1} and ζ_k and, therefore, we can introduce the notations

$$-\mathrm{i}P_{0k}(\nabla\Phi_k \times \boldsymbol{e}_3) =: A_{k,m+k-1}\zeta_{k-1} + A_{k,m+k}\zeta_k, \qquad k = 1, 2, \ldots, m. \tag{3.63}$$

Based on (3.63), we build the $m \times (m-1)$-dimensional matrix operator $\hat{A}_{12}$ that has only the following nonzero elements

$$(\hat{A}_{12})_{k,k-1} := A_{k,m+k-1}, \qquad k = 2,3,\dots,m,$$
$$(\hat{A}_{12})_{kk} := A_{k,m+k}, \qquad k = 1,2,\dots,m-1. \tag{3.64}$$

Then we can get the shorter version,

$$\left\{-2\omega_0 \frac{\mathrm{d}}{\mathrm{d}t} P_{ok}(\nabla\Phi_k \times \boldsymbol{e}_3)\right\}_{k=1}^{m} = -2\omega_0 \hat{A}_{12}\frac{\mathrm{d}\hat{\zeta}}{\mathrm{d}t}, \qquad \hat{\zeta} = \{\zeta_j\}_{j=1}^{m-1}. \tag{3.65}$$

To perform similar changes to all the terms with time derivatives of the first order in (3.54), we note that the group of terms $\{(\rho_j\psi_j - \rho_{j+1}\psi_{j+1})_{\Gamma_j}\}_{j=1}^{m-1}$ is uniquely determined by the element $\hat{\boldsymbol{v}} = \{\boldsymbol{v}^k\}_{k=1}^m \in \hat{\boldsymbol{J}}_0(\Omega)$, and the similar group $\{(\rho_j\Psi_j - \rho_{j+1}\Psi_{j+1})_{\Gamma_j}\}_{j=1}^{m-1}$ is determined by the element $\widehat{\nabla\Phi} = \{\nabla\Phi_k\}_{k=1}^m \in \hat{\boldsymbol{G}}_{h,S}(\Omega)$. Consequently, for the first group we can denote

$$-i(\psi_j)_{\Gamma_j} =: \sum_{l=1}^{m} A'_{m+j,l}\boldsymbol{v}^l \qquad j = 1,2,\dots,m-1,$$
$$i(\psi_{j+1})_{\Gamma_j} =: \sum_{l=1}^{m} A''_{m+j,l}\boldsymbol{v}^l \qquad j = 1,2,\dots,m-1. \tag{3.66}$$

With regard to (3.66), we build the $(m-1)\times m$-dimensional matrix block $\hat{A}_{21}$ with the following entries

$$(\hat{A}_{21})_{jl} := \rho_j A'_{m+j,l} + \rho_{j+1}A''_{m+j,l}. \tag{3.67}$$

Then, the collection

$$\left\{-2\omega_0 \frac{\mathrm{d}}{\mathrm{d}t}(\rho_j\psi_j - \rho_{j+1}\psi_{j+1})_{\Gamma_j}\right\}_{j=1}^{m-1}$$

can be written down as $-2\mathrm{i}\omega_0\hat{A}_{21}\mathrm{d}\hat{\boldsymbol{v}}/\mathrm{d}t$. It easily follows that $(\hat{A}_{21})_{kk}$ and $(\hat{A}_{21})_{k,k+1}$, $k = 1,2,\dots,m-1$ are the only nonzero elements in the matrix $\hat{A}_{21}$.

Finally, the group of terms

$$\left\{-2\omega_0 \frac{\mathrm{d}}{\mathrm{d}t}(\rho_j\Psi_j - \rho_{j+1}\Psi_{j+1})_{\Gamma_j}\right\}_{j=1}^{m-1}$$

can be written down as $-2\mathrm{i}\omega_0\hat{A}_{22}\mathrm{d}\hat{\zeta}/\mathrm{d}t$. The elements $(\hat{A}_{22})_{jl}$ of the $(m-1)\times(m-1)$-dimensional matrix $\hat{A}_{22}$ are defined by the following formulas

$$-\mathrm{i}(\Psi_j)_{\Gamma_j} =: \sum_{l=1}^{m-1} A'_{m+j,m+l}\zeta_l,$$

$$\mathrm{i}(\Psi_{j+1})_{\Gamma_j} =: \sum_{l=1}^{m-1} A''_{m+j,m+l}\zeta_l, \tag{3.68}$$

hence

$$(\rho_j\Psi_j - \rho_{j+1}\Psi_{j+1})_{\Gamma_j} = \sum_{l=1}^{m-1} \left(\rho_j A'_{m+j,m+l} + \rho_{j+1}A''_{m+j,m+l}\right)\zeta_l, \tag{3.69}$$

and, therefore,

$$(\hat{A}_{22})_{jl} := \rho_j A'_{m+j,m+l} + \rho_{j+1}A''_{m+j,m+l}, \qquad j,l = 1,2,\ldots,m-1. \tag{3.70}$$

The above introduced matrix blocks make it possible to write down the equations (3.53) and (3.54) in the form of a differential equation,

$$\begin{pmatrix} \hat{I} & 0 \\ 0 & \hat{C} \end{pmatrix} \frac{\mathrm{d}^2}{\mathrm{d}t^2}\begin{pmatrix} \hat{\boldsymbol{v}} \\ \hat{\zeta} \end{pmatrix} - 2k\omega_0 \begin{pmatrix} \hat{A}_{11} & \hat{A}_{12} \\ \hat{A}_{21} & \hat{A}_{22} \end{pmatrix} \frac{\mathrm{d}}{\mathrm{d}t}\begin{pmatrix} \hat{\boldsymbol{v}} \\ \hat{\zeta} \end{pmatrix} + \begin{pmatrix} 0 & 0 \\ 0 & \hat{B} \end{pmatrix}\begin{pmatrix} \hat{\boldsymbol{v}} \\ \hat{\zeta} \end{pmatrix} = \begin{pmatrix} \hat{\boldsymbol{f}}_0 \\ \hat{F} \end{pmatrix}, \tag{3.71}$$

considered in the space $\hat{H} := \hat{\boldsymbol{J}}_0(\Omega) \oplus \hat{L}_{2,\Gamma}$, with $\hat{\boldsymbol{f}}_0 := \{\boldsymbol{f}_{0k}\}_{k=1}^m$ and $\hat{F} := \{(\rho_j F_j - \rho_{j+1}F_{j+1})_{\Gamma_j}\}_{j=1}^{m-1}$.

To deduct the final equation, we proceed in a similar way to the reasoning developed at the end of Section 6.3.3 in connection with the transition from (3.24) and (3.15) to equation (3.28). Namely, we perform in (3.71) the substitution (similar to (3.22))

$$\hat{\eta} = \hat{C}^{1/2}\hat{\zeta} = \hat{C}^{-1/2}(\gamma_\Gamma\hat{\Phi}), \tag{3.72}$$

where the trace operator γ_Γ is defined by the formulas (3.6.21). Then we apply the operator $\mathrm{diag}(\hat{I};\hat{C}^{-1/2})$ to both sides of (3.71) to obtain the following Cauchy problem

$$\begin{aligned}
&\frac{\mathrm{d}^2y}{\mathrm{d}t^2} - 2\mathrm{i}\omega_0 A\frac{\mathrm{d}y}{\mathrm{d}t} + By = f, \qquad y(0) = y^0,\ y'(0) = y^1, \\
&A = \left(\tilde{A}_{ij}\right)_{i,j=1}^2 = \begin{pmatrix} \hat{A}_{11} & \hat{A}_{12}\hat{C}^{-1/2} \\ \hat{C}^{-1/2}A_{21} & \hat{C}^{-1/2}A_{22}\hat{C}^{-1/2} \end{pmatrix}, \\
&B = \mathrm{diag}(0;\hat{B}_0), \quad \hat{B}_0 := \hat{C}^{-1/2}\hat{B}\hat{C}^{-1/2}, \\
&y = \begin{pmatrix} \hat{\boldsymbol{v}} \\ \hat{\eta} \end{pmatrix}, \quad f(t) = \begin{pmatrix} \hat{\boldsymbol{f}}_0 \\ \hat{C}^{-1/2}\hat{F} \end{pmatrix}, \\
&y^0 = \begin{pmatrix} \hat{\boldsymbol{v}}^0 \\ \hat{C}^{-1/2}\hat{\zeta}^0 \end{pmatrix} = \begin{pmatrix} \hat{P}_0\{\boldsymbol{w}^{k,0}\}_{k=1}^m \\ \hat{C}^{-1/2}\{(\boldsymbol{w}^{j,0}\cdot\boldsymbol{n}_j)_{\Gamma_j}\}_{j=1}^{m-1} \end{pmatrix}, \\
&y^1 = \begin{pmatrix} \hat{\boldsymbol{v}}^{0,1} \\ \hat{C}^{-1/2}\hat{\zeta}^{0,1} \end{pmatrix} = \begin{pmatrix} \hat{P}_0\{\boldsymbol{u}^{k,0}\}_{k=1}^m \\ \hat{C}^{-1/2}\{(\boldsymbol{u}^{j,0}\cdot\boldsymbol{n}_j)_{\Gamma_j}\}_{j=1}^{m-1} \end{pmatrix}.
\end{aligned} \tag{3.73}$$

Comparing the problem (3.73) for a system of nonmixing fluids with the similar problem (3.27) on small oscillations in a single fluid rotating in a partially filled container, we observe that they have the same form. Let us next prove that the general properties of the operators involved in these problems are identical, too.

First, we note that $\hat{B}_0$ in (3.73) is an unbounded positive definite operator with discrete spectrum. Indeed, since $\hat{B} = \mathrm{diag}(\sigma_j B_j)_{j=1}^{m-1}$ and all the operators $\sigma_j B_j$ are positive definite according to the assumption on static stability of the system, then all $(\sigma_j B_j)^{-1}$ are bounded. Therefore, operator $\hat{B}^{-1}$ is also bounded and positive (this operator is even compact for capillary fluids). Then, operator $\hat{C}^{1/2}\hat{B}^{-1}\hat{C}^{1/2}$ is also positive and compact, because $\hat{C}$ is positive and compact. Hence, one gets that the inverse operator $\hat{B}_0 = \hat{C}^{-1/2}\hat{B}\hat{C}^{-1/2}$ defined on the set $\mathcal{D}(\hat{B}_0) = \mathcal{R}(\hat{B}_0^{-1})$ dense in $L_{2,\Gamma}$, is positive definite and has a discrete positive spectrum with the limit point $+\infty$. Its eigenvalues, $\{\lambda_n(\hat{B}_0)\}_{n=1}^{\infty}$, can be found as consequtive minima of the variational ratio

$$\frac{\left(\hat{B}_0^{1/2}\hat{\eta}, \hat{B}_0^{1/2}\hat{\eta}\right)_{\hat{L}_{2,\Gamma}}}{(\hat{\eta}, \hat{\eta})_{\hat{L}_{2,\Gamma}}} = \frac{\left(\hat{B}^{1/2}\hat{\zeta}, \hat{B}^{1/2}\hat{\zeta}\right)_{\hat{L}_{2,\Gamma}}}{\left(\hat{C}\hat{\zeta}, \hat{\zeta}\right)\hat{L}_{2,\Gamma}}. \tag{3.74}$$

The variation in (3.74) should be performed in the class of functions $\hat{\zeta} = \{\zeta_j\}_{j=1}^{m-1} \in \mathcal{D}(\hat{B}^{1/2}) \subset \hat{L}_{2,\Gamma}$, which are connected with $\hat{\nabla}\Phi = \{\nabla\Phi_k\}_{k=1}^{m} \in \hat{\boldsymbol{G}}_{h,S}(\Omega)$ by solutions of the Neumann problem (3.62). Then, (3.74) becomes

$$\frac{\displaystyle\sum_{j=1}^{m-1}\left\{\int_{\Gamma_j}\left[\sigma_j \nabla_{\Gamma_j}(\zeta_j, \zeta_j) + a_{\Gamma_j}|\zeta_j|^2\right] \mathrm{d}\Gamma_j + \sigma_j \oint_{\partial\Gamma_j} \chi_j|\zeta_j|^2 \mathrm{d}s_j\right\}}{\displaystyle\sum_{k=1}^{m}\rho_k \int_{\Omega_k}|\nabla\Phi_k|^2 \mathrm{d}\Omega_k}, \tag{3.75}$$

with

$$a_{\Gamma_j}(x) := -\sigma_j(k_{1j}^2 + k_{2j}^2) + (\Delta\rho)_j\left[g\cos(\widehat{n_j, \mathrm{e}_3}) - \omega_0^2 r\cos(\widehat{n_j, \mathrm{e}_r})\right], \quad j = 1, 2, \ldots, m-1.$$

Variational problems such as (3.75) and (3.62) have been already considered in Section 3.6 for a system of heavy nonrotating fluids (see (3.6.26)) and in Section 4.1 for a system of capillary nonrotating fluids (see (4.1.22)). For example, in the case of capillary fluids when $\sigma_j > 0$, the asymptotic formulas for the eigenvalues $\lambda_n = \lambda_n(\hat{B}_0)$ still have the form (4.1.55),

$$\lambda_n = (n/K_\sigma)^{3/2}[1 + \mathrm{o}(1)](n \to \infty),$$
$$K_\sigma := \frac{1}{4\pi}\sum_{j=1}^{m-1}\left(\frac{\rho_j + \rho_{j+1}}{\sigma_j}\right)^{2/3} \mathrm{mes}\,\Gamma_j > 0. \tag{3.76}$$

In the case of a system of heavy rotating fluids, that is, for $\sigma_j = 0$, $j = 1, 2, \ldots, m-1$, the asymptotic formula for $\lambda_n(\hat{B}_0)$ does not coincide with the formula (3.6.28) due to the new form of the function in (3.75),

$$\lambda_n = (n/K_0)^{1/2}[1 + \mathrm{o}(1)], \qquad n \to \infty,$$
$$K_0 := \frac{1}{4\pi}\sum_{j=1}^{m-1}(\rho_j + \rho_{j+1})^2 \int_{\Gamma_j} |a_{\Gamma_j}(x)|^{-2} \mathrm{d}\Gamma_j. \tag{3.77}$$

For $\omega_0 \to 0$, that is, when the surfaces Γ_j become horizontal planes, the constant K_0 becomes the constant $K_{C,B}$ in (3.6.28), which corresponds to the case of a system of heavy nonrotating fluids.

Now, let us prove that operator A in (3.73), which can be naturally called the *gyroscopic operator* for the system of fluids, is a bounded self-adjoint operator with the norm equal to 1. For the block $\tilde{A}_{11} = \hat{A}_{11} = \operatorname{diag}(A_{kk})_{k=1}^m$ this is the case because all operators A_{kk} are self-adjoint and the property (3.61) holds true. Therefore, it is sufficient to prove, that $A = A^*$ and $\|A\| \leq 1$.

To prove these properties, let us note that the following relations take place,

$$\hat{P}_0(\hat{\boldsymbol{v}} \times \boldsymbol{e}_3) =: \hat{A}_{11}\hat{\boldsymbol{v}}, \quad \hat{P}_0(\nabla\Phi \times \boldsymbol{e}_3) = \mathrm{i}A_{12}\hat{\zeta},$$
$$\gamma_\Gamma\hat{\psi} := \{(\rho_j\psi_j - \rho_{j+1}\psi_{j+1})_{\Gamma_j}\}_{j=1}^{m-1} = \mathrm{i}\hat{A}_{21}\hat{\boldsymbol{v}},$$
$$\gamma_\Gamma\hat{\Psi} := \{(\rho_j\Psi_j - \rho_{j+1}\Psi_{j+1})_{\Gamma_j}\}_{j=1}^{m-1} = \mathrm{i}\hat{A}_{22}\hat{\zeta}. \tag{3.78}$$

Here, γ_Γ is the trace operator that results after applying the abstract scheme of Section 1.8 to the pair of spaces $F = \hat{H}^1_\Gamma(\Omega)$ and $G = L_{2,\Gamma}$. Let us recall that the range of γ_Γ is the space $G^{1/2} = \hat{H}^{1/2}_\Gamma(\Gamma) = \mathcal{D}(\hat{C}^{-1/2})$, where $\hat{C}^{-1}$ is the generating operator of the Hilbert pair $(G^{1/2}; G)$, $\hat{C} = \gamma_\Gamma\hat{T}$. The operator $\hat{C}^{-1}$ isometrically maps the space $\mathcal{D}(\hat{C}^{-1/2})$ into $G^{-1/2} = (\hat{H}^{1/2}_\Gamma(\Gamma))^*$.

We next prove that operator $\hat{A}_{12}$ acts boundedly from $G^{-1/2}$ to $\hat{\boldsymbol{J}}_0(\Omega)$. Indeed, $\hat{\Phi} = \{\Phi_k\}_{k=1}^m = \hat{T}\hat{\zeta}$, where $\hat{T}$ is an isometrical operator, which makes it possible to build a solution $\hat{\Phi}$ of problem (3.62) in the space $\hat{H}^1_{h,S}(\Omega)$ by using the collection

$$\{\zeta_j\}_{j=1}^{m-1} = \hat{\zeta} \in G^{-1/2} = \left(\bigoplus_{j=1}^{m-1} H^{1/2}_{\Gamma_j}(\Gamma_j)\right)^*.$$

Then, the collection $\widehat{\nabla\Phi} = \{\nabla\Phi_k\}_{k=1}^m$ belongs to the space $\hat{\boldsymbol{G}}_{h,S}(\Omega)$, and the operator that maps the element $\hat{\zeta}$ into the solution $\widehat{\nabla\Phi}$ is an isometry from $G^{-1/2}$ to $\hat{\boldsymbol{G}}_{h,S}(\Omega)$. The boundedness of operator $\hat{A}_{12}$ from $G^{-1/2}$ to $\hat{J}_0(\Omega)$ follows now from the latter observation combined with the equation $\hat{P}_0(\widehat{\nabla\Phi} \times \boldsymbol{e}_3) = \mathrm{i}A_{12}\hat{\zeta}$ and from the properties of the projector and the operation of vector product (the componentwise multiplication is assumed).

Similarly, it can be proven, using the properties of operator $\hat{C}^{-1}$ mentioned above, that $\hat{A}_{21}$ acts boundedly from $\hat{\boldsymbol{J}}_0(\Omega)$ to $G^{1/2} = \mathcal{D}(\hat{C}^{-1/2})$ and the operator $\hat{A}_{22}$ acts boundedly from $G^{-1/2}$ to $\mathcal{D}(\hat{C}^{-1/2})$. From all these facts and formulas (3.73) for the matrix blocks $\tilde{A}_{ij}$ we obtain that operator $A = (\tilde{A}_{ij})_{i,j=1}^2$ is bounded on $\hat{H}$.

Now, we prove that $A = A^*$ and $\|A\| \leq 1$. Let $y_1 = (\hat{\boldsymbol{v}}_1; \hat{\eta}_1)^{\mathrm{t}}$ and $y_2 = (\hat{\boldsymbol{v}}_2; \hat{\eta}_2)^{\mathrm{t}}$ be two arbitrary elements from $\hat{H}$. Then

$$
\begin{aligned}
(Ay_1, y_2)_{\hat{H}} &= \left(\hat{A}_{11}\hat{\boldsymbol{v}}_1, \hat{\boldsymbol{v}}_2\right)_{\hat{\boldsymbol{L}}_2(\Omega)} + \left(\hat{A}_{12}\hat{C}^{-1/2}\hat{\eta}_1, \hat{\boldsymbol{v}}_2\right)_{\hat{\boldsymbol{L}}_2(\Omega)} \\
&\quad + \left(\hat{C}^{-1/2}\hat{A}_{21}\hat{\boldsymbol{v}}_1, \hat{\eta}_2\right)_{\hat{L}_{2,\Gamma}} + \left(\hat{C}^{-1/2}\hat{A}_{22}\hat{C}^{-1/2}\hat{\eta}_1, \hat{\eta}_2\right)_{\hat{L}_{2,\Gamma}} \\
&= \left(\hat{A}_{11}\hat{\boldsymbol{v}}_1, \hat{\boldsymbol{v}}_2\right)_{\hat{\boldsymbol{L}}_2(\Omega)} + \left(\hat{A}_{12}\hat{\zeta}_1, \hat{\boldsymbol{v}}_2\right)_{\hat{\boldsymbol{L}}_2(\Omega)} \\
&\quad + \left(\hat{A}_{21}\hat{\boldsymbol{v}}_1, \hat{\zeta}_2\right)_{\hat{L}_{2,\Gamma}} + \left(\hat{A}_{22}\hat{\zeta}_1, \hat{\zeta}_2\right)_{\hat{L}_{2,\Gamma}} \\
&= \left(-\mathrm{i}\hat{P}_0\left(\hat{\boldsymbol{v}}_1 \times \boldsymbol{e}_3\right), \hat{\boldsymbol{v}}_2\right)_{\hat{\boldsymbol{L}}_2(\Omega)} + \left(-\mathrm{i}\hat{P}_0\left(\widehat{\nabla\Phi}^1 \times \boldsymbol{e}_3\right), \hat{\boldsymbol{v}}_2\right)_{\hat{\boldsymbol{L}}_2(\Omega)} \\
&\quad + \left(-\mathrm{i}\left(\gamma_\Gamma\hat{\psi}_1 + \gamma_\Gamma\hat{\Psi}_1\right), \hat{\zeta}_2\right)_{\hat{L}_{2,\Gamma}} \\
&= \left(-\mathrm{i}\hat{P}_0\left(\left(\hat{\boldsymbol{v}}_1 + \widehat{\nabla\Phi}^1\right) \times \boldsymbol{e}_3\right), \hat{\boldsymbol{v}}_2\right)_{\hat{\boldsymbol{L}}_2(\Omega)} \\
&\quad + \left(-\mathrm{i}\hat{P}_S\left(\left(\hat{\boldsymbol{v}}_1 + \widehat{\nabla\Phi}^1\right) \times \boldsymbol{e}_3\right), \widehat{\nabla\Phi}^2\right)_{\hat{\boldsymbol{L}}_2(\Omega)} \\
&= \left(-\mathrm{i}\left(\hat{P}_0 + \hat{P}_S\right)\left(\left(\hat{\boldsymbol{v}}_1 + \widehat{\nabla\Phi}^1\right), \hat{\boldsymbol{v}}_2 + \widehat{\nabla\Phi}^2\right)\right)_{\hat{\boldsymbol{L}}_2(\Omega)} \\
&= \left(\left(\hat{\boldsymbol{v}}_1 + \widehat{\nabla\Phi}^1\right), -\mathrm{i}\left(\hat{P}_0 + \hat{P}_S\right)\left(\left(\hat{\boldsymbol{v}}_2 + \widehat{\nabla\Phi}^2\right) \times \boldsymbol{e}_3\right)\right)_{\hat{\boldsymbol{L}}_2(\Omega)} \\
&= \ldots = (y_1, Ay_2)_{\hat{H}},
\end{aligned}
\tag{3.79}
$$

that is, A is a self-adjoint operator.

The following general formula, just used in the deduction of (3.79),

$$(\gamma v, \psi)_G = (v, T\psi)_F,$$

was introduced in the abstract scheme of Section 1.8 and, in the case we considered it, has the form

$$\left(\gamma_\Gamma\left(\hat{\psi}_1 + \hat{\Psi}_1\right), \hat{\zeta}_2\right)_{\hat{L}_{2,\Gamma}} = \left(\widehat{\nabla\psi}_1 + \widehat{\nabla\Psi}_1, \widehat{\nabla\Phi}^2\right)_{\hat{\boldsymbol{L}}_2(\Omega)}. \tag{3.80}$$

Setting $y_1 = y_2 = y$ in (3.79) and using the elementary inequality

$$|(\hat{\boldsymbol{w}} \times \boldsymbol{e}_3, \hat{\boldsymbol{w}})_{\hat{\boldsymbol{L}}_2(\Omega)}| \leq \|\hat{\boldsymbol{w}}\|^2_{\hat{\boldsymbol{L}}_2(\Omega)},$$

we next obtain

$$|(Ay, y)_{\hat{H}}| \leq \|\hat{v} + \widehat{\nabla\Phi}\|^2_{\hat{L}_2(\Omega)} = \|\hat{v}\|^2_{\hat{L}_2(\Omega)} + \|\widehat{\nabla\Phi}\|^2_{\hat{L}_2(\Omega)}$$
$$= \|\hat{v}\|^2_{\hat{L}_2(\Omega)} + \|\hat{\eta}\|^2_{\hat{L}_{2,\Gamma}} = \|y\|^2_{\hat{H}}. \tag{3.81}$$

Here we used the orthogonality of $\hat{v}$ and $\widehat{\nabla\Phi}$ in $\hat{L}_2(\Omega)$ and the above mentioned isometric property $\|\widehat{\nabla\Phi}\|_{\hat{L}_2(\Omega)} = \|\eta\|_{L_{2,\Gamma}}$. From (3.81) it follows that $\|A\| \leq 1$, which along with the equality $\|\hat{A}_{11}\| = 1$ proves all the required properties of operator A.

Hence, the operator equations (3.27) for oscillations of a single fluid and the ones in (3.73) for a system of fluids are identical. That is why it is reasonable to consider an abstract Cauchy problem of the same form and the corresponding spectral problem.

6.4 Solving the Initial Boundary Value Problem

In this section we consider the problem of correct solvability of an abstract evolution equation generated by the problem on oscillations of an ideal rotating fluid, and its hydrodynamic concretization.

6.4.1 Generalized Solution of the Operator Equation

In the Hilbert space H, which is the orthogonal sum of the Hilbert spaces H_1 and H_2, let us consider the following evolution problem:

$$\frac{\mathrm{d}^2u}{\mathrm{d}t^2} - \mathrm{i}A\frac{\mathrm{d}u}{\mathrm{d}t} + Bu = f(t), \tag{4.1}$$

with the initial conditions

$$u(0) = u^0, \qquad u'(0) = u^1. \tag{4.2}$$

We assume that the following conditions hold true for the operator matrices $A = (A_{ij})^2_{i,j=1}$ and B from (4.1)

$$A = A^* \in \mathcal{L}(H), \quad B = \operatorname{diag}(0, B_0), \qquad B_0 \gg 0. \tag{4.3}$$

To transform the problem (4.1)–(4.2) into an equation of the desired type, let us perform in (4.1) the substitution

$$\frac{\mathrm{d}u_3}{\mathrm{d}t} = \mathrm{i}B_0^{1/2}u_2, \quad u_3(0) = 0, \qquad u = (u_1; u_2)^{\mathrm{t}}. \tag{4.4}$$

Assuming that differentiation with respect to t is possible in (4.4), we have

$$\frac{\mathrm{d}^2 u_3}{\mathrm{d}t^2} = \mathrm{i}B_0^{1/2}\frac{\mathrm{d}u_2}{\mathrm{d}t}, \qquad u_3'(0) = \mathrm{i}B_0^{1/2}u_2(0). \tag{4.5}$$

Performing the substitution (4.4) in (4.1) and taking into account (4.5), we obtain the problem

$$\frac{\mathrm{d}^2\tilde{u}}{\mathrm{d}t^2} = \mathrm{i}\mathcal{A}\frac{\mathrm{d}\tilde{u}}{\mathrm{d}t} + h(t), \quad \tilde{u}(0) = \tilde{u}^0, \quad \tilde{u}'(0) = \tilde{u}^1, \tag{4.6}$$

where $\tilde{u} = (u_1; u_2; u_3)^{\mathrm{t}}$, $h(t) = (f_1(t); f_2(t); 0)^{\mathrm{t}}$, $\tilde{u}(0) = (u_1^0; u_2^0; 0)^{\mathrm{t}}$, $\tilde{u}^1 = (u_1^1; u_2^1; iB_0^{1/2}u_2^0)^{\mathrm{t}}$, and the operator matrix $\mathcal{A}$ has the form

$$\mathcal{A} := \begin{pmatrix} A_{11} & A_{12} & 0 \\ A_{21} & A_{22} & B_0^{1/2} \\ 0 & B_0^{1/2} & 0 \end{pmatrix}. \tag{4.7}$$

$\mathcal{A}$ is a self-adjoint operator that acts on the space $\tilde{H} = H_1 \oplus H_2 \oplus H_2$, and it is unbounded if the operator B_0 is unbounded.

According to Section 1.5.4, the form of equation (4.6) makes it possible to represent the solution $\mathrm{d}\tilde{u}/\mathrm{d}t$ by using the group of unitary operators

$$U(t) := \exp(it\mathcal{A}), -\infty < t < \infty. \tag{4.8}$$

We obtain

$$\frac{\mathrm{d}\tilde{u}}{\mathrm{d}t} = U(t)\tilde{u}^1 + \int_0^t U(t-\tau)h(\tau)\mathrm{d}\tau. \tag{4.9}$$

Integrating with respect to t, we finally get

$$\tilde{u}(t) = \tilde{u}^0 + \int_0^t U(\tau)\mathrm{d}\tau \cdot \tilde{u}^1 + \int_0^t \mathrm{d}\tau \int_0^\tau U(\tau - \xi)h(\xi)\mathrm{d}\xi. \tag{4.10}$$

Let us next assume that $f(t)$ in (4.1) is a continuous function of t with values in H, and u^0, u^1 are elements from H, with $u_2^0 \in \mathcal{D}(B_0^{1/2})$. Then, in the Cauchy problem (4.6), we have that $\tilde{u}^0 \in \tilde{H}$, $\tilde{u}^1 \in \tilde{H}$, and $h(t)$ is a continuous function with values in $\tilde{H}$. Hence, from (4.9) it follows that the function $\mathrm{d}\tilde{u}/\mathrm{d}t$ is a continuous function of t with values in $\tilde{H}$ and, therefore, the solution $\tilde{u}(t)$, which is represented by formula (4.10), is a continuously differentiable function of t.

Considering the initial problem (4.1) and (4.2), we are now in a position to formulate the final conclusion on its solvability.

If $f(t)$ is a continuous function of t with values in $H, u^i \in H$, $i = 0,1$, and $u_2^0 \in \mathcal{D}(B_0^{1/2})$, then the Cauchy problem (4.1)–(4.2) has a generalized solution $u(t)$, which is continuously differentiable and such that the function $u_2(t)$ belongs to $\mathcal{D}(B_0^{1/2}) \subset H_2$ for every t and the function $B_0^{1/2}u_2(t)$ is continuous in t.

For this generalized solution, twice the full energy of the system,

$$\left\|\frac{d\tilde{u}}{dt}\right\|_{\tilde{H}}^2 = \left\|\frac{du}{dt}\right\|_H^2 + \left\|B^{1/2}u\right\|_H^2 = \left\|\frac{du_1}{dt}\right\|_{H_1}^2 + \left\|\frac{du_2}{dt}\right\|_{H_2}^2 + \left\|B_0^{1/2}u_2\right\|_{H_2}^2, \tag{4.11}$$

is a continuous function of t, and the full energy balance law takes place for this function,

$$\left\|\frac{du}{dt}\right\|_H^2 + \left\|B^{1/2}u\right\|_H^2 = \|u^1\|_H^2 + \left\|B_0^{1/2}u_2^0\right\|_{H_2}^2 + 2Re\int_0^t \left(f(\tau), \frac{du}{d\tau}\right)_H d\tau. \tag{4.12}$$

For $f(t) \equiv 0$, we get the full energy conservation law.

The proof of the equality (4.12) can be carried out in the following way. According to (4.11) and the representation (4.9) we have

$$\left\|\frac{du}{dt}\right\|_H^2 + \left\|B^{1/2}u\right\|_H^2 = \left\|\frac{d\tilde{u}}{dt}\right\|_{\tilde{H}}^2 = \left(U(t)\tilde{u}^1 + \int_0^t U(t-\tau)h(\tau)d\tau, \frac{d\tilde{u}}{d\tau}\right)_{\tilde{H}}, \tag{4.13}$$

where $U(t) = \exp(it\mathcal{A})$.

The scalar product in the right side of (4.13) equals the expression

$$\begin{aligned}
&\left(\tilde{u}^1, U(-t)\frac{d\tilde{u}}{du}\right)_{\tilde{H}} + \int_0^t \left(h(\tau), U(\tau - t)\frac{d\tilde{u}}{dt}(t)\right)_{\tilde{H}} d\tau \\
&= \left(\tilde{u}^1, \tilde{u}^1 + \int_0^t \exp(i\tau A)h(\tau)d\tau\right)_{\tilde{H}} + \int_0^t (h(\tau), \exp(-i\tau\mathcal{A})\tilde{u}^1)_{\tilde{H}}\, d\tau \\
&+ \int_0^t \left(h(\tau), \int_0^t \exp(i(\tau - \xi)\mathcal{A})h(\xi)d\xi\right)_{\tilde{H}} d\tau.
\end{aligned} \tag{4.14}$$

In $\int_0^t \exp(i(\tau - \xi)\mathcal{A})h(\xi)d\xi$, we divide the integration interval into two parts, from 0 to τ and from τ to t, and change the order of integration in the so-obtained double integral.

Finally, the right side of (4.14) takes the form

$$\|\tilde{u}^1\|_{\tilde{H}}^2 + 2Re\int_0^t \left(h(\tau), \frac{d\tilde{u}}{d\tau}\right)_{\tilde{H}} d\tau,$$

which coincides with the right side of (4.12) in virtue of the definitions of $\tilde{u}^1$ and $h(\tau)$.

6.4.2 Small Movements of Fluid in a Partially Filled Container

The general line of reasoning carried out in Section 6.4.1 can be applied to the initial boundary value problem (3.28) on oscillations of a fluid, which is rotating in a partially filled container. Comparing these two problems, one can see that operator A in equation (4.1) should be replaced by $2\omega_0 A$, with $A = A^* \in \mathcal{L}(\boldsymbol{H})$, $\|A\| = 1$, and operator B_0 in (4.1) and (4.3) should be replaced by operator $\sigma\rho^{-1}B_0$. Hence, problem (3.28) for y^0, y^1 from $\boldsymbol{J}_0(\Omega) \oplus L_{2,\Gamma}$, $\eta^0 \in \mathcal{D}(B_0^{1/2})$ and a continuous function $f(t)$ with values in $\boldsymbol{J}_0(\Omega) \oplus L_{2,\Gamma} = \boldsymbol{H}$ is univalently solvable and has a generalized solution with continuous full energy. For this solution the full energy balance law holds true.

For the initial boundary value problem (3.6)–(3.11), this reasoning proves that whenever the initial fields $\boldsymbol{w}^0$ and $\boldsymbol{u}^0$ in (3.11) belong to the space $\boldsymbol{L}_2(\Omega)$, $(\boldsymbol{w}^0 \cdot \boldsymbol{n})_\Gamma \in \mathcal{D}(B_1^{1/2}) \subset L_{2,\Gamma}$ (these requirements guarantee the finiteness of kinetic and potential energies when $t = 0$), and the field of external forces $\boldsymbol{f}(t,x)$ is a continuous function of t with values in $\boldsymbol{L}_2(\Omega)$, then the problem is univalently solvable and has a generalized solution $\boldsymbol{w}(t,x)$, for which the energy balance law (4.1.40) holds true.

6.4.3 On the Structure of the Spectrum of a Vortical Operator

Further, based on equation (3.28), we will consider some special cases of free movements of a fluid for $f(t) \equiv 0$.

First of all, let us determine the structure of the kernel of operator A_{11} in a certain typical situation. The operator A_{11} can be naturally called a *vortical operator.* We suppose that the container and the region Ω are axially symmetric relatively to the axis Ox_3, and the equations of the equilibrium surface Γ and the solid boundary S in the cylindrical coordinate system $Or\theta x_3$ are defined by continuously differentiable functions, namely,

$$x_3 = \varphi_\Gamma(r), \quad x_3 = \varphi_S(r), \quad r \in [0, r_0],$$

where $r = \sqrt{x_1^2 + x_2^2}$. Therefore, if the following condition holds true,

$$\varphi_\Gamma'(r) \neq \varphi_S'(r), \qquad r \in (0, r_0), \tag{4.15}$$

then the kernel $\ker A_{11}$ of operator A_{11} has the form

$$\ker A_{11} = \left\{ \boldsymbol{w} \in \boldsymbol{J}_0(\Omega) : w_3 \equiv w_r \equiv 0,\ w_\theta = r\omega(r),\ \int_0^{r_0} r^3\omega^2(r)\mathrm{d}r < \infty \right\}. \tag{4.16}$$

Let us prove (4.16) for smooth fields $\boldsymbol{w}$ from $\boldsymbol{J}_0(\Omega)$. According to (3.26), $A_{11}\boldsymbol{w} := -\mathrm{i}P_0(\boldsymbol{w} \times \boldsymbol{e}_3)$. Thus, in virtue of the orthogonal decomposition $\boldsymbol{L}_2(\Omega) = \boldsymbol{J}_0(\Omega) \oplus \boldsymbol{G}(\Omega)$, the equality $A_{11}\boldsymbol{w} = \boldsymbol{0}$ gives $\boldsymbol{w} \times \boldsymbol{e}_3 = \nabla\psi \in G(\Omega)$. Hence we obtain

$$w_1 = -\frac{\partial\psi}{\partial x_3}, \quad w_2 = \frac{\partial\psi}{\partial x_1}, \quad \psi = \psi(x_1, x_2). \tag{4.17}$$

From the condition $\operatorname{div} \boldsymbol{w} = 0$ in Ω and (4.17) we can deduce that

$$\frac{\partial w_3}{\partial x_3} = 0, \qquad w_3 = f(x_1, x_2). \tag{4.18}$$

For the axially symmetric container Ω, let us write down the boundary condition on $\partial\Omega$ for the field $\boldsymbol{w} \in \boldsymbol{J}_0(\Omega)$ on the surfaces Γ and S in the cylindrical coordinate system $Or\theta x_3$. On the surface Γ it has the form

$$w_n \equiv w_r \cos(\widehat{\boldsymbol{n}, \boldsymbol{e}_r}) + w_\theta \cos(\widehat{\boldsymbol{n}, \boldsymbol{e}_\theta}) + w_3 \cos(\widehat{\boldsymbol{n}, \boldsymbol{e}_3}) = 0.$$

We note that

$$\begin{aligned}
&\cos(\widehat{\boldsymbol{n}, \boldsymbol{e}_\theta}) = 0, \\
&\cos(\widehat{\boldsymbol{n}, \boldsymbol{e}_3}) = \left(1 + (\varphi_\Gamma'(r))^2\right)^{-1/2}, \\
&\cos(\widehat{\boldsymbol{n}, \boldsymbol{e}_r}) = -\varphi_\Gamma'(r) \left(1 + (\varphi_\Gamma'(r))^2\right)^{-1/2},
\end{aligned}$$

and obtain, with regard to (4.17) and (4.18), that

$$w_3 = \varphi_\Gamma'(r) w_r = \frac{1}{r}\frac{\partial \psi}{\partial \theta}\varphi_\Gamma'(r) = f(r, \theta). \tag{4.19}$$

The similar condition on S gives

$$w_3 = \frac{1}{r}\frac{\partial \psi}{\partial \theta}\varphi_S'(r). \tag{4.20}$$

Equations (4.19) and (4.20) do not depend on x_3 and are true for all $r \in (0, r_0)$, $\theta \in [0, 2\pi]$, that is, they are valid for all the points $x \in \Omega$. Therefore, by equating the right sides of (4.19) and (4.20) we get

$$\frac{\partial \psi}{\partial \theta}\left[\varphi_\Gamma'(r) - \varphi_S'(r)\right] \equiv 0.$$

From the latter and (4.15), it follows that $\partial\psi/\partial\theta \equiv 0$, that is, $\psi = \psi(r)$. Then from (4.17) and (4.19) we obtain

$$w_3 \equiv 0, \quad w_r \equiv 0, \quad w_\theta = -\frac{\mathrm{d}\psi}{\mathrm{d}r} =: r\omega(r). \tag{4.21}$$

Here, $\omega(r)$ is an arbitrary function, which for $\boldsymbol{w} \in \boldsymbol{J}_0(\Omega) \subset \boldsymbol{L}_2(\Omega)$ satisfies only the following finiteness condition

$$\int_0^{r_0} |w_\theta|^2 r \mathrm{d}r = \int_0^{r_0} r^3 \omega^2(r) \mathrm{d}r < \infty. \tag{4.22}$$

The property (4.16) is proved.

Hence, in an axially symmetric container Ω, the kernel $\ker A_{11}$ of the vortical operator A_{11} does not equal zero. Moreover, this kernel is infinite-dimensional because the function $\omega(r)$ in (4.22) is arbitrarily chosen. Here, the natural yet unsolved question would be the following: Does the operator A have any nonzero eigenvalues?

Let us show that the end-points of the spectrum of operator A_{11}, that is, the points ± 1, are not eigenvalues.

We will prove this fact for the point $\omega = +1$. Let us consider the equation

$$A_{11}\boldsymbol{u} := -\mathrm{i}P_0(\boldsymbol{u} \times \boldsymbol{e}_3) = \boldsymbol{u}, \qquad \boldsymbol{u} \in \boldsymbol{J}_0(\Omega), \tag{4.23}$$

and prove that it only has the trivial solution. Since $P_0\boldsymbol{u} = \boldsymbol{u}$, then from (4.23) it follows that

$$-\mathrm{i}(\boldsymbol{u} \times \boldsymbol{e}_3) - \boldsymbol{u} = \nabla p, \qquad \nabla p \in \boldsymbol{G}(\Omega), \tag{4.24}$$

and the orthogonality of the subspaces $\boldsymbol{J}_0(\Omega)$ and $\boldsymbol{G}(\Omega)$ gives the identity

$$\mathrm{i}\int_\Omega (\boldsymbol{u} \times \boldsymbol{e}_3) \cdot \nabla\varphi \mathrm{d}\Omega = -\int_\Omega \nabla p \cdot \nabla\varphi \mathrm{d}\Omega, \tag{4.25}$$

which holds true for any element $\nabla\varphi \in \boldsymbol{G}(\Omega)$. Now, we show that $\nabla p(x) \equiv \boldsymbol{0}$. Indeed, let us set $\varphi(x) \equiv \overline{p(x)}$ in (4.25). Hence,

$$\int_\Omega |\nabla p|^2 \mathrm{d}\Omega = -\mathrm{i}\int_\Omega (\boldsymbol{u} \times \boldsymbol{e}_3) \cdot \overline{\nabla p} \mathrm{d}\Omega. \tag{4.26}$$

However, from (4.24) it follows that

$$\frac{\partial p}{\partial x_1} = i\frac{\partial p}{\partial x_2} = -(u_1 + \mathrm{i}u_2), \ \frac{\partial p}{\partial x_3} = -u_3, \tag{4.27}$$

and from (4.26) we obtain

$$\int_\Omega |\nabla p|^2 \mathrm{d}\Omega = \int_\Omega |u_1 + \mathrm{i}u_2|^2 \mathrm{d}\Omega = \int_\Omega \left|\frac{\partial p}{\partial x_1}\right|^2 \mathrm{d}\Omega = \int_\Omega \left|\frac{\partial p}{\partial x_2}\right|^2 \mathrm{d}\Omega. \tag{4.28}$$

Hence, it can be concluded that $\nabla p \equiv \boldsymbol{0}$. Then, from (4.27) it follows that

$$u_3 \equiv 0, \quad u_1 + iu_2 \equiv 0 \quad \text{in } \Omega. \tag{4.29}$$

From (4.25) and the equality $\nabla p \equiv \boldsymbol{0}$ we obtain

$$\int_\Omega (\boldsymbol{u} \times \boldsymbol{e}_3) \cdot \nabla\varphi \mathrm{d}\Omega = 0, \qquad \text{for any } \varphi \in H^1_{\partial\Omega}(\Omega). \tag{4.30}$$

From (4.30) we have

$$\int_\Omega \left(u_2\frac{\partial\varphi}{\partial x_1} - u_1\frac{\partial\varphi}{\partial x_2}\right) \mathrm{d}\Omega = \int_\Omega \left(\frac{\partial u_1}{\partial x_2} - \frac{\partial u_2}{\partial x_1}\right)\varphi \mathrm{d}\Omega = 0, \qquad \text{for any } \varphi \in H_0^1(\Omega). \tag{4.31}$$

It means that

$$\operatorname{div}(\boldsymbol{u} \times \boldsymbol{e}_3) = \frac{\partial u_1}{\partial x_2} - \frac{\partial u_2}{\partial x_1} = 0. \tag{4.32}$$

By (4.32), the existence of the next potential follows,

$$\Phi(x_1, x_2, x_3^0) = \int_{(x_1^0, x_2^0, x_3^0)}^{(x_1, x_2, x_3^0)} (u_1 \mathrm{d}x_1 + u_2 \mathrm{d}x_2), \qquad \boldsymbol{u} = \nabla_2\Phi.$$

This potential describes the movements of the fluid in the plane $x_3 = x_3^0 = \text{const}$. Since $\boldsymbol{u} \in \boldsymbol{J}_0(\Omega)$ and $u_3 \equiv 0$, then $\Delta_2\Phi = \operatorname{div}\boldsymbol{u} = 0$. If condition $\cos(\widehat{\boldsymbol{n}, \boldsymbol{e}_3}) \neq 1$ holds true on the boundary $\partial\Omega \cap \{x_3 = x_3^0\}$, then on this boundary, in virtue of the nonleaking condition $u_n = 0$, we obtain $\partial\Phi/\partial n_2 = 0$. Therefore, $\Phi = \text{const}(x_3^0)$, that is, $\boldsymbol{u} \equiv \boldsymbol{0}$. To conclude, for a wide class of cavities Ω, the point $\omega = 1$ is not an eigenvalue of the operator A_{11}. The case when $\omega = -1$ can be treated similarly.

As for the points in the spectrum of operator A_{11} included in the interval $[-1, 1]$ that are neither equal to $\omega = 0$ nor $\omega = \pm 1$, they belong to the essential spectrum $\sigma_{\mathrm{e}}(A_{11})$ as it was previously mentioned. In this case, for certain shapes of the containers, like for example, for a cylindrical container with an arbitrary section, the point spectrum $\sigma_{\mathrm{p}}(A_{11})$ of the vortical operator forms a dense set in $[-1, 1]$. The point spectrum in an elipsoid-of-rotation shaped region Ω has the same property. For differently shaped containers Ω, in particular, for a toroidal container, there are parts of the interval $[-1, 1]$ that are filled with the continuous spectrum.

6.4.4 Classes of Free Movements

Let us consider now some important special cases of free movements of the system based on the properties of the spectrum of the vortical operator A_{11}. We assume in (3.28) that $f(t) \equiv 0$. In this case, whenever the full energy conservation law holds true, the field $\boldsymbol{u}$ of small displacements of fluid particles relative to a uniform rotation cannot have anything but a linear growth in t. We would like to find out the characteristic special features of the solutions of the problem

$$\frac{\mathrm{d}^2 y}{\mathrm{d}t^2} - 2\mathrm{i}\omega_0 A \frac{\mathrm{d}y}{\mathrm{d}t} + \sigma\rho^{-1} B y = 0, \qquad y = (\boldsymbol{v}; \eta)^{\mathrm{t}}, \tag{4.33}$$

that are given by functions $y(t)$ with linear growth in t, that is,

$$y(t) = t\varphi + \psi. \tag{4.34}$$

A substitution of (4.34) into (4.33) leads to the conditions

$$\begin{aligned} \sigma\rho^{-1}B\varphi &= \left(0; \sigma\rho^{-1}B_0\varphi_2\right)^{\mathrm{t}} = \mathbf{0}, \\ 2\mathrm{i}\omega_0 A\varphi &= \sigma\rho^{-1}B\psi = \left(0; \sigma\rho^{-1}B_0\psi_2\right)^{\mathrm{t}}, \end{aligned}$$

hence we conclude that

$$\begin{aligned} B_0\varphi_2 &= 0, \\ 2\mathrm{i}\omega_0(A_{11}\varphi_1 + A_{12}\varphi_2) &= 0, \\ 2\mathrm{i}\omega_0(A_{21}\varphi_1 + A_{22}\varphi_2) &= \sigma\rho^{-1}B_0\psi_2. \end{aligned}$$

Since $B_0 \gg 0$, then $\varphi_2 = 0$ and, therefore,

$$A_{11}\varphi_1 = 0, \quad -2\mathrm{i}\omega_0 A_{21}\varphi_1 + \sigma\rho^{-1}B_0\psi_2 = 0.$$

If the kernel $\ker A_{11}$ of the vortical operator A_{11} is not equal to zero, then it can be shown that $\psi_2 = 2\mathrm{i}\omega_0\rho\sigma^{-1}B_0^{-1}A_{21}\varphi_1$, where $\varphi_1 \in \ker A_{11} \neq \{\mathbf{0}\}$, and the element ψ_1 in the representation $\psi = (\psi_1; \psi_2)^{\mathrm{t}}$ can be an arbitrary element.

This line of reasoning allows us to derive the following physical conclusions, which separate the movements that are linear by t into classes.

1. Let $\varphi = 0$, $\psi = (\boldsymbol{v}^0, 0)^{\mathrm{t}}$, where $\boldsymbol{v}^0$ is an arbitrary element from $\boldsymbol{J}_0(\Omega)$. Then, the solution of problem (4.33) has the form $y(t) = y^0 = \psi$. To this solution (let us call it a *trivial movement of the first type*) there corresponds a field of displacements from the vortical subspace $\boldsymbol{J}_0(\Omega)$ that is constant in time and the zero field from the potential harmonic subspace $\boldsymbol{G}_{h,S}(\Omega)$. Since the normal components of the fields in $\boldsymbol{J}_0(\Omega)$ equal zero on $\partial\Omega$, it follows that the free surface Γ does not move away from this solution, and the perturbation of the field of pressures equals zero.

2. Now, let the functions φ and ψ in (4.34) be of the form

$$\psi = (0; \eta^0)^{\mathrm{t}}, \quad \varphi = (\boldsymbol{v}^1; 0)^{\mathrm{t}}, \qquad \eta^0 = 2\mathrm{i}\omega_0\rho\sigma^{-1}B_0^{-1}A_{21}\boldsymbol{v}^1, \tag{4.35}$$

with

$$\boldsymbol{v}^1 \in \boldsymbol{J}_0(\Omega), \qquad \boldsymbol{v}^1 = \tilde{\omega}_0 P_0(\boldsymbol{e}_3 \times \boldsymbol{r}), \tag{4.36}$$

where $\boldsymbol{r}$ is the radius-vector of the point $x \in \Omega$. Physically, conditions (4.35) and (4.36) mean that at the initial moment of time $t = 0$ the fluid rotates as a rigid body with the additional angular velocity $\tilde{\omega}_0$, that is, the initial field of velocities equals $\tilde{\omega}_0(\boldsymbol{e}_3 \times \boldsymbol{r})$.

Obviously, for such initial data, the field of deviations $\boldsymbol{v}(t)$ of the fluid particles from the vortical subspace $\boldsymbol{J}_0(\Omega)$ should grow linearly with time (in the nonstationary coordinate system $Ox_1x_2x_3$, which rotates relatively to the stationary system with the angular velocity ω_0), and this field should be constant in the potential harmonic subspace $\boldsymbol{G}_{h,S}(\Omega)$. Indeed, for the above mentioned rotation of the whole fluid, its free surface should occupy a new stationary position $\tilde{\Gamma}$, corresponding to the angular velocity of the system, which equals $\omega_0 + \tilde{\omega}_0$. This new form $\tilde{\Gamma}$ of the free surface is defined by the deviation $\zeta = \zeta^0 = C^{-1/2}\eta^0 = 2\mathrm{i}\omega_0\rho\sigma^{-1}C^{-1/2}B_0^{-1}A_{21}\boldsymbol{v}^1$ of the surface $\tilde{\Gamma}$ from Γ.

Hence, in the second case, the solution of the problem has the form (4.34) and (4.35), where $\boldsymbol{v}^1$ is an element from $\boldsymbol{J}_0(\Omega)$. Let us call such solutions *trivial movements of the second type.*

3. Physically, it is obvious that there exists a sufficiently large set of initial functions $\boldsymbol{v}^1 \in \boldsymbol{J}_0(\Omega)$ for which the solution $y(t)$ is a trivial movement of the second type. Indeed, since the kernel of the vortical operator A_{11} can be infinite-dimensional, then all the elements $\boldsymbol{v}^1 \in \ker A_{11}$ generate trivial movements of the second type according to the formulas (4.34) and (4.35). In particular, the following movement can be considered as a movement of the second type: The layers of fluid of thickness dr, which are located at the distance r from the axis of rotation, have an arbitrary angular velocity $\omega(r)$ at the initial moment of time, and the integral (4.16) converges for that angular velocity. Here, the free surface of the fluid takes the stationary position $\tilde{\Gamma}$ described by the function

$$\zeta^0 = C^{-1/2}\eta^0 = 2i\omega_0\rho\sigma^{-1}C^{-1/2}B_0^{-1}A_{21}\boldsymbol{v}^1, \tag{4.37}$$

with $\boldsymbol{v}^1 = P_0\left[\omega(r)(\boldsymbol{e}_3 \times \boldsymbol{r})\right]$, and $\boldsymbol{r} = \sqrt{x_1^2 + x_2^2}$.

6.4.5 Free Movements of a System of Fluids

All the above mentioned classes of free movements also occur in the evolution problem for a system of fluids subject to certain modifications and additional requirements. First, let us determine the structure of the kernel of the vortical operator $\hat{A}_{11}$ in (3.71).

Since $\hat{A}_{11} := \operatorname{diag}(A_{kk})_{k=1}^m$ and the operator A_{kk}, for $k = 1, 2, \ldots, m$, has a nonzero kernel $\ker A_{kk}$ of the form (4.16) in the axially symmetric case, then $\ker A_{kk}$ is also infinite-dimensional. The following analysis gives more specific information.

In the axially symmetric case, we assume that the equations of the solid boundary S and the equilibrium surfaces Γ_j, $j = 1, 2, \ldots, m-1$, have the form

$$x_{3j} = \varphi_j(r), \qquad 0 \le r \le r_j,\ j = 0, 1, \ldots, m, \tag{4.38}$$

in the cylindrical coordinate system $Or\theta x_3$, where the lower and upper generatrixes of S correspond to the indexes $j = 0$, $j = m$, and the functions $\varphi_j(r)$ are continuously differentiable. If the following conditions of the form (4.15) hold true,

$$\varphi_j'(r) \neq \varphi_{j+1}'(r), \qquad j = 0, 1, \ldots, m-1, \tag{4.39}$$

then the kernel $\ker \hat{A}_{11}$ of the vortical operator for the system of fluids has the form

$$\ker \hat{A}_{11} = \Big\{ \hat{\boldsymbol{w}} := \{\boldsymbol{w}^k\}_{k=1}^m \in \hat{\boldsymbol{J}}_0(\Omega) :$$
$$w_3^k \equiv w_r^k \equiv 0,\ w_\theta^k = r\omega_k(r),\ \sum_{k=1}^m \int_0^{r_k} r^3 \omega_k^2(r) \mathrm{d}r < \infty \Big\}; \tag{4.40}$$

for the parts of the interval $(0, r_j)$ where

$$\varphi_j'(r) \equiv \varphi_{j+1}'(r), \tag{4.41}$$

the functions $\hat{\boldsymbol{w}} \in \ker \hat{A}_{11}$ have the following components

$$\begin{aligned} w_r^k &= \frac{1}{r}\frac{\partial \psi_k}{\partial \theta}, \\ w_\theta^k &= -\frac{\partial \psi_k}{\partial r}, \\ w_3^k &= \frac{1}{r}\frac{\partial \psi_k}{\partial \theta}\varphi_k'(r) = \frac{1}{r}\frac{\partial \psi_k}{\partial \theta}\varphi_{k-1}'(r), \end{aligned} \tag{4.42}$$

where $\psi_k = \psi_k(r, \theta)$ are arbitrary functions that satisfy only the general requirement

$$\sum_{k=1}^m \int_0^{2\pi} \mathrm{d}\theta \int_0^{r_k} \left\{ \left| \frac{\partial \psi_k}{\partial r} \right|^2 + \frac{1}{r^2} \left| \frac{\partial \psi_k}{\partial \theta} \right|^2 \left[1 + (\varphi_k'(r))^2 \right] \right\} r \mathrm{d}r < \infty. \tag{4.43}$$

The proof of this assertion is completely analogous to the proof of the similar result in the case of a single fluid given in Section 6.4.3. Namely, after obtaining the relations

$$w_3^k = \frac{1}{r}\frac{\partial \psi_k}{\partial \theta}\varphi_{k-1}'(r), \quad w_3^k = \frac{1}{r}\frac{\partial \psi_k}{\partial \theta}\varphi_k'(r), \qquad k = 1, 2, \ldots, m,$$

which are similar to formulas (4.19) and (4.20), we have the equations

$$\frac{\partial \psi_k}{\partial \theta}\left[\varphi'_{k-1}(r)-\varphi'_k(r)\right]\equiv 0, \qquad \text{for any } \theta\in[0,2\pi],\ k=1,2,\dots,m,$$

where $\psi_k = \psi_k(r,\theta)$. Hence, if conditions (4.39) hold true, then we obtain the first assertion on the structure of the kernel $\hat{A}_{11}$, that is, (4.40). For those k and parts of the interval $(0,r_k)$ where one of the conditions (4.41) holds true, the functions $\psi_k(r,\theta)$ can be arbitrary. Hence, the second part of the proved assertion follows.

Now, let us note the following important fact. For a system of capillary fluids that rotate in an axially symmetrical container, the situation when conditions (4.39) holds true is typical: ker $\hat{A}_{11}$ is described by the set (4.40). If a system of heavy fluids is being considered, that is, when all $\sigma_j = 0$, $j = 1,2,\dots,m-1$, then the equilibrium surfaces Γ_j are paraboloids of rotation. In this case, according to formulas (3.42), the functions $\varphi_j(r)$, $j = 1,2,\dots,m-1$, are equal to

$$\varphi_j(r) = (2g)^{-1}\omega_0^2 r^2 + c_j$$

and, therefore, conditions (4.41) hold true. The functions in ker $\hat{A}_{11}$ are described by the relations (4.42) and (4.43).

Considering the trivial free movements of a system of nonmixing fluids, that is, the solutions of problem (3.73) for $f(t)\equiv 0$ that depend linearly on t, let us note that here, for the solutions of the first type, $y(t) = y^0 = (\hat{\boldsymbol{v}}^0;0)^{\mathrm{t}}$ for any $\hat{\boldsymbol{v}}^0 \in \hat{\boldsymbol{J}}_0(\Omega)$. For such a solution all the equilibrium surfaces Γ_j do not move, and the perturbed field of pressures equals zero in Ω_k. As an example of trivial movements of the second type, one can consider the movement of the system, when the fluid in region Ω_k is being rotated with the angular velocity $\tilde{\omega}_{0k}$. Then, the boundaries Γ_j occupy the stationary position during the whole time of movement,

$$\hat{\zeta}_0 = \{\zeta_j\}_{j=1}^{m-1} = \hat{C}^{-1/2}\hat{\eta}^0 = 2i\omega_0\hat{C}^{-1/2}\hat{B}_0^1\hat{A}_{21}\hat{\boldsymbol{v}},$$

with

$$\hat{\boldsymbol{v}} = \hat{P}_0\{\boldsymbol{v}^k\}_{k=1}^m = \{\tilde{\omega}_{0k}P_{0k}(\boldsymbol{e}_3\times\boldsymbol{r})\},$$

and the field of deviations $\boldsymbol{v}^k$ in the region Ω_k grows linearly in time. Finally, the family of functions $\hat{\boldsymbol{v}}$ that belong to ker $\hat{A}_{11}$ makes it possible to obtain the next set of more complicated trivial movements of the second type,

$$y(t) = \varphi t + \psi,$$

with $\varphi = (\hat{\boldsymbol{v}};0)^{\mathrm{t}}$, $\hat{\boldsymbol{v}}\in$ ker $\hat{A}_{11}$ and $\psi = (0;\hat{\beta})^{\mathrm{t}}$, $\hat{\beta}^0 = 2i\omega_0\hat{B}_0^{-1}A_{21}\hat{\boldsymbol{v}}$. Here, the totality of such movements for a system of heavy fluids turns out to be a larger set than the one for a system of capillary fluids.

6.5 Self-adjoint Operator Pencils Generated by Problems on Oscillations of a Rotating Ideal Fluid

In this section we consider the spectral problem corresponding to the evolution problem studied in Section 6.4. In the next section, based on this problem, we will investigate the properties of surface and interval waves that occur in a rotating fluid which partially fills the container.

6.5.1 The Main Operator Pencil

Let us assume that operators A and B of the evolution problem (4.1) possess the additional properties

$$\|A\| = 1, \quad \sigma(A_{11}) = [-1, 1], \quad B_0^{-1} \in \mathfrak{S}_\infty, \tag{5.1}$$

and look for some proper oscillations, that is, solutions of the problem (4.1) for $f(t) \equiv 0$ that depend on time according to the law $\exp(\mathrm{i}\omega t)$. Then, instead of (4.1) we obtain the following spectral problem,

$$L(\omega)v := (B + \omega A - \omega^2 I)v = 0, \qquad v \in H = H_1 \oplus H_2. \tag{5.2}$$

Since B and A are self-adjoint operators, the quadratic pencil $L(\omega)$ with parameter ω is self-adjoint, too. Let us show that the spectrum of the resulting spectral problem (5.2) is real. Indeed, for $\omega \neq 0$ a substitution similar to the one in (4.4) can be performed in (5.2), that is, we can assume that

$$B_0^{1/2} v_2 := \omega v_3, \qquad v_3 \in H_2. \tag{5.3}$$

Then, problem (5.2) is equivalent to the spectral problem

$$\mathcal{A}\tilde{v} = \omega\tilde{v}, \qquad \tilde{v} = (v_1; v_2; v_3)^{\mathrm{t}} \in \tilde{H}, \tag{5.4}$$

where the operator matrix $\mathcal{A}$ is defined as in (4.7) and $\tilde{H} = H_1 \oplus H_2 \oplus H_2$. The operator $\mathcal{A}$ is self-adjoint and also unbounded due to the unboundedness of operator B_0 and it is defined in the dense set $\mathcal{D}(\mathcal{A}) = H_1 \oplus \mathcal{D}\left(B_0^{1/2}\right) \oplus \mathcal{D}\left(B_0^{1/2}\right) \subset \tilde{H}$. Its spectrum, $\sigma(\mathcal{A})$, is real and, therefore, the spectrum $\sigma(L(\omega))$ of the pencil $L(\omega)$ is real, too. Since the eigenelements of the operator $\mathcal{A} = \mathcal{A}^*$ do not have any associated elements, the pencil $L(\omega)$ has the same property for $\omega \neq 0$. However, the pencil $L(\omega)$ can have associated elements for $\omega = 0$.

It is obvious that the eigenspace Z_0 of the pencil $L(\omega)$ that corresponds to the eigenvalue $\omega = 0$ is infinite-dimensional and consists of elements of the form $v^0 = (v_1^0; 0)^t$, with $v_1^0 \in H_1$.

Let us consider now the associated elements that correspond to the eigenvalue $\omega = 0$. The following equations should hold true,

$$L(0)v^0 = 0, \qquad L(0)v^1 + L'(0)v^0 = 0,$$

or,

$$A_{11}v_1^0 = 0, \qquad B_0 v_2^1 + A_{21}v_1^0 = 0,$$

for the first element associated to the element $v^0 = (v_1^0, 0)^t$, namely, $v^1 = (v_1^1; v_2^1)^t$. Hence, only eigenelements of the form $(v_1^0; 0)^t$, where $v_1^0 \in \ker A_{11}$, can have associated elements. Here, $v_2^1 = -B_0^{-1}A_{21}v_1^0$, and the component v_1^1 is arbitrary. It turns out that there are no second associated elements. Indeed, the equation

$$L(0)v^2 + L'(0)v^1 + \frac{1}{2}L''(0)v^0 = 0$$

reduces to the following,

$$A_{11}v_1^1 + A_{21}v_2^1 - v_1^0 = 0, \quad B_0 v_2^2 + A_{21}v_1^1 + A_{22}v_2^1 = 0. \tag{5.5}$$

From the first equation in (5.5), after a scalar multiplication by v_1^0 we have

$$\|v_1^0\|_{H_1}^2 + \left\|B_0^{-1/2}A_{21}v_1^0\right\|_{H_2}^2 = 0.$$

Hence, it follows that $v_1^0 = 0$, a contradiction to the assumption that $v^0 = (v_1^0; 0)^t$ is an eigenelement.

Let us note that by using the eigen- and associated elements

$$v^0 = (v_1^0; 0)^t, \qquad v_1^0 \in \ker A_{11} \neq \{0\},\ v^1 = (0; -B_0^{-1}A_{21}v_1^0)^t \perp Z_0,$$

the solutions of the second type of the Cauchy problem (4.1) and (4.2) can be expressed as

$$u(t) = v^0 t - iv^1.$$

6.5.2 On the Spectrum of the Operator Pencil

Let us write down equation (5.4) in the vector-matrix form

$$\begin{pmatrix} A_{11} & A_{12} & 0 \\ A_{21} & A_{22} & B_0^{1/2} \\ 0 & B_0^{1/2} & 0 \end{pmatrix} \begin{pmatrix} v^1 \\ v^2 \\ v^3 \end{pmatrix} = \omega \begin{pmatrix} v^1 \\ v^2 \\ v^3 \end{pmatrix}. \tag{5.6}$$

In order to understand the type of spectrum in problem (5.6), let us assume that operators A_{12} and $A_{21} = A_{12}^*$ in the operator matrix $\mathcal{A}$ equal zero. Then, problem (5.6) splits into two separate problems,

$$A_{11}v_1 = \omega v_1, \qquad \begin{pmatrix} A_{22} & B_0^{1/2} \\ B_0^{1/2} & 0 \end{pmatrix} \begin{pmatrix} v^2 \\ v^3 \end{pmatrix} = \omega \begin{pmatrix} v_2 \\ v_3 \end{pmatrix}, \tag{5.7}$$

which can be easily investigated.

Indeed, according to (5.1), the spectrum of the first problem in (5.7) consists of the interval $[-1, 1]$. The second problem in (5.7) is the eigenvalue problem for an unbounded self-adjoint operator, which has a compact inverse operator,

$$D := \begin{pmatrix} 0 & B_0^{-1/2} \\ B_0^{-1/2} & -B_0^{-1/2}A_{22}B_0^{-1/2} \end{pmatrix}, \tag{5.8}$$

acting on $H_2^2 := H_2 \oplus H_2$. Hence, one gets that the second problem (5.7) is equivalent to the problem

$$D\varphi = \omega^{-1}\varphi, \qquad \varphi = (v_2; v_3)^{\mathrm{t}} \in H_2^2, \tag{5.9}$$

and, according to the Hilbert-Schmidt theorem, this problem has a real discrete spectrum that consists of eigenvalues with finite multiplicities and the limit points $\omega = \pm\infty$.

Therefore, when $A_{12} = A_{21}^* = 0$, the problems (5.6) and (5.2) have both their discrete and essential spectra in the interval $[-1, 1] =: I_0$. Let us show that this is also the case for problem (5.2) when $A_{12} = A_{21}^* \neq 0$.

6.5.3 Operator Pencils with Analytic Perturbations

With regard to the discrete spectrum of problem (5.2), let us consider the region on the real axis outside the interval $I_0 = [-1, 1]$ and, for $\omega \in \mathbb{R} \setminus I_0$, perform the substitution of parameter $\lambda = 1/\omega$ in equation (5.2), which is next written down in a vector-matrix form. The latter form leads to the equations

$$(I - \lambda A_{11})v_1 = \lambda A_{12}v_2, \qquad v_2 = \lambda(A_{21}v_1 + A_{22}v_2) + \lambda^2 B_0 v_2. \tag{5.10}$$

Since $|\lambda| < 1$ and $\|A\| = 1$, the operator-valued function $I - \lambda A_{11}$ is invertible, and the inverse operator-valued function, which is holomorphic in λ for $|\lambda| < 1$, is expressed by the Neumann series

$$\tilde{R}(\lambda) := (I - \lambda A_{11})^{-1} = \sum_{k=0}^{\infty} \lambda^k A_{11}^k. \tag{5.11}$$

Therefore, from the first equation in (5.10) we have $v_1 = \lambda\tilde{R}(\lambda)A_{12}v_2$. Substituting v_1 into the second equation we obtain

$$v_2 = \lambda A_{22}v_2 + \lambda^2 B_0 v_2 + \lambda^2 A_{21}\tilde{R}(\lambda)A_{12}v_2, \qquad |\lambda| < 1. \tag{5.12}$$

Equation (5.12) contains the unbounded operator $B_0 \gg 0$ and the holomorphic operator-valued function $\lambda^2 A_{21}\tilde{R}(\lambda)A_{12}$, which is self-adjoint. We eliminate the unbounded operator B_0 by introducing a new unknown quantity and consider a self-adjoint operator pencil that is an analytic perturbation of the linear pencil $D - \lambda I$. We perform the substitution (5.3) in (5.12) for $\omega = \lambda^{-1}$, that is, let us assume

$$v_3 = \lambda B_0^{1/2} v_2. \tag{5.13}$$

Applying operator $B_0^{-1/2}$ to (5.12), from (5.12) and (5.13) we obtain the system of equations

$$\begin{pmatrix} 0 & B_0^{-1/2} \\ B_0^{-1/2} & -B_0^{-1/2}A_{22}B_0^{-1/2} \end{pmatrix}\begin{pmatrix} v_2 \\ v_3 \end{pmatrix} = \lambda\begin{pmatrix} v_2 \\ v_3 \end{pmatrix} + \begin{pmatrix} 0 & 0 \\ 0 & B_0^{-1/2}A_{21}\tilde{R}(\lambda)A_{12}B_0^{-1/2} \end{pmatrix}\begin{pmatrix} v_2 \\ v_3 \end{pmatrix}, \tag{5.14}$$

where on the left side we have the compact operator D that was investigated above in (5.8) and (5.9). Let us note that for $A_{21} = A_{12}^* = 0$, problem (5.14) reduces to the nonperturbed problem (5.9).

Denoting the operator matrix in the right side of (5.14) by

$$\Phi(\lambda) := \operatorname{diag}(0; B_0^{-1/2}A_{21}\tilde{R}(\lambda)A_{12}B_0^{-1/2}), \qquad \tilde{R}(\lambda) = (I - \lambda A_{11})^{-1}, \tag{5.15}$$

we write down (5.14) as the following equation,

$$D\varphi = \lambda\varphi + \lambda\Phi(\lambda)\varphi, \qquad \varphi = (v_2; v_3)^{\mathrm{t}} \in H_2^2, \tag{5.16}$$

in the space $H_2^2 = H_2 \oplus H_2$. To equation (5.16), there corresponds a self-adjoint analytic pencil of operators,

$$A(\lambda) := \lambda I - D + \lambda\Phi(\lambda), \qquad |\lambda| < 1. \tag{5.17}$$

The self-adjointness of $A(\lambda)$ follows from the fact that operator D is self-adjoint and compact and, moreover, has a zero kernel. Further, the operator-valued function $\Phi(\lambda)$ defined according to (5.15) is also self-adjoint, because $A_{11}^* = A_{11}$, $A_{12} = A_{21}^*$, and $B_0^{-1/2}$ is a self-adjoint positive operator. For $|\lambda| < 1$, the function $\Phi(\lambda)$ takes compact values, because the operator-valued function $A_{21}\tilde{R}(\lambda)A_{12}$ takes values in $\mathcal{L}(H_2^2)$ and $B_0^{-1/2} \in \mathfrak{S}_\infty$.

Therefore, the operator-valued function $A(\lambda)$ is an analytic Fredholm operator pencil for $\lambda \in \mathbb{C} \setminus \{(-\infty, -1], [1, +\infty)\}$. Consequently, it has a discrete spectrum in this region, which is located on $\mathbb{R}$, with possible limit points $\lambda = 0$ and $\lambda = \pm 1$. Accordingly, the eigenvalues $\omega = \lambda^{-1}$ of problem (5.2) are of finite multiplicity and can have the points $\omega = \pm\infty$ and $\omega = \pm 1$ as limit points.

Subsequent investigations of solutions of the problem (5.2) for $|\omega| > 1$ are based on the study of properties of the operator pencil $A(\lambda)$.

6.5.4 Factorization of the Operator Pencil

We will use sufficient conditions for the factorization of the pencil $A(\lambda)$. Let us begin by estimating the norm of operator D.

According to the notation used in (5.8), for any $\varphi = (\varphi_1; \varphi_2)^{\mathrm{t}}$ from H_2^2 and any $\varepsilon > 0$ we have

$$\begin{aligned}
|(D\varphi, \varphi)_{H_2^2}| &= \left| \left(B_0^{-1/2}\varphi_2, \varphi_1\right)_{H_2} + \left(B_0^{-1/2}\varphi_1, \varphi_2\right)_{H_2} - \left(A_{22}B_0^{-1/2}\varphi_2, B_0^{-1/2}\varphi_2\right)_{H_2} \right| \\
&\leq 2\left\|B_0^{-1/2}\right\| \cdot \|\varphi_1\|_{H_2} \cdot \|\varphi_2\|_{H_2} + \left\|B_0^{-1/2}\right\|^2 \|\varphi_2\|^2_{H_2} \\
&\leq \left\|B_0^{-1/2}\right\| \left(\varepsilon^{-1}\|\varphi_1\|^2_{H_2} + \left(\varepsilon + \left\|B_0^{-1/2}\right\|\right) \|\varphi_2\|^2_{H_2}\right). \qquad (5.18)
\end{aligned}$$

To obtain the last inequality in (5.18) we used the estimate $\|A_{22}\| \leq 1$. Let us choose $\varepsilon > 0$ from the condition $\varepsilon^{-1} = \varepsilon + \left\|B_0^{-1/2}\right\|$ that gives

$$\varepsilon = \frac{\left(\sqrt{\|B_0^{-1/2}\|^2 + 4} - \|B_0^{-1/2}\|\right)}{2}.$$

Then, from (5.18) we have

$$|(D\varphi, \varphi)_{H_2^2}| \leq \frac{1}{2}\|B_0^{-1/2}\| \left(\sqrt{\|B_0^{-1/2}\|^2 + 4} + \|B_0^{-1/2}\|\right) \|\varphi\|^2_{H_2^2},$$

hence,

$$\|D\| \leq \|B_0^{-1/2}\| \left(\sqrt{4 + \|B_0^{-1/2}\|^2} + \|B_0^{-1/2}\|\right)/2. \qquad (5.19)$$

Now, we can use the sufficient conditions 2°—formulated at the end of Section 1.6.9—for the factorization of a self-adjoint operator pencil holomorphic in a neighborhood of the interval $[a, b] \subset \mathbb{R}$. In this case, we choose as $[a, b]$ the interval $[-q, q] \subset [-1, 1]$ for some q, $0 < q < 1$. Then, according to assertion 2° in Section 1.6.9, for the factorization of $A(\lambda)$ on the interval $[-q, q]$ it is sufficient to have

$$A(-q) \ll 0, \quad A(q) \gg 0, \quad A'(\lambda) \gg 0, \qquad \lambda \in [-q, q]. \qquad (5.20)$$

Let us first check that the first condition (5.20) holds true for any $q \in (0,1)$. Indeed, since

$$(\lambda(I-\lambda A_{11})^{-1})'_\lambda = (I-\lambda A_{11})^{-1} + \lambda A_{11}(I-\lambda A_{11})^{-2} = (I-\lambda A_{11})^{-2} \gg 0 \quad (5.21)$$

for any $\lambda \in [-q,q]$, then from (5.21) and (5.15) it follows that

$$(\lambda\Phi(\lambda))'_\lambda = \operatorname{diag}\left(0; B_0^{-1/2}A_{21}(I-\lambda A_{11})^{-2}A_{12}B_0^{-1/2}\right) \geq 0. \quad (5.22)$$

Therefore,

$$A'(\lambda) = I + (\lambda\Phi(\lambda))'_\lambda \geq I \gg 0. \quad (5.23)$$

Further, for $\lambda = q > 0$ we have the estimate

$$A(q) = qI - D + q\Phi(q) \geq (q - \|D\|)I, \quad (5.24)$$

where the property $\Phi(q) \geq 0$ was used, because $(I - qA_{11})^{-1} \gg 0$ for $0 < q < 1$ due to the property $\|A_{11}\| = 1$. Hence, it appears that if $\|D\| < q < 1$, then the second condition (5.20) holds true. Similarly, one can prove that these inequalities are sufficient for the first condition (5.20) as well.

The following conclusion is the final result of the previous reasoning:

If the condition

$$\|B_0^{-1/2}|\left(\sqrt{4+\|B_0^{-1/2}\|^2} + \|B_0^{-1/2}\right) < 2 \quad (5.25)$$

is satisfied, then $\|D\| < 1$, and there exists $q \in (\|D\|, 1)$ such that the pencil $A(\lambda)$ admits a factorization of the form

$$A(\lambda) = A_+(\lambda)(\lambda I - Z), \quad (5.26)$$

on the interval $[-q,q]$. Here, $\sigma(Z) \subset [-q,q] \subset [-1,1]$, the operator-valued function $A_+(\lambda)$ is holomorphic and holomorphically invertible in a neighborhood of the interval $[-q,q]$ and, moreover, in any open disk $|\lambda| < r \in (q,1)$.

It can be proved that inequality (5.25) is equivalent to the inequality $\|B_0^{-1/2}\|^2 = \|B_0^{-1}\| < 1/2$, that is, to the inequality

$$\lambda_{\min}(B_0) = \lambda_1(B_0) > 2. \quad (5.27)$$

Let us assume that the operator B_0 in (5.2), with $B = \operatorname{diag}(0; B_0)$, satisfies condition (5.27).

From (5.16), (5.17), and (5.26) for $\lambda = 0$, it follows that $-D = -A_+(0)Z$. In virtue of the invertibility of operator $A_+(\lambda)$, we obtain that the operator $Z = A_+^{-1}(0)D$ is compact.

Hence, if condition (5.27) holds true, then instead of the spectral problem (5.16) for the pencil $A(\lambda)$ one can consider the following problem, for $|\lambda| < 1$,

$$Z\varphi = \lambda\varphi, \qquad \varphi = (\varphi_1; \varphi_2)^{\mathrm{t}} \in H_2^2. \tag{5.28}$$

Since for the operator Z there exists a positive definite symmetrizator, then Z is similar to a self-adjoint operator. Further, from the fact that $\ker D = \{0\}$, it follows that $\ker Z = \{0\}$ and, therefore, according to the Hilbert-Schmidt theorem, the problem (5.28) has a discrete spectrum $\{\lambda_k\}_{k=1}^{\infty} \subset (-q, q)$ that consists of eigenvalues $\lambda_k \to 0$ as $k \to \infty$ with finite multiplicities. To these eigenvalues there correspond some eigenelements $\{\varphi_k\}_{k=1}^{\infty}$ of problem (5.28) that form a Riezs basis in the space H_2^2.

Finally, from the above reasoning it follows that if condition (5.27) holds true, then outside the interval $I_0 = [-1, 1]$ problem (5.2) has a discrete spectrum $\{\omega_k\}_{k=1}^{\infty} = \{1/\lambda_k\}_{k=1}^{\infty}$ that consists of eigenvalues ω_k with finite multiplicities and the limit point $\omega = \infty$.

Later on we will prove that the set of eigenvalues of operator Z can be split up into two infinite sets $\{\lambda_n^+\}_{n=1}^{\infty}$ and $\{\lambda_n^-\}_{n=1}^{\infty}$ of positive and negative values. To these values there correspond the eigenvalues $\omega_n^{\pm}$ of problem (5.2), for which $\omega_n^+ > 1$, $\omega_n^- < -1$ and $\lim_{n\to\infty} \omega_n^{\pm} = \pm\infty$.

6.5.5 Systems of Eigenelements With Defect Basicity

Up until now, it has been assumed that condition (5.27) holds true. That led to the inequality $\|D\| < 1$ based on (5.25) and (5.19). Whenever property (5.27) is not valid, the operator pencil $A(\lambda)$ in (5.17) may not admit the factorization (5.26) and the former constructions do not make any sense. In this case, the eigenelements $\{\varphi_k\}_{k=1}^{\infty}$ of problem (5.16) form a Riezs basis with finite defect in H_2^2.

To prove this fact, let us note that the following conditions hold true for the operator-valued function $A(\lambda) = \lambda I - D + \lambda\Phi(\lambda)$,

$$A(0) = -D \in \mathfrak{S}_\infty, \qquad A'(0) \gg 0. \tag{5.29}$$

The first condition is obvious and the second one follows from (5.22),

$$A'(0) = I + \operatorname{diag}\left(0; B_0^{-1/2} A_{21} A_{12} B_0^{-1/2}\right) \geq I \gg 0.$$

Therefore, according to property 3° in Section 1.6.10, we obtain that for any ε, $0 < \varepsilon < 1$, the eigenelements of problem (5.16) corresponding to the eigenvalues $\{\lambda_k\}$ from the interval $(-\varepsilon, \varepsilon)$ form a Riezs basis of a subspace that has a finite defect in $H_2^2 = H_2 \oplus H_2$.

Since the eigenvalues $\lambda_k \in (-\varepsilon, \varepsilon)$ converge to zero for $k \to \infty$, then their reciprocals $\omega_k = 1/\lambda_k$, which are located outside the interval $(-1/\varepsilon, 1/\varepsilon)$ and are the eigenvalues of the problems (5.6) and (5.2), converge to infinity. Consequently, for the normalized solutions $v_k = (v_{1k}; v_{2k})^t$ of problem (5.2) that correspond to the eigenvalues ω_k, the properties

$$\|v_{1k}\|_{H_1} \to 0, \quad \|v_{2_k} - \omega_k^{-2} B_0 v_{2k}\|_{H_2} \to 0, \qquad k \to \infty, \tag{5.30}$$

are true. These relations follow from the equations (5.2) written down as

$$v_{1k} = \omega_k^{-1}(A_{11}v_{1k} + A_{12}v_{2k}), \quad v_{2k} - \omega_k^{-2} B_0 v_{2k} = \omega_k^{-1}(A_{21}v_{1k} + A_{22}v_{2k}), \tag{5.31}$$

and from the fact that $\omega_k \to \infty$ as $k \to \infty$, and $\|Aij\| \le 1$, $i, j = 1, 2$.

In conclusion, let us remark that whenever condition (5.27) is not valid, then in contrast to Section 6.5.4, the eigenvalues $\omega_n^{\pm}$ no longer have the points $\omega = \pm 1$ as accumulation points. This fact follows from the discreteness property of operator B_0, equation (5.6), and some general results in finite dimensional perturbation theory of unbounded self-adjoint operators (see, for example, [BS], Section 9.3).

We also note that if $B_0^{-1/2} \in \mathfrak{S}_{p_0}$, then from property 4° of Section 1.6.10 it follows that the eigenelements of problem (5.16) that correspond to the eigenvalues $\omega_n^{\pm}$ from $\mathbb{R} \setminus [-1, 1]$ form a p-basis (or defect p-basis) in H_2^2 for $p = p_0$.

6.5.6 Double-Sided Estimates of Positive and Negative Eigenvalues

Let us assume that condition $\lambda_1(B_0) > 2$ holds true, which is sufficient for the factorization of the analytic operator-valued function $A(\lambda)$ from (5.17). Then, the eigenvalues $\{\lambda_n^{\pm}\}_{n=1}^{\infty}$ of problem (5.16) are located in the interval $(-q, q)$, $q \in (\|D\|, 1)$, and their reciprocals, $\omega_n^{\pm}$, that are eigenvalues of the operator pencil (5.2), are located outside the interval $(-1/q, 1/q) \supset [-1, 1]$.

Using the variational methods from Section 1.6.11, we will next obtain double-sided estimates for the eigenvalues $\omega_n^{\pm}$ of problem (5.2). First, we perform a transition from (5.2) to an operator pencil with bounded operator coefficients.

We write down (5.2) in the vector-matrix form as

$$\omega^2 \begin{pmatrix} v_1 \\ v_2 \end{pmatrix} - \omega \begin{pmatrix} A_{11} & A_{12} \\ A_{21} & A_{22} \end{pmatrix} \begin{pmatrix} v_1 \\ v_2 \end{pmatrix} - \begin{pmatrix} 0 & 0 \\ 0 & B_0 \end{pmatrix} \begin{pmatrix} v_1 \\ v_2 \end{pmatrix} = \begin{pmatrix} 0 \\ 0 \end{pmatrix}, \tag{5.32}$$

and then use the substitution

$$B_0^{1/2} v_2 = w_2. \tag{5.33}$$

Applying operator $B_0^{-1/2}$ to the equation that results from (5.32), we obtain

$$\omega^2 \begin{pmatrix} I & 0 \\ 0 & B_0^{-1/2} \end{pmatrix} \begin{pmatrix} v_1 \\ w_2 \end{pmatrix} - \omega \begin{pmatrix} A_{11} & A_{12} B_0^{-1/2} \\ B_0^{-1/2} A_{21} & B_0^{-1/2} A_{22} B_0^{-1/2} \end{pmatrix} \begin{pmatrix} v_1 \\ w_2 \end{pmatrix} - \begin{pmatrix} 0 & 0 \\ 0 & I \end{pmatrix} \begin{pmatrix} v_1 \\ w_2 \end{pmatrix}$$
$$= \begin{pmatrix} 0 \\ 0 \end{pmatrix}, \tag{5.34}$$

that is, the equation

$$\tilde{L}(\omega) w := (\omega^2 C - \omega A_0 - P) w = 0, \qquad w = (v_1; w_2)^{\mathrm{t}} \in H_2^2. \tag{5.35}$$

Taking the scalar product of (5.35) and w (in H_2^2), we obtain that the numbers ω, for which there exist nonzero solutions of problem (5.32), belong to the set of Rayleigh functionals

$$p_{\pm}(w) = \frac{(A_0 w, w) \pm \sqrt{(A_0 w, w)^2 + 4(Cw, w)(Pw, w)}}{2(Cw, w)}. \tag{5.36}$$

Let us next consider the interval $(a, b) := (1/q, b)$, for $b > 1/q$, $q \in (\|D\|, 1)$, and the functional $p_+(w)$ in this interval; for $b = \infty$, all the eigenvalues ω_n^+ of problem (5.2) are located in $(1/q, \infty)$. The properties $p_+(tw) = p_+(w)$ for any $t > 0$, and $(\tilde{L}(p_+(w)) w, w) = 0$ are obvious. Besides, for $w \neq 0$ we have the inequality

$$\left(\tilde{L}'(p_+(w)) w, w \right) > 0. \tag{5.37}$$

Indeed, the left side of (5.37) is equal to

$$\begin{aligned} 2p_+(w)(Cw, w) - (A_0 w, w) &= [(Cw, w) - (A_0 w, w)] + (2p_+(w) - 1) C(w, w) \\ &\geq (2p_+(w) - 1)(Cw, w) > 0. \end{aligned} \tag{5.38}$$

In deriving (5.38), we used the property $p_+(w) > 1$ and the following inequality

$$|(A_0 w, w)| \leq (Cw, w), \tag{5.39}$$

which follows from the obvious inequality

$$|(Av, v)| \leq \|v\|^2, \qquad \|A\| = 1.$$

From the proved properties of the Rayleigh functional, it follows that $\tilde{L}(\omega)$ and $p_+(w)$ satisfy conditions 1°–6° in Section 1.6.11 in the interval $(1/q, b)$, $b \leq \infty$. Since on this interval problem (5.2) and the equivalent problem (5.35) for the pencil $\tilde{L}(\omega)$ have a discrete spectrum $\{\omega_n^+\}_{n=1}^{\infty}$ with the limit point $\omega = +\infty$, then the variational principles from Section 1.6.11 of the type (1.10) and (1.11) hold true for the numbers ω_n^+. Namely, taking into account that $\omega_n^+ \to \infty$ as $n \to \infty$ and, therefore, $\lambda_n^+ = 1/\omega_n^+ \to 0$, instead of (1.6.11) we obtain the Poincare–Ritz principle in the following form,

$$\omega_n^+ = \min_{\dim M = n} \max_{0 \neq w \in M} p_+(w), \qquad n = 1, 2, \ldots, \tag{5.40}$$

where M is an arbitrary n-dimensional subspace of H_2^2. Similarly, it can be proved that for negative eigenvalues in the interval $(-\infty, 1/q)$ the following is valid,

$$\omega_n^- = \max_{\dim M = n} \min_{0 \neq w \in M} p_-(w), \qquad n = 1, 2, \ldots. \tag{5.41}$$

Now, let us obtain double-sided inequalities for $p_+(w)$. According to the definition of $p_+(w)$ we have

$$[p_+(w)]^2 (Cw, w) - p_+(w)(A_0 w, w) - (Pw, w) = 0,$$

whence

$$[p_+(w)]^2 = (Pw, w)/(Cw, w) + p_+(w)(A_0 w, w)/(Cw, w).$$

From the latter equation and (5.39) we obtain

$$(Pw, w)/(Cw, w) - p_+(w) \leq [p_+(w)]^2 \leq (Pw, w)/(Cw, w) + p_+(w). \tag{5.42}$$

The inequality on the right side leads to

$$\left(p_+(w) - \frac{1}{2}\right)^2 \leq (Pw, w)/(Cw, w) + \frac{1}{4},$$

and from the inequality on the left we obtain

$$(Pw, w)/(Cw, w) + \frac{1}{4} \leq \left(p_+(w) + \frac{1}{2}\right)^2.$$

Since $p_+(w) > 1$, by taking square roots we finally have

$$\sqrt{(Pw, w)/(Cw, w) + \frac{1}{4}} - \frac{1}{2} \leq p_+(w) \leq \sqrt{(Pw, w)/(Cw, w) + \frac{1}{4}} + \frac{1}{2}. \tag{5.43}$$

Now, we use the Poincare–Ritz principle for the numbers ω_n^+ in the form (5.40). Then, the inequality in the right side of (5.43) gives

$$\begin{aligned}\omega_n^+ &\le \frac{1}{2} + \sqrt{\frac{1}{4} + \min_{\dim M=n} \max_{0\neq w\in M} (Pw,w)/(Cw,w)} \\ &= \frac{1}{2} + \sqrt{\frac{1}{4} + \min_{\dim M=n} \max_{0\neq w\in M} \|w_2\|^2 / \left(\|v_1\|^2 + \left\|B_0^{-1/2} w_2\right\|^2\right)} \\ &= \frac{1}{2} + \sqrt{\frac{1}{4} + \min_{\dim M=n} \max_{0\neq w\in M} \|w_2\|^2 / \left\|B_0^{-1/2} w_2\right\|^2} \\ &= \frac{1}{2} + \sqrt{\frac{1}{4} + \lambda_n(B_0)}, \qquad n = 1, 2, \ldots. \end{aligned} \tag{5.44}$$

In the last step we used a principle of the type (5.40) for the eigenvalues $\lambda_n(B_0)$ of the unbounded operator $B_0 \gg 0$, $B_0^{-1} \in \mathfrak{S}_\infty$.

From the left inequality in (5.43) we obtain another estimate of the type (5.44), which in the end leads to the inequalities

$$\sqrt{1 + 4\lambda_n(B_0)} - 1 \le 2\omega_n^+ \le \sqrt{1 + 4\lambda_n(B_0)} + 1, \qquad n = 1, 2, \ldots. \tag{5.45}$$

By reasoning similarly for the Rayleigh functional $p_-(w)$ in (5.36) and using the principle (5.41), one obtains estimates of the type (5.45) for $|\omega_n^-|$. Hence, we finally have

$$\sqrt{1 + 4\lambda_n(B_0)} - 1 \le 2|\omega_n^\pm| \le \sqrt{1 + 4\lambda_n(B_0)} + 1, \qquad n = 1, 2, \ldots. \tag{5.46}$$

The asymptotic formula

$$\omega_n^\pm = \pm \left(\lambda_n(B_0)\right)^{1/2} + O(1), \qquad n \to \infty, \tag{5.47}$$

which is a corollary to the inequalities (5.46), allows us to estimate the eigenvalues $\omega_n^\pm$ by eigenvalues $\lambda_n(B_0)$ of the operator B_0.

6.5.7 On the Essential Spectrum of the Problem

Up until now, the spectrum of problem (5.2), (5.4), or (5.6), which is the same for everyone of them, has been investigated outside the interval $I_0 = [-1, 1]$. It has been proved that whenever the condition $\lambda_1(B_0) > 2$ is satisfied, the spectrum is discrete and has a countable set of positive and negative eigenvalues $\{\omega_n^\pm\}_{n=1}^\infty$, $\lim_{n\to\infty} \omega_n^\pm = \pm\infty$. Next, we consider the case when in the problem (5.2) and (5.6) the spectral parameter ω lies in $I_0 = [-1, 1]$.

If in problem (5.6) $A_{12} = A_{21}^* = 0$, then the whole interval $[-1, 1]$ is part of the spectrum of the first problem (5.7), according to (5.7) and (5.1). Let us show that in the general situation when $A_{12} = A_{21}^* \neq 0$, the interval $I_0 = [-1, 1]$ is part of the essential spectrum $\sigma_e(\mathcal{A})$ of problem (5.6), or part of the essential spectrum $\sigma_e(L(\omega))$ of the pencil $L(\omega)$ from (5.2), which are equal spectra.

We transform equations (5.10), which are equivalent to equation (5.2), by using the substitution $\lambda = 1/\omega$ and get

$$A_{11}v_1 + A_{12}v_2 = \omega v_1, \qquad \Psi(\omega)v_2 = -\omega A_{21}v_1,$$
$$\Psi(\omega) := B_0 + \omega A_{22} - \omega^2 I. \tag{5.48}$$

First, let us assume that the basic condition $\lambda_1(B_0) > 2$ holds true. Then, for all $\omega \in [-1, 1] = I_0$, the operator $\Psi(\omega)$ is positive definite, because

$$(\Psi(\omega)v_2, v_2)_{H_2} \geq (B_0 v_2, v_2)_{H_2} - |\omega| \, \|v_2\|_{H_2}^2 - |\omega|^2 \|v_2\|_{H_2}^2$$
$$\geq (\lambda_1(B_0) - 2) \, \|v_2\|_{H_2}^2. \tag{5.49}$$

Therefore, for any $\omega \in I_0$, operator $\Psi(\omega)$ is invertible and has a compact inverse operator,

$$[\Psi(\omega)]^{-1} = B_0^{-1/2} \left[I + \omega B_0^{-1/2} A_{22} B_0^{-1/2} - \omega^2 B_0^{-1} \right]^{-1} B_0^{-1/2}. \tag{5.50}$$

Solving the second equation (5.48) for the element v_2 and substituting the solution into the first equation we obtain the next equation for v_1,

$$M(\omega)v_1 := \left\{ A_{11} - \omega A_{12} [\Psi(\omega)]^{-1} A_{21} \right\} v_1 = \omega v_1. \tag{5.51}$$

Here, $A_{11} = A_{11}^* \in \mathcal{L}(H_1)$, $\sigma(A_{11}) = I_0$, and the term $V(\omega) := -\omega A_{12} [\Psi(\omega)]^{-1} A_{21}$ is an analytic self-adjoint perturbation that takes compact operator values for $\omega \in I_0$.

Let us fix an arbitrary $\omega_1 \in I_0$ and consider the operator $M(\omega_1)$ that equals the sum of the bounded operator A_{11} and the compact self-adjoint operator $V(\omega_1)$. According to Weyl Theorem (Section 1.1.19) the essential spectrum $\sigma_e(M(\omega_1))$ of operator $M(\omega_1)$ is similar to the spectrum of operator A_{11}, that is, in virtue of the property $\sigma(A_{11}) = \sigma_e(A_{11}) = [-1, 1] = I_0$, it coincides with I_0.

According to the definition of the essential spectrum, for any $\omega_2 \in I_0$ there exists a Weyl sequence $\{v_{1n}\}_{n=1}^{\infty}$, $\|v_{1n}\|_{H_1} = 1$, depending on ω_1 and ω_2, such that

$$(M(\omega_1) - \omega_2 I) \, v_{1n} \to 0, \qquad n \to \infty.$$

Choosing $\omega_2 = \omega_1$ and the corresponding Weyl sequence $\{v_{1n}\}_{n=1}^{\infty}$, we conclude that for this sequence

$$\lim_{n\to\infty} (M(\omega_1) - \omega_1 I)\, v_{1n} = 0.$$

This equality means that every arbitrarily chosen point $\omega_1 \in I_0$ belongs to the essential spectrum of problem (5.51).

Now, let us leave aside the constraint $\lambda_1(B_0) > 2$ for the operator $\Psi(\omega)$. Since the spectral problem $\Psi(\omega)v_2 = 0$ is transformed after the substitution (5.3) into the second problem (5.7), which is equivalent to problem (5.9) with an operator D as in (5.8), the operator pencil $\Psi(\omega)$ has a discrete spectrum $\{\omega_k\}_{k=1}^{\infty} = \|\lambda_k^{-1}(D)\|_{k=1}^{\infty}$ with the limit point $\omega = \infty$.

Consequently, the interval $I_0 = [-1, 1]$ includes no more than a finite number of points ω_k at which the operator $\Psi(\omega)$ does not have an inverse. At all the other points ω in the interval I_0, the above proof can be carried out once again, and we obtain that at each of them there exists a Weyl sequence for the operator-valued function $M(\omega) - \omega I$. Since the essential spectrum is a closed set, the proved assertion is true for all points in the interval $I_0 = [-1, 1]$.

Summing up the results of this section, we note that the spectrum of the problem investigated here consists of the essential spectrum $I_0 = [-1, 1]$ and, the points of the discrete spectrum that are only located outside this interval.

6.6 Proper Oscillations of a Rotating Fluid

In this section, based on the general conclusions obtained in Section 6.5, we investigate the problem on proper oscillations of a fluid rotating in a partially filled container.

6.6.1 Surface and Internal Waves

Let us consider the spectral problem that corresponds to the Cauchy problem (3.28) for small movements of a single fluid. Assuming that $f(t) \equiv 0$, $y(t) = u \exp(i\omega t)$, $u = (\boldsymbol{v}; \eta)^{\mathrm{t}} \in \boldsymbol{J}_0(\omega) \oplus L_{2,\Gamma} = H$, we have

$$L_{\omega_0}(\omega)u := (B + 2\omega_0\omega A - \omega^2 I)u = 0, \tag{6.1}$$

with

$$A = (A_{ij})_{i,j=1}^{2} = A^*, \quad \|A\| = 1, \quad \sigma(A_{11}) = [-1, 1],$$
$$B = \operatorname{diag}(0, \sigma\rho^{-1}B_0), \qquad B_0 \gg 0,\ B_0^{-1} \in \mathfrak{S}_\infty.$$

Here, the operator pencil $L_{\omega_0}(\omega)$ depends on the parameter ω_0 (the angular velocity of rotation of the system) both in an explicit and implicit form. Indeed, by changing ω_0, the region Ω occupied by fluid in the state of relative equilibrium changes, too; hence, the operators A and B, the surface Γ, and the space $H = \boldsymbol{J}_0(\Omega) \oplus L_{2,\Gamma}$ will change as well. These changes are continuous in the parameter ω_0, provided the changes of all the other parameters of the considered system are continuous in ω_0.

Problem (6.1) makes the abstract problem (5.2) considered in Section 6.5 more concrete. To perform the transition from (5.2) to (6.1), it is necessary to substitute $A \mapsto 2\omega_0 A$ and $B_0 \mapsto \sigma\rho^{-1}B_0 =: \tilde{B}_0$ in (5.1), and assume that $\boldsymbol{v} \in \boldsymbol{J}_0(\Omega) =: H_1$ and $\eta \in L_{2,\Gamma} =: H_2$. Conditions (5.1) are valid.

Further, in Section 6.5, we performed a transition from (5.2) to (5.6) by means of the substitution (5.3), therefore, we make in (6.1) the following substitution,

$$(\sigma\rho^{-1}B_0)^{1/2}\eta = \tilde{B}_0^{1/2}\eta = \omega\xi, \qquad \xi \in L_{2,\Gamma}. \tag{6.2}$$

Then, instead of (5.6), for $\omega \neq 0$ we obtain the problem

$$\mathcal{A}_{\omega_0}\tilde{u} = \omega\tilde{u}, \qquad \tilde{u} = (\boldsymbol{v};\eta;\xi)^t \in \boldsymbol{J}_0(\Omega) \oplus L_{2,\Gamma} \oplus L_{2,\Gamma} =: \tilde{H}, \tag{6.3}$$

with

$$\mathcal{A}_{\omega_0} := \begin{pmatrix} 2\omega_0 A_{11} & 2\omega_0 A_{12} & 0 \\ 2\omega_0 A_{21} & 2\omega_0 A_{22} & \tilde{B}_0^{1/2} \\ 0 & \tilde{B}_0^{1/2} & 0 \end{pmatrix} = \mathcal{A}_{\omega_0}^*, \tag{6.4}$$

which is equivalent to problem (6.1).

As examples show, in a rotating (capillary or heavy) ideal fluid partially filling the container, the totality of all proper oscillations is divided into two classes, namely, the surface and the internal waves.

The surface waves are caused by the presence of the free surface of the fluid and the gravitation, capillar, and centrifugal forces acting on the system. These waves correspond to the discrete spectrum of the problem on proper oscillations. The eigenfunctions associated with the modes of surface waves fade quickly (exponentially, in the interesting examples) along the normal $\boldsymbol{n}$ going from the equilibrium surface into fluid, that is, these oscillations are of the same kind as the ordinary gravitation-capillary waves in the nonrotating fluid. For $\omega_0 \to 0$, the characteristics of surface waves are converted into the corresponding characteristics of gravitation-capillary waves.

The internal inertial waves, caused by the Coriolis force acting on the system represent the second class of proper oscillations in a rotating ideal fluid. The mentioned type of waves is similar to the waves in a wholly filled rotating container and is close to the pure vortical movements in a fluid that were considered in Section 6.1. The spectrum of frequencies of internal waves fills the interval $[-2\omega_0, 2\omega_0]$ and the vortical operator $2\omega_0 A_{11}$ acting in $\boldsymbol{J}_0(\Omega)$ is "responsible" for this sectrum. As examples show, the modes of internal waves have the following property: For them, the free surface of the fluid is almost unperturbed during the process of oscillations, and the proper movements take place mainly inside the region Ω.

In problem (6.3), (6.4), let us take the limit as $\omega_0 \to 0$. Assuming that all the characteristics of the system depend continuously on ω_0, then by taking the limit we obtain the following problem,

$$\mathcal{A}_0 \tilde{u} = \omega \tilde{u}, \quad \mathcal{A}_0 = \begin{pmatrix} 0 & 0 & 0 \\ 0 & 0 & \tilde{B}_{00}^{1/2} \\ 0 & \tilde{B}_{00}^{1/2} & 0 \end{pmatrix}, \quad \tilde{B}_{00} := \tilde{B}_0\big|_{\omega_0=0}. \tag{6.5}$$

Obviously, the spectrum of this problem consists of the point $\omega = 0$, which is an eigenvalue with infinite multiplicity corresponding to eigenelements of the form $\tilde{u}_0 = (\boldsymbol{v}; 0; 0)^{\mathrm{t}}$ with arbitrary $\boldsymbol{v} \in \boldsymbol{J}_0(\Omega)$, and of the discrete spectrum of frequencies $\{\tilde{\omega}_n^{\pm}\}_{n=1}^{\infty}$ of gravitation-capillary waves, $\tilde{\omega}_n^{\pm} = \pm\lambda_n(\tilde{B}_{00}^{1/2})$, for which the eigenelements $\tilde{u}_n^{\pm}$ have the form $\tilde{u}_n^{\pm} = (0; \eta_n; \pm\eta_n)^{\mathrm{t}}$, where $\eta_n = \eta_n\left(\tilde{B}_{00}^{1/2}\right)$ are eigenelements of operator $\tilde{B}_{00}^{1/2}$. Here, the elements $(\eta_n; \pm\eta_n)^{\mathrm{t}}$ form an orthogonal basis in the space $L_{2,\Gamma} \oplus L_{2,\Gamma}$.

If ω_0 increases from $\omega_0 = 0$ up to a given value, then (as it will be proved later) the eigenvalue $\omega = 0$ with infinite multiplicity is converted into the interval $[-2\omega_0, 2\omega_0]$, and the internal waves in the rotating fluid correspond to this interval. Outside this interval, the problems (6.1), (6.3) have a discrete spectrum of frequencies, which corresponds to the surface waves.

Let us consider these phenomena in detail.

6.6.2 Properties of the Surface Waves

We assume in (6.1) that $\omega \in \mathbb{R}\setminus I_0$, with $I_0 : [-2\omega_0, 2\omega_0]$, and perform the substitution $\omega = 2\omega_0/\lambda$. Then $|\lambda| < 1$, and according to the scheme developed in Section 5.3, we can perform a transition to an equation of the form (5.16), where now the operator $\sigma(4\omega_0^2\rho)^{-1}B_0$ stands for B_0. Hence, it follows that the general conclusions derived in Sections 5.3–5.11, with the obvious modifications, remain essentially unchanged for the case of the surface waves in problem (6.1). We formulate these conclusions without any detailed explanation.

1°. For $\omega \in \mathbb{R}\setminus I_0$, problem (6.1) has a discrete spectrum $\{\omega_n^{\pm}\}_{n=1}^{\infty}$ of frequencies that consists of normal eigenvalues $\omega_n^{\pm}$, with $\omega_n^{+} > 2\omega_0$ and $\omega_n^{-} < -2\omega_0$.

2°. The condition

$$8\rho\omega_0^2 < \sigma\lambda_1(B_0) \tag{6.6}$$

is sufficient for a spectral factorization of the operator pencil in problem (6.1), which is analogous to (5.17).

Either when (6.6) holds true, or in the opposite case, the points $\omega = \pm 2\omega_0$ cannot be limit points of the discrete spectrum $\{\omega_n^\pm\}_{n=1}^\infty$ located outside the interval I_0.

In the case of a heavy fluid, it is necessary to take the limit in the right side of (6.6) as $\sigma \to 0$, which leads to the replacement of σB_0 with the operator $(B_0)_{gr} \gg 0$, where again $(B_0)_{gr}^{-1} \in \mathfrak{S}_\infty$. This remark will apply to all the subsequent conclusions and formulas, where the operator σB_0 takes part.

3°. Let us assume that the fluid partially filling the container does not rotate, and the equilibrium state of the fluid in container is statically stable in linear approximation, that is, $\sigma B_0 \gg 0$ for $\omega_0 = 0$, that is, $\sigma\lambda_1(B_0) > 0$. By increasing the rotation of the system, that is, by increasing ω_0, the right side of (6.6) changes continuously from a positive value and the left side changes starting from zero. Therefore, there exists a diapason $(0, \omega_{0*})$ of values ω_0, such that for $\omega_0 \in (0, \omega_{0*})$ condition (6.6)t holds true by any means. This conclusion is valid under the assumption that for changes in the parameter ω_0, the region Ω changes continuously and preserves its type of connectiveness, and the equilibrium surface Γ together with its boundary $\partial\Gamma$ remain in certain given classes of smoothness (for a capillary fluid). In particular, this takes place in a convex container with a piecewise smooth boundary.

4°. Using the conclusions 1°–3° above, we obtain the following main result. Suppose that for $\omega_0 = 0$ the considered system is statically stable in linear approximation. Then, for a sufficiently small angular velocity ω_0, that is, in the case when condition (6.6) holds true, problem (6.1) has a discrete spectrum of frequencies $\{\omega_k^\pm\}_{k=1}^\infty$ with the limit point $\omega = \infty$, to which there correspond the proper modes of oscillations $\{u_k\}_{k=1}^\infty = \{(\boldsymbol{v}_k; \eta_k)^t\}_{k=1}^\infty \subset \boldsymbol{J}_0(\Omega) \oplus L_{2,\Gamma}$. These oscillations have the properties of surface waves, namely, by assuming that $\|\boldsymbol{v}_k\|^2_{\boldsymbol{L}_2(\Omega)} + \|\eta_k\|^2_{L_{2,\Gamma}} = 1$ we get

$$\|\boldsymbol{v}_k\|_{\boldsymbol{L}_2(\Omega)} \to 0, \quad \|\eta_k - \sigma(\rho\omega_k^2)^{-1} B_0\eta_k\|_{L_{2,\Gamma}} \to 0, \tag{6.7}$$

as $k \to \infty$.

Indeed, if the condition $\sigma B_0 \gg 0$ holds true for $\omega_0 = 0$, then conclusions 1° and 2° are valid for $\omega_0 \in (0, \omega_{0*})$ and, therefore, conclusion 3° takes place, too. Further, from equation (6.1), written down in the vector-matrix form, for the solutions ω_k, $u_k = (\boldsymbol{v}_k; \eta_k)^t$, we have

$$\boldsymbol{v}_k = \frac{2\omega_0}{\omega_k}(A_{11}\boldsymbol{v}_k + A_{12}\eta_k),$$

$$\eta_k - \sigma(\rho\omega_k^2)^{-1} B_0\eta_k = \frac{2\omega_0}{\omega_k}(A_{21}\boldsymbol{v}_k + A_{22}\eta_k),$$

whence (6.7) follows.

The first relation in (6.7) shows that for a sufficiently large k, the vortical component $\boldsymbol{v}_k \in \boldsymbol{J}_0(\Omega)$ of the solution $u_k = (\boldsymbol{v}_k; \eta_k)^{\mathrm{t}}$ is small in the norm of $\boldsymbol{L}_2(\Omega)$. Here, the potential component η_k that determines the potential harmonic field $\nabla\Phi_k = V\eta_k \in \boldsymbol{G}_{h,S}(\Omega)$ satisfies, as we take the limit $k \to \infty$, the same equation as the equation $\sigma B_0\eta = \rho\omega^2\eta$ of potential oscillations of a nonrotating fluid (here, the operator B_0 is changed due to the condition $\omega_0 > 0$). In other words, relations (6.7) show that in this case the proper oscillations $u_k = (\boldsymbol{v}_k; \eta_k)^{\mathrm{t}}$ have the characteristic of the surface waves.

5°. If condition (6.6) holds true, then for the frequencies $\omega_n^{\pm}$ of the oscillations of surface waves, the following inequalities are satisfied

$$\sqrt{\sigma\rho^{-1}\lambda_n(B_0) + \omega_0^2} - \omega_0 \leq |\omega_n^{\pm}| \leq \sqrt{\sigma\rho^{-1}\lambda_n(B_0) + \omega_0^2} + \omega_0, \qquad n = 1, 2, \ldots. \tag{6.8}$$

Indeed, the Rayleigh functionals for the pencil (6.1) have the form (instead of (5.6))

$$p_{\pm}(w) = \frac{\omega_0(A_0w, w) \pm \sqrt{\omega_0^2(A_0w, w)^2 + (Cw, w)(Pw, w)}}{(Cw, w)},$$
$$w = \left(\boldsymbol{v}; (\sigma\rho^{-1}B_0)^{1/2}\eta\right)^{\mathrm{t}} \in \boldsymbol{J}_0(\Omega) \oplus L_{2,\Gamma}, \tag{6.9}$$

where the operator matrices A_0, C, and P are defined by formulas (5.35). Therefore, we obtain the formulas (6.8) by repeating for (6.9) the same reasoning we used to derive inequalities (5.43) in Section 6.5.10.

6°. The following asymptotic formula for frequencies of oscillations of surface waves is a corollary of inequalities (6.8),

$$\omega_n^{\pm} = \pm(\sigma\rho^{-1})^{1/2}\lambda_n\left(B_0^{1/2}\right) + \mathrm{O}(1), \qquad n \to \infty, \tag{6.10}$$

where, according to the formulas (3.39) and (3.40), the numbers $(\sigma\rho^{-1})^{1/2}\lambda_n(B_0^{1/2})$ have the following asymptotic behavior as $n \to \infty$,

$$(\sigma\rho^{-1})^{1/2}\lambda_n\left(B_0^{1/2}\right)$$
$$= \begin{cases} (\sigma\rho^{-1})^{1/2}K_\sigma^{-3/4}n^{3/4}[1 + \mathrm{o}(1)], & K_\sigma = \mathrm{mes}\Gamma/(4\pi),\ \sigma > 0, \\ \rho^{-1/2}K_0^{-1/4}n^{1/4}[1 + \mathrm{o}(1)], & K_0 = \int_\Gamma |a_\Gamma^0|^2 d\Gamma/(4\pi),\ \sigma = 0. \end{cases} \tag{6.11}$$

Formulas (6.10) and (6.11) show that the frequencies of oscillations of surface waves depend asymptotically only on characteristics of the region Ω occupied by fluid in the state of relative equilibrium and the gravitation, centrifugal, and capillary forces, and do not depend at all on the Coriolis forces.

7°. If condition (6.6) holds true, then the spectral problem on finding the frequencies and the modes of surface waves admits a factorization of the form (5.26), where the factor Z can be written as $Z = A_+^{-1}(0)D$, with $A_+^{-1}(0) \in \mathcal{L}\left(L^2_{2,\Gamma}\right)$ and operator D has the form

$$D := \begin{pmatrix} 0 & 2\omega_0(\sigma\rho^{-1}B_0)^{-1/2} \\ 2\omega_0(\sigma\rho^{-1}B_0)^{-1/2} & -4\omega_0^2\rho\sigma^{-1}B_0^{-1/2}A_{22}B_0^{-1/2} \end{pmatrix}. \tag{6.12}$$

From the latter property and from the existence of a symmetrizator for the factor Z, it can be proved (as in Section 6.5.4) that the eigenelements $\{(\eta_n^{\pm};\xi_n^{\pm})^{\mathrm{t}}\}_{n=1}^{\infty}$, with $\xi_n^{\pm} = (\omega_n^{\pm})^{-1}(\sigma\rho^{-1}B_0)^{1/2}\eta_n^{\pm}$, which correspond to the eigenfrequencies $\omega_n^{\pm}$, form a Riesz basis in the space $L_{2,\Gamma} \oplus L_{2,\Gamma}$. These elements also form a p-basis for $p > p_0$, where $p_0 = 2$ for a heavy fluid ($\sigma = 0$) and $p_0 = 2/3$ for a capillary fluid ($\sigma = 0$, see formulas (3.39) and (3.40)).

8°. If condition (6.6) is not valid, then the eigenelements $\{(\eta_n^{\pm};\xi_n^{\pm})^{\mathrm{t}}\}$ that correspond to the eigenvalues $\omega_n^{\pm}$ outside the interval $[-(2\omega_0+\varepsilon), 2\omega_0+\varepsilon]$ for any given $\varepsilon > 0$, form a Riesz basis with finite defect in $L_{2,\Gamma} \oplus L_{2,\Gamma}$ (and a defect p-basis for the above mentioned values $p > p_0$). This assertion follows from the conclusions of Section 6.5.5.

6.6.3 On Existence and Properties of Internal Waves

Similarly to Section 6.5.7, it can be proved that the interval $I_0 = [-2\omega_0, 2\omega_0]$ is part of the essential spectrum of the operator pencil (6.1). From a physical point of view, this means that in a rotating ideal fluid that partially fills a container, besides surface waves there exist internal waves whose spectrum of frequencies fills the interval $[-2\omega_0, 2\omega_0]$.

The study of some examples, in particular, the problems on oscillations of a heavy fluid in a cylindrical container with an arbitrary section and with a small angular velocity ω_0, and the problems on oscillations of a capillary fluid located in a circular cylindrical container under conditions of complete low-gravity, shows that the spectrum of oscillation frequencies that correspond to internal waves is a pure point spectrum. In other words, the point spectrum of the corresponding spectral problem of the form (5.32) contains a set dense in the interval $[-2\omega_0, 2\omega_0]$. Let us show that in this case, the eigenelements of the spectral problem can be naturally called *internal waves*.

For problem (6.1), let us perform a transition similar to the one described in Section 6.5.7 from equation (5.2) to equation (5.51). Then, instead of (5.51) we obtain the following spectral problem

$$M_{\omega_0}(\omega)\boldsymbol{v} := \left\{2\omega_0 A_{11} - 4\omega_0^2\omega A_{12}\left[\Psi_{\omega_0}(\omega)\right]^{-1} A_{21}\right\}\boldsymbol{v} = \omega\boldsymbol{v},$$
$$\Psi_{\omega_0}(\omega) := \sigma\rho^{-1}B_0 + 2\omega_0\omega A_{22} - \omega^2 I, \qquad \boldsymbol{v} \in \boldsymbol{J}_0(\Omega). \tag{6.13}$$

Let us assume that the point spectrum of problem (6.13) is dense in the interval $I_0 = [-2\omega_0, 2\omega_0]$. Then, for any $\omega' \in I_0$ that does not belong to the possible finite set of points of the discrete spectrum of the pencil Ψ_{ω_0} in I_0, and for any sequence $\{\omega_k\}_{k=1}^{\infty}$ of eigenfrequencies of oscillations converging to ω', the corresponding eigenelements $\tilde{u}_k = (\boldsymbol{v}_k; \eta_k; \xi_k)^t$ of problem (6.3) that are normalized by the conditions

$$\|\tilde{u}_k\|^2_{\tilde{\boldsymbol{H}}} = \|\boldsymbol{v}_k\|^2_{\boldsymbol{L}_2(\Omega)} + \|\eta_k\|^2_{L_{2,\Gamma}} + \|\xi_k\|^2_{L_{2,\Gamma}} = 1, \qquad k = 1, 2, \dots, \tag{6.14}$$

possess the properties of internal waves, namely, as $k \to \infty$ we have

$$\|\eta_k\|^2_{L_{2,\Gamma}} = \|\nabla\Phi_k\|_{\boldsymbol{L}_2(\Omega)} \to 0,$$
$$\|\xi_k\|^2_{L_{2,\Gamma}} = \|\omega_k^{-1}(\sigma\rho^{-1}B_0)^{1/2}\eta_k\|_{L_{2,\Gamma}} \to 0,$$
$$\|(\sigma B_1)^{1/2}\zeta_k\|_{L_{2,\Gamma}} = \|(\sigma\rho^{-1}B_0)^{1/2}\eta_k\|_{L_{2,\Gamma}} \to 0,$$
$$\|\boldsymbol{v}_k\|_{\boldsymbol{L}_2(\Omega)} \to 1,$$
$$\|(2\omega_0 A_{11} - \omega' I)\boldsymbol{v}_k\|_{\boldsymbol{L}_2(\Omega)} \to 0. \tag{6.15}$$

We next prove the properties (6.15). Let ω' be an arbitrary point in the interval I_0, for which the operator $\Psi_{\omega_0}(\omega')$ is invertible. Then, in virtue of the representation

$$[\Psi_{\omega_0}(\omega)]^{-1} = \rho\sigma^{-1}B_0^{-1/2}\left[I + 2\omega_0\omega\rho\sigma^{-1}B_0^{-1/2}A_{22}B_0^{-1/2} - \rho\omega^2\sigma^{-1}B_0^{-1}\right]^{-1}B_0^{-1/2}$$
$$=: B_0^{-1/2}\psi(\omega)B_0^{-1/2}, \tag{6.16}$$

and the compactness of $B_0^{-1/2}$, we obtain that $[\Psi_{\omega_0}(\omega)]^{-1}$ takes compact operator values. Now, let $\{\omega_k\}_{k=1}^{\infty}$ be an arbitrary sequence of eigenvalues for problem (6.13) that is equivalent to problem (6.3), such that $\omega_k \to \omega'$ as $k \to \infty$. The eigenelements $\tilde{u}_k = (\boldsymbol{v}_k; \eta_k; \xi_k)^t$ of problem (6.3) corresponding to the numbers ω_k are orthonormal because the operator matrix $\mathcal{A}_{\omega_0}$ is self-adjoint, that is, for these eigenelements we have

$$(\tilde{u}_k, \tilde{u}_j)_{\tilde{\boldsymbol{H}}} = (\boldsymbol{v}_k, \boldsymbol{v}_j)_{\boldsymbol{L}_2(\Omega)} + (\eta_k, \eta_j)_{L_{2,\Gamma}} + (\xi_k, \xi_j)_{L_{2,\Gamma}} = \delta_{kj}. \tag{6.17}$$

In particular, conditions (6.14) hold true.

By the properties (6.17), the sequence $\{\tilde{u}_k\}_{k=1}^{\infty}$ converges weakly to zero in the space $\tilde{\boldsymbol{H}} = \boldsymbol{J}_0(\Omega) \oplus L_{2,\Gamma} \oplus L_{2,\Gamma}$. Hence it follows that the sequence $\{\boldsymbol{v}_k\}_{k=1}^{\infty}$ converges weakly to zero in the space $\boldsymbol{J}_0(\Omega)$. Since $\omega_k \to \omega'$ and the operator $\Psi_{\omega_0}(\omega')$ is invertible, the operators $\Psi_{\omega_0}(\omega_k)$ are invertible too for any sufficiently large k. Moreover,

$$[\Psi_{\omega_0}(\omega_k)]^{-1} \in \mathfrak{S}_\infty, \quad [\Psi_{\omega_0}(\omega_k)]^{-1} - [\Psi_{\omega_0}(\omega')]^{-1} \to 0, \qquad \text{as } k \to \infty.$$

Let us next write down the equation of problem (6.13) and the connections between the solutions $\boldsymbol{v} \in \boldsymbol{J}_0(\Omega)$, $\eta \in L_{2,\Gamma}$ and $\xi \in L_{2,\Gamma}$ of problem (6.3) for the chosen sequence $\omega_k \to \omega'$:

$$\begin{aligned}(2\omega_0 A_{11} - \omega_k I)\boldsymbol{v}_k &= 4\omega_0^2\omega_k A_{12}[\Psi_{\omega_0}(\omega_k)]^{-1}A_{21}\boldsymbol{v}_k,\\ \eta_k &= -2\omega_0\omega_k B_0^{-1/2}\psi(\omega_k)B^{-1/2}A_{21}\boldsymbol{v}_k,\\ \xi_k &= \omega_k^{-1}(\sigma\rho^{-1})^{1/2}B_0^{1/2}\eta_k.\end{aligned} \tag{6.18}$$

Since $\boldsymbol{v}_k$ converges weakly to zero in $\boldsymbol{J}_0(\Omega)$ and the operator $B_0^{-1/2}$ is compact, the sequence $w_k := B_0^{-1/2}A_{21}\boldsymbol{v}_k$ from $L_{2,\Gamma}$ converges strongly to zero in $L_{2,\Gamma}$. From the latter property and the continuity of the operator-valued function $\psi(\omega)$ from (6.16), we obtain that in a sufficiently small neighborhood of the point ω', the sequence of elements $\varphi_k := \psi(\omega_k)w_k$ converges strongly in $L_{2,\Gamma}$ as well. But then, the elements η_k defined by $\eta_k = -2\omega_0\omega_k B_0^{-1/2}\varphi_k$ belong to $\mathcal{D}(B_0^{1/2})$, as it follows from the representation (6.18) and, moreover, $B_0^{1/2}\eta_k = -2\omega_0\omega_k\varphi_k \to 0$ as $k \to \infty$. Hence, by using the property proved in Section 6.3.4 we obtain

$$\|(\sigma\rho^{-1}B_0)^{1/2}\eta_k\|_{L_{2,\Gamma}} = \|(\sigma B_1)^{1/2}\zeta_k\|_{L_{2,\Gamma}} \to 0, \qquad \text{as } k \to \infty.$$

Further, from $\xi_k = \omega_k^{-1}(\sigma\rho^{-1}B_0)^{1/2}\eta_k$ and $\eta_k = -2\omega_0\omega_k B_0^{-1/2}\varphi_k$ we get $\xi_k = -2\omega_0(\sigma\rho^{-1})^{1/2}\varphi_k \to 0$. Now, let us note that since $\varphi_k \to 0$, then $\eta_k = -2\omega_0\omega_k B_0^{-1/2}\varphi_k \to 0$ as $k \to \infty$. As it has been proved in Section 6.3.3 (see (3.23)), the element $V\eta = \nabla\Phi \in \boldsymbol{G}_{h,S}(\Omega)$ that represents the potential-harmonic component of the solution of the problem on oscillations can be univalently found from the element $\eta \in L_{2,\Gamma}$. Here, the operator V is an isometry. The first relation in (6.15) follows from the last remark and the property $\eta_k \to 0$ (in $L_{2,\Gamma}$).

Since, as it has been already proved, $\eta_k \to 0$ and $\xi_k \to 0$, then from (6.14) we obtain that $\|\boldsymbol{v}_k\|_{\boldsymbol{L}_2(\Omega)} \to 1$ as $k \to \infty$. The last equality in (6.15) follows from the first equality in (6.18), if we note that its right side equals $-2\omega_0 A_{12}\eta_k$ and $\eta_k \to 0$ and $\omega_k \to \omega'$ as $k \to \infty$. Thus, all relations in (6.15) are proved.

Formulas (6.15) show that for arbitrarily chosen eigensolutions of problem (6.1) or of the equivalent problem (6.3) for $\omega \neq 0$, for which $\omega_k \to \omega'$ and the normalizing conditions (6.14) hold true, the potential component $\nabla\Phi_k$ of the solutions converges to zero as $k \to \infty$ and, at the same time, the norm of the vortical component $\boldsymbol{v}_k$ in $\boldsymbol{J}_0(\Omega)$ converges to 1. Here, for any arbitrarily chosen number $\omega' \in I_0$ the relation $2\omega_0 A_{11}\boldsymbol{v} = \omega'\boldsymbol{v}$ holds true by taking the limit. This relation corresponds to the spectral problem on oscillations of a rotating fluid in a wholly filled container considered in Section 6.1. Since for that problem the deviation of the boundary $\partial\Omega$ during the process of oscillations equals zero, it is natural to expect that in the currently considered problem the deviation of the free surface of the fluid should asymptotically—that is, as $k \to \infty$—converge to zero. This very fact is expressed by the first formula in (6.15). Moreover, the second line of formulas in (6.15) shows that in this process, the potential energy of the system converges to zero, that is, the deviations ζ_k of the free surface from the equilibrium surface Γ converge to zero not only in the norm of $L_{2,\Gamma}$, but also in the norm of $H^1_\Gamma(\Gamma)$. These conclusions demonstrate once again that to oscillations of a rotating fluid with frequencies located in the interval $I_0 = [-2\omega_0, 2\omega_0]$, there correspond internal wave movements, for which the free surface of the fluid does not deviate almost at all from the equilibrium state. Obviously, the sources of such oscillations are the Coriolis forces. The vortical operator A_{11} is used in problem (6.1) precisely because of their action on the given system.

6.6.4 Oscillations of a System of Nonmixing Fluids

The general conclusions already obtained in this section for the problem on proper oscillations of a single fluid that partially fills an arbitrary rotating container, hold for the problem on oscillations of a system of nonmixing fluids, too. The general evolution equation of this problem was obtained in Section 6.3.

Indeed, since the evolution equations of the above-mentioned problems are identical, then the spectral problem for a system of fluids has the form (6.1), where now one should formally assume that $\sigma = 1$, $\rho = 1$, $\boldsymbol{u} = \hat{\boldsymbol{u}} = (\hat{\boldsymbol{v}}; \hat{\eta})^t \in \hat{\boldsymbol{H}} = \hat{\boldsymbol{J}}_0(\Omega) \oplus \hat{L}_{2,\Gamma}$. Further, the operators A and B from (3.73) that have the properties described in Section 6.3.6, are to be understood as A and B in (6.1). Hence, all the general conclusions obtained in Sections 6.5 and 6.6 are still valid in the case of the problem on proper oscillations of a system of nonmixing fluids.

In particular, in such systems there are waves whose spectrum of frequencies is discrete and located outside the interval $[-2\omega_0, 2\omega_0]$. Here, these waves are naturally called *boundary waves* instead of "surface" waves. Indeed, examples and the previous way of reasoning show that the modes of surface waves that correspond to the discrete spectrum of frequencies, are described by functions concentrated mainly near the boundaries between fluids. The amplitude of these waves fades quickly (in many examples—exponentially) while going from the equilibrium boundary Γ_j between two fluids into one of them (with density ρ_j) or into the other (with density $\rho_{j+1} < \rho_j$) along the normal $\boldsymbol{n}_j$.

For the boundary waves appearing in a system of fluids, the general conclusions 1°–8° from Section 6.6.2, with the corresponding substitution (6.6)–(6.8) are still valid. Here, in the asymptotic formulas (6.11) one should assume that $\rho = 1$, $\sigma = 1$, $B_0 \mapsto \hat{B}_0$, and the constants K_0, K_σ are defined by the formulas (3.76) and (3.77).

For $\omega \in [-2\omega_0, 2\omega_0]$ there exist internal waves in the system of fluids caused by the influence of Coriolis forces. For these waves, the relations (6.15), which in this case have the following form as $k \to \infty$,

$$\begin{aligned} \|\hat{\eta}^k\|_{\hat{L}_{2,\Gamma}} &= \|\widehat{\nabla\Phi}_k\|_{\hat{\boldsymbol{L}}_2(\Omega)} \to 0, \\ \|\hat{\xi}^k\|_{\hat{L}_{2,\Gamma}} &= \|\omega_k^{-1}\hat{B}_0^{1/2}\hat{\eta}^k\|_{\hat{L}_{2,\Gamma}} \to 0, \\ \|\hat{B}^{1/2}\hat{\zeta}^k\|_{\hat{L}_{2,\Gamma}} &= \|\hat{B}_0^{1/2}\hat{\eta}^k\|_{\hat{L}_{2,\Gamma}} \to 0, \\ \|\hat{\boldsymbol{v}}^k\|_{\hat{\boldsymbol{L}}_2(\Omega)} &\to 1, \\ \|(2\omega_0\hat{A}_{11} - \omega'\hat{I})\hat{\boldsymbol{v}}^k\|_{\hat{\boldsymbol{L}}_2(\Omega)} &\to 0, \end{aligned}$$

where the notations are the ones used in Section 6.3.6, still hold true.

Examples and a heuristic analysis of problem (6.1) show that presumably, the following assertions formulated as hypotheses, are true for the boundary and internal waves.

Hypothesis 1. For the problem on oscillations of a system of m nonmixing fluids filling an arbitrary rotating container, the frequencies and modes of oscillations that correspond to the discrete spectrum $\{\omega_k\}_{k=1}^\infty$ outside the interval $[-2\omega_0, 2\omega_0]$ are asymptotically divided into $m-1$ classes

$$\{\omega_k\}_{k=1}^\infty = \bigcup_{j=1}^{m-1} \{\omega_{jn}\}_{n=1}^\infty, \quad \{\hat{\eta}_k\}_{k=1}^\infty = \bigcup_{j=1}^{m-1} \{\hat{\eta}^{jn}\}_{n=1}^\infty,$$

as $k \to \infty$, in the following way. The modes $\hat{\eta}^{jn}$ in j-th class for large n describe oscillations of the system that occur mainly near the boundary Γ_j. The maximal amplitude of the oscillations $\hat{\eta}^{jn}$ is reached on Γ_j and, at the same time, the boundaries Γ_q for $q \neq j$ are almost not perturbed. The amplitude is fading while going away from Γ_j, and the effect is stronger if the number n is larger, that is, we are dealing with the so-called "skin-effect" phenomenon.

Hypothesis 2. The modes of internal waves that correspond to oscillation frequencies ω from $[-2\omega_0, 2\omega_0]$ can be divided into m classes for the system of m fluids as follows. The fluid oscillations in the k-th class take place mainly in the region Ω_k occupied by the fluid with the density ρ_k and the corresponding frequencies are dense in the interval $[-2\omega_0, 2\omega_0]$. Here, the perturbations of the boundaries Γ_k, Γ_{k-1}, and the other boundaries are not present, and there are no internal wave movements in the regions Ω_p, $p \neq k$.

Finally, let us recall one more situation that was not investigated in this section so far. If condition (6.6) is not satisfied, then by increasing the parameter ω_0, that is, by increasing the angular velocity of the system (consisting of a single or several fluids), the points of the discrete spectrum corresponding to boundary waves can be located in the interval $[-2\omega_0, 2\omega_0]$. In this situation, it is possible to deal with the internal resonance of two types of waves because of the internal influence of Coriolis and potential forces.

The questions formulated in Hypotheses 1 and 2 and the phenomenon of internal resonance are still waiting to be investigated.

Appendix B

Remarks and Reference Comments to Part II

This Appendix includes references to books and articles used by the authors while writing some sections in Part II, and additional information about problems closely related to those discussed in the main text.

B.1 Chapter 3

3.1. The material presented in this section has an auxiliary character and is discussed mainly for the reader's convenience. It can be found in several standard courses of theoretical mechanics and classical fluid dynamics. While writing this section, the authors used the monographs of N. E. Kochin, I. A. Kibel, and N. V. Roze [1], G. N. Mikishev and B. I. Rabinovich [1], and N. M. Moiseev and V. V. Rumyantsev [1].

3.2. The system of equations of fluid dynamics is not a system of partial differential equations of Cauchy–Kovalevskaya type, because the time derivative of one of the unknown functions—the pressure—is not present in these equations. To eliminate the pressure, in classical fluid dynamics one applies the operator $\boldsymbol{rot}$ to both sides of the equation and obtain as a result the equation in the Helmholtz form. However, even in this case, the system is not a Cauchy–Kovalevskaya one. E. Hopf [1], S. G. Krein [1], and S. L. Sobolev [2], have applied independently the method of orthogonal projections onto the subspace $\boldsymbol{J}_0(\Omega)$ (using the projector P_0), in which the solenoidality condition and the boundary conditions take the time into account automatically, and there is no pressure gradient in this space. Various variants of the method of orthogonal projection are being used in all parts of the book.

We want to point out that this method, in which the pressure is eliminated, is equivalent to the method used in the works of E. Hopf [1] and O. A. Ladyuzhenskaya [1,3], while determining generalized solutions by means of integral identities.

3.3. The operator of kinetic energy of potential movements in an ideal fluid was used in the works of N. N. Moiseev [1,2,4]. Later on, it has been investigated in details by N. D. Kopachevsky [1,3]. The asymptotic formula (3.24) is based on results in the works of I. L. Vulis and M. Z. Solomyak [1,2], and also the work of N. A. Karazeeva and M. Z. Solomyak [1].

3.4. In a classical framework, the solutions of problem (4.1)–(4.3) for the case of a potential field of mass forces have been obtained by N. E. Zhukovsky [1].

3.5. The problem of the motion of a body with a cavity partially filled with an ideal fluid was extensively considered in the early 1950s and later on, in the works of L. N. Sretenskii [1], N. N. Moiseev [2,4], G. S. Narimanov [1], B. I. Rabinovich [1], D. E. Okhotsimskii [1] and others. A great number of problems on cavities and motions of a special type have been investigated and numerous calculations of the hydrodynamic coefficients were made over the years (see, for example, the books of N. N. Moiseev and A. A. Petrov [1], S. F. Feschenko, I. A. Lukovskii, B. I. Rabinovich, and L. V. Dokuchaev [1], the article by B. I. Rabinovich, L. V. Dokuchaev, and Z. M. Polyakova [1] and others). A general approach from the point of view of mechanics and applied mathematics to the above-mentioned problems is described in the monographs by G. N. Mikishev and B. I. Rabinovich [1], N. N. Moiseev, and V. V. Rumyantsev [1], and F. L. Chernousko [1].

An operator approach to these problems has been carried out by N. N. Moiseev [2,4]. Further, this approach was more thoroughly applied in the work of S. G. Krein and N. N. Moiseev [1], In this particular work, Krein and Moiseev used the results of L. S. Pontryagin on self-adjoint operators in spaces with indefinite metrics (see Section 1.3). A hefty part of our Section 3.5 is being stated following the work of S. G. Krein and N. N. Moiseev [1]. However, Krein and Moiseev came short in noting that the operator B always equals zero on the element corresponding to a slow stationary rotation of the body. That is why, for example, in this book we do not ask for the operator B to be positive definite; taking this into account (see Section 1.5) makes the investigation more complicated.

3.6. This section is written based on the work of N. D. Kopachevsky [9] with particular regard to using the abstract scheme from Section 1.8.

The asymptotic formula (6.28) follows from the work of N. A. Karazeeva and M. Z. Solomyak [1].

B.2 Chapter 4

4.1. The works of N. N. Moiseev and F. L. Chernousko [1] and N. D. Kopachevsky [1,3] were the first ones to deal with small oscillations of an ideal capillary fluid in a container. Further investigations of these issues were carried out by N. D. Kopachevsky [4–8], and later by M. Y. Barnyak [1–3], M. Y. Barnyak and A. N. Komarenko [1], I. A. Lukobskii [1], and others. Early contributions and preliminary results are illustrated in the monograph by N. N. Moiseev and V. V. Rumyantsev [1], and in the selection of works edited by N. N. Moiseev [1]. In more details, these results are stated in the monograph by V. G. Babskii, N. D. Kopachevsky, A. D. Myshkis, L. A. Slobozhanin, and A. D. Tyuptsov [1] (see also [2] and the monograph by A. D. Myskis, V. G. Babskii, M. Y. Zhukov, N. D. Kopachevsky, L. A. Slobozhanin, and A. D. Tyuptsov [1]). In the same works, a detailed bibliography on the mentioned issues is indicated.

We should note that besides the two variational principles (1.31) and (1.34), there exists a third principle which allows one to calculate the eigenvalues $\lambda = \rho\omega^2/\sigma$ and the frequencies of oscillations ω. This principle is formulated and justified in the papers of M. Y. Barnyak [1–3] and M. Y. Barnyak and A. N. Komarenko [1–2].

One can deduce (by referring to Section 4.1) that the dynamic and kinematic conditions on the equilibrium surface Γ lead to the system of equations

$$\sigma\rho^{-1}Bh = -\mathrm{i}\omega\varphi, \qquad \frac{\partial\varphi}{\partial n} = \mathrm{i}\omega h,$$

for the amplitude functions $h(\xi^1,\xi^2)$ and $\varphi(x)$ (see Section 4.1.6). This system can be written down in the vector-matrix form

$$\begin{pmatrix} \sigma\rho^{-1}B & 0 \\ 0 & C^{-1} \end{pmatrix} \begin{pmatrix} h \\ \varphi \end{pmatrix} = \omega \begin{pmatrix} 0 & -iI \\ iI & 0 \end{pmatrix} \begin{pmatrix} h \\ \varphi \end{pmatrix},$$

or,

$$\mathcal{B}y = \omega Jy, \qquad y = (h;\varphi)^{\mathrm{t}} \in \mathcal{D}(\mathcal{B}) \subset H \oplus H, \ J = J^* = J^{-1}.$$

Here, $C^{-1} \gg 0$ stands for an unbounded operator, the inverse to the compact positive kinetic energy operator C.

Since $\mathcal{B} \gg 0$, one can do the substitution $\mathcal{B}^{1/2}y = z$ in the above mentioned equation, and get for the new unknown z the next equation with a compact self-adjoint operator,

$$z = \omega\mathcal{B}^{-1/2}J\mathcal{B}^{-1/2}z.$$

From this we deduce that the numbers $(\omega_k^+)^{-1}$ are consecutive maxima of the variational ratio

$$\int_\Gamma \varphi h \mathrm{d}\Gamma \Big/ \left[\sigma\rho^{-1}\|h\|_B^2 + \int_\Omega |\nabla\varphi|^2 \mathrm{d}\Omega\right], \qquad \int_\Gamma \varphi \mathrm{d}\Gamma = 0,$$

and $\omega_k^- = -\omega_k^+$. Such a variational principle is very convenient for computer calculations by Ritz method. In particular, one does not need to find the inverse operator B^{-1} (as it is necessary in the variational principle (1.34)), and also one does not require the additional conditions

$$\Delta\varphi = 0 \quad \text{in } \Omega, \qquad \frac{\partial\varphi}{\partial n} = 0 \quad \text{on } S,$$

(as in (1.31)). These conditions are the natural ones from the point of view of variational calculus.

The asymptotic formulas (1.35) and (1.55) were obtained by N. D. Kopachevsky using an heuristic method and proved by T. A. Suslina [1–3].

4.2. The problem of small oscillations of an ideal fluid completely filling a container with rigid walls and one or two horizontal plane ends, has been considered in the works of A. V. Andronov [1,2] and A. V. Andronov and N. D. Kopachevsky [1]. The asymptotic formulas (2.47), (2.60), (2.62), (2.83) were proved by T. A. Suslina [2,3] at the request of N. D. Kopachevsky.

4.3. Nowadays, the problem of oscillations of a fluid in a container with elastic walls is considered to be classic; for example, a detailed bibliography on this issue can be found in the review report of E. I. Grigolyuk [1]. For the case of an elastic plane bottom and a partially filled container, this problem was investigated by operator methods in the works of Ngo Zui Kan [4–6]. It was Ngo Zui Kan who considered the corresponding problem of oscillations of a system of nonmixing fluids, when the fluids can be either capillary or heavy.

The present section is written based on the works of Ngo Zui Kan [4–6] and on subsequent investigations carried out by A. V. Andronov [1,2] and by A. V. Andronov and N. D. Kopachevsky [1]. Here one used the abstract scheme from Section 1.8. The asymptotic formulas (3.17), (3.36), (3.41) and other related results were proved by T. A. Suslina [1–3].

Let us point out that the works of L. S. Srubschik and V. I. Yudovich [1], A. D. Tyuptsov [1], L. A. Tyomkin and V. S. Tyomkina [1] are also connected with the subject matter of Section 4.1, and the works of R. Y. Amenzade [1], M. P. Petrenko [1,2] and the monograph by I. M. Rapoport [1] have direct relationship with Sections 4.2, 4.3.

The problem of small oscillations of an ideal fluid located in an immovable container and enclosed by a membrane, and the similar problem for a system of fluids with membrane boundaries in between, are close—in the mathematical sense—to the problems considered in Sections 3.3, 3.6, 4.1, and 4.3. These issues were investigated by Y. S. Pashkova [1,2] (with advice and help from N. D. Kopachevsky).

Let $\rho_0 > 0$ be the surface density of a horizontally located membrane Γ, and let $T > 0$ be its tension in the immovable state of the fluid. The membrane Γ and the solid boundary S of the container limit the region Ω filled with an ideal fluid of density ρ_0. The membrane is fixed along the boundary $\partial\Gamma$, i.e. its deviation during the process of oscillations equals zero on $\partial\Gamma$.

For the potential of displacements $\Phi(x)$ of fluid particles and for the vertical deviation $\zeta(x)$, $x \in \Gamma$, in the problem of proper oscillations of the given hydromechanical system we have the following spectral problem,

$$\Delta\Phi = 0 \quad \text{in } \Omega, \qquad \frac{\partial\Phi}{\partial n} = 0 \quad \text{on } S, \qquad \frac{\partial\Phi}{\partial n} = \zeta \quad \text{on } \Gamma,$$
$$\int_\Gamma \zeta d\Gamma = \int_\Gamma \Phi d\Gamma = 0, \qquad B\zeta = \lambda(\rho_0\zeta + \rho\Phi) \quad \text{on } \Gamma,$$

where

$$B\zeta := P_\Gamma(-T\Delta_2\zeta + \rho g\zeta)P_\Gamma, \quad \mathcal{D}(b) = H_0^2(\Gamma) \cap H,$$

P_Γ is the orthoprojector onto $H := L_2(\Gamma) \ominus \{1\}$, and $\lambda = \omega^2$ is a squared frequency of proper oscillations.

Similar to Sections 3.3, 4.1 it can be proved that the given spectral problem is equivalent to the operator equation $B\zeta = \lambda(\rho_0 I + \rho C)\zeta$, where $\zeta \in \mathcal{D}(B) \subset H$, $B \gg 0$, $B^{-1} \in \mathfrak{S}_\infty$, $0 < C \in \mathfrak{S}_\infty$. Therefore, according to the conclusions derived in Section 1.4, this problem has a discrete spectrum with the accumulation point $+\infty$, and its eigenvalues λ_n are consecutive minima of the variational ratio

$$\int_\Gamma [T|\nabla_2\zeta|^2 + \rho g|\zeta|^2] d\Gamma \Big/ \left(\rho_0 \int_\Gamma |\zeta|^2 d\Gamma + \rho \int_\Omega |\nabla\Phi|^2 d\Omega\right),$$

subject to the above-mentioned relations for Φ and ζ. Using the variational principles from Section 1.4 and the properties $0 < C \in \mathfrak{S}_\infty$, it can be proved that $\lambda_n \le \mu_n + g\rho/\rho_0$, where μ_n are eigenvalues of the membrane without fluid, and the asymptotic formula $\lambda_n = (\mu_n + g\rho/\rho_0)[1 + o(1)]$ as $n \to \infty$ can be obtained.

If the region Ω, similar to the problems addressed in Sections 3.6 and 4.3, is filled with a system of m nonmixing fluids with the densities $\rho_1 > \rho_2 > \ldots > \rho_m > 0$ and the horizontal membranes Γ_j, $j = 1, 2, \ldots, m-1$ with surface densities $\rho_{0,j}$ are located between the fluid layers, then the corresponding problem of small proper oscillations can be reduced to determining the spectrum of the variational ratio

$$\frac{\sum_{j=1}^{m-1}\left\{\int_{\Gamma_j}\left[T_j|\nabla_2\zeta_j|^2+g(\rho_j-\rho_{j+1})|\zeta_j|^2\right]d\Gamma j\right\}}{\sum_{j=1}^{m-1}\rho_{0,j}\int_{\Gamma_j}|\zeta_j|^2 d\Gamma_j+\sum_{k=1}^{m}\rho_k\int_{\Omega_k}|\nabla\Phi_k|^2 d\Omega_k},$$

where ζ_j and Φ_k satisfy the additional conditions

$$\Delta\Phi_k=0 \quad \text{in } \Omega_k, \qquad \frac{\partial\Phi_k}{\partial n}=0 \quad \text{on } S_k, \qquad k=1,2\ldots,m;$$

$$\frac{\partial\Phi_j}{\partial n_j}=\frac{\partial\Phi_{j+1}}{\partial n_j}=\zeta_j \quad \text{on } \Gamma_j, \qquad j=1,2,\ldots,m-1;$$

$$\int_{\Gamma_j}(\rho_j\Phi_j-\rho_{j+1}\Phi_{j+1})\,d\Gamma_j=0, \quad j=1,2,\ldots,m-1, \quad \int_{\Gamma_{m-1}}\rho_m\Phi_m d\Gamma_{m-1}=0.$$

The spectrum of this problem is also discrete, and for its eigenvalues one can prove properties similar to the ones mentioned above for a system that consists of one fluid and one membrane.

B.3 Chapter 5

5.1. The content of this section is based on the work of N. G. Kuznetsov [1] (see also [2]). The work of D. V. Fox and J. R. Kuttler [1] is closely related to these issues, too.

5.2. Here, one states the problem considered by N. D. Kopachevsky and A. V. Sigal [1]. It should be pointed out that the theory of shallow water for spectral problems of fluid dynamics was used very rarely in the past. For the case of a capillary fluid, some problems are considered in the monograph by N. N. Moiseev and V. V. Rumyantsev [1], and in the work of N. D. Kopachevsky [4].

5.3. The problem of small oscillations of the system "ideal fluid-gas" was formulated by N. D. Kopachevsky and investigated by B. M. Vronskii [1]. Here, it is presented in the form that has been slightly modified by N. D. Kopachevsky.

B.4 Chapter 6

The problem of movements of an ideal fluid, which are close to a rotation with constant angular velocity, has drawn attention of many researchers; the bibliography on this issue is quite extensive. Here, we will mention only the monographs by H. Greenspan [1], T. I. Zelenyak [1], and F. L. Chernousko [1]. The application of operator methods in investigating small perturbations of the motion of a top filled with fluid has been worked out in the classic work of S. L. Sobolev (1943) and in the doctorate thesis of

S. G. Krein (1950). Later on, problems related to the S. L. Sobolev's problem have been considered by R. A. Alexandryan [1], T. I. Zelenyak and his scholars, etc. Let us point out that S. L. Sobolev's investigations lead to the creation of a new branch in the theory of abstract evolution equations, dealing with equations that are not solvable relatively to a derivative (see the review by S. G. Krein and M. I. Khazan [1]).

In the part corresponding to fluid dynamics, the above-mentioned works dealt with the case when the fluid completely occupies a cavity in a rigid body. The first work where small perturbations of the stationary rotation of a symmetric cavity filled with an ideal fluid were investigated, with the assumption of small thickness of the fluid layer, was the work of G. S. Narimanov [2]. Later on, these investigations were developed in the works of K. Stewarson [1] and B. A. Kostandyan [1,2]. In the works of J. V. Mails [1–3], the condition of small thickness of the fluid layer was omitted, but the free surface of the fluid was replaced by a weightless cover, which eliminates the surface waves. The later limitation was overcome in the works of M. P. Dyachenko [1,2].

The operator approach to problems of small oscillations of an ideal rotating fluid in an open container based on applying a projection to the equations of motion onto orthogonal subspaces was developed in the doctorate thesis of N. D. Kopachevsky (1980). Further investigations, which took into account this general scheme for problems of oscillations of an ideal incompressible fluid in a rotating container with elastic walls, have been carried out by A. Garadzhaev [1–3].

6.1. The simplest properties of the Coriolis operator (1.4) were found out by S. L. Sobolev [2]. The general theorem on the spectrum structure of this operator was proved by J. V. Ralstone [1]. Also, let us note the work of S. A. Gabov [1], in which it was shown that in the case of a cylindrical cavity of an arbitrary section, the set of eigenvalues of the Coriolis operator is countable and, in view of the result of J. V. Ralstone [1], dense in the interval $[-2i\omega_0, 2i\omega_0]$. The above-mentioned work of S. A. Gabov also proved the theorem on two-multiple basicity of the system of eigenfunctions.

Among the numerous works devoted to the investigation of spectral issues appearing in connection with the problem of S. L. Sobolev, let us point out only the works of R. Denchev [1,2], Y. N. Grigoriyev [1], and S. N. Sklyar [1].

6.2. The theorem on solvability of the boundary value problem for a symmetrical top was obtained by S. L. Sobolev [2]. Here, this theorem has been proved in a general situation.

The spectral problem corresponding to the problem (2.1)–(2.3) for the case of a symmetrical top was investigated in the same work of S. L. Sobolev. In it, S. L.

Sobolev introduced a quadratic form that generates a Hilbert subspace whenever the following magnitude is positive,

$$L := C_1 + C_2 - A_1 - A_2 - \omega_0^{-2} g(l_1 M_1 + l_2 M_2).$$

Here, M_1 and M_2 are the masses of the top and the fluid, C_1 and C_2 are the inertia moments of the body and the fluid relatively to a common symmetry axis, A_1 and A_2 are the moments of inertia relatively to the perpendicular axis that goes through the stationary point, and l_1 and l_2 are distances from the centers of mass of the top and the fluid to the stationary point. In this case, all the eigenvalues are real, the normal oscillations of the system are periodic, and the motion is stable.

In the case when $L < 0$, the mentioned theorem generates a space with indefinite metric (the form has one negative square). The corresponding operator is self-adjoint relatively to the indefinite form. Here, for the first time in a problem of mathematical physics, a linear operator self-adjoint in a space with an indefinite metric has been introduced. In connection with this problem, L. S. Pontryagin, while answering a question raised by of S. L. Sobolev, created the theory of self-adjoint operators in spaces with indefinite metrics (the spaces Π_κ, see Section 1.3). However, for this particular problem, independent of the general theory, S. L. Sobolev investigated the spectrum of the problem and showed that the corresponding operator has no more than two complex conjugate eigenvalues, with a multiplicity that does not exceed 1, and all the real eigenvalues are simple except for, maybe, one eigenvalue.

Hence, in the case when $L < 0$, the motion of the top can become unstable.

In the case of an ellipsoidal cavity, the space of the velocity fields of the fluid can be decomposed into a direct sum of subspaces; one of them is three-dimensional, and only the movements from this subspace interact with the movements of the rigid body. The corresponding frequencies of proper oscillations of the top can be obtained from an algebraic equation of the third order.

In the case of a cylindrical cavity, the corresponding characteristic equation has a countable number of solutions. The motion of the body is connected with a countable set of oscillations of the fluid. If the angular velocity changes, then an infinite set of zones of unstability and stability of the motion can appear.

A. Y. Ishlinskii and M. E. Temchenko [1] used another (more simple) deduction of the characteristic equations of the problem in the case of ellipsoidal and cylindrical cavities.

In the general case, this problem was also considered by F. L. Chernousko [2]. He introduced into consideration the solutions of three auxiliary boundary value problems in the region filled with fluid, which become the Zhukovsky potential when $\omega_0 \to 0$. For $\omega_0 \neq 0$, the fluid velocity and the gyrostatic moment can be expressed in terms of these solutions. In the case of the motion of a free body, a characteristic equation was obtained that has been later investigated in some particular cases. The same method is being applied for studying the motion of a slightly viscous rotating fluid.

6.3. The formulation and consideration of this problem belong to N. D. Kopachevsky [10–12]. Here, both the method of orthogonal projection of the equations of motion and the general reasoning similar to the scheme from Section 1.8 are applied.

The asymptotic formulas (3.39), (3.40), (3.76), (3.77) were obtained by N. D. Kopachevsky (using a heuristic method) and proved by T. A. Suslina [1-3].

6.4. This section was written based on the works by N. D. Kopachevsky [10,11].

6.5. The abstract spectral problem (5.1), (5.2) is discussed based on the works of N. D. Kopachevsky [10,14,15,17,19], and on the work of A. V. Andronov and N. D. Kopachevsky [1]. Here, the results in the works of A. S. Markus and V. I. Matsaev [1], A. S. Markus, V. I. Matsaev, and G. I. Russu [1], and in the monograph by Y. S. Abramov [1] have been considerably used.

The assertion that whenever conditions (5.27) are not valid, then the points $\omega = \pm 1$ are not accumulation points of eigenvalues outside the interval $[-1, 1]$, that is, of the frequencies of surface waves, was proven by M. Y. Tsarkov and later on by A. Fragela. For example, the mentioned fact follows from Theorem 3 in Section 9.3 of the monograph by M. S. Birman and M. Z. Solomyak [BS], p. 206.

6.6. The results of this section belong to N. D. Kopachevsky [13].

Let us point out that the works of N. D. Kopachevsky and B. M. Vronskii [1] and A. V. Andronov, N. D. Kopachevsky, T. P. Krasnoslobodtseva (Temchenko), and B. M. Vronskii [1] address the problems of small movements and proper oscillations of a system "fluid-gas" that rotates uniformly with an angular velocity ω_0, and of the same system in the case when an ideal incompressible fluid partially fills a container with an elastic bottom.

The sophisticated structure of the spectrum of internal waves in an arbitrary partially filled rotating container was used in the work of A. M. Gomilko [1]. In the work of R. V. Rvalov [1] (also see the work of R. V. Rvalov and V. M. Rogovoy [1]), oscillations of a rotating fluid were considered for particular types of partially filled cavities.

Standard Reference Texts

[AI] Azizov, T. Y. and Iokhvidov, I. S. (1986). *Foundations of the Theory of Linear Operators in Spaces with Indefinite Metrics*, pp. 352. Moscow: Nauka.

[AG] Akhiezer, N. I. and Glasman, I. M. (1966). *Theory of Linear Operators in Hilbert Spaces*, pp. 554. Moscow: Nauka.

[BS] Birman, M. S. and Solomyak, M. Z. (1980). *Spectral Theory of Self-adjoint Operators in Hilbert Spaces*, pp. 264. Leningrad: Izd-vo LGU.

[F] Krein, S. G., ed. (1972). *Functional Analysis*, 2nd edition, pp. 544. Moscow: Nauka.

[GG] Gohberg, I. and Goldberg, S. (1981). *Operator Theory*, pp. 285. Basel: Birkhäuser Verlag.

[GGK] Gohberg, I., Goldberg, S., and Kaashoek, M. A. (1990; 1993). *Classes of Linear Operators*, Vol. 1, pp. 468; Vol. 2, pp. 556. Basel: Birkhäuser Verlag.

[GK] Gohberg, I. and Krein, M. G. (1965). *Introduction to the Theory of Nonselfadjoint Operators in Hilbert Spaces*, pp. 448. Moscow: Nauka.

[KKN] Kopachevsky, N. D., Krein, S. G., and Ngo Zui Kan (1989). *Operator Methods in Linear Fluid Dynamics: Evolution and Spectral Problems*, pp. 416. Moscow: Nauka.

[K] Krein, S. G. (1967). *Linear Differential Equations in Banach Spaces*, pp. 464. Moscow: Nauka.

[M] Markus, A. S. (1986). *Intoduction to the Spectral Theory of Polynomial Operator Pencils*, pp. 260. Kishinev: Ştiinţa.

Bibliography

ABRAMOV, Y. S.

[1] *Variational Methods in the Theory of Operator Pencils. Spectral Optimization,* 1983, pp. 178. Leningrad: Izd-vo LGU.

ABRAMSON, H. N.

[1] Dynamic behavior of fluids in a moving container. *Appl. Mech. Reviews,* 16(7), 1973, 501–505.

AZIZOV, T. Y.

[1] On completeness and basicity of eigen- and associated vectors of J-self-adjoint operators of class $\mathcal{K}(\mathcal{H})$. *DAN SSSR,* 253(5), 1980, 1033–1036.

[2] Spectral theory and the theory of extension of operators in spaces with indefinite metrics. PhD Thesis, Voronezh University, Voronezh, 1985, pp. 275.

ALEXANDRIAN, R. A.

[1] Spectral properties of operators generated by Sobolev-type systems of differential equations. *Trudy MMO,* 9, 1960, 455–505.

AMENZADE, R. Y.

[1] Small oscillations of an ideal fluid with regard to mass forces in an elastic hull. *DAN SSSR,* 241(1), 1978, 44–47.

ANDRONOV, A. V.

[1] On small oscillations of an ideal fluid in a container with elastic ends. Preprint No. 1478, SGU, Simferopol, 1993, pp. 26.

[2] On oscillations of a system of rotating ideal fluids in a container with elastic ends. Preprint No. 687, SGU, Simferopol, 1985, pp. 32.

ANDRONOV, A. V. AND KOPACHEVSKY, N. D.

[1] Self-adjoint operator pencils generated by problems on oscillations of a rotating ideal fluid. Preprint No. 1667, SGU, Simferopol, 1986, pp. 19.

ANDRONOV, A. V., KOPACHEVSKY, N. D., KRASNOSLOBODTSEVA (TEMCHENKO), T. P., AND VRONSKI, B. M.

[1] Spectral problems of mathematical physics generated by problems of hydrodynamics and hydroelasticity. *Proceedings of the All-Union Conference "Differential Equations and Their Applications," Ashkhabad, 1986: Abstracts*, p. 17–18.

BABSKII, V. G., KOPACHEVSKY, N. D., MYSHKIS, A. D., SLOBOZHANIN, L. A., AND TYUPTSOV, A. D.

[1] *Gidromehanika Nevesomosty*. Moscow: Nauka, 1976, pp. 504.

[2] Low-gravity fluid mechanics: Some unsolved problems. In *Proceedings of the Inst. of Problems of Mechanics of AN SSSR: Fluid Mechanics and Heat-Mass Transfer in Low-Gravity*. Moscow: Nauka, 1982, 53–59.

[3] *Low-Gravity Fluid Mechanics*. Springer Verlag, Heidelberg, 1987, pp. 583.

BALAKRISHNAN, A. V.

[1] *Applied Functional Analysis*. Moscow: Nauka, 1980, pp. 384.

BARNYAK, M. Y.

[1] On determining frequencies and shapes of proper oscillations of fluid in a container under conditions close to low-gravity. In *Linear and Nonlinear Boundary Value Problems*, Y. A. Mitropolskii (editor). Kiev: Inst. Mat. AN USSR, 1971, 31–45.

[2] Determining eigenfrequencies and shapes of small oscillations of an ideal fluid in a container located in a weak gravitation field. In *Mathematical Physics*. Vol. 9. Kiev: Naukova Dumka, 1971, 3–10.

[3] Variational formulations of the problem of small oscillations of fluid in a container under conditions close to low-gravity. I. A. Lukovskii (editor). Kiev: Inst. Mat. AN USSR, 1977, 74–83.

BARNYAK, M. Y. AND KOMARENKO, A. N.

[1] Variational method of solving the problems on fluid oscillations in a weak gravitation field. Reports at the Seminar on Differential Equations, Volume 1. Kiev: Inst. Mat. AN USSR, 1969, 166–175.

[2] Variational method of determining eigenfrequencies of fluid oscillations in a weak gravitation field. *DAN USSR, Series A*, 2, 1972, 108–111.

BASKAKOV, A. G.

[1] Theorem on splitting of an operator and some adjacent questions from the analitic theory of perturbations. *Izv. AN SSSR. Series Mat.*, 49(3), 1986, 435–457.

[2] *Harmonic Analysis of Linear Operators: A Textbook.* Voronezh: Izd-vo VGU, 1987, pp 164.

BELONOS, S. M. AND CHERNOUS, K. A.

[1] *Boundary Value Problems for Navier-Stokes Equations.* Moscow: Nauka, 1985, pp. 312.

BEREZANSKII, Y. M.

[1] *Decomposition by Eigenfunctions of Self-Adjoint Operators.* Kiev: Naukova Dumka, 1965, pp. 800.

[2] Spaces with negative norm. *UMN*, 18(1), 1963, 63–96.

BESOV, O. V., ILJIN, V. P., AND NIKOLSKII, S. M.

[1] *Integral Representations of Functions and Embedding Theorems.* Moscow: Nauka, 1975, pp. 480.

BIRMAN, M. S. AND SOLOMYAK, M. Z.

[1] Quantitative analysis of Sobolev's theorems and applications to spectral theory. In *Proceedings of the 10th Mathematical School.* Kiev: Inst. Mat. AN USSR, 1974, 5–189.

[2] Asymptotics of the spectrum of differential equations. In *Achievements of Science and Technology: Mathematical Analysis.* Moscow: VINITI, Vol. 14, 1977, 5–52.

[3] Asymptotics of the spectrum of variational problems on solutions of elliptic equations. *SMJ*, 1, 1979, 3–22.

BYHOVSKII, E. B. AND SMIRNOV, N. V.

[1] On orthogonal decompositions of spaces of vector-valued functions square summable in a given region, and on operators of vector analysis. *Trudy MI AN SSSR*, 59, 1960, 5–35.

CHERNOUSKO, F. L.

[1] Movements of a rotating rigid body with a cavity filled with fluid. *PMM*, 31, 1967, 416–432.

COOPER, R. M.

[1] Dynamics of liquids in moving containers. *ARSJ*, 30(8), 1960, 1506–1513.

DALETSKII, Y. L. AND KREIN, M. G.

[1] *Stability of Solutions of Differential Equations in a Banach Space.* Moscow: Nauka, 1970, pp. 534.

DENCHEV, R.

[1] On spectrum of an operator. *DAN SSSR*, 126, 1959, 256–252.

[2] On spectrum of an operator. *DAN SSSR*, 127, 1959, 501–504.

DENY, J. AND LIONS, J. L.

[1] Les espaces du type de Beppo Levi. *Ann. Inst. Fourier*, 5, 1953–1954, 305–370.

DE-RAM, G.

[1] *Differentiable Manifolds.* Moscow: IL, 1956, pp. 271.

DOKUCHAEV, L. V. AND RVALOV, R. V.

[1] On stability of stationary rotation of a rigid body with a cavity containing a fluid. *Izv. AN SSSR. MJG*, 4, 1972, 3–9.

DYACHENKO, M. P.

[1] On the theory of oscillations of a nonviscous incompressible uniformly rotating fluid. *DAN USSR, Series A*, 8, 1970, 748–752.

[2] On oscillation of a gyrostate with a cavity partially filled with a nonviscous incompressible fluid. *DAN USSR, Series A*, 10, 1971, 915–919.

FESCHENKO, S. F., LUKOVSKII, I. A., RABINOVICH, B. I., AND DAKUCHAEV, L. V.

[1] *Methods for Determining Adjoint Masses of Fluid in Moving Cavities.* Kiev: Naukova Dumka, 1969, pp. 250.

FOX, D. W. AND KUTTLER, J. R.

[1] Sloshing frequencies. *Z. Angew. Math. Phys.*, 34(5), 1983, 668–696.

FRIEDMAN, A. AND SHINBROB, M.

[1] The initial value problem for the linearized equations of water waves. *J. Math. Mech.*, 17, 1967, 107–180.

FRIEDRICHS, K. O.

[1] On the boundary value problems of the theory of elasticity and Korn's inequality. *Ann. of Math.*, 48(2), 1947, 347–354.

GABOV, S. A.

[1] On the spectrum of one of S. L. Sobolev's problems. *DAN SSSR*, 253(3), 1980, 521–524.

GARADZHAEV, A.

[1] On the problem of normal oscillations of a heavy viscous fluid in a container. *Sib. Mat. J.*, 25(2), 1984, 213–216.

[2] Some model problems connected with problems on oscillations of a viscous fluid in an elastic container. *Diff. Equations*, 20(1), 1984, 14–19.

[3] Spectral theory of the problem on small oscillations of an ideal fluid in a rotating elastic container. *Diff. Equations*, 23(1), 1987, 38–47.

GATAULLIN, I. G., IMSHENETSKII, A. B., AND RABINOVICH, B. I.

[1] On calculation of frequencies and associated masses of fluid in moving cavities using the method of random search. *Automatics and Calculation Technique*, 4, 1967, 37–47.

GLUSHKO, V. P. AND KREIN, S. G.

[1] Inequalities for norms of derivatives in weighted L_p spaces. *Sib. Mat. J.*, 1(3), 1960, 343–382.

GOMILKO, A. M.

[1] On the continuous spectrum of a problem of fluid mechanics. *UMN*, 36(5), 1981, 169–170.

GOBERT, I.

[1] Une inegalite fondamentale de la theorie de l'elasticite. *Bull. Soc. Roy. Sci. Liege*, 3–4, 1962, 182–191.

GOLDSTEIN, V. M. AND RESHETNYAK, Y. S.

[1] *Introduction to the Theory of Functions with Generalized Derivatives and Quasi-Conformal Mappings.* Moscow: Nauka, 1969, pp. 283.

GREENSTEIN, V. A.

[1] On p-basicity of the system of eigen- and associated vectors of a polinomial self-adjoint operator pencil. Preprint No. 890, Simferopol Univ., Simferopol, 1990, pp. 8.

[2] Basicity of part of the system of eigenvectors of a holomorphic operator-valued function. *Mathem. Notes*, 50(1), 1991, 142–143.

GREENSTEIN, V. A. AND KOPACHEVSKY, N. D.

[1] On the spectrum problem of a bounded self-adjoint operator. Preprint No. 109, Simferopol Univ., Simferopol, 1989, pp. 25.

[1] On the p-basicity of a system of eigenelements for a self-adjoint operator-valued function. In *Proceedings of the 15th All-Union School on Operator Theory in Functional Spaces, Ulyanovsk, 1990, Part 1: Abstracts*, 1990, pp. 72.

GREENSPAN, H.

[1] *Theory of Rotating Fluids.* Leningrad: Gidrometeoizdat, 1975, pp. 304.

GRIGOLYUK, E. I.

[1] Problems on the interaction of hulls with fluids. In *Proceedings of the 8th All-Union Conference on the Theory of Hulls and Plates.* Moscow: Nauka, 1970, 755–778.

GRIGORIEV, Y. N.

[1] On the spectrum of the operator pencil of the S. L. Sobolev's problem. In *Dynamics of Continuous Media.* L. V. Ovsyannikov (editor). Novosibirsk, Vol. 17, 1974, 12–18.

HENRY, D.

[1] *Geometric Theory of Semilinear Parabolic Equations.* Berlin, Heidelberg, New York: Springer Verlag, 1981.

ISHLINSKII, A. Y. AND TEMCHENKO, M. E.

[1] On small oscillations of the vertical axis of a top with a cavity completely filled with an ideal incompressible fluid. *Journ. of Appl. Mech. and Tech. Phys.*, 3, 1960, 65–75.

JOSEPH, D.

[1] *Stability of Motions of a Fluid.* Moscow: Mir, 1981, pp. 245.

KARAZEEVA, N. A. AND SOLOMYAK, M. Z.

[1] Asymptotics of the spectrum of Steklov-type problems in component regions. In *Problems of Mathematical Analysis*, Vol. 8. Leningrad: Izd-vo LGU, 1981, 36–48.

KATO, T.

[1] *Theory of Perturbations of Linear Operators.* Moscow: Mir, 1972, pp. 740.

KELDYSH, M. V.

[1] On eigenvalues and eigenfunctions of some classes of nonself-adjoint operators. *DAN SSSR*, 77(1), 1951, 11–14.

[2] On completeness of eigenfunctions of some classes of nonself-adjoint linear operators. *UMN*, 24(4), 1971, 15–41.

KELDYSH, M. V. AND LIDSKII, V. B.

[1] Issues on the spectral theory of nonself-adjoint operators. In *Proceedings of the 4th All-Union Mathematics Congress*, Vol. 1. Leningrad: Izd-vo AN SSSR, 1963, 102–120.

KHARLAMOV, P. V. AND SAVCHENKO, A. Y.

[1] On integration of equations of motion of a body with a cavity filled with fluid (in Ukranian). *DAN URSR, Series A*, 2, 1968.

KOCHIN, N. E., KIBEL, I. A., AND ROZE, N. V.

[1] *Theoretical Fluid Mechanics.* Moscow: Fizmatgiz, Part 1, 1963, pp. 584; Part 2, 1963, pp. 604.

KOPACHEVSKY, N. D.

[1] Fluid dynamics in weak force fields. On small oscillations of an ideal fluid. *Izv. AN SSSR. MJG*, 2, 1966, 141–144.

[2] On the Cauchy problem for small oscillations of a viscous fluid in a weak field of mass forces. *Journ. of Comp. Math. and Math. Phys.*, 7(1), 1967, 129–146.

[3] On small oscillations of an ideal fluid in a container under conditions close to low-gravity. N. N. Moiseev (editor). Moscow: BT AN SSSR, 1968, 98–134.

[4] On the influence of capillary forces on the frequencies of proper oscillations of an ideal fluid in a slightly filled spheric container. *Journ. of Comp. Math. and Math. Phys.*, 9(6), 1969, 1347–1356.

[5] On the free oscillations of a fluid rotating in a cylindrical container under conditions of low-gravity. *Izv. AN SSSR. MJG*, 4, 1972, 3–9.

[6] Fluid dynamics in weak gravitation fields: On plane oscillations of an ideal fluid in a rectangular channel. *Izv. AN SSSR. MJG*, 5, 1972, 3–13.

[7] Fluid dynamics in weak gravitation fields: On small oscillations of an ideal fluid in a cylindrical container. *Izv. AN SSSR. MJG*, 2, 1973, 3–13.

[8] Fluid dynamics in weak gravitation fields: Plane problems on oscillations of an ideal fluid in a container. *Journ. of Comp. Math. and Math. Phys.*, 13(4), 1973, 952–970.

[9] On oscillation of nonmixing fluids. *Journ. of Comp. Math. and Math. Phys.*, 13(5), 1973, 1249–1263.

[10] The Cauchy problem for small movements of an ideal capillary rotating fluid. *DAN SSSR*, 219(6), 1974, 1310–1313.

[11] The Cauchy problem for small movements of a rotating fluid in a container. In *Proceedings of the 8th Voronezh Winter Mathematics School: Abstracts.* Voronezh: Izd-vo VGU, 1974, 52–53.

[12] Using S. L. Sobolev's method in the problem on oscillations of an ideal capillary rotating fluid. *Journ. Comp. Math. and Math. Phys.*, 16(2), 1976, 426–439.

[13] Small movements and proper oscillations of an ideal rotating fluid. Preprint FTINT AN USSR. Kharkov, 1978, pp. 54.

[14] On the existence of surface waves in the problem on normal oscillations of an ideal fluid rotating in a partially filled container. *Functional Analysis and its Applications*, 12(2), 1978, 84–85.

[15] On the basicity of the system of eigen- and associated vectors of a self-adjoint operator pencil. *Functional Analysis and its Applications*, 15(2), 1981, 77–78.

[16] *Problems of the Theory of Fluid Oscillations in Low-Gravity: A Textbook.* Kharkov: Izd-vo KAI, 1981, pp. 100.

[17] On small oscillations of a capillary fluid rotating in a partially filled container. In *Proceedings of the Inst. of Problems of Mechanics of AN SSSR: Fluid Mechanics and Heat-Mass Transfer in Low-Gravity.* Moscow: Nauka, 1982, 59–60.

[18] On properties of the system of modes of surface waves in a rotating ideal fluid. In *Functional Analysis and Applied Mathematics.* Kiev: Naukova Dumka, 1982, 43–55.

[19] On p-basicity of the system of root vectors of the self-adjoint operator pencil $I-\lambda A-\lambda^{-1}B$. In *Functional Analysis and Applied Mathematics.* Kiev: Naukova Dumka, 1982, 55–70.

[20] On the spectral theory of self-adjoint operator-valued functions and its applications to fluid dynamics. In *Proceedings of the 8th All-Union School on Operator Theory in Functional Spaces, Minsk: Abstracts.* Minsk, 1982, 89–90.

KOPACHEVSKY, N. D. AND MOROZOVSKAYA, N. N.

[1] Determining the first frequency of oscillations of an ideal fluid in a cylindrical container in a weak gravitation field. In *Introduction to the Dynamics of a Body with Fluid in Low-Gravity*, N. N. Moiseev (editor). Moscow: BT AN SSSR, 1968, 135–151.

KOPACHEVSKY, N. D., MYSHKIS, A. D., AND TYUPTSOV, A. D.

[1] On fluid oscillations under conditions of complete or partial low-gravity. In *Mathematical Physics and Functional Analysis*, Vol. 1. Kharkov: FTINT AN USSR, 1969, 5–23.

KOPACHEVSKY, N. D., MYSHKIS, A. D., SLOBOZHANIN, L. A., AND TYUPTSOV, A. D.

[1] New results in the hydromechanics of fluids in low-gravity. In *Proceedings of the 4th All-Union Congress in Theoretical and Applied Mechanics: Abstracts.* Kiev: Naukova Dumka, 1976, pp. 62.

KOPACHEVSKY, N. D. AND NEMIRSKAYA, O. I.

[1] On p-basicity of the system of eigenelements of a self-adjoint operator-valued function. Preprint No. 1969, Simferopol Univ., Simferopol, 1992, pp. 10.

KOPACHEVSKY, N. D. AND RADYAKIN, N. K.

[1] On small oscillations of an ideal capillary fluid rotating in an axially symmetric container. In *Questions of Comput. Mathematics and Technology: Proceedings of Scientific Seminars FTINT AN USSR.* Kiev: Naukova Dumka, 1976, 3–25.

[2] Free oscillations of two capillary fluids rotating in a cylindrical container. *Izv. AN SSSR. MJG*, 5, 1976, 97–104.

KOPACHEVSKY, N. D. AND SIGAL, A. V.

[1] The theory of shallow water in the problem on oscillations of an ideal fluid in a bounded container. In *Proceedings of the All-Union Conference "Problems of Complex Automatization of Hydrophysical Research," Sevastopol: Abstracts.* Sevastopol, 1989, 100–101.

KOPACHEVSKY, N. D., TYOMKIN, L. A., AND TYOMKINA, V. S.

[1] On oscillations of a fluid drop located on a plane in low-gravity. In *Mathematical Physics and Functional Analysis*, Vol. 1. Kharkov: FTINT AN USSR, 1969, 37–48.

KOPACHEVSKY, N. D. AND VRONSKII, B. M.

[1] Self-adjoint operator pencils generated by a problem on oscillations of the system "fluid-gas." In *Proceedings of the 9th All-Union School on the Theory of Operators in Function Spaces, Ternopol: Abstracts.* Ternopol, 1984, 25.

KOSTANDYAN, B. A.

[1] On stability of movements of a rotating top with a cavity not completely filled with fluid. *J. Appl. Mech. and Tech. Phys.*, 3, 1960, 56–64.

[2] Influence of oscillations of the free surface on the stability of movements of a rotating top containing a fluid. *PMM*, 25(4), 1961, 646–659.

KOSTYUCHENKO, A. G. AND ORAZOV, M. B.

[1] On some properties of the roots of a self-adjoint quadratic pencil. *Functional Analysis and its Applications*, 9(4), 1975, 28–40.

[2] The problem on oscillations of an elastic half-cylinder and related self-adjoint quadratic pencils. *Proceedings of the Seminar in Honor of I. G. Petrovskii*, No. 6, 1981, 97–146.

KOSTYUCHENKO, A. G. AND SHKALIKOV, A. A.

[1] Self-adjoint quadratic pencils of operators and elliptic problems. *Functional Analysis and its Applications*, 17(2), 1983, 38–61.

[2] On the theory of self-adjoint pencils of operators. *Vestn. MGU, Series 1: Mathematics, Mechanics*, 6, 1983, 40–51.

KREIN, M. G.

[1] Introduction to the geometry of indefinite J-spaces and to the theory of operators in these spaces. In *Proceedings of the Second Summer Mathematical School*, Vol. 1. Kiev: Naukova Dumka, 1965, 15–92.

KREIN, M. G. AND LANGER, G. K.

[1] On the theory of quadratic pencils of nonself-adjoint operators. *DAN SSSR*, 154(6), 1964, 1258–1261.

[2] On some mathematical principles of linear theory of differential oscillations of continua. In *Proceedings of the International Symposium on Application of the Theory of Functions of Complex Variable in the Mechanics of Continuous Medium*, Vol. 2. Moscow: Nauka, 1965, 283–322.

KREIN, S. G.

[1] On functional properties of the operators of vector analysis and fluid mechanics. *DAN SSSR*, 93(6), 1953, 969–972.

[2] Mathematical questions of the theory of motion of a rigid body with a cavity filled with fluid. In *Proceedings of the 3rd All-Union Mathematics Congress*, Vol. 1. Moscow: Izd-vo AN SSSR, 1956, pp. 205.

[3] Differential equations in a Banach space and their applications to fluid mechanics. *UMN*, 12(1), 1957, 208–211.

[4] On an interpolation theorem in operator theory. *DAN SSSR*, 130(3), 1960, 491–494.

[5] Asymptotic splitting of operator equations. In *Lyapunov's Functions and Their Applications*, L. V. Ovsyannikov (editor). Novosibirsk: Nauka, 1986, 206–214.

KREIN, S. G. AND KHAZAN, M. I.

[1] Differential equations in a Banach space. In *Achievements of Science and Technology: Mathematical Analysis.* M: Viniti, Vol. 21, 1983, 130–264.

KREIN, S. G. AND MOISEEV, N. N.

[1] On oscillations of a rigid body containing a fluid with free surface. *PMM*, 21(2), 1957, 169–174.

KREIN, S. G., PETUNIN, Y. I., AND SEMYONOV, E. M.

[1] *Interpolation of Linear Operators.* Moscow: Nauka, 1978, pp. 400.

KREIN, S. G. AND PETUNIN, Y. I.

[1] Scales of Banach spaces. *YMN*, 21(2), 1968, 39–168.

KREIN, S. G. AND TROFIMOV, V. P.

[1] On some operators holomorphically depending on a parameter. In *Proceedings of Math. Dept. of VGU.* Voronewh: Izd-vo VGU, 1970, 63–85.

KUZNETSOV, N. G.

[1] On a variational method of determining proper frequencies of oscillations of fluid in a channel. *Applied Math. and Mechanics*, 54(4), 1990, 553–561.

[2] Lower estimates of proper frequencies of plane oscillations of a fluid in a channel. *Applied Math. and Mechanics*, Vol. 56(2), 1992, 342–346.

LADYZHENSKAYA, O. A.

[1] *Mixed Problem for a Hyperbolic Equation.* Moscow: Gostechizdat, 1953, pp. 280.

[2] On nonstationary operator equation and their applications to linear problems of mathematical physics. *Mat. Sb.*, 45(2), 1968, 123–158.

[3] *Mathematical Issues on Dynamics of a Viscous Incompressible Fluid.* Moscow: Nauka, 1970, pp. 288.

LAMB, H.

[1] *Fluid Dynamics.* Moscow, Leningrad: Gostechizdat, 1947, pp. 928.

LANGER, H.

[1] Zur spectral Theorie J-selbstadjugierter Operatoren. *Math Ann.*, 146, 1962, pp. 60.

LARIONOV, E. A.

[1] On bases made of root vectors of an operator pencil. *DAN SSSR*, 206(2), (1972), 283–286.

LAX, P. D.

[1] On Cauchy problem for hyperbolic equations. *Comm. Pure and Appl. Math.*, 8(4), 1955, 615–633.

LERAY, J.

[1] *Lectures on Hyperbolic Equations with Variable Coefficients.* Inst. for Adv. Stud., Princeton, Fall, 1952, pp. 125.

LIMARCHENKO, P. S.

[1] Influence of fluid contents on the motion of a container. *Applied Mech.*, 16(3), 1960, 131–133.

LIONS, J. L.

[1] Espaces intermediares entre espaces hilbertiens et applications. *Bull. Math. Soc. Math. Phys. R. P. Roumanie*, 2(4), 1958, pp. 419.

LIONS, J. L. AND MAGENES, E.

[1] *Problemes aux Limites Nonhomogenes et Applications*, Vol. 1. Paris: Dunod, 1968.

LUKOVSKII, I. A.

[1] Determining dynamics characteristics of a fluid in a moving container under conditions of weak gravity by the method of eigenfunction decomposition. In *Mathematical Physics*, Vol. 11. Kiev, 1972, 57–64.

LYANTSE, V. E. AND STOROZH, O. G.

[1] *Methods of the Theory of Unbounded Operators.* Kiev: Naukova Dumka, 1983, pp. 210.

LYUSTERNIK, L. A. AND SOBOLEV, V. I.

[1] *Brief Course on Functional Analysis.* Moscow: Vysshaya shkol., 1982, pp. 271.

MARKUS, A. S. AND MATSAEV, V. I.

[1] On basicity of some part of eigen- and associated vectors of a self-adjoint operator pencil. *Math. Selection,* 133(3), 1987, 293–313.

MARKUS, A. S., MATSAEV, V. I., AND RUSSU, G. I.

[1] On some generalizations of the theory of strongly damped pencils for the case of pencils of arbitrary order. *Acta Sci. Math. Szeged,* 34, 1973, 245–271.

MATSAEV, V. I. AND PALANT, Y. A.

[1] *Spectral Theory of Operator Pencils.* Donetsk: Izd-vo DGU, 1975, pp. 39.

MAILS, J. U. AND TRESH, B. A.

[1] Surface oscillations of a rotating fluid. In *Proc. Amer. Soc. Eng.-Mech. Applied Mech., Series E,* 4, 1961.

MAZJA, V. G.

[1] *Spaces of S. L. Sobolev.* Leningrad: Izd-vo LGU, 1985, pp. 416.

MIKHAILOV, V. P.

[1] On the mixed boundary value problem for the system of equations of Navier-Stokes. *Sib. Math. J.,* 9(5), 1968, 1221–1237.

MIKISHEV, G. N. AND RABINOVICH, B. I.

[1] *Dynamics of a Rigid Body with Cavities Partially Filled with Fluid.* Moscow: Mashinostroenie, 1968, pp. 532.

MILES, I. W.

[1] Free surface oscillations in a rotating liquid. *Phys. Fluid.,* 2(3), 1959, 297–305.

[2] The Cauchy-Poisson problem for a rotating fluid. *J. Fluid Mech.,* 17(1), 1963, 89–97.

[3] Free-surface oscillations in a slowly rotating fluid. *J. Fluid Mech.,* 18(2), 1964, 187–194.

MILES, I. W. AND TROESCH, B. A.

[1] Surface oscillations of a rotating liquid. *J. Appl. Mech.,* 83(4), 1961, 491–496.

MILOSLAVSKII, A. I.

[1] On the stability of some classes of evolution equations. *Sib. Math. J.,* 36(5), 1985, 118–132.

MOGULSKII, E. Z.

[1] On completeness of the system of eigen- and associated vectors of an operator pencil. *DAN SSSR,* 183(4), 1968, 775–778.

MOISEEV, G. A.

[1] Motion of a rigid body with a cavity completely filled with two nonmixing fluids. In *Mathematical Physics*, Vol. 13. Kiev, 1973, pp. 66–73.

MOISEEV, N. N.

[1] On oscillations of a heavy ideal incompressible fluid in a container. *DAN SSSR*, 85(5), 1952, 963–965.

[2] Motion of a rigid body with a cavity partially filled with ideal dropped fluid. *DAN SSSR*, 85(4), 1952, 719–722.

[3] Problems of small oscillations of an open container with fluid under the influence of an elastic force. *Ukr. Math. J.*, 4(2), 1952, 168–173.

[4] Problems of motion of a rigid body containing fluid masses, which have a free surface. *Math. Selection*, 32(74:1), 1953, 61–96.

[5] *Variational Methods in the Problem of Oscillations of a Fluid and a Body with Fluid.* N. N. Moiseev (editor). Moscow: Izd-vo BT AN SSSR, 1962, pp. 248.

MOISEEV, N. N. (EDITOR)

[1] *Introduction to the Dynamics of a Body with Fluid under Conditions of Low-Gravity.* Moscow: Izd-vo BT AN SSSR, 1968, pp. 280.

MOISEEV, N. N. AND CHERNOUSKO, F. L.

[1] Problems on oscillations of fluids influenced by forces of surfaces tension. *J. Comp. Math. and Math. Phys.*, 5(6), 1965, 1017–1035.

MOISEEV, N. N. AND PETROV, A. A.

[1] *Numerical Methods for Calculating Eigenfrequencies of Oscillations of a Bounded Volume of Fluid.* Moscow: Izd-vo BT AN SSSR, 1966, pp. 270.

MOISEEV, N. N. AND RUMYANTSEV, V. V.

[1] *Dynamics of a Body with Cavities Containing Fluid.* Moscow: Nauka, 1965, pp. 440.

MYSHKIS, A. D., BABSKSII, B. G., ZHUKOV, M. Y., KOPACHEVSKY, N. D., SLOBOZHANIN, L. A., AND TYUPTSOV, A. D.

[1] *Methods of Solving Problems of Fluid Mechanics for Conditions of Low-Gravity.* Kiev: Naukova Dumka, 1992, pp. 592.

NARIMANOV, G. S.

[1] On the motion of a rigid body with a cavity partially filled with fluid. *PMM*, 20(1), 1956, 21–38.

[2] On the motion of a symmetric gyroscope with a cavity partially filled with fluid. *PMM*, 21(5), 1957, 696–708.

NGO ZUY KAN

[1] Ve Chuyn dong quau cua vat co lo hong chua chat long nhot. *Tap San Toan Hoc.*, 1(1), 1973, 36–44.

[2] Dao dong cua con lac co lo hong chua chat long nhot. *Tap San Toan Hoc.*, 2(1–2), 1973, 53–57.

[3] Ve Viec Tehn he so tat dan cua dao dong trong chat long nhot khong nen duok. *Tap San toan Hoc.*, 3(2), 1975, 7–11.

[4] On the motion of nonmixing fluids in a container with a plane elastic end. *Izv. AN SSSR. MTT*, 5, 1979.

[5] On small oscillations of an elastic container that has the form of a rectangular parallelepiped, filled with ideal fluid with surface tension. In *Operator Methods in Differential Equations*, S. G. Krein (editor). Voronezh: Izd-vo VGU, 1979, 56–62.

[6] On the motion of an ideal fluid with surface tension in a container with a plane elastic end. *Izv. AN SSSR. MTT*, 3(4), 1980, 143–154.

NGO ZUY KAN AND NGUYEN TAT DAC

[1] Dac dong cua binh dan hoi chua knong day chat long ly tnong. *Tap Chi Jean Hoc.*, 6(4), 1978, 1–10.

NGO ZUY KAN, NGUYEN TAT DAC, AND PHAM VAN NIHN

[1] On the oscillations of an elastic reservoir containing fluid. *Acta Scientiarum Vietnamicarum, Sectio Scientiarum Mathematicarum et Physicarum*, 8(1), 1972, 109–123.

NGO ZUY KAN, NGUYEN XUAN HUY, AND NGUEN CHI THANH

[1] Ve daodong cua chat long nhot khong nen duec chua khong day trong binh hing tru nam ngang. *Tom tat bao cao hoi nghi toan hoc Viet nam lan thu 2*, Hanoi, 1977, 32–39.

NICODYM, O.

[1] Sur une classe de founction considerees dans le probleme de Dirichlet. *Fundam. Mat.*, 21, 1933, 129–150.

NIKOLSKII, S. M.

[1] *Approximation of Functions of Several Variables and Embedding Theorems.* Moscow: Nauka, 1977, pp. 455.

OHOTSIMSKII, D. E.

[1] On the theory of motion of a rigid body with cavities partially filled with fluid. *PMM*, 20(1), 1956, 3–20.

Paraska, V. I.

[1] On asymptotics of eigen- and singular values of linear operators with increasing smoothness. *Math Selection*, 68(4), 1965, 621–631.

Pashkova, J. S.

[1] Oscillations of a hydrosystem with a membrane boundary. In *Proceedings of the First Crimean All-Union School—Symposium on Special and Evolution Problems, Sevastopol-Laspi: Abstracts.* Kiev, 1990.

[2] Oscillations of a hydrodynamic system with membrane boundaries. Preprint No. 1506, Simferopol Univ., Simferopol, 1992, pp. 28.

Petrenko, M. P.

[1] Proper oscillations of a fluid with free surface and an elastic end in a cylindrical cavity. *Applied Mech.*, 5(6), 1969, 44–50.

[2] Forced joint oscillations of a fluid and an elastic end in a cylindrical container. *Applied Mech.*, 6(6), 1970. 127–131.

Petrov, A. A., Popov, Y. P., and Pukhnachev, Y. V.

[1] Calculation by the variational method of proper oscillations of a fluid in immovable containers. *J. Comp. Math. and Math. Phys.*, 4(5), 1964, 880–895.

Pontryagin, L. S.

[1] Hermitian operators in spaces with indefinite metrics. *Izv. AN SSSR, Series Math.*, 8(6), 1944, 243–280.

Prigorskii, V. A.

[1] On some classes of bases of a Hilbert space. *UMN*, 20(5:125), 1965, 231–236.

Pshenichnov, G. I.

[1] Free fluid oscillations in rigid containers. *PMM*, 36(2), 1972, 248–254.

Rabinovich, B. I.

[1] On the equations of perturbed motion of a rigid body with a cylindrical cavity partially filled with fluid. *PMM*, 20(1), 1956, 39–50.

Rabinovich, B. I., Dokuchaev, L. V., and Polyakova, Z. M.

[1] On calculation of coefficients of equations of perturbed motion of a rigid body with cavities partially filled with fluid. *Space Research*, 3(2), 1965, 179–207.

Radzievskii, G. V.

[1] Multiple completeness of root vectors of the Keldish pencil perturbed by an operator-valued function analytic in a circle. *Math. Selection*, 91(133:3), 1973, 310–335.

[2] Completeness of root vectors of the Keldysh pencil perturbed by an analytic operator-valued function $S(\lambda)$ with $S(\infty) = 0$. *Math. Notes*, 21(3), 1977, 391–398.

[3] The problem on completeness of root vectors in the spectral theory of operator-valued functions. *UMN*, 5, 1982, 81–145.

RALSTON, J. V.

[1] On stationary modes in inviscid rotating fluids. *J. Math. Analysis and Appl.*, 44, 1973, 366–383.

RAPOPORT, I. M.

[1] *Oscillations of an Elastic Hull Partially Filled with Fluid.* Moscow: Mashinostroenie, 1967, pp. 357.

RIESZ, F. AND SZ.-NAGY, B.

[1] *Lectures on Functional Analysis.* Moscow: Mir, 1979, pp. 588.

RUMYANTSEV, V. V.

[1] On the stability of motion of a rigid body with surface tension. *PMM*, 28(4), 1964, 746–753.

RVALOV, R. V.

[1] Boundary value problems of free oscillations of a rotating ideal fluid. *Izv. AN SSSR. MJG*, 4, 1973, 81–88.

RVALOV, R. V. AND ROGOVOY, V. M.

[1] On the movement of a rotating body with a cavity containing fluid. *Izv. AN SSSR. MJG*, 3, 1972, pp. 98.

SKALAK, R. AND GONLY J.

[1] Surface waves on a rotating liquid. *AIAA J.*, 2(2), 1964.

SKLYAR, S. N.

[1] On the eigenfunction basis of an operator related to S. L. Sobolev's problem. In *Dynamics of Continuous Media*, Vol. 17. Novosibirsk: Izd-vo Inst. Gidrodinamiki SO AN SSSR, 1974, 81–88.

SMIRNOVA, E. P.

[1] Stabilization of the free rotation of an assymetrical top with cavities completely filled with fluid. *PMM*, 38(6), 1974, 980–985.

SOBOLEV, S. L.

[1] The Cauchy problem in spaces of functionals. *DAN SSSR*, 3, 1935, 291–294.

[2] On a new problem for a system of partial differential equations. *DAN SSSR*, 81(6), 1951, 1007–1009.

[3] On a new problem of mathematical physics. *Izv. AN SSSR. Series Math.*, 18(1), 1954, 3–50.

[4] On the motion of a symmetric top with a cavity filled with fluid. *J. Appl. Mech. and Tech. Phys.*, 3, 1960, 20–55.

[5] *Introduction to the Theory of Cubature Formulas.* Moscow: Nauka, 1974, pp. 808.

[6] *Some Applications of Functional Analysis in Mathematical Physics.* Moscow: Nauka, 1988, pp. 336.

SOLONNIKOV, V. A. AND URALTSEVA, N. N.

[1] Sobolev's Spaces. In *Selected Chapters on Analysis and Higher Algebra, Part 3*, M. Z. Solomyak (editor). Leningrad: Izd-vo LGU, 1981, 129–197.

SRETENSKII, L. N.

[1] Oscillations of fluid in a moving container. *Izv. AN SSSR, Otdel Teh. Nauk.*, 10, 1951, 1483–1494.

SRUBSCHIK, L. S. AND YUDOVICH, V. I.

[1] Asymptotic shape of the free surface of a uniformly rotating fluid with large Bond numbers. *Izv. AN SSSR. MJG*, 6, 1973, 3–12.

STEWARSON, K.

[1] On almost rigid rotations, Part 2. *J. Fluid. Mech.*, 26, 1966, 131–144.

SUSLINA, T. A.

[1] Asymptotics of the spectrum of variational problems for solutions of a homogeneous elliptic equation if there are connections on part of the boundary. In *Problems of Math. Analysis*, Vol. 9. Leningrad: Izd-vo LGU, 1984, 84–97.

[2] Asymptotics of the spectrum of variational problems for solutions of an elliptic equations in a region with piecewise smooth boundary. *Zap. Nauk. Semin. LOMI AN SSSR*, 147, 1985, 179–183.

[3] Asymptotics of the spectrum of some problems connected with fluid oscillations. Preprint No. 8058-B, Leningrad Electromeh. Inst. Svyazi.-L., 1985, pp. 79.

TEMAM, R.

[1] *Navier-Stokes Equations.* Moscow: Mir, 1981, pp. 408.

TROESH, B. A.

[1] Free oscillations of a fluid in a container. Boundary problems. In *Differential Equations.* Madison: Univ. of Wisconsin Press, 1960, 279–299.

TROFIMOV, V. P.

[1] On root subspaces of operators depending analytically on a parameter. *Mathematical Research*, 3(3), Kishinev: RIO AN MSSR, 1968, 117–125.

TYOMKIN, L. A. AND TYOMKINA, V. S.

[1] On the calculation of proper oscillations of an ideal fluid in an axially symmetric container with regard to surface forces. *Izv. AN SSSR. MJG*, 5, 1972, 14–22.

TYUPTSOV, A. D.

[1] Hydrostatics in weak force fields. Stability of equilibrium shapes of fluid surface. *Izv. AN SSSR. MJG*, 2, 1966, 78–85.

USPENSKII, S. V., DEMIDENKO, G. V., AND PEREPYOLKIN, V. G.

[1] *Embedding Theorems and Appendices to Differential Equations.* Novosibirsk: Nauka, 1984, pp. 224.

VASILIJEVA, A. B. AND BUTUZOV, V. F.

[1] *Asymptotic Decompositions of Solutions of Singularly Perturbed Equations.* Moscow: Nauka, 1973, pp. 272.

WEYL, H.

[1] Method of orthogonal projection in potential theory: Selected works. In *Mathematics and Theoretical Physics*, V. I. Arnold (editor). Moscow: Nauka, 1984, 275–307.

VISHIK, M. I. AND LYUSTERNIK, L. A.

[1] Regular degeneration and boundary layer for linear differential equations with small parameter. *UMN*, 12(5), 1957, 3–122.

VRONSKII, B. M.

[1] Oscillations of the system "fluid-gas" in a bounded region. Preprint No. 1811, Simferopol, 1986, pp. 26.

VULIS, I. L. AND SOLOMYAK, M. Z.

[1] Spectral asymptotics of degenerating Steklov's problem. *Vestnik LGU*, 19, 1973, 148–150.

[2] Spectral asymptotics of degenerating elliptic operators of the second order. *Izv. AN SSSR, Series Mat.*, 38(6), 1974, 1362–1392.

YUDOVICH, V. I.

[1] *Method of Linearization in Hydrodynamics Theory of Stability.* Rostov-on-Don: Izd-vo RGU, 1984, pp. 192.

ZABCZYK, J.

[1] A note on C_0-semigroups. *Bull. Acad. Pol. Sci., Series Sci. Math., Astron. et Phys.*, 23(8), 895–898.

ZELENYAK, T. I.

[1] *Selected Issues on Qualitative Theory of Partial Equations.* Novosibirsk: Izd-vo NGU, 1970, pp. 164.

ZHUKOVSKY, N. E.

[1] On the motion of a rigid body with cavities filled with a homogeneous dropped fluid. *Izbr. Soch.*, Vol. 1. Moscow-Leningrad: Gostehizdat, 1948, 31–152.

List of Symbols

$\emptyset$, $\mathbb{N}$, $\mathbb{R}$, $\mathbb{C}$	the empty set and the sets of natural, real, and complex numbers
$\in$, $\subset$, $\cup$, $\cap$	membership, inclusion, union, and intersection
$A := B$ ($B =: A$)	equality by definition; the colon sign is placed on the side of the object to be defined
E, F, G	abstract Hilbert or Banach spaces (Chapter 1)
$\dim E$	dimension of space E
$(\cdot,\cdot)_E$ ($[\cdot,\cdot]_E$)	definite (indefinite) scalar product in E (Sections 1.1, 1.3)
$\|\cdot\|_E$	norm in E (Section 1.1)
$L \dot{+} M$	direct sum of subspaces L and M (Section 1.8.14)
$L \oplus M$ ($L \ominus M$)	orthogonal sum (difference) of subspaces L and M (Sections 1.1.3, 1.1.14)
$F \hookrightarrow E$	embedding of space F into space E (Section 1.1.6)
F^*	space dual to space F (Section 1.1.6)
Π_κ	Pontryagin space (Section 1.3.9)
$M[+]N$	direct J-orthogonal sum of subspaces (Section 1.3.1)
E^α ($-\infty < \alpha < \infty$)	Hilbert scale of spaces (Section 1.8.3)
$N^\perp$ ($N^{[\perp]}$)	orthogonal (J-orthogonal) complement of N in E (Sections 1.1.3, 1.3.1)
A, B, C	linear operators acting from E into F, or from E into E (Sections 1.1.3, 1.3.1)
$\mathcal{D}(A)$, $\mathcal{R}(A)$	domain (range) of operator A (Section 1.1.8)
A^*	adjoint of operator A in a space with a definite scalar product (Section 1.1.9)

A^+	adjoint of the operator A in in a space with an indefinite scalar product (Section 1.3.4)
$A \geq 0$ $(A > 0,\ A \gg 0)$	nonnegativeness, positiveness, and positive definiteness of operator A, respectively (Section 1.1.10)
$A, L, \gamma, T = \gamma^* = \partial^{-1}$, $C = \gamma T$	operators from the general scheme for solving boundary value problems (Section 1.8)
$\mathcal{L}(E,F)$ $(\mathcal{L}(E))$	linear bounded operators from E into F (from E into E) (Section 1.1.8)
$\mathfrak{S}_\infty(E,F)$, $\mathfrak{S}_\infty$	compact operators from E into F, or from E into E (Section 1.1.12)
$\mathfrak{S}_p$	operators of a finite order (Section 1.1.12)
P, P_0, P_Γ, P_L	orthoprojectors on subspaces from E (Section 1.1.10)
$\operatorname{diag}(A_1, A_2)$	diagonal matrix with operator entries on the main diagonal (Section 1.1.14)
$A_1 \oplus A_2$	orthogonal sum of operators A_1 and A_2 (Section 1.1.14)
$\operatorname{Ker} A$	kernel (null space) of operator A (Section 1.6.5)
$W(A)$	numerical range of operator A (Section 1.1.17)
$\lambda_j(A)$, $j \in \mathbb{N}$	eigenvalues of operator A (Section 1.1.15)
$s_j(A)$, $j \in \mathbb{N}$	s-numbers of operator A (Section 1.1.12)
$r(A)$	spectral radius of operator A (Section 1.1.17)
$A(\lambda)$ $(L(\lambda),\ M(\lambda))$	operator pencil, operator-valued function (Section 1.6.2)
$\sigma(A)$ $(\sigma(A(\lambda)))$	spectrum of operator A (operator-valued function $A(\lambda)$) (Sections 1.1.17, 1.6.2)
$\sigma_p(A)$, $\sigma_c(A)$, $\sigma_r(A)$, $\sigma_e(A)$, $\sigma_d(A)$	subsets of $\sigma(A)$ (Sections 1.1.18, 1.1.9)
$\rho(A)$, $\rho(A(\lambda)))$	resolvent set of operator A (of operator pencil $A(\lambda)$) (Sections 1.1.17, 1.6.2)
$n(r,A)$, $n^\pm(r,A)$ $(n(r,A(\lambda)),$ $n^\pm(A(\lambda)))$	distribution functions of eigenvalues of operator A (operator pencil $A(\lambda)$) (Section 1.6.11)
$\mathcal{Z}_{\lambda_0}(A) = \operatorname{Ker}(A - \lambda_0 I)$	eigensubspace of operator A corresponding to eigenvalue λ_0 (Section 1.1.15)
α_{λ_0} (ν_{λ_0})	eigen (algebraic) multiplicity of operator A corresponding to eigenvalue λ_0 (Section 1.1.15)
$\varphi_0, \varphi_1, \ldots, \varphi_{p-1}$	chain of eigen- and associated elements corresponding to eigenvalue λ_0 (Sections 1.1.15, 1.6.2)
$\varphi(\lambda)$	root function corresponding to an eigenvalue of pencil $A(\lambda)$ (Section 1.6.3)
e^{tA} $(\exp(tA))$	exponential function of operator A (Section 1.5.1)

$\mathcal{U}(t)$	semigroup of operators (Section 1.5.2)
$\cos(tA)$, $\sin(tA)$	trigonometric functions of operator A (Section 1.5.8)
$L_2(\Omega)$, $H^l(\Omega)$, $L_2^1(\Omega)$, $H_0^1(\Omega)$, $H_h^1(\Omega)$, $H_{\partial\Omega}^1(\Omega)$, $H_\Gamma^1(\Omega)$, H_Ω^1, $H_{0,\Gamma}^1(\Omega)$, $L_2(\Gamma)$, $L_{2,\Gamma}$, $H_{h,S}^1(\Omega)$, $H_{\partial\Omega}^{1/2}$, ...	the Hilbert spaces of square-integrable scalar-valued functions and their subspaces (Sections 1.1, 1.8)
$\boldsymbol{L}_2(\Omega)$, $\boldsymbol{G}_0(\Omega)$, $\boldsymbol{J}(\Omega)$, $\boldsymbol{J}_0(\Omega)$, $\boldsymbol{G}(\Omega)$, $\boldsymbol{J}_{0,S}(\Omega)$, $\boldsymbol{G}_{0,\Gamma}(\Omega)$, ...	spaces and subspaces of hydrodynamics for an ideal fluid (Section 2.1)
$\boldsymbol{H}^1(\Omega)$, $\boldsymbol{J}^1(\Omega)$, $\boldsymbol{J}_0^1(\Omega)$, $\boldsymbol{M}_1(\Omega)$, $\boldsymbol{N}_1(\Omega)$	spaces and subspaces of hydrodynamics for a viscous fluid (Section 2.2)
$\boldsymbol{u} = \boldsymbol{u}(t,x)$	velocity field of fluid (Section 2.2.1)
ρ, ν, $\mu = \rho\nu$	density and the kinematic and dynamic viscosities of a fluid (Sections 2.1.1, 2.2.1)
$p = p(t,x)$, $P = P(t,x)$	dynamic and full pressure (Section 3.1.1)
σ	surface tension coefficient of a capillary fluid (Section 3.4.1)
$\boldsymbol{g}$	acceleration of the gravitational field (Section 3.1.5)
$\operatorname{div}\boldsymbol{u}$	divergence of field $\boldsymbol{u}$ (Section 2.1.3)
$\Delta p := \operatorname{div}\nabla p$	Section 2.1.5
$\gamma_n\boldsymbol{u} = u_n = \boldsymbol{u}\cdot\boldsymbol{n}$	normal component of field $\boldsymbol{u}$ on the boundary (Section 2.1.6)
$\tau_{ij}(\boldsymbol{u}) := \dfrac{\partial u_i}{\partial x_j} + \dfrac{\partial u_j}{\partial x_i}$	Section 2.2.1
$\tilde{\tau}_{ij}(\boldsymbol{u}) := -P\delta_{ij} + \rho\nu\tau_{ij}(\boldsymbol{u})$	tensor of tensions (Section 2.2.1)
$\boldsymbol{J}$	inertia tensor (Section 3.1.3)

Subject Index

Sections

A

absolute velocity 3.1.1
abstract Cauchy problem 1.5.1
abstract Green formula 1.8.7
acceleration 3.1.1
adjoint operator 1.1.9
adjoint space 1.1.5
algebraic multiplicity of an
  eigenvalue
    for operators 1.1.15
    for operator-valued
      functions 1.6.1
angular operator 1.3.1
angular velocity 3.1.1
associated elements 1.1.15, 1.6.1
asymptotic expansion 1.7.4
asymptotically stable solution 1.5.6

B

basis
  canonical 1.6.1
  orthonormal 1.1.7
  Riesz (p-basis) 1.1.13
bounded
  functional 1.1.5
  operator 1.1.8

C

canonical factorization 1.6.6
Cartesian coordinate system
  stationary 3.1.1
  nonstationary 3.1.1
  semistationary 3.1.1
Cauchy-Lagrange integral 3.3.2
Cauchy problem 1.5.1
closable operator 1.1.8
closed
  extension 1.1.16
  operator 1.1.16
codimension 1.1.3
compact operator 1.1.8
  of finite order 1.1.12
complete
  space 1.1.1

system of eigen- and
associated elements 1.6.4
completeness
with a finite defect 1.6.7
continuous spectrum 1.1.16
contractive semigroup 1.5.4
Coriolis acceleration 3.1.1

D

defect 1.1.2
definite subspace 1.3.1
degenerate pencil 1.6.4
densely embedded space 1.8.1
dimension 1.1.3
direct decomposition 1.1.14
Dirichlet
problem 1.8.2
norm 1.2.4
discrete spectrum 1.1.8
displacement
acceleration 3.1.1
velocity 3.1.1
dissipation of the energy 2.2.1
dissipative operator 1.5.4
distribution function of
eigenvalues 1.6.8
divergence of a field 2.1.3
domain of definition 1.1.16
Dupré-Young condition 4.1.1

E

eigenmultiplicity 1.1.15
eigenelement
for an operator 1.1.15
for an operator-valued
function 1.6.1
eigenvalue
for an operator 1.1.15
for an operator-valued
function 1.6.1
embedding
compact 1.2.6, 1.8.1
continuous 1.2.6, 1.8.1
energetic space 4.1.5
equation
of hyperbolic type 1.5.8
Navier-Stokes 3.1.2
Poisson 1.8.2
Sophie Germaine 4.2.2
Volterra 1.5.9
equipped spaces 1.2.5
equivalent norms 1.1.4
exponential stability 1.5.6
essential spectrum 1.1.19
extended spectrum 1.1.17
extension of a linear operator 1.1.16

F

factor-space 1.1.23
field of mass external forces 3.1.2
field with a finite kinetic energy 2.1.1
finite function 1.2.1
Fisher-Courant-Weyl principle 1.6.11
fluid
capillary 3.1.3
ideal 2.1.1
incompressible 2.1.1
heavy (noncapillary) 3.1.3
homogeneous 2.1.1
viscous 2.2.1
form of an operator 1.1.10
Fourier coefficient 1.1.7
Fredholm
holomorphic operator-valued
function 1.6.3

pencil 1.6.3
Friedrichs extension 1.8.4
function of
operator 1.1.20
self-adjoint operator 1.1.22

G

generalized
derivative 1.2.2
solution 1.8.4
generating operator 1.8.1
gyroscopic operator 6.3.4
gyrostate 3.1.4
gyrostatical moment 3.1.4
Green's formula 2.1.7, 2.2.3

H

harmonic field 2.1.7
Hilbert
pair 1.8.2
scale 1.8.3
space 1.1.1
Hilbert-Schmidt operator 1.1.12
holomorphic operator-valued
function 1.6.1

I

indefinite metrics 1.3.1
inequality of
Bessel 1.1.7
Buniakowsky-Schwarz 1.1.1
Korn 2.2.4
Poincare 1.2.4
inertia tensor 3.1.4
integral Volterra equation 1.5.9
interlacing operator 1.7.2
invariant subspace 1.1.14
inverse operator 1.1.8
isometric operator 1.1.8
isomorphic spaces 1.1.8
isotropic
element 1.3.1
subspace 1.3.1

J

Jordan chain 1.6.1

K

kernel
function 1.6.2
lineal 1.1.15
of an operator 1.1.15
subspace 1.1.15
kinetic energy 2.1.1
energy operator 3.3.3
moment 3.1.4
Korn identity 2.2.4
inequality 2.2.4

L

Laplace-Beltrami operator 4.1.2
length of chain 1.6.1
limiting spectrum 1.1.19
lineal
degenerate 1.3.1
isotropic 1.3.1
negative 1.3.1
neutral 1.3.1
nonnegative 1.3.1
nonpositive 1.3.1
positive 1.3.1
linear
functional 1.1.5
hull 1.1.3
operator 1.1.8
operator pencil 1.6.4

M

maximal
  dissipative operator 1.5.4
  subspace 1.3.1
minimax principle 1.4.1
min-max theorem 1.4.2
multiple completeness by
  Keldysh 1.6.5
multiplicity
  algebraic 1.1.15
  eigen 1.1.15

N

negative subspace 1.3.1
Neumann
  problem 1.8.2
  series 1.1.17
neutral subspace 1.3.1
nonleaking condition 3.1.1
nonnegative
  operator 1.1.10
  subspace 1.3.1
nonpositive subspace 1.3.1
norm of
  Dirichlet 1.2.4
  linear operator 1.1.8
  functional 1.1.5
normal
  eigenvalue 1.1.18
  component of a field on the
    boundary 2.1.6
nuclear operator 1.1.12
numerical range 1.1.17

O

operator
  bounded from below 1.1.10
  of kinetic energy 3.3.3
  of potential energy 4.1.5
  pencil 1.6.1
orthocompletable subspace 1.3.1
orthogonal
  complement 1.1.3
  decomposition 1.1.3
  elements 1.1.3
  subspaces 1.1.3
  sum 1.1.3
orthonormal system 1.1.7
orthoprojector 1.1.10

P

Parseval identity 1.1.7
perturbed operator 1.5.5
Poincare-Ritz principle 1.6.1
Pontryagin space 1.3.6
point spectrum 1.1.18
positive
  definite operator 1.1.10
  operator 1.1.10
  subspace 1.3.1
potential
  energy 4.1.5
  field 2.1.2
  of a field 2.1.2
power of a self-adjoint
  operator 1.1.22
pressure 3.1.2
  dynamical 3.2.1
  statical 3.2.1
problem
  well posed 1.5.2
  of completeness 1.6.4
  uniformly well posed 1.5.2
projection onto a subspace 1.1.3
projector 1.1.10

Q

quadratic form 1.1.10
quite subordinated operator 1.5.5

R

range of operator 1.1.8
Rayleigh functional 1.6.11
region
  of the first type 1.2.4
  of the second type 1.2.6
  of the third type 1.2.7
regular point 1.1.17
relative
  acceleration 3.1.2
  velocity 3.1.1
resolution
  of the identity 1.1.22
  for a bounded self-adjoint operator 1.1.22
  for an unbounded self-adjoint operator 1.1.22
resolvent of an operator 1.1.17
  pencil 1.6.1
restriction of an operator 1.1.16
residual spectrum 1.1.18
Riesz projector 1.1.20
root
  elements 1.1.15
  function 1.6.2

S

scalar product 1.1.1
  degenerate 1.1.23
  indefinite 1.3.1
self-adjoint operator 1.1.10
  pencil 1.6.9
separable space 1.1.7
set
  closed 1.1.1
  compact 1.1.1
  dense 1.1.3
  resolvent 1.1.17
symmetrizator 1.6.9
semigroup 1.5.2
  of C_0-class 1.5.2
  holomorphic 1.5.3
  strongly continuous 1.5.2
seminorm 1.1.23
skew-symmetric operator 6.1.2
s-numbers 1.1.12
Sobolev space 1.2.1
solenoidal field 2.1.4
space
  of harmonic functions 1.8.6
  of potentials 2.1.1
spectral
  decomposition 1.1.22
  factorization 1.6.6
  radius 1.1.17
spectrum of
  a bounded operator 1.1.17
  an operator 1.1.17
  an operator-valued function 1.6.1
  an operator pencil 1.6.1
  an unbounded operator 1.1.18
stability 1.5.6
statistically stable in linear approximation 4.1.5
stickiness condition 3.1.1
Stokes operator 2.2.5
subordinated operator 1.5.5
subspace 1.1.1
support of a function 1.2.1
symmetric operator 1.1.21

T

tension tensor 3.1.3
Theorem of
  Banach 1.1.4
  Hilber-Schmidt 1.1.11
  Keldysh 1.6.4
  Riesz 1.1.5, 1.1.6
  Weyl 1.1.19
trace operator 1.8.6
transformer 1.7.3

U

unbounded operator 1.1.16
uniformly definite subspace 1.3.2
unitary operator 1.1.8

V

vector Laplace operator 2.2.3
Volterra operator 1.5.9

W

weak
  Cauchy problem 1.5.2
  solution 1.5.2, 1.8.4
weakly
  converging sequence 1.1.8
  perturbed operator 1.6.4
Weyl
  decomposition 2.1.8
  sequence 1.1.19

Z

Zaremba problem 1.8.2
Zhukovsky potential 3.4.4

MIX
Papier aus verantwortungsvollen Quellen
Paper from responsible sources
FSC® C105338

If you have any concerns about our products,
you can contact us on
ProductSafety@springernature.com

In case Publisher is established outside the EU,
the EU authorized representative is:
Springer Nature Customer Service Center GmbH
Europaplatz 3, 69115 Heidelberg, Germany

Printed by Libri Plureos GmbH
in Hamburg, Germany